Prentice-Hall Computer Applications in Electrical Engineering Series

FRANKLIN F. KUO, *Editor*

CADZOW AND MARTENS   *Discrete-Time and Computer Control Systems*
DAVIS   *Computer Data Displays*
JENSEN AND LIEBERMAN   *IBM Circuit Analysis Program: Techniques and Analysis*
KUO AND MAGNUSON   *Computer Oriented Circuit Design*
LIN   *An Introduction to Error-Correcting Codes*
STOUTEMEYER   *PL/1 Programming for Engineering and Science*

DISCRETE-TIME
AND
COMPUTER CONTROL
SYSTEMS

PRENTICE-HALL INTERNATIONAL, INC., *London*
PRENTICE-HALL OF AUSTRALIA, PTY. LTD., *Sydney*
PRENTICE-HALL OF CANADA, LTD., *Toronto*
PRENTICE-HALL OF INDIA PRIVATE LIMITED, *New Delhi*
PRENTICE-HALL OF JAPAN, INC., *Tokyo*

DISCRETE-TIME AND COMPUTER CONTROL SYSTEMS

JAMES A. CADZOW

Associate Professor
Department of Electrical Engineering
State University of New York at Buffalo

HINRICH R. MARTENS

Professor
Departments of Electrical and Mechanical Engineering
State University of New York at Buffalo

PRENTICE-HALL, INC.

ENGLEWOOD CLIFFS, NEW JERSEY

© 1970 by The Research Foundation of State University of New York

Except for the right to materials reserved by others, the Publisher and the Author hereby grant permission to domestic persons of the United States and Canada for use of this work without charge in the English language in the United States and Canada after four years from date of publication. For conditions of use and permission to use materials contained herein for foreign publication or publication in other than the English language, apply to the Publisher or the copyright owner.

The copyright owner will give permission for use of this work without charge after four years from date of publication. For conditions of use, permission to use, and for other permissions, apply to the Publisher or the copyright holder.

Current printing (last digit):

10 9 8 7 6 5 4 3

13-216036-6

Library of Congress Catalog Card Number: 76127549

Printed in the United States of America

To our mothers

Preface

This book has been written with the intent of exposing the advanced undergraduate student to the theories and techniques of systems operating in discrete time, including computer control systems, discrete data processing, computer simulations, and digital data systems.

The level of the book is appropriate for students with a background in analog and digital computation, automatic control systems, and state variable techniques. The material has been used in note form for the past three years at the State University of New York at Buffalo in a senior elective course. It is addressed to the interest of electrical and mechanical engineering students.

The motivation leading to the preparation of this text is based on the belief that the graduating engineering student should avail himself of a knowledge and understanding of digital systems and data processes. With this objective the course is structured to include a study of discrete-time systems, their analysis and design through time-domain and z-domain techniques; an exposure to time-domain synthesis and optimal control; and a presentation of mathematical and physical aspects of selected classes of computer control systems. Accompanying the course lectures is a laboratory program offering hands-on experience with a small-scale digital computer, digital logic modules and other hardware typical of digital systems.

The book is organized into nine chapters. Chapter 1 is essentially an introduction to discrete-time systems concepts for orientation and perspective. By way of example, a number of definitions are presented. A variety of discrete system topics are explored, many of which are discussed in detail in later chapters. The modeling and analysis through time-domain techniques are the subject of Chapters 2 and 3. The weighting sequence, the nth order difference equation, and the state variable representation are used to characterize a discrete system. Typical discrete systems are analyzed; techniques for calculating the response and determining stability are presented.

Both chapters have appendices describing computer programs suitable for aiding in the solution of the chapters' problems.

Chapters 4 and 5 contain additional background for the mathematical processing of discrete systems: the z-transform and state variable representations. Both chapters give detailed coverage of these topics. While Chapter 4 has the conventional coverage of z-transform topics, i.e., definition, properties, transform pairs, inverse transform, transfer function, etc., Chapter 5 is devoted to the derivation of a variety of state variable representations of discrete systems and their relation to other forms of discrete system models.

The z-transform analysis of discrete-time and sampled-data systems is the subject of Chapter 6. Efforts are made throughout to correlate results obtained in the z domain and the time domain. Chapter 7 contains a number of approaches to the analytical design of discrete system. Optimum controllers are developed for the minimum time case using both z-domain and time-domain techniques. The concepts of controllability and observability are introduced to support the discussion of other optimal control problems, such as the minimum energy problem and the discrete linear regulator problem. An effort is made here to present introductory optimization concepts at a reasonable mathematical level. Computer programs implementing some of the optimization algorithms are contained in the chapter's appendix.

Physical design characteristics of systems involving a discrete processor are presented in Chapter 8. This chapter is addressed to the student's need to know about information flow through a computer, data conversion, number conversion, digital logic elements, and selected system hardware, such as stepper motors and shaft encoders. The presence of this chapter provides an appropriate balance between the mathematical modeling of discrete systems components and their physical description and operation.

The use of computer methods in the solution and simulation of discrete systems problems is discussed in Chapter 9. The general objective is the development of facility and understanding in the use of digital, analog, and hybrid computers. The chapter begins with a section on the numerical integration of differential equations. This is followed by the presentation of alternate ways of simulating systems by digital computer. Analog computer simulation and digital simulation languages are the next topics. A brief description of MIMIC is given. Hybrid computation is introduced and several examples of hybrid systems simulation are presented. The chapter concludes with a detailed presentation of three computer control problems. The chapter's appendix has a computer program implementing the Runge-Kutta-Merson integration routine.

In teaching from the material of this text, we have found it very helpful to sprinkle the contents of Chapters 8 and 9 into the lectures throughout the course. Computer solutions and simulations are appropriate for homework and laboratory assignments in connection with material of all chapters.

Similarly useful are discussions of the engineering characteristics of discrete systems and components. The course, as we have taught it, is accompanied by a laboratory. The central component of this laboratory is a small-scale digital computer, a DEC PDP-8. Most of the laboratory studies are designed around this computer. The availability of any small computer for data acquisition, data processing, and control function generation will prove most helpful in understanding many of the discrete system concepts.

Most of the material of this book, including the computer programs and laboratory studies, were produced under an NSF course development grant for which H. R. Martens was project director. Acknowledgment is gratefully extended to the three project consultants, Professors R. C. Dorf, H. E. Koenig, and J. L. Shearer, for their most valuable assistance throughout the project. The talented efforts of W. M. Horner and G. Piosenka, Ph.D. students in our department, are highly recognized in the development of computer programs and laboratory studies. The diligence and patience of Mrs. LaVerne Christie is gratefully acknowledged for typing most of the manuscript.

J. A. CADZOW
H. R. MARTENS

Contents

1 Introduction to Discrete Systems — 1

1.1 Introduction — 1
1.2 System Classification — 2
1.3 Signal Conversion, Sampling and Holding — 12
1.4 Simulation — 15
1.5 Modern Control Theory — 19
1.7 Sampled-data Systems — 22
1.6 Radar Tracking — 27
1.8 Conclusion — 29

2 Time-domain Representations of Linear Discrete Systems — 33

2.1 Introduction — 33
2.2 The Linear Difference Equation — 34
2.3 Weighting Sequence of Linear Discrete Systems — 35
2.4 Weighting Sequence for Cascaded Systems — 45
2.5 State Variable Representations — 47
2.6 Continuous Systems with Piecewise Constant Inputs — 51
2.7 Relationships between Discrete System Representations — 55

3 The Analysis of Discrete-time Systems: Time-domain Approach — 75

3.1 Introduction — 75
3.2 Data-hold Techniques — 77
3.3 Open-loop Sampled-data Systems — 80

3.4	Discrete State Equations of Closed-loop Sampled-data Systems	86
3.5	The Discrete State Analysis of Computer Control Systems	91
3.6	Stability of Discrete Systems	97
3.7	Analysis of a Digital Process Controller	100
3.8	Response of Sampled Systems between Sampling Instants	106

4 z-Transformation and Linear Discrete Systems 118

4.1	Introduction	118
4.2	z-Transform	119
4.3	Linearity of z-transform	122
4.4	Determination of z-Transform Pairs	122
4.5	Properties of the One-sided z-Transform	132
4.6	Transfer Function of a Linear Discrete System	139
4.7	Inverse z-Transform of One-sided z-Transforms	142
4.8	Linear Difference Equations	146
4.9	Response of a Linear Discrete System	150
4.10	z-Transform of a Product of Two Functions	157
4.11	Conclusion	159

5 State Variable Representation 172

5.1	Introduction	172
5.2	Concept of State	173
5.3	State Variable Representations of Linear Discrete Systems	178
5.4	Discrete State Equations	194
5.5	Transfer Matrix for Discrete Systems	199

6 Analysis of Linear Discrete-time System: z-Domain Approach 203

6.1	Introduction	203
6.2	Artifice of Impulse Sampling	204
6.3	Systems with Impulse Sampling	209
6.4	The Transfer Function of a Digital Computer	215
6.5	Typical Sampled-data Systems	216
6.7	Stability Analysis	223
6.6	Relationship Between z Domain and s Domain	224
6.8	Effect of Pole-zero Configuration of $C(a)$ in the z-Plane upon System Transient Response	226
6.9	The Root Locus Technique	229

7 The Analytical Design of Discrete Systems — 246

- 7.1 Introduction — 246
- 7.2 Time-domain Synthesis with Minimum Settling Time — 247
- 7.3 Minimal Prototype Design Using z-Transform Method — 268
- 7.4 Controllability and Observability — 274
- 7.5 Regulator Problem — 283
- 7.6 Minimum Energy Control — 286
- 7.7 Tracking Test Inputs — 298
- 7.8 Controller with a Quadratic Performance Index — 302
- 7.9 DC Gain of a Discrete System — 308

8 Engineering Characteristics of Computer Control Systems — 323

- 8.1 Number Systems — 324
- 8.2 Digital Encoding — 331
- 8.3 Shaft Encoders — 334
- 8.4 Digital Circuit Modules — 339
- 8.5 Data Converters — 350
- 8.6 The Stepper Motor — 357
- 8.7 The Stepper Motor in Control Applications — 365
- 8.8 Summary — 373

9 Computer Methods in Systems Studies — 376

- 9.1 Introduction — 376
- 9.2 Numerical Methods of Simulating System Dynamics — 377
- 9.3 Use of the State Transition Method in Simulation Studies — 395
- 9.4 Digital Computer Simulation of a Digital Control System — 398
- 9.5 Analog Computer Simulation of Systems — 404
- 9.6 Digital Analog-system Simulators — 408
- 9.7 Hybrid Computer Techniques and Applications in Simulation — 415
- 9.8 Computer Control — 437
- 9.9 Summary — 461

Appendices:

- 2A Computer Program for $\mathbf{A}(T)$ and $\mathbf{B}(T)$ — 64
- 2B The Transition Matrix — 67

3A	Computer-generated Printer Plot	110
3B	Eigenvalues of a Matrix	116
4A	Power Series	162
6A	Proof of Theorem 6.3-1	243
7A	Computer Program for Design of Deadbeat Controller	314
7B	Computer Program for Discrete Linear Regulator	319
9A	Program for the Numerical Integration of State Variable Equations	464

Index   ***469***

1

Introduction to Discrete Systems

1.1 Introduction

In recent years, computer technology has made revolutionary advances. Many sophisticated engineering applications are now made possible by the incorporation of a digital computer in a system design. Such designs place a high demand on a basic understanding of computer system characteristics. In the chapters that follow, material is presented with the objective of preparing the reader for the analysis and design of systems incorporating digital devices, called *discrete-time systems*.

Attention will be focused, in the main, on systems that are linear. This restriction is justified because many systems are linear or approximately linear. Furthermore, the mathematical tools available for the study of discrete-time systems are primarily restricted to linear systems.

It will be shown that for linear systems in which a digital computer is an element, linear difference equations arise in a natural manner, thus providing a motivation for the study of linear difference equations. This is analogous to the importance attached to systems governed by linear differential equations in continuous-time systems.

This chapter introduces some of the basic definitions of discrete systems such as linearity, time-invariance, and nonanticipativeness. Then some typical engineering applications in which difference equations arise are presented. These applications are treated in more detail later and are presented here only to offer some orientation and perspective.

1.2 System Classification

Before an analysis of a system may be attempted, a mathematical representation (or model) of the system must be given or obtained. This mathematical representation must adequately characterize the behavior of the system, but it also should be sufficiently simple to make the resultant analysis tractable. Unless otherwise specified, the mathematical model is assumed to be given.

Usually, the model is constructed so that there exist two types of variables called the *input* and the *output variables*. The input variables, through the dynamical behavior of the system, influence the output variables in some manner. The objective of system analysis is that of determining the nature of this input-output relation. In Figure 1.2-1, a block diagram representation of a system is shown. The set of variables $\{r_1, r_2, \ldots, r_m\}$ is the system input, while the set $\{c_1, c_2, \ldots, c_p\}$ is the system output (responses). It is common to designate the input and output variables as the input and output signals. In many practical situations, there exist only one input (r_1) signal and one output (c_1) signal.

1.2-1 Continuous and Discrete Signals

In most cases of interest, the input and output signals are functions of the independent variable time t. By introducing the independent variable t, the ability to change the input signals as functions of t is obtained. If the independent variable t can take on a continuum of values, the signal is said to be a *continuous signal*. To denote a continuous signal, the argument t is incorporated whenever the signal is formally written (e.g., $\sin t$). Therefore, if the set of inputs appearing in Figure 1.2-1 is continuous, it will be written as

$$\{r_1(t), r_2(t), \ldots, r_m(t)\} \qquad (1.2\text{-}1)$$

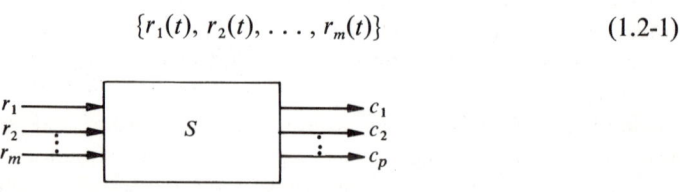

Figure 1.2-1. Block diagram representation of a system.

If the independent variable t takes on only a finite or at most an infinitely countable number of values, the time variable t is said to be a *discrete-time variable*. The values for which the discrete-time variable is defined are denoted by t_k with $k = 0, \pm 1, \pm 2, \ldots$. A signal whose independent variable t is discrete is said to be a *discrete signal*. A discrete signal will be written as $r(t_k)$. Examples of continuous and discrete signals are illustrated in Figure 1.2-2.

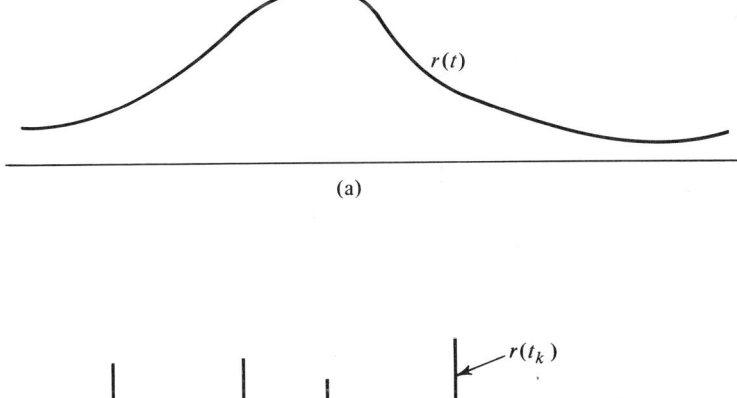

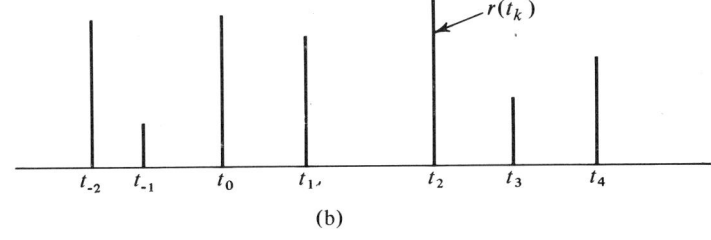

Figure 1.2-2. Example of (a) continuous signal; (b) discrete signal.

1.2-2 Sampling

The process of sampling is a very basic concept in digital control and data processing theory. In a very idealistic sense, a sampler is a device that transforms a continuous signal into a discrete signal. Let $r(t)$ be the continuous input signal to a sampler as shown in Figure 1.2-3(a). The output of this sampler will be a sequence of numbers, spaced in time, which appear at the sampling instances t_k and have the values equal to the continuous input signal at the sampling instances. A simple model of the sampler is shown in Figure 1.2-3(b), where the switch is thought of as being open for all time except at the sampling instances t_k, when it closes instantaneously to pass the input signal $r(t)$. Normally, the sampling intervals are spaced equidistant in time so that $t_k = kT$, where T is the sampling time.

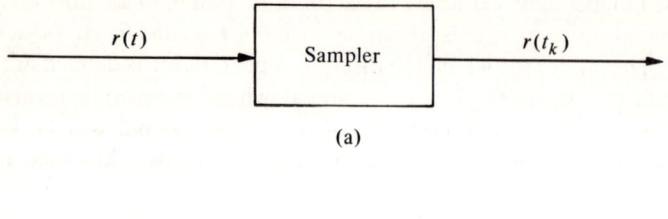

(a)

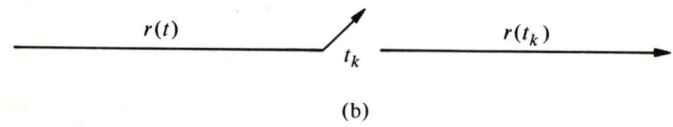

(b)

Figure 1.2-3. Process of sampling.

EXAMPLE 1.2-1

Let the input to the sampler depicted in Figure 1.2-3(a) be given by

$$r(t) = 2 + 3t + \sin 5t$$

Then its output will be a sequence of numbers that appear at times t_k and have values

$$r(t_k) = 2 + 3t_k + \sin 5t_k$$

1.2-3 Continuous and Discrete Systems

A system in which the characterizing signals (e.g., input and output signals) are all continuous is called a *continuous system*. Most continuous systems of interest are governed by differential equations.

When the characterizing signals are all discrete-type signals, the system is said to be a *discrete system*. In general, a discrete system is characterized by difference equations that relate the various input and output signals. As an example, any dynamical process implemented by a digital computer is discrete.

Some systems possess both continuous and discrete signals. Such systems are referred to as *hybrid systems*. Sections 1.4 and 1.6 give illustrative examples of such systems.

1.2-4 Difference Equations

We shall concentrate our efforts toward studying systems that are characterized by difference equations. A true discrete system is inherently representable by difference equations. Hybrid systems may, in many cases, be characterized by difference equations. The remainder of this chapter will illustrate a variety of typical engineering systems in which such a representation is possible. For the purposes of this book, any system that may be adequately described by difference equations will be considered a discrete system, and the analysis and synthesis techniques developed will apply to such systems.

In order to obtain a "feel" for what exactly a difference equation is, let us consider the following simple example. Suppose that it is desired to perform the operation of integration numerically, that is, to evaluate

$$c(t) = \int_0^t r(\tau)\, d\tau \tag{1.2-2}$$

by some iterative technique. This may be desirable if the function to be integrated, $r(\tau)$, has no closed form integral, for example. A standard technique is first to approximate the function $r(\tau)$ by a piecewise constant function $r_a(\tau)$ as shown in Figure 1.2-4. The piecewise constant function $r_a(\tau)$ is given by

$$r_a(\tau) = r(nT) \quad \text{for } nT \leq \tau < nT + T$$

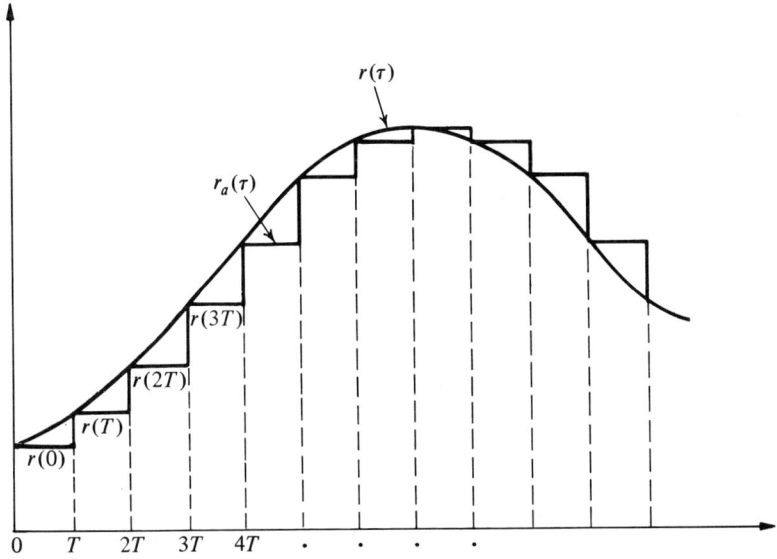

Figure 1.2-4. Numerical integration process.

The reader will recall from calculus that the integral (1.2-2) is approximated by

$$c(kT) = \int_0^{kT} r(\tau)\, d\tau \approx \sum_{m=0}^{k-1} Tr(mT) \tag{1.2-3}$$

In fact, this is the ordinary definition of the integration process as T approaches zero. This relationship requires the retention of all past values of $r(mT)$ (k in number) in order to determine the value of the integral at $t = kT$. A much simpler and systematic method is obtained by first determining

$$c(kT + T) = \int_0^{kT+T} r(\tau)\, d\tau \approx \sum_{m=0}^{k} Tr(mT)$$

and subtracting (1.2-3) from this expression, which gives

$$c(kT + T) - c(kT) = Tr(kT) \qquad (1.2\text{-}4)$$

Now, it is necessary only to retain the previous value of the integral $c(kT)$ and the present sampled value of $r(\tau)$ [i.e., $r(kT)$] in order to determine the value of the integral at $t = (k + 1)T$. Incidentally, this represents substantial savings in storage space if a digital computer is used in evaluating (1.2-3) and k is large.

The dynamics of this first-order difference equation work as follows:
1. One first notes the initial condition

$$c(0) = \int_0^0 r(\tau)\, d\tau = 0$$

2. Next, one evaluates (1.2-4) first for $k = 0$, then $k = 1$, and so forth; that is,

$$c(T) = Tr(0) + c(0) = Tr(0)$$
$$c(2T) = Tr(1) + c(1)$$
$$c(3T) = Tr(2) + c(2)$$
$$\cdots \cdots \cdots$$
$$c(mT) = Tr(mT - T) + c(mT - T)$$

At each state of the iteration, the new value of $c(mT)$ is evaluated by adding its previous value to the new input $r(mT - T)$ multiplied by T.

A more general nth-order, linear difference equation has the form

$$c(n + k) + a_1 c(n + k - 1) + \cdots + a_n c(k)$$
$$= b_0 r(n + k) + b_1 r(n + k - 1) + \cdots + b_n r(k) \qquad (1.2\text{-}5)$$

In this case, one computes $c(n + k)$ by retaining its previous values $c(n + k - 1), c(n + k - 2), \ldots, c(k)$ and the input sequence $r(n + k), r(n + k - 1), \ldots, r(k)$ and then performs the multiplications and additions prescribed by (1.2-5).*

---

*In comparing the general linear difference equation (1.2-5) with the numerical integration expression 1.2-4, one observes the explicit appearance of computation time T in one and not the other. In what follows, the appearance of T will be entered only when directly applicable. It will normally be implied as in (1.2-5), but not formally written.

It is seen that the input signal [sequence of numbers ..., $r(-1), r(0), r(1), ...$] influences in a definite way the output sequence of numbers $c(k)$. This has its obvious parallel in continuous systems, where the input signal [e.g., $r(t) = \sin t$] influences the system's output in a similar manner. The coefficients $a_0, a_1, ..., a_{n-1}, b_0, b_1, ..., b_{n-1}$ determine the characteristic of this input-output dynamical behavior. How one selects these coefficients is one of the main topics of this text.

1.2-5 Linear Systems

In introducing the concept of linear systems, it will be convenient first to consider systems with one input and one output. Conceptually, suppose that any input r_a is applied to such a system with the resultant response being denoted by c_a; similarly, let c_b be the response of this same system to the general input r_b. Then, the system is said to be linear if the system's response to the input $\alpha r_a + \beta r_b$ is

$$c = \alpha c_a + \beta c_b \tag{1.2-6}$$

for all real numbers α and β. The characteristic of the independent time variable (continuous or discrete) for the input signal r and output signal c has not been explicitly shown, since the same definition of linearity applies in either case. In the language of transformation theory, the relationship between the input and output signals is given by

$$c = S[r]$$

for which (1.2-6) can be rewritten as

$$c = S[\alpha r_a + \beta r_b] = \alpha S[r_a] + \beta S[r_b] \tag{1.2-7}$$

This transformation S is to be interpreted as how the system transforms (changes) the input signal r into the output signal c. Examples of continuous and discrete linear and nonlinear systems will now be given.

EXAMPLE 1.2-2

The simple RC network shown in Figure 1.2-5 is characterized by the first-order differential equation

$$\dot{c}(t) + c(t) = r(t)$$

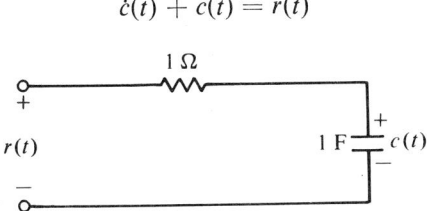

Figure 1.2-5. Simple RC network.

Let $c_a(t)$ and $c_b(t)$ be the response to the inputs $r_a(t)$ and $r_b(t)$, respectively; that is,

$$\dot{c}_a(t) + c_a(t) = r_a(t)$$
$$\dot{c}_b(t) + c_b(t) = r_b(t) \tag{1.2-8}$$

The response to the input $\alpha r_a(t) + \beta r_b(t)$ will be defined by $c(t)$ so that

$$\dot{c}(t) + c(t) = \alpha r_a(t) + \beta r_b(t) \tag{1.2-9}$$

Equation (1.2-9) is satisfied by letting $c(t) = \alpha c_a(t) + \beta c_b(t)$, as can be shown by substituting (1.2-8) into (1.2-9). The system is therefore linear.

EXAMPLE 1.2-3

Consider the system shown in Figure 1.2-6. This system is characterized by the first-order differential equation

$$\dot{c}(t) = r^2(t)$$

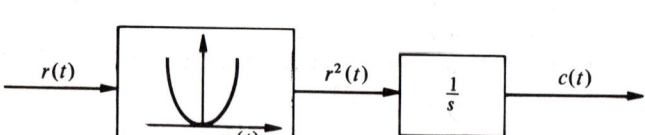

Figure 1.2-6. Simple nonlinear system.

Let $c_a(t)$ and $c_b(t)$ be the responses to the inputs $r_a(t)$ and $r_b(t)$, respectively. Therefore,

$$\dot{c}_a(t) = r_a^2(t)$$
$$\dot{c}_b(t) = r_b^2(t)$$

If the input $\alpha r_a(t) + \beta r_b(t)$ is applied to the system, the response $c(t)$ is given by

$$\dot{c}(t) = [\alpha r_a(t) + \beta r_b(t)]^2$$
$$= \alpha^2 r_a^2(t) + \beta^2 r_b^2(t) + 2\alpha\beta r_a(t) r_b(t)$$
$$= \alpha^2 \dot{c}_a(t) + \beta^2 \dot{c}_b(t) + 2\alpha\beta r_a(t) r_b(t)$$
$$\neq \alpha \dot{c}_a(t) + \beta \dot{c}_b(t)$$

Thus, the system is not linear.

EXAMPLE 1.2-4

A discrete approximation to numerical integration is given by

$$c[(k+1)T] - c(kT) = Tr(kT)$$

where T is a constant. If $c_a(kT)$ and $c_b(kT)$ are the responses of this system to the inputs $r_a(kT)$ and $r_b(kT)$, respectively, we have

$$c_a[(k+1)T] - c_a(kT) = Tr_a(kT)$$
$$c_b[(k+1)T] - c_b(kT) = Tr_b(kT)$$

When the input $\alpha r_a(kT) + \beta r_b(kT)$ is applied to the system, the response $c(kT)$ satisfies the difference equation

$$c[(k+1)T] - c(kT) = T[\alpha r_a(kT) + \beta r_b(kT)]$$

However, $c(kT) = \alpha c_a(kT) + \beta c_b(kT)$ satisfies this difference equation, as may be verified by substitution. The system is linear.

By analogous methods, it may be shown that the discrete system

$$c(k+1) - c(k) = r^2(k)$$

is nonlinear.

In general, a system may have more than one input signal and one output signal. Let the set of m input signals be denoted by the $(m \times 1)$ vector **r**, where

$$\mathbf{r} = \begin{bmatrix} r_1 \\ r_2 \\ \vdots \\ r_m \end{bmatrix}$$

and the set of p output signals by the $(p \times 1)$ vector **c**, where

$$\mathbf{c} = \begin{bmatrix} c_1 \\ c_2 \\ \vdots \\ c_p \end{bmatrix}$$

A system with the vector input **r** and vector output **c** is said to be linear if the relationship

$$\mathbf{c} = S[\alpha \mathbf{r}_a + \beta \mathbf{r}_b] = \alpha S[\mathbf{r}_a] + \beta S[\mathbf{r}_b] \qquad (1.2\text{-}10)$$

is satisfied for any set of real numbers (α, β) and input vectors $\mathbf{r}_a, \mathbf{r}_b$. The transformation S transforms (changes) an m-dimensional input vector **r** into a p-dimensional output vector **c**. A system in which the relationship (1.2-10) does not hold is said to be nonlinear.

1.2-6 Time-invariant System

A system in which the relationship (dynamical behavior) between the input and output signals is unchanging with respect to time is called a *time-invariant system*. For example, if $\mathbf{c}(t)$ is the output vector in response to the input vector $\mathbf{r}(t)$ for a continuous system, that is,

$$\mathbf{c}(t) = S[\mathbf{r}(t)]$$

then a time-invariant system is characterized by the relationship

$$\mathbf{c}(t - \tau) = S[\mathbf{r}(t - \tau)] \qquad (1.2\text{-}11)$$

for all τ and $\mathbf{r}(t)$. In words, if $\mathbf{c}(t)$ is the response to $\mathbf{r}(t)$, then the system's response to the same input signal applied τ seconds later is $\mathbf{c}(t)$ delayed by τ seconds [i.e., $\mathbf{c}(t - \tau)$].

A discrete system for which the dynamical behavior between the input and output vectors is given by

$$\mathbf{c}(t_n) = S[\mathbf{r}(t_n)] \qquad (1.2\text{-}12)$$

is said to be time invariant if the relationship

$$\mathbf{c}(t_n - t_k) = S[\mathbf{r}(t_n - t_k)] \qquad (1.2\text{-}13)$$

holds for all t_n, t_k, and $\mathbf{r}(t_k)$.

EXAMPLE 1.2-5

Consider the system characterized by

$$\dot{c}(t) = tr(t)$$

Let $c_a(t)$ be the response to the input $r_a(t)$, i.e., $\dot{c}_a = tr_a(t)$; and let $c_b(t)$ be the response to $r_a(t - \tau)$, i.e.,

$$\dot{c}_b(t) = \tau r_a(t-\tau) = (t-\tau)r_a(t-\tau) + \tau r_a(t-\tau)$$
$$= \dot{c}_a(t-\tau) + \tau r_a(t-\tau)$$
$$\neq \dot{c}_a(t-\tau)$$

The system is not time invariant.

EXAMPLE 1.2-6

Consider the system of Example 1.2-2

$$\dot{c}_a(t) + c_a(t) = r_a(t)$$

If $c_b(t)$ is the response to $r_a(t-\tau)$, i.e.,

$$\dot{c}_b(t) + c_b(t) = r_a(t-\tau) \qquad (1.2\text{-}14)$$

it is seen that $c_b(t) = c_a(t-\tau)$ satisfies (1.2-14). The system is time invariant.

It may be shown that any system governed by differential equations is time invariant if the variable t does not explicitly appear in these equations. A system that is not time invariant is called time varying.

EXAMPLE 1.2-7

The discrete system under investigation is governed by the difference equation

$$c(k+1) - c(k) = kr(k)$$

Let $c_a(k)$ be the response to the input sequence $r_a(k)$. Therefore,

$$c_a(k+1) - c_a(k) = kr_a(k)$$

If the input $r_b(k) = r_a(k-m)$ is applied, then the system's response $c_b(k)$ satisfies

$$c_b(k+1) - c_b(k) = kr_a(k-m)$$
$$= (k-m)r_a(k-m) + mr_a(k-m)$$
$$= c_a(k-m+1) - c_a(k-m) + mr_a(k-m)$$

so that $c_b(k) \neq c_a(k-m)$, and the system is not time invariant. It may be shown that the discrete system

$$c(k+1) - c(k) = r(k)$$

is time invariant.

Any discrete system that is governed by difference equations is time invariant if the time variable k does not appear explicitly in the difference equations.

1.2-7 Nonanticipative Systems

A system in which the present output c does not depend on the values of the input r in future time is *nonanticipative*. Nonanticipative systems are characterized by the output's being dependent completely on present and previous values of the input signals. Anticipative systems are physically nonrealizable, since the present output depends on future values of the input.

1.3 Signal Conversion, Sampling and Holding

One of the most interesting aspects of discrete systems is the presence of signals in different physical forms. For instance, signals may be continuous in one part of a system and discrete in another part. In many cases of interest a digital processor, such as a digital computer, will be the receiver or transmitter of signals. Then we must not only distinguish between continuous and discrete signals, but also between analog and digital signals. Let us use an illustration to bring this aspect into focus and discuss related mathematical and physical considerations.

Consider the system block diagram shown in Figure 1.3-1. In this system a continuous signal enters a sampler and is transformed into a discrete signal consisting of a sequence of sampled values of the original signal. The sampled signal has a common characteristic with the continuous signal: they are both *analog* signals.

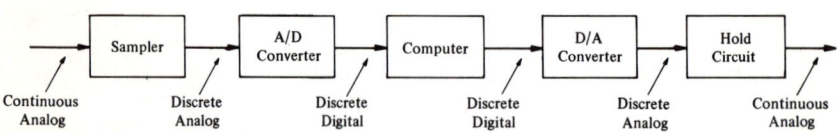

Figure 1.3-1. Sequential conversion of signal passing through a digital computer.

An analog signal is a signal (continuous or discrete) whose amplitude is not restricted in any sense. That is, at any instant of time, the amplitude of the signal may take on any value in a given range. On the other hand, a *digital* signal is a signal whose amplitude is restricted to a given set of values. For a demonstration of the distinction between analog and digital signals, consider the following example.

EXAMPLE 1.3-1

Let it be desired to read the voltage at some point in an electronic circuit. It is known beforehand that the voltage will be in the range of $+10$ volts to -10 volts.

If one were to use a standard voltmeter whose scale range was $+10$ volts to -10 volts, the resultant voltage measured would be considered an analog signal because "any" level in this range could be measured on the scale. On the other hand, if one were to use a digital voltmeter with fixed range $+9.99$ volts to -9.99 volts (i.e., the meter reads polarity and three decimal figures with fixed decimal point) it is possible to read only 2000 voltage levels. This signal would be considered digital, since it can take on only a limited number of values.

In order for the sampled, or discrete, signal to be processable by a digital computer its physical format must be changed from an analog to a digital signal. A digital computer can accept only digital (usually binary) signals. The indicated conversion is carried out by an *analog-to-digital* converter (A/D). The operational principles of an A/D converter are explained in Chapter 8. Suffice it to say that an A/D converter determines a binary equivalent to an analog input.

Whether or not the conversion process is mathematically significant (to be included in the system's description) depends on the conversion speed (normally a few microseconds) and the bit capacity of the resulting binary signal. A signal that passes through an A/D converter becomes a *discrete digital* signal. In this form it is quantized with respect to amplitude. Since such a signal is normally represented in binary form, the number of quantization levels is determined by the range of signal divided by 2^N, where N is the number of binary digits, or "bits." If N is small (below 8 to 10) the quantization may be too coarse to adequately represent all values that the signal may assume. As an illustration of the quantization problems, consider Example 1.3-2.

EXAMPLE 1.3-2

Determine the quantization levels of a continuous signal of range ± 10 volts passing through a sampler and A/D converter with a bit capacity of 4.

The total number of binary positions is given by 2^4, or 16. Hence, there are 16 quantization levels of 20/16, or 1.25 volts magnitude. Typically, the signal may be assigned binary numbers according to Table 1.3-1.

Table 1.3-1

Decimal	Binary	Decimal	Binary
+9.375	0111	− .625	1000
+8.125	0110	−1.875	1001
+6.875	0101	−3.125	1010
+5.625	0100	−4.375	1011
+4.375	0011	−5.625	1100
+3.125	0010	−6.875	1101
+1.875	0001	−8.125	1110
+ .625	0000	−9.375	1111

There are eight positive and eight negative positions. Notice that the binary numbers are so divided as to use the most significant bit as the sign bit. Thus, a 0xxx denotes a positive number, while a 1xxx denotes a negative number. Such an assignment of bits is called *digital encoding*.

An incoming signal of +5.325 volts would be recognized by the converter as binary 0100.

Since the bit capacity of most computers is at least 12, and more frequently at least 18, the quantization error resulting from A/D conversion is mathematically insignificant. Therefore, we will usually ignore the presence of an A/D converter in making an analysis of hybrid systems.

The signal is now fed into the digital computer in binary format at a rate corresponding to the sampling rate. After it leaves the digital computer the signal may be converted back to a continuous analog signal by passing it through a *digital-to-analog* converter. As the name implies, the digital-to-analog converter, or D/A converter, changes the physical format of a signal from digital to analog. The resulting analog signal will appear quantized, although only indirectly so. Its magnitude will be in a direct correspondence with the originating binary number. But all real numbers may be used to represent it. A detailed discussion of a D/A converter is presented in Chapter 8. D/A conversion is normally insignificant from a mathematical point of view, and its presence will be ignored when an analysis of a hybrid system is made.

At the output of the D/A converter the signal is classified as a *discrete analog* signal. It may be changed into a continuous analog signal by passing it through a data hold circuit. The operation of a typical data hold (zero-order hold) may be mathematically defined by

$$r(\tau) = r(nT) \quad \text{for } nT \leq \tau < (n+1)T \qquad (1.3\text{-}1)$$

A zero-order data hold generates a "staircase" function such as those shown in Figures 1.2-4 and 1.5-2.

There are five blocks in Figure 1.3-1, each one indicating a separate but important operation. Three of these, the sampler, the digital computer, and the hold circuit, are mathematically significant; they affect the dynamic description of a signal passing through them. The remaining two, the A/D conversion and D/A conversion, are only of physical significance, and their characteristics do not normally have to be included in the mathematical description of the overall system, provided that the bit capacity of the binary process is sufficiently large to keep the quantization error at an acceptable level.

Whenever a digital computer is present in a system, an information flow like that indicated by Figure 1.3-1 is involved. Sampling and A/D conversion appear together preceding the digital computer, and D/A conversion and holding appear together following the digital computer. Because of the relative mathematical significance, only the operations of sampling and holding are indicated; the operations of data conversion are implied. Thus, the system shown in Figure 1.3-1 is shown in the simplified version of Figure 1.3-2.

Figure 1.3-2. Mathematically significant operations.

1.4 Simulation

The digital computer was first developed primarily to carry out computations that were exceedingly lengthy or impracticable for the individual to do by hand. Such an example is that of determining the response of a system to a given input when the system's dynamics are characterized by linear, time-invariant differential equations. If the input waveform can be approximated by a convenient function such as a step, ramp, etc., it is possible, in principle, to use known techniques to analytically determine the system's response. However, even in this case, if the system is of high order, then these techniques become complex at best for determining the response. In a typical situation, the input signal has no simple functional representation, so one is forced to use numerical techniques in evaluating the system's response.

Two techniques for simulating the dynamical characteristics of a system, characterized by a differential equation, on a digital computer will now be presented. In Chapter 9, a more detailed study of simulation will be conducted.

1.4-1 First Difference Method

The first difference method is a most appealing and intuitive method for digital simulation. It will be demonstrated by means of an example. Suppose that the system under consideration is governed by the transfer function

$$H(s) = \frac{C(s)}{R(s)} = \frac{1}{s(s+a)} \tag{1.4-1}$$

where $R(s)$ and $C(s)$ are the Laplace transforms of the system's input and output signals, respectively. The differential equation governing this system is found by first multiplying both sides of (1.4-1) by $s(s+a)R(s)$, which gives

$$R(s) = s(s+a)C(s) = (s^2 + as)C(s)$$

and by noting that the inverse transform of $s^m C(s)$ is $d^m c(t)/dt^m$ (where initial conditions have been taken to be zero). This results in

$$\frac{d^2 c(t)}{dt^2} + a\frac{dc(t)}{dt} = r(t) \tag{1.4-2}$$

It is desired to simulate the dynamical characteristics of this differential equation, which, in turn, necessitates the determination of the first two derivatives of the time function $c(t)$. Since the simulation is to be carried out by means of a digital computer, it is natural to solve (1.4-2) at computation times that are taken to be spaced at T-second intervals, that is,

$$\left.\frac{d^2 c(t)}{dt^2}\right|_{t=nT} + a\left.\frac{dc(t)}{dt}\right|_{t=nT} = r(nT) \tag{1.4-3}$$

In order to determine $dc(t)/dt|_{t=nT}$, we shall employ the following approximation from differential calculus.

$$\dot{c}(nT) = \left.\frac{dc(t)}{dt}\right|_{t=nT} \approx \frac{c(nT) - c(nT-T)}{T} \tag{1.4-4}$$

and similarly

$$\ddot{c}(nT) = \left.\frac{d^2 c(t)}{dt^2}\right|_{t=nT} = \frac{\dot{c}(nT) - \dot{c}(nT-T)}{T}$$
$$= \frac{c(nT) - 2c(nT-T) + c(nT-2T)}{T^2}$$

so that (1.4-3) becomes

$$\frac{c(nT) - 2c(nT-T) + c(nT-2T)}{T^2} + a\frac{c(nT) - c(nT-T)}{T} = r(nT)$$

After simplification, we have

$$c(nT) = \frac{T^2 r}{1+aT}(nT) + \frac{2 + aT}{1 + aT}c(nT - T) - \frac{1}{1 + aT}c(nT - 2T) \quad (1.4\text{-}5)$$

Relationship (1.4-5) is a linear, second-order difference equation that may be programmed on a digital computer to determine the system's response $c(t)$ to a general input $r(t)$ at the computation times nT. The computation time interval T must be selected to be small enough so that the approximation (1.4-4) is good. However, if the computation time interval is selected too small, the number of iterations required to obtain a given time history of $c(t)$ increases, thus requiring more computer time and, therefore, a more expensive simulation. Some compromise value of T must be selected. The extension to more complex system simulations may be carried out by using this technique in an obvious manner.

1.4-2 Impulse Response Method

An alternate simulation method will now be presented.

If the linear system shown in Figure 1.4-1 has an impulse response $h(t)$, then, using the convolution integral, we find that the response of the system $c(t)$ to the input $r(t)$ is

$$c(t) = \int_0^t r(t - \tau)h(\tau)\,d\tau \quad (1.4\text{-}6)$$

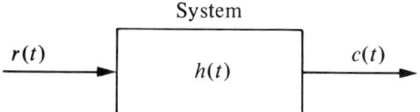

Figure 1.4-1. System with impulse response $h(t)$.

where it has been assumed that $r(t) = 0$ for $t < 0$. In order to convert this integral (1.4-6) into a form convenient for use on a digital computer, it will be convenient to approximate the impulse response $h(t)$ by a piecewise constant function, as illustrated in Figure 1.4-2. The impulse response is approximated by

$$h(t) \approx \sum_{j=0}^{N} h_j[u(t - jT) - u(t - jT - T)] \quad (1.4\text{-}7)$$

where the time interval T has been selected small enough so that the approximation to $h(t)$ is adequate and N is chosen so that

$$h(t) \approx 0, \quad t \geq NT$$

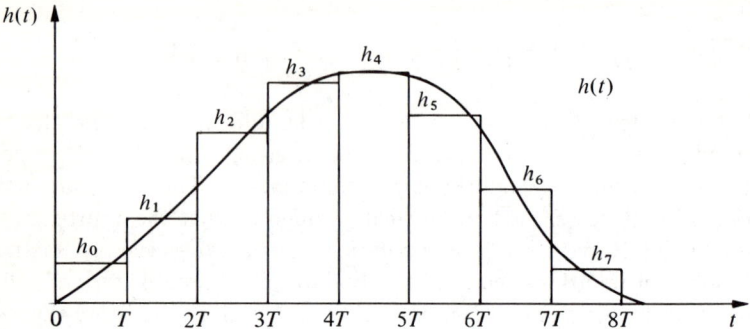

Figure 1.4-2. Approximation to impulse response.

and $u(t)$ denotes the unit step function

$$u(t) = \begin{cases} 1 & \text{for } t \geq 0 \\ 0 & \text{for } t < 0 \end{cases}$$

Inserting (1.4-7) into (1.4-6) and evaluating the result at $t = kT$ gives

$$c(kT) = c_k = \int_0^{kT} r(kT - \tau) \sum_{j=0}^{N} h_j[u(\tau - jT) - u(\tau - jT - T)]\, d\tau \quad (1.4\text{-}8)$$

$$c_k = \sum_{j=0}^{N} h_j \int_{jT}^{jT+T} r(kT - \tau)\, d\tau \quad (1.4\text{-}9)$$

Making the change of variables $\alpha = kT - \tau$, we find that Equation (1.4-9) becomes

$$c_k = \sum_{j=0}^{N} h_j r_{k-j} \quad (1.4\text{-}10)$$

where

$$r_{k-j} = \int_{(k-j-1)T}^{(k-j)T} r(\tau)\, d\tau \quad (1.4\text{-}11)$$

If T is selected small enough so that the waveform $r(\tau)$ remains essentially constant in the T second interval, then (1.4-11) can be approximated by

$$r_{k-j} = Tr(kT - jT) \quad (1.4\text{-}12)$$

Equation (1.4-10) is a difference equation with which, for a given input $r(t)$, the approximate output $c(t)$ at the sampling times kT may be determined.

1.5 Modern Control Theory

In many control applications, a digital computer is used to generate the control input that serves to control a given system in some desirable manner. The digital computer emits a sequence of numbers that are fed into a digital-to-analog converter. A digital-to-analog converter is a device that changes a discrete set of numbers (digital information) into a function of continuous time.* The digital computer in conjunction with the digital-to-analog converter is sometimes referred to as a digital controller in control applications. The output of the digital controller serves as the control input to the system under control. Because of the basic discrete nature of digital computers, these control inputs remain of a fixed form, such as steps, ramps, etc., during the sampling or computation period of the computer. Usually, the control inputs are held at a fixed value during each sampling interval of T seconds.

If the system to be controlled is characterized by a linear, time-invariant differential equation and its inputs are held at constant values for intervals of T seconds, the system's dynamics may be represented by linear difference equations. To illustrate this, consider the system shown in Figure 1.5-1. It is assumed that the controller, acting on the output signal $c(t)$, generates a piecewise constant signal of the form

$$r(t) = r(kT), \qquad kT \leq t < [k+1]T$$

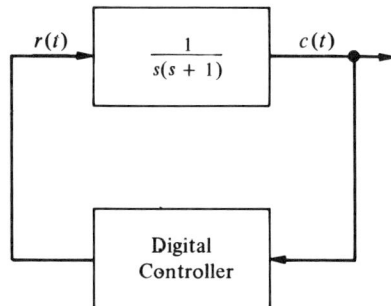

Figure 1.5-1. Controlled system.

where $r(kT)$ is a number and T is the sampling period of the controller. How the controller generates the control input $r(t)$ depends on the particular control problem under consideration and will be studied in detail in later chapters. Typically, the control input is like that shown in Figure 1.5-2.

*Normally, a digital-to-analog converter has an integrally built-in zero-order data-hold circuit.

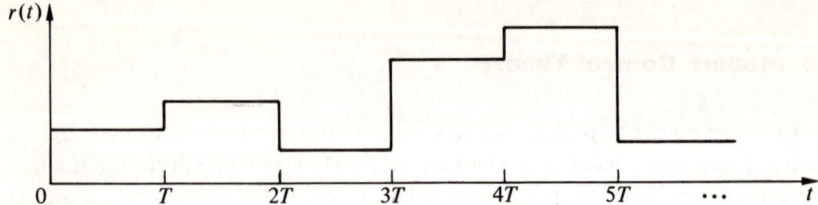

Figure 1.5-2. Piecewise constant input signal.

A state model characterizing the system shown in Figure 1.5-1 is

$$\frac{d}{dt}\begin{bmatrix} x_1 \\ x_2 \end{bmatrix} = \begin{bmatrix} -1 & 0 \\ 1 & 0 \end{bmatrix}\begin{bmatrix} x_1 \\ x_2 \end{bmatrix} + \begin{bmatrix} 1 \\ 0 \end{bmatrix} r(t) \quad (1.5\text{-}1)$$

with

$$c(t) = x_2$$

For $0 < t \leq T$, $r(t) = r_0$, a constant, (1.5-1) becomes

$$\frac{d}{dt}\begin{bmatrix} x_1 \\ x_2 \end{bmatrix} = \begin{bmatrix} -1 & 0 \\ 1 & 0 \end{bmatrix}\begin{bmatrix} x_1 \\ x_2 \end{bmatrix} + \begin{bmatrix} 1 \\ 0 \end{bmatrix} r(0) \quad \text{for } 0 < t \leq T \quad (1.5\text{-}2)$$

$$c(t) = x_2$$

which has the solution

$$\mathbf{x}(t) = e^{\mathbf{F}t}\mathbf{x}(0) + \int_0^t e^{\mathbf{F}(t-\tau)}\mathbf{G}r(0)\, d\tau \quad \text{for } 0 < t \leq T \quad (1.5\text{-}3)$$

$$c(t) = [0 \quad 1]\mathbf{x}(t)$$

where

$$\mathbf{F} = \begin{bmatrix} -1 & 0 \\ 1 & 0 \end{bmatrix}, \quad \mathbf{G} = \begin{bmatrix} 1 \\ 0 \end{bmatrix}$$

Since this is basically a hybrid system, it would be reasonable to agree that the output should be calculated only at time instants when the input changes its value. At the end of the first time period $t = T$, the state vector is

$$\mathbf{x}(T) = e^{\mathbf{F}T}\mathbf{x}(0) + \int_0^T e^{\mathbf{F}(T-\tau)}\mathbf{G}r(0)\, d\tau$$

$$= \mathbf{A}(T)\mathbf{x}(0) + \int_0^T e^{\mathbf{F}(T-\tau)}\, d\tau\, \mathbf{G}r(0) \quad (1.5\text{-}4)$$

where $\mathbf{A}(T) = e^{\mathbf{F}T}$.

Sec. 1.5   Modern Control Theory

Now the integral in (1.5-4) may be evaluated directly by making the change of variables $t = T - \tau$.

$$\int_0^T e^{\mathbf{F}(T-\tau)} \, d\tau = \int_0^T e^{\mathbf{F}t} \, dt = \int_0^T \mathbf{A}(t) \, dt \tag{1.5-5}$$

Furthermore, if we introduce the notation

$$\int_0^T \mathbf{A}(\tau) \, d\tau \, \mathbf{G} = \mathbf{B}(T)$$

then (1.5-4) may be simplified to

$$\mathbf{x}(T) = \mathbf{A}(T)\mathbf{x}(0) + \mathbf{B}(T)r(0) \tag{1.5-6}$$

The matrices $\mathbf{A}(T)$ and $\mathbf{B}(T)$ are composed of constant elements. Hence, (1.5-6) is a simple algebraic equation in which the state vector of the system at time $t = T$ is related to the initial condition vector and the input for the period $0 < t \leq T$.

The numerical characteristics given in (1.5-1) lead to

$$\mathbf{A}(T) = \begin{bmatrix} e^{-T} & 0 \\ 1 - e^{-T} & 1 \end{bmatrix} \tag{1.5-6a}$$

and

$$\mathbf{B}(T) = \int_0^T \begin{bmatrix} e^{-\tau} & 0 \\ 1 - e^{-\tau} & 1 \end{bmatrix} \begin{bmatrix} 1 \\ 0 \end{bmatrix} d\tau = \begin{bmatrix} 1 - e^{-T} \\ T - 1 + e^{-T} \end{bmatrix} \tag{1.5-6b}$$

For convenience T is selected as 1. Then

$$\mathbf{A}(T) = \begin{bmatrix} .368 & 0 \\ .632 & 1 \end{bmatrix}$$

and

$$\mathbf{B}(T) = \begin{bmatrix} .632 \\ .368 \end{bmatrix}$$

With these numerical values (1.5-6) becomes

$$\begin{bmatrix} x_1(1) \\ x_2(1) \end{bmatrix} = \begin{bmatrix} .368 & 0 \\ .632 & 1 \end{bmatrix} \begin{bmatrix} x_1(0) \\ x_2(0) \end{bmatrix} + \begin{bmatrix} .632 \\ .368 \end{bmatrix} r(0) \tag{1.5-7}$$

$$c(1) = x_2(1)$$

Equation (1.5-7) represents the response of the system to the first of the sequence of control inputs (which we view as computer outputs) at time $t = 1$ second. The next instant at which an observation of the system is to be made is at time $t = 2$ seconds, up to which time the input consists of the sequence $\{r_0, r_1\}$. The output of the system at $t = 2$ may be very simply obtained by taking $x(1)$ as the initial conditions for the second interval and r_1 as the driving input. Thus, (1.5-7) may be applied again, i.e.,

$$\begin{bmatrix} x_1(2) \\ x_2(2) \end{bmatrix} = \begin{bmatrix} .368 & 0 \\ .632 & 1 \end{bmatrix} \begin{bmatrix} x_1(1) \\ x_2(1) \end{bmatrix} + \begin{bmatrix} .632 \\ .368 \end{bmatrix} r(1) \qquad (1.5\text{-}8)$$

$$c(2) = x_2(2)$$

with no additional calculations for the matrices $\mathbf{A}(T)$ and $\mathbf{B}(T)$. Indeed, (1.5-7) holds for the nth period, i.e.,

$$\begin{bmatrix} x_1(k+1) \\ x_2(k+1) \end{bmatrix} = \begin{bmatrix} .368 & 0 \\ .632 & 1 \end{bmatrix} \begin{bmatrix} x_1(k) \\ x_2(k) \end{bmatrix} + \begin{bmatrix} .632 \\ .368 \end{bmatrix} r(k) \qquad (1.5\text{-}9)$$

$$c(k+1) = x_2(k+1)$$

This equation allows us to compute response of the system for any time period, providing the state of the system at the beginning of the period and the input during the period are known. Such an equation is of great importance in the study of discrete-time systems and is generally known as the *discrete state equation* (1.5-9). Equation (1.5-9) is the difference equation that gives the system's output at the discrete times $t = kT$ when the control input $r(t)$ changes its level. This difference equation arises in a natural way, because the digital controller generates the piecewise constant input.

The study of linear discrete-time systems is further supported by the z-transform, which will be developed in Chapters 4, 5, and 6.

1.6 Sampled-data Systems

In the previous decade, much effort has been made toward the analysis and synthesis of sampled-data systems. Historically, sampled-data theory is the forerunner of discrete optimal control theory.

A sampled-data system is defined to be a collection of dynamical subsystems in which some of the signals appear at one or more points in the form of a sequence of numbers. The control systems of Section 1.5 are examples of sampled-data systems. Figure 1.6-1 depicts a portion of a typical sampled-data system.

Sec. 1.6 — Sampled-data Systems

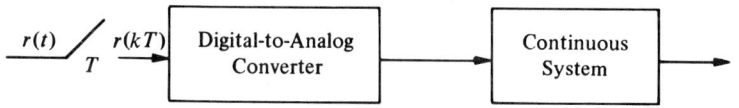

Figure 1.6-1. Sampled-data system.

The switch shown in Figure 1.6-1 is assumed to close instantaneously every T seconds and to generate the number sequence

$$\{\ldots r(-2T), r(-T), r(0), r(T), \ldots\} \quad (1.6\text{-}1)$$

which is used as the digital-to-analog converter's input. The function of the digital-to-analog converter is to change this number sequence (1.6-1) into a continuous signal $c(t)$, which may be used to influence (control) the continuous system. Most modern sampled-data analysis is based on the assumption that the digital-to-analog converter is characterized by

$$c(t) = r(kT) \quad \text{for } kT \leq t < (k+1)T \quad (1.6\text{-}2)$$

This is the well-known characteristic of a zero-order hold.

In order to make an analysis of sampled-data systems, most investigators have utilized the concept of impulse sampling. Impulse sampling is introduced in order that the resultant mathematical relationships describing the system can be generated in a straightforward manner. The beginning student is cautioned to realize that impulse sampling is a fictitious concept introduced only for mathematical convenience. A pictorial representation of an impulse sampler is shown in Figure 1.6-2.

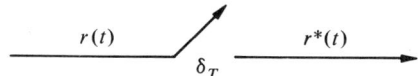

Figure 1.6-2. Impulse sampler.

By definition, the output of the impulse sampler is given by

$$r^*(t) = \sum_{k=-\infty}^{\infty} r(kT)\delta(t - kT) \quad (1.6\text{-}3)$$

where $\delta(t)$ is the dirac delta function (impulse). A typical function $r(t)$ and its impulse sample are shown in Figure 1.6-3. The sampled function $r^*(t)$ is a time sequence of appropriately weighted impulses. Taking the Laplace transform of (1.6-3), assuming $r(t) = 0$ for $t < 0$, we have

$$R^*(s) = \mathscr{L}[r^*(t)] = \int_0^\infty \sum_{k=0}^{\infty} r(kT)\delta(t - kT)e^{-st}\, dt \quad (1.6\text{-}4)$$

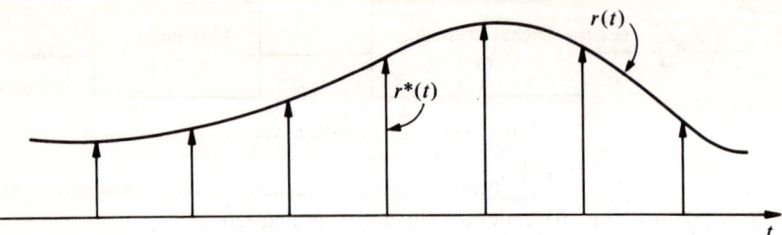

Figure 1.6-3. Input $r(t)$ and output sequence $r^*(t)$ of impulse sampler.

and under the assumption that interchanging the operations of summation and integration is proper, $R^*(s)$ becomes

$$R^*(s) = \sum_{k=0}^{\infty} r(nT)e^{-ksT} \qquad (1.6\text{-}5)$$

If the sequence of impulses in (1.6-3) serves as an input to a system with the transfer function (later shown to be related to the zero-order data hold)

$$G_0(s) = \frac{C(s)}{R^*(s)} = \frac{1 - e^{-sT}}{s} \qquad (1.6\text{-}6)$$

or its equivalent impulse response

$$g_0(t) = u(t) - u(t - T) \qquad (1.6\text{-}7)$$

it follows that the output $c(t)$ of such a system is given by

$$c(t) = r(kT) \quad \text{for } kT \leq t < (k+1)T \qquad (1.6\text{-}8)$$

Comparison of (1.6-2) with (1.6-8) reveals the equivalence of the two systems. The first, in which a physically realizable digital-to-analog converter is used, and the second, in which a fictitious impulse sampler in conjunction with the system of (1.6-6) is used, are shown to give identical responses to the same input. The equivalent systems are shown in Figure 1.6-4. Because of this equivalence, whenever the digital-to-analog converter of Figure 1.6-4(a) appears in an actual system, it may be mathematically replaced by the impulse sampler of Figure 1.6-4(b) without affecting the signals resulting in the remaining portion of the overall system. This is a mathematical equivalence only, since an impulse sampler is not physically realizable. A typical response of the system shown in Figure 1.6-4 is shown in Figure 1.6-5.

The response of the system in Figure 1.6-4(b) to the input $r^*(t)$ is by (1.6-5) and (1.6-6)

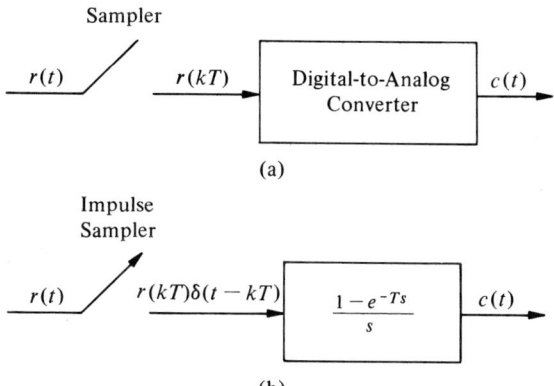

Figure 1.6-4. Equivalent systems.

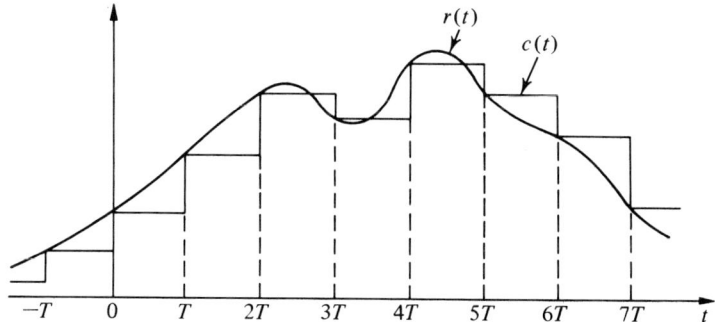

Figure 1.6-5. Response with zero-order hold D-A converter.

$$C(s) = \frac{1 - e^{-sT}}{s} \sum_{k=0}^{\infty} r(kT) e^{-ksT} \tag{1.6-9}$$

For inputs $r(t)$ whose Laplace transforms are ratios of polynomials in s, this infinite summation (1.6-9) has a convenient closed form representation in the variable e^{-ksT}.

EXAMPLE 1.6-1

For the systems shown in Figures 1.6-4(a) and 1.6-4(b) determine the Laplace transform of the output $c(t)$ to the step input

$$r(t) = u(t) = \begin{cases} 1 & \text{for } t \geq 0 \\ 0 & \text{for } t < 0 \end{cases} \tag{1.6-10}$$

Evaluating (1.6-9) with $r(kT) = 1$, we obtain

$$C(s) = \frac{1 - e^{-sT}}{s} \sum_{k=0}^{\infty} e^{-ksT} \qquad (1.6\text{-}11)$$

The infinite summation is a geometric series that is equal to

$$\sum_{k=0}^{\infty} e^{-ksT} = \frac{1}{1 - e^{-sT}} \quad \text{for } |e^{-sT}| \leq 1 \qquad (1.6\text{-}12)$$

$C(s)$ is therefore given by

$$C(s) = \frac{1}{s} \qquad (1.6\text{-}13)$$

This implies that the response $c(t)$ is

$$c(t) = \begin{cases} 1 & \text{for } t \geq 0 \\ 0 & \text{for } t < 0 \end{cases} \qquad (1.6\text{-}14)$$

which, based on the characteristics assumed for a digital-to-analog converter [i.e., Equation (1.6-1)], is the response expected.

The mathematical conveniences that result by introducing the impulse modulator concept are self-evident. In the original system, both continuous and discrete signals are present, whereas in the fictitious mathematical equivalent system no discrete signals explicitly appear. An analysis using familiar Laplace transform methods is then possible, as is reflected by (1.6-9). From (1.6-9), it is noted that the variable e^{sT} occurs. This variable occurs so frequently in sampled-data systems that it is written more compactly as

$$z = e^{sT} \qquad (1.6\text{-}15)$$

Equation (1.6-15) reflects a change of variables from the s variable to the z variable. By (1.6-15) Equation (1.6-5) can be rewritten in the new variable z as

$$R(z) = R^*(s)\Big|_{z=e^{sT}} = \sum_{k=0}^{\infty} r(kT) z^{-k} \qquad (1.6\text{-}16)$$

Analyses of sampled-data systems are then made with the new z variable. It is shown in many classical texts on sampled-data systems how the z-transform leads in a natural manner to difference equations. This is an important point. In essence, an analysis of systems characterized by linear difference equations is equivalent to an analysis of a linear sampled-data system and vice versa. A more detailed study of sampled-data systems is given in Chapter 5.

1.7 Radar Tracking

The previous two sections have shown how difference equations occur in a natural way when the systems under consideration are basically continuous. There do exist systems that are inherently represented by difference equations. One such example is that of a pulsed radar tracking system. In this system, radar pulses are sent out periodically and the return signals are processed in order to determine the position of airborne objects (targets). The returned information is basically discrete (i.e., a sequence of numbers), because the emitted radar pulses themselves are discrete rather than continuous. Since the returned information is discrete, the processing of it is best accomplished by using discrete processing techniques that are carried out by special-purpose digital computers. This processing usually takes the form of a system of difference equations.

In the information processing, two conflicting requirements are normally made.

1. Good noise-suppressing capability.
2. Fast maneuver-following capability.

The noise-suppression property is desirable, since the input information to the processing computer is usually contaminated with noise; therefore, one of the processor's functions is to remove this noise without seriously degrading the useful information content of the input signal. In tracking some airborne object, it is also desirable that the tracker be able to track the object when it makes evasive movements. This is why requirement (2) is postulated. Unfortunately, as one makes the processing more noise-insensitive, the maneuvering capabilities suffer and vice versa.

In many pulsed radar tracker systems, the radar processor seeks to generate

3. A good estimate of the target's present position.
4. A good estimate of the target's present velocity.
5. A good estimate of the target's future position.

These estimates are made by processing the raw input data, which are the noise-contaminated positions of the target obtained from the returned radar pulses. Basically, one may think of the radar processor as a discrete filter. To illustrate how the information processing may take place, let

$$c_n = \text{noise-contaminated target position obtained from radar pulse } n$$
$$\bar{c}_n = \text{estimate of the actual position of the target at the } n\text{th pulse.}$$

(1.7-1)

If a linear processing is desired, one method for generating $\bar{c}_n$ is by letting it be a linear combination of present and past values of c_n; that is,

$$\bar{c}_n = \sum_{k=0}^{\infty} h_k c_{n-k} \tag{1.7-2}$$

The numbers $\{h_k\}$ are to be selected so that requirements (1) and (2) are satisfied. In many cases, as will be shown in Chapter 2, (1.7-2) is equivalent to the difference equation

$$\bar{c}_n = \sum_{i=0}^{N} \alpha_i c_{n-i} + \sum_{j=1}^{M} \beta_j \bar{c}_{n-j} \tag{1.7-3}$$

A set of equations similar to (1.7-2) or (1.7-3) may be formulated to give an estimate of

$$\bar{\dot{c}}_n = \text{estimate of the uncontaminated velocity of the target}$$

such as

$$\bar{\dot{c}}_n = \sum_{k=0}^{\infty} g_k c_{n-k} \tag{1.7-4}$$

or, equivalently,

$$\bar{\dot{c}}_n = \sum_{i=0}^{N_1} \gamma_i c_{n-i} + \sum_{i=1}^{M_1} \xi_i \bar{\dot{c}}_{n-i} \tag{1.7-5}$$

By combining the difference equations for position and velocity estimates, (1.7-3) and (1.7-5), a set of processing equations is evolved. As an example, the so-called "α-β" tracker equations [6]* have the form

$$\bar{c}_n = c_{pn} + \alpha(c_n - c_{pn}) \tag{1.7-6}$$

$$\bar{\dot{c}}_n = \bar{\dot{c}}_{n-1} + \frac{\beta}{T}(c_n - c_{pn}) \tag{1.7-7}$$

$$c_{pn-1} = \bar{c}_n + T\bar{\dot{c}}_n \tag{1.7-8}$$

where $T = $ radar repetition time

$c_{pn} = $ predicted position of target at nth radar pulse.

This system of equations, (1.7-6), (1.7-7), and (1.7-8), is a very simple, but effective, form of data processing. The parameters α and β are selected so to

*Numbers in square brackets refer to the references at the end of the chapter.

satisfy requirements (1) and (2). Methods for making this selection have been studied in detail (e.g. [6]).

1.8 Conclusion

The importance of the linear difference equation with regard to discrete systems has been demonstrated by the examples considered in this chapter. An understanding of the characteristics of linear difference equations and linear discrete systems is of prime importance in order properly to utilize such systems. The next few chapters will be directed toward this goal.

REFERENCES

1. Ragazzini, J. R. and G. F. Franklin, *Sampled-Data Control Systems*, McGraw-Hill, New York, 1958.
2. Dorf, R.C., *Time Domain Analysis and Design of Control Systems*, Addison-Wesley, Reading, Mass., 1966.
3. Timothy, L. K. and B. E. Bona, *State Variable—An Introduction*, McGraw-Hill, New York, 1967.
4. Schwarz, R. J. and B. Friedland, *Linear Systems*, McGraw-Hill, New York, 1965.
5. DeRusso, P. M., R. J. Roy, and C. M. Close, *State Variables for Engineers*, Wiley, New York, 1965.
6. Benedict, T. R. and G. W. Bordner, "Synthesis of an Optimal Set of Radar Track-While Scan Smoothing Equations," *IRE Transactions on Automatic Control*, Vol. AC-7, No. 4, July 1962, pp. 27–32.

PROBLEMS

1.1 For the single input–single output continuous and discrete system characterized by the following equations, determine which coefficients must be zero for the systems to be

    (a) Linear.
    (b) Time invariant.

    (i) $\alpha_1 \left(\dfrac{d^3 c}{dt^3}\right)^2 + \alpha_2 \dfrac{d^2 c}{dt^2} + (\alpha_3 + \alpha_4 c + \alpha_5 \sin t)\dfrac{dc}{dt} + \alpha_6 c = \alpha_7 r$

    (ii) $\alpha_1 c(k+3)^2 + \alpha_2 c(k+2) + [\alpha_3 + \alpha_4 c(k) + \alpha_5 \sin k]c(k+1) + \alpha_6 c(k) = \alpha_7 r(k)$, where $c(\ )$ is the output signal and $r(\ )$ the input signal.

1.2 Determine if the systems below are

(a) Linear.
(b) Time invariant.

(i) $S[r(\ )] = c(\ ) = r^2(\ )$

(ii) $S[r(\ )] = c(\ ) = \begin{cases} r(\ ), & \text{argument} \geq 0 \\ 0, & \text{argument} < 0 \end{cases}$

(iii) $S[r(\ )] = c(\ ) = \begin{cases} r(\ ), & r(\ ) \geq 0 \\ 0, & r(\ ) < 0 \end{cases}$

The time variables ( ) can be either continuous or discrete.

1.3 What are suitable quantization levels for the purpose of converting an analog signal of range ± 5 volts into a 5-bit digital signal?

1.4 Which of these combinations are valid signals?

Analog—continuous
Analog—discrete
Digital—continuous
Digital—discrete

1.5 Let an initial amount of money P_0 be invested in a savings account having a compound interest rate of $r\%$ per annum. Thus, after any given year, the amount on deposit is equal to the amount on deposit one year earlier plus the interest obtained from the previous year's deposit. If P_n represents the amount on deposit after the nth year, determine the relationship between P_{n+1} and P_n. Demonstrate that

$$P_n = P_0\left(1 + \frac{r}{100}\right)^n$$

satisfies (is a solution to) this first-order difference equation.

1.6 Let i_n be the current in the nth loop of the latter network shown in Figure P1.6. Write the difference equations that one obtains by using Kirchhoff's voltage law. Are these equations linear? Time invariant?

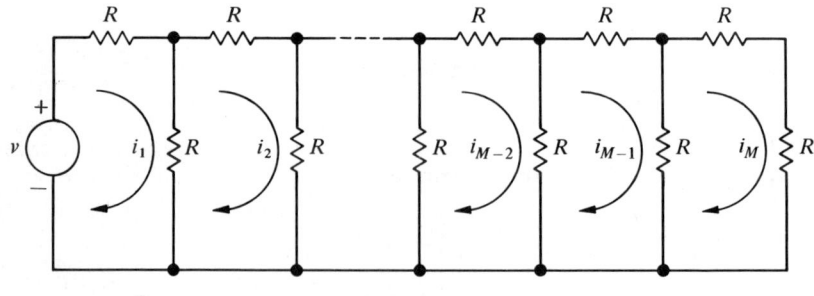

Figure P1.6.

1.7 If i_n denotes the current in the nth loop of the latter network shown in Figure P1.7, write the differential-difference equation for the nth loop.

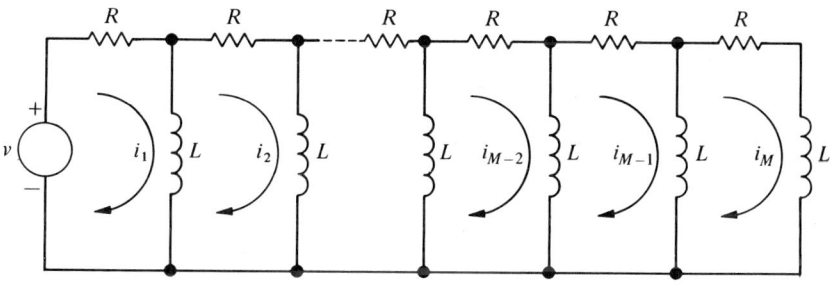

Figure P1.7.

1.8 For the system shown in Figure P1.8, the switch opens when the system reaches equilibrium (i.e., $i_L = V/R$) and closes when $i_L = 0$. V_n denotes the voltage across the capacitor at the nth closing of the switch; determine V_n as a function of R, L, C, and E. The diode is assumed perfect and the switch is initially closed at $t = 0$ with all initial conditions being zero. *Hint:* Use conservation of energy concepts.

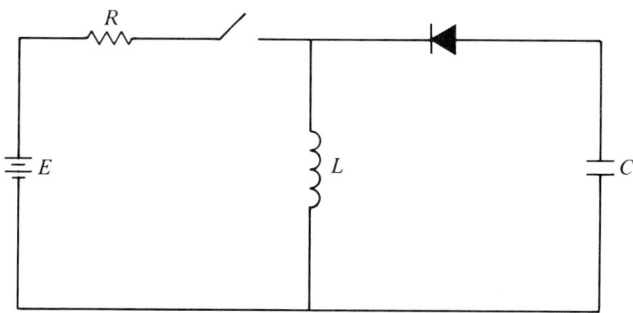

Figure P1.8.

1.9 It is desired to simulate on a digital computer a system whose Laplace transfer function is given by

$$H(s) = \frac{b-a}{(s+a)(s+b)}$$

(a) Using the two techniques of Section 1.4, determine two alternate difference equations that simulate this system.

(b) If a step input $r(t) = u(t)$ is applied to this system,
  (i) Compute its actual response $c_a(t)$.
  (ii) Compute the response one obtains by using the digital simulation $c_{s_1}(kT)$, $c_{s_2}(kT)$.
  (iii) As a comparison of how good the simulations are for the step input evaluate

$$\sum_{k=0}^{\infty} [c_a(kT) - c_{s_1}(kT)]^2 \quad \text{and} \quad \sum_{k=0}^{\infty} [c_a(kT) - c_{s_2}(kT)]^2$$

1.10 If the input to the system shown in Figure P1.10 is piecewise constant, that is,

$$r(t) = r_k \quad \text{for} \quad kT \leq t < (k+1)T$$

determine the difference equation that gives $c(kT)$ as a function of r_k. Put the result into matrix form.

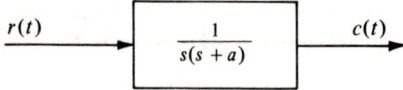

Figure P1.10

2

Time-domain Representations of Linear Discrete Systems

2.1 Introduction

It is apparent from the examples given in Chapter 1 that difference equations in various forms are the mathematical language utilized in the description of discrete-time systems. Some systems operate inherently in discrete time, whereas others, such as computer-controlled systems, operate partially in discrete time and partially in continuous time. The common characteristic of these seemingly different systems is that their dynamical behavior may be analyzed by applying discrete-time mathematical concepts.

A linear discrete system may be represented in many different, but equivalent, forms. Three such representations are (1) the *nth-order linear difference equation*, (2) the *weighting sequence*, and (3) the *discrete state equation*. Each representation method has its counterpart in linear continuous systems. The nth-order linear difference equation plays the same role as the nth-order linear differential equation; the weighting sequence is analogous to the impulse response; finally, the discrete state equations are analogous to the continuous state equations.

These three methods are so-called time-domain representations of linear discrete systems, since the discrete-time variable k appears in each. This chapter will investigate these three representation methods and show how they are related to one another. Another method for characterizing linear discrete systems, the transfer function, is introduced in Chapter 4.

2.2 The Linear Difference Equation

A linear, time-invariant, discrete system with one input and one output is characterized by the linear difference equation

$$c[(n+k)T] + a_1 c[(n+k-1)T] + \cdots + a_n c(kT)$$
$$= b_0 r[(n+k)T] + b_1 r[(n+k-1)T] + \cdots + b_n r(kT) \qquad (2.2\text{-}1)$$

where $c(kT)$ and $r(kT)$ are the system's output and input at time kT (kth iteration), respectively. The order of the difference equation as represented by (2.2-1) is the difference between the highest and lowest argument of $c(\ )$, in this case $(n+k) - (k) = n$. The coefficients $a_n, \ldots, a_1, b_n, \ldots, b_0$ are fixed numbers that characterize the dynamical behavior of the system. If these coefficients are not fixed but depend on the discrete-time parameter k, the system is said to be time varying in nature. We shall, in general, treat only time-invariant systems. Figure 2.2-1 illustrates a block diagram representation of such a system.

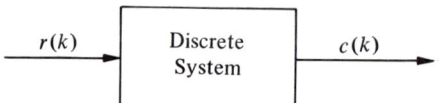

Figure 2.2-1. Block diagram representation of a discrete system.

The input is the sequence of numbers

$$\{\ldots r(-2T), r(-T), r(0), r(T), r(2T), \ldots\}$$

and the output is the sequence

$$\{\ldots c(-2T), c(-T), c(0), c(T), c(2T), \ldots\}$$

Although time appears as an explicit argument in these sequences it is not uncommon to delete T from the argument so that $r(kT)$ and $c(kT)$ may be written as $r(k)$ and $c(k)$, respectively. Whenever it is advisable, this abbreviated notation will be employed.

Equation (2.2-1) is typically used in describing the input-output relationship of a digital computer. Solving (2.2-1) for $c(n+k)$, the output, yields

$$c(n+k) = b_0 r(n+k) + b_1 r(n+k-1) + \cdots + b_n r(k)$$
$$- a_1 c(n+k-1) - \cdots - a_n c(k) \qquad (2.2\text{-}2)$$

Quite frequently, this equation is also written in the equivalent form

$$c(k) = b_0 r(k) + b_1 r(k-1) + \cdots + b_n r(k-n)$$
$$- a_1 c(k-1) - a_2 c(k-2) - \cdots - a_n c(k-n) \qquad (2.2\text{-}3)$$

Equation (2.2-3) is usually used in connection with digital computers. The number $c(k)$ represents the output at discrete time k, while $c(k-1)$, $c(k-2)$, ... are the past outputs that have been stored in the computer's memory. Similarly, $r(k)$, $r(k-1)$, ... represent the input numbers at discrete times k, $k-1$, etc. Equation (2.2-3) is thus a recursion equation that is applied every iteration. A FORTRAN program implementing this recursion equation is as follows:

```
        DIMENSION  C(N), R(N)
C       UPDATE PAST VALUES OF INPUT
        R(K-N)   = R(K-N+1)
        R(K-N+1) = R(K-N+2)
         .
         .
         .
        R(K-2)   = R(K-1)
        R(K-1)   = R(K)
        R(K)     = XIN
C       UPDATE PAST VALUES OF OUTPUT
        C(K-N)   = C(K-N+1)
        C(K-N+1) = C(K-N+2)
         .
         .
         .
        C(K-1)   = C(K)
C       CALCULATE NEW OUTPUT
        C(K)     = B0*R(K)+B1*R(K-1)+ ... +BN*R(K-N)
                   -A1*C(K-1)-A2*(K-2)- ... -AN*C(K-N)
```

The above program is, of course, not complete, but it effectively demonstrates the storage updating required to apply (2.2-3) iteratively.

2.3 Weighting Sequence of Linear Discrete Systems

For systems governed by linear differential equations, the concept of impulse response is very valuable. If one knows the impulse response for such a system, which is initially at rest, then by using the convolution integral, the response to any input may be evaluated. There exists a similar concept for linear discrete systems, the weighting sequence.

The impulse response is defined to be the continuous system's response (mathematical) to a Dirac delta function input. To calculate the impulse response of a linear, continuous system, one conceptually applies the Dirac delta function to the system's input and in some manner determines the resultant response. This response is called the *impulse response*. In discrete systems, we are working with a sequence of numbers rather than a function of continuous time; therefore, a sequence that has properties analogous to the Dirac delta function (impulse) is desired. The Kronecker delta sequence is such a sequence. It is defined as

$$r(k) = \delta_j(k) = \begin{cases} 1 & \text{for } k = j \\ 0 & \text{for } k \neq j \end{cases}$$

It is a sequence that is zero for all discrete time except $k = j$, where it equals one. Using the analogy from continuous systems, we apply the input

$$r(k) = \delta_0(k) = \begin{cases} 1 & \text{for } k = 0 \\ 0 & \text{for } k \neq 0 \end{cases}$$

to the system governed by the difference equation

$$c(k) = b_0 r(k) + b_1 r(k-1) + \cdots + b_n r(k-n) \\ - a_1 c(k-1) - a_2 c(k-2) - \cdots - a_n c(k-n) \quad (2.3\text{-}1)$$

This system is assumed to be at rest prior to the application of the input [i.e., $c(k) = 0$ for $k = -1, -2, -3, \ldots$].

Denote the resulting response $c(k)$ by $h(k)$; that is,

$$h(k) = c(k) \quad \text{for all } k$$

This sequence $h(k)$ is called the *weighting sequence*. It plays the same role for discrete systems as does the impulse response for continuous systems. Figure 2.3-1(a) shows a typical weighting sequence.

To demonstrate how the sequence $h(k)$ evolves, let us apply the Kronecker delta input $\delta_0(k)$ to the system (2.3-1). Since the system is initially at rest, we have $c(k) = 0$ for $k < 0$. Now we evaluate (2.3-1) at $k = 0$, obtaining

$$c(0) = b_0 r(0) = b_0$$

To determine $c(1)$, we let $k = 1$ in (2.3-1), obtaining

$$c(1) = b_1 r(0) - a_1 c(0) = b_1 - a_1 b_0$$

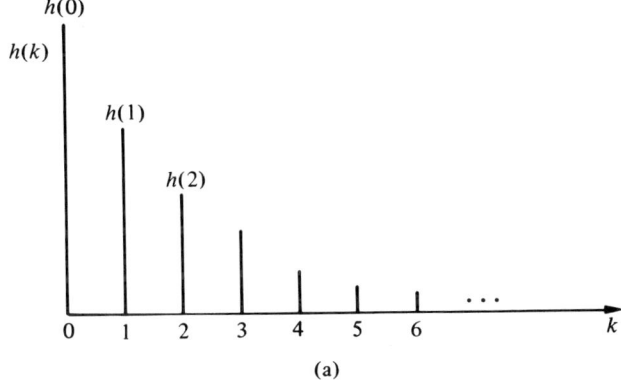

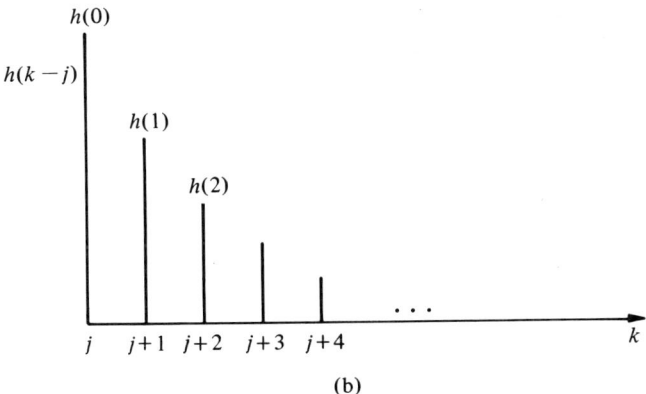

Figure 2.3-1. Weighting sequence.

Similarly,

$$c(2) = b_2 r(0) - a_1 c(1) - a_2 c(0)$$
$$= b_2 - a_1 b_1 + a_1^2 b_0 - a_2 b_0$$

and so forth. The weighting sequence $h(k)$ for this system is therefore

$$\begin{aligned} h(k) &= 0 \quad \text{for } k < 0 \\ h(0) &= c(0) = b_0 \\ h(1) &= c(1) = b_1 - a_1 b_0 \\ h(2) &= c(2) = b_2 - a_1 b_1 + a_1^2 b_0 - a_2 b_0 \end{aligned} \qquad (2.3\text{-}2)$$

$$\cdot \ \cdot \ \cdot \ \cdot \ \cdot \ \cdot \ \cdot \ \cdot \ \cdot \ \cdot \ \cdot \ \cdot$$

More systematic methods for determining $h(k)$ will be introduced later.

Since the system under consideration is linear, it follows that if the input

$$r(k) = \alpha_0 \delta_0(k) = \begin{cases} \alpha_0 & \text{for } k = 0 \\ 0 & \text{for } k \neq 0 \end{cases} \qquad (2.3\text{-}3)$$

is applied then the resultant response is

$$c^0(k) = \alpha_0 h(k) \qquad (2.3\text{-}4)$$

The superscript on $c(k)$ is used to denote the weighted (by α_0) Kronecker delta response applied at $k = 0$. To demonstrate the validity of (2.3-4), apply the input (2.3-3) to system (2.3-1). For $k = 0, 1, 2, \ldots$ we obtain

$$c(0) = b_0 r(0) = \alpha_0 b_0$$
$$c(1) = b_1 r(0) - a_1 c(0) = \alpha_0 [b_1 - a_1 b_0]$$
$$c(2) = b_2 r(0) - a_1 c(1) - a_2 c(0) = \alpha_0 [b_2 - a_1 b_1 + a_1^2 b_0 - a_2 b_0]$$
$$\cdot \quad \cdot \quad \cdot \quad \cdot \quad \cdot \quad \cdot \quad \cdot \quad \cdot \quad \cdot \quad \cdot \quad \cdot \quad \cdot \quad \cdot$$

A comparison of this response with (2.3-2) indicates the correctness of the assertion.

To further generalize the characteristics of this system, let us apply the weighted Kronecker delta sequence

$$r(k) = \alpha_1 \delta_1(k) = \begin{cases} \alpha_1 & \text{for } k = 1 \\ 0 & \text{for } k \neq 1 \end{cases} \qquad (2.3\text{-}5)$$

to (2.3-1). Again we assume that the system is initially at rest prior to application of the input. In this case, this implies that $c(k) = 0$ for $k < 1$. Therefore, evaluating (2.3-1) at $k = 1, 2, 3, \ldots$ results in

$$c(1) = b_0 r(1) = \alpha_1 b_0$$
$$c(2) = b_1 r(1) - a_1 c(1) = \alpha_1 [b_1 - a_1 b_0]$$
$$c(3) = b_2 r(1) - a_1 c(2) - a_2 c(1) = \alpha_1 [b_2 - a_0 b_1 + a_0^2 b_0 - a_1 b_0]$$
$$\cdot \quad \cdot \quad \cdot \quad \cdot \quad \cdot \quad \cdot \quad \cdot \quad \cdot \quad \cdot \quad \cdot \quad \cdot \quad \cdot \quad \cdot$$

A comparison of this result with (2.3-2) reveals that the response of the system to the input given by (2.3-5) is

$$c^1(k) = \alpha_1 h(k - 1)$$

More generally, let the input sequence

$$r(k) = \alpha_j \delta_j(k) = \begin{cases} \alpha_j & \text{for } k = j \\ 0 & \text{for } k \neq 1 \end{cases} \qquad (2.3\text{-}6)$$

be applied to system (2.3-1). Since the system is at rest prior to the application of (2.3-6), we have $c(k) = 0$ for $k < j$. Evaluating (2.3-1) at $k = j, j+1, j+2, \ldots$ gives

$$c(j) = b_0 r(j) = \alpha_j b_0$$
$$c(j+1) = b_1 r(j) - a_1 c(j) = \alpha_j [b_1 - a_1 b_0]$$
$$c(j+2) = b_2 r(j) - a_1 c(j+1) - a_2 c(j) = \alpha_j [b_2 - a_1 b_1 + a_1^2 b_0 - a_2 b_0]$$
$$\cdot \quad \cdot \quad \cdot \quad \cdot \quad \cdot \quad \cdot \quad \cdot \quad \cdot \quad \cdot \quad \cdot \quad \cdot \quad \cdot \quad \cdot \quad \cdot \quad \cdot \quad \cdot \quad \cdot$$

so that the general response to input (2.3-6) is

$$c^j(k) = \alpha_j h(k-j) \tag{2.3-7}$$

This result could have been immediately obtained by appealing to the linearity and time-invariance properties of system (2.3-1). The response $h(k-j)$ is shown in Fig. 2.3-1(b). It is simply the weighting sequence $h(k)$, delayed j units of discrete time.

If the more general input

$$r(k) = \begin{cases} 0 & \text{for } k < 0 \\ \alpha_k & \text{for } k \geq 0 \end{cases}$$

is applied, the response $c(k)$ can be found by invoking the linearity property. First, we shall rewrite this input as

$$r(k) = \alpha_0 \delta_0(k) + \alpha_1 \delta_1(k) + \alpha_2 \delta_2(k) + \cdots$$

Because of the system's linearity, the output at time $c(k)$ is equal to the sum of the individual responses caused by the inputs $r(0), r(1), r(2), \ldots, r(k)$, taken separately; that is,

$$c(k) = \sum_{j=0}^{k} c^j(k)$$

which by (2.3-7) becomes

$$c(k) = \sum_{j=0}^{k} \alpha_j h(k-j)$$

The following result has been demonstrated: If the general input sequence $\{r(0), r(1), \ldots\}$ is applied to the linear system of (2.3-1), which is initially at rest, the output at time k can be obtained by evaluating the expression

$$c(k) = \sum_{j=0}^{k} h(k-j) r(j) \qquad k = 0, 1, 2, \ldots \tag{2.3-8}$$

Equation (2.3-8) is called the *convolution summation* of the system (2.3-1). The similarity between it and the convolution integral for continuous systems is evident.

A change of variables will now be made to illustrate an alternate form for the convolution summation. Let the integers m and j be related by $m = k - j$ so that we have the following equivalences

j	0	1	2	...	k
m	k	$k-1$	$k-2$	...	0

and

$$r(j) = r(k - m)$$
$$h(k - j) = r(m)$$

Using these equivalences, we may rewrite (2.3-8) as

$$c(k) = \sum_{m=k}^{0} h(m) r(k - m)$$

and, since we are adding $k + 1$ numbers, it is possible to reverse the order of summation. We obtain

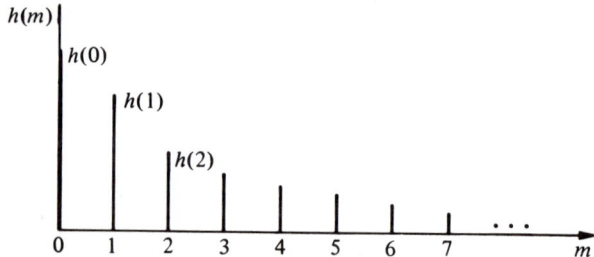

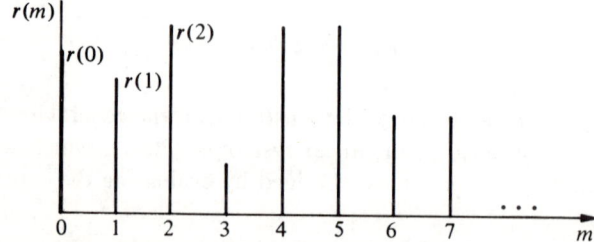

Figure 2.3-2. Typical weighting sequence and input.

$$c(k) = \sum_{m=0}^{k} h(m)r(k-m) \qquad (2.3\text{-}9)$$

To illustrate how the convolution summation (2.3-9) may be interpreted graphically, consider the input and weighting sequence shown in Figure 2.3-2.

To obtain $r(k - m)$, we simply take the mirror image of $r(m)$ and shift it to the right k discrete-time intervals, as shown in Figure 2.3-3(a) for $k = 3$.

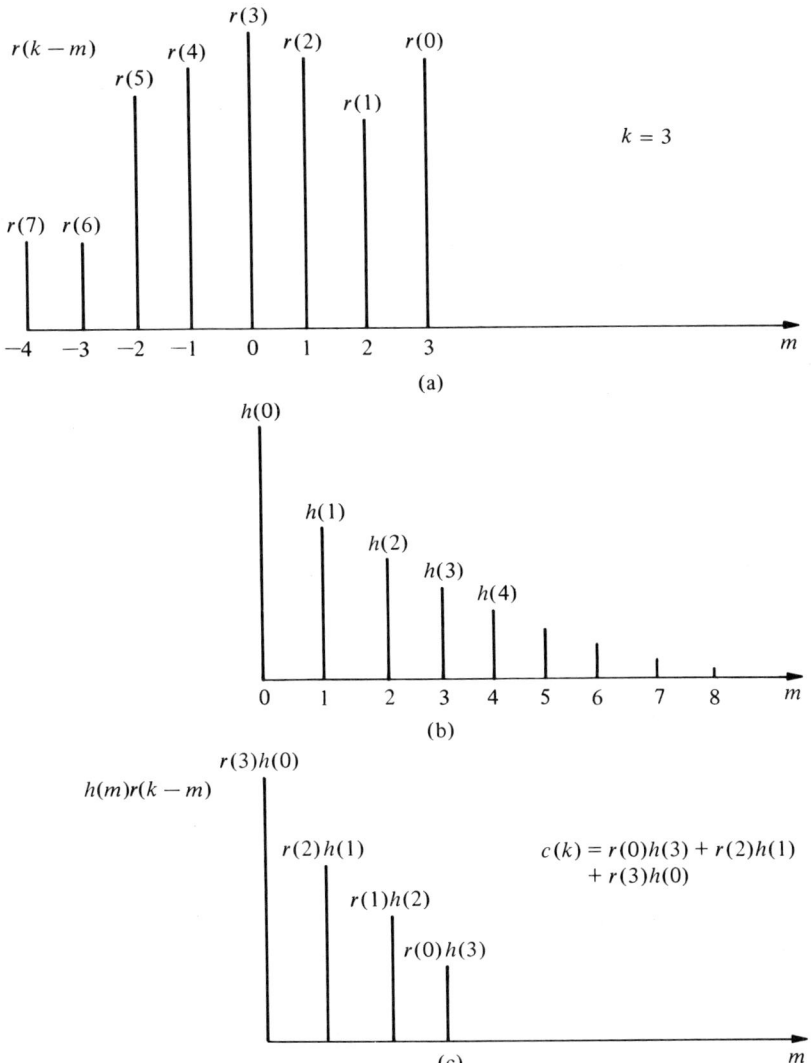

Figure 2.3-3. Illustration of convolution process.

Next we plot $h(m)$ directly below the plot of $r(k-m)$, as shown in Figure 2.3-3(b). Then we perform the multiplication of $r(k-m)$ with $h(m)$ by multiplying the magnitudes of the overlapping magnitudes of Figures 2.3-3(a) and (b), as shown in Figure 2.3-3(c). Finally, we sum the results of the individual products in Figure 2.3-3(c) to obtain $c(k)$.

The number $h(m)$ weights the contribution that the input m iterations in the past, $r(k-m)$, makes to the present output $c(k)$. For this reason, the sequence $\{h(0), h(1), \ldots\}$ is called the *weighting sequence*.

EXAMPLE 2.3-1

Consider a system represented by the difference equation

$$c(k) - \alpha c(k-1) = r(k) \qquad (2.3\text{-}10)$$

Determine its weighting sequence.

For the Kronecker delta input sequence $r(0) = 1, r(k) = 0$ for $k \neq 0$, the output sequence obtained by applying (2.3-10) repeatedly is

$$c(0) = 1, \quad c(1) = \alpha, \quad c(2) = \alpha^2, \quad \ldots, \quad c(k) = \alpha^k, \quad \ldots$$

The weighting sequence is therefore

$$h(k) = \begin{cases} \alpha^k & \text{for } k \geq 0 \\ 0 & \text{for } k < 0 \end{cases}$$

EXAMPLE 2.3-2

For the system of Example 2.3-1, determine the system's response to the discrete step input sequence

$$r(k) = \begin{cases} 1 & \text{for } k \geq 0 \\ 0 & \text{for } k < 0 \end{cases} \qquad (2.3\text{-}11)$$

Applying the convolution summation (2.3-9) with the weighting sequence obtained in Example 2.1-1 gives

$$c(k) = \sum_{m=0}^{k} \alpha^m$$

which is a finite geometric series. Its closed-form expression is found by subtracting $\alpha c(k)$ from $c(k)$; that is,

$$c(k) - \alpha c(k) = \sum_{m=0}^{k} \alpha^m - \alpha \sum_{m=0}^{k} \alpha^m$$

$$c(k)[1-\alpha] = \sum_{m=0}^{k} \alpha^m - \sum_{m=1}^{k+1} \alpha^m = 1 - \alpha^{k+1}$$

which, after we have solved for $c(k)$, gives

$$c(k) = \frac{1 - \alpha^{k+1}}{1 - \alpha}$$

It should be noted that if the parameter α has a magnitude less than one, then the response $c(k)$ approaches the constant $1/(1 - \alpha)$ for large values of discrete time k. The discrete step response of such a system is a discrete step in the steady state (i.e., k large). On the other hand, if α is greater than one in magnitude, then the response becomes unbounded for large k, a characteristic associated with an unstable system. Figure 2.3-4 shows the response plotted for $\alpha = \frac{1}{2}$.

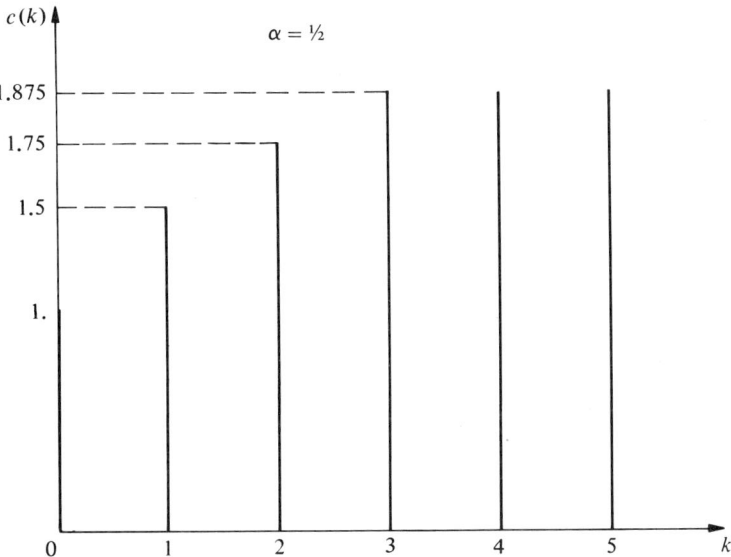

Figure 2.3-4. Discrete step response for system of (2.3-10).

In developing the convolution summation that characterizes a linear discrete system, it was assumed that the input was applied at $k = 0$. In the general case, the input may be initially applied for any value of k, either positive or negative. Since in (2.3-9) $h(m)$ weights the contribution that $r(k - m)$ makes to $c(k)$, this more general case is treated by reformulating (2.3-9) so that

$$c(k) = \sum_{m=0}^{\infty} h(m) r(k - m), \qquad -\infty < k < \infty \qquad (2.3\text{-}12)$$

By making the change of variables $j = k - m$, we obtain

$$c(k) = \sum_{j=-\infty}^{k} h(k - j) r(j), \qquad -\infty < k < \infty \qquad (2.3\text{-}13)$$

which corresponds to (2.3-8). Equations (2.3-12) and (2.3-13) are the general convolution summations that characterize a linear, time-invariant, discrete system with one input and one output as governed by the difference equation (2.3-1).

For linear continuous time systems, a necessary requirement for the physical realizability of a system is that its impulse response be zero for negative time. One might, therefore, naturally expect that a physically realizable linear discrete system be characterized by its weighting sequence's being zero for negative values of discrete time k. There exist many linear discrete systems for which $h(k) \neq 0$ for negative k. To demonstrate this, consider the following difference equation.

$$c(k) = \alpha c(k-1) + r(k+1) \tag{2.3-14}$$

To determine the weighting sequence of this system, we apply the input $r(k) = \delta_0(k)$ and calculate the response. At $k = -1$, (2.3-14) becomes

$$c(-1) = \alpha c(-2) + r(0)$$

and since the system is initially at rest, $c(-2) = 0$, it follows that

$$c(-1) = r(0) = 1$$

At $k = 0, 1, 2, \ldots$ we have

$$c(0) = \alpha c(-1) = \alpha$$
$$c(1) = \alpha c(0) = \alpha^2$$
$$c(2) = \alpha c(1) = \alpha^3$$
$$\cdot \quad \cdot \quad \cdot \quad \cdot \quad \cdot \quad \cdot$$
$$c(k) = \alpha^{k+1}$$

so that the weighting sequence for (2.3-14) is given by

$$h(k) = \begin{cases} \alpha^{k+1} & \text{for } k \geq -1 \\ 0 & \text{for } k < -1 \end{cases}$$

In this example, we have $h(-1) \neq 0$, which clearly demonstrates that a linear discrete system need not have $h(k) = 0$ for $k < 0$. This system is an anticipative system, as we can appreciate by noting that the present output $c(k)$ depends on the input one discrete time in the future [i.e., $r(k+1)$]. Since it is not possible to implement discrete systems whose present output depends on future values of input (as yet unknown), such systems are said to be physically unrealizable systems. Thus a necessary condition for a physically realizable system is that the weighting sequence $h(k)$ be zero for all negative k.

2.4 Weighting Sequence for Cascaded Systems

The reason that this anticipativeness property occurs is that (2.3-14) is not in the form given by (2.3-1). Nevertheless, even for anticipative systems, the weighting sequence plays the same role as it does for a nonanticipative system. In the anticipative situation, the output at time k depends both on previous and future values of inputs; that is,

$$c(k) = \sum_{m=-\infty}^{\infty} c^m(k)$$

where $c^m(k)$ is the consequence of the input $r(k-m)$ on the present output. Therefore,

$$c(k) = \sum_{m=-\infty}^{\infty} h(m)r(k-m) \qquad (2.3\text{-}15)$$

The only distinction between (2.3-12) and (2.3-15) is in the lower limit of summation. In making the change of variables $j = k - m$, (2.3-15) becomes

$$c(k) = \sum_{j=-\infty}^{\infty} h(k-j)r(j) \qquad (2.3\text{-}16)$$

Convolution summations (2.3-15) and (2.3-16) are the most general weighting sequence representations of a linear, time-invariant, discrete system. If the system is nonanticipative, then $h(k) = 0$ for $k < 0$, and (2.3-15) and (2.3-16) simplify to (2.3-12) and (2.3-13), respectively.

2.4 Weighting Sequence for Cascaded Systems

Many interesting properties exist for linear, time-invariant discrete systems connected in cascade. Two systems are said to be connected in cascade if the output of one system serves as the input to the second system. Figure 2.4-1 is a block diagram representation of two systems connected in cascade.

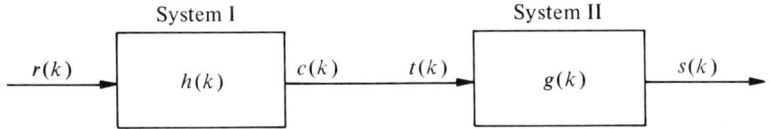

Figure 2.4-1. Systems in cascade.

Let the weighting sequences of systems I and II be $h(k)$ and $g(k)$, respectively. The output $c(k)$ of system I is related to its input $r(j)$ by

$$c(k) = \sum_{j=-\infty}^{\infty} h(k-j)r(j) \qquad (2.4\text{-}1)$$

and for system II the output $s(m)$ is

$$s(m) = \sum_{k=-\infty}^{\infty} g(m-k)t(k) \qquad (2.4\text{-}2)$$

where the signals $r(k)$, $c(k)$, $t(k)$, and $s(k)$ are located as indicated in Figure 2.4-1. For the purpose of deriving the weighting sequence of the two systems in cascade, we shall apply the input

$$r(k) = \delta_0(k) = \begin{cases} 1 & \text{for } k = 0 \\ 0 & \text{for } k \neq 0 \end{cases}$$

to system I. The response of system I to this input, by definition, is its weighting sequence. Therefore,

$$c(k) = h(k)$$

Now, the output of system I is equal to the input of system II; that is,

$$t(k) = c(k) = h(k) \qquad (2.4\text{-}3)$$

Therefore, the output of system II is given by

$$s(m) = \sum_{k=-\infty}^{\infty} g(m-k)h(k) \qquad (2.4\text{-}4)$$

which is a convolution summation. It is the weighting sequence of the cascaded system. By this technique, it is possible to replace two linear, time-invariant discrete systems connected in cascade by an equivalent linear, time-invariant system. Another property of interest is evident if one makes the change of variables $i = m - k$ in (2.4-4), giving

$$s(m) = \sum_{i=-\infty}^{\infty} h(m-i)g(i) \qquad (2.4\text{-}5)$$

A comparison of (2.4-4) with (2.4-5) reveals that the response of two linear, time-invariant discrete systems connected in cascade is independent of the order in which the systems are cascaded.

These cascading properties are easily generalized for the case when more than two linear, time-invariant discrete systems are cascaded.

EXAMPLE 2.4-1

Determine the equivalent weighting sequence for two systems connected in cascade when each system is characterized by the same difference equation

$$c(k) - \alpha c(k-1) = r(k) \qquad (2.4\text{-}6)$$

In Example 2.3-1, it was shown that the weighting sequence for the system of (2.4-6) is

$$h(k) = \begin{cases} \alpha^k & \text{for } k \geq 0 \\ 0 & \text{for } k < 0 \end{cases}$$

so that the equivalent weighting sequence for two such systems connected in cascade is, by (2.4-4),

$$f(k) = \sum_{j=0}^{k} \alpha^{k-j}\alpha^j = (k+1)\alpha^k \quad k = 0, 1, 2, \ldots \quad (2.4\text{-}7)$$

Equation (2.4-7) is the weighting sequence of the equivalent system.

2.5 State Variable Representations

A general state variable representation for a system governed by an nth-order, linear, time-invariant differential equation consists of a set of n simultaneous first-order differential equations. If such a system has one input and one output, this representation takes on the general form

$$\frac{d}{dt}\begin{bmatrix} x_1 \\ x_2 \\ \vdots \\ x_n \end{bmatrix} = \begin{bmatrix} f_{11} & f_{12} & \cdots & f_{1n} \\ f_{21} & & & \\ \vdots & & & \\ f_{n1} & & & f_{nn} \end{bmatrix} \begin{bmatrix} x_1 \\ x_2 \\ \vdots \\ x_n \end{bmatrix} + \begin{bmatrix} g_1 \\ g_2 \\ \vdots \\ g_n \end{bmatrix} r(t) \quad (2.5\text{-}1)$$

where $x_1, x_2, \ldots, x_n$ are the selected state variables, $r(t)$ is the single input, and the system's output $y(t)$ is given by

$$y(t) = [c_1 \ c_2 \ \cdots \ c_n] \begin{bmatrix} x_1 \\ x_2 \\ \vdots \\ x_n \end{bmatrix} + dr(t) \quad (2.5\text{-}2)$$

or in its matrix form

$$\dot{\mathbf{x}} = \mathbf{Fx} + \mathbf{G}r$$
$$y = \mathbf{Cx} + dr$$

The underlying principle of a state variable representation lies in the fact that if one is given the initial conditions

$$\begin{bmatrix} x_1(t_0) \\ x_2(t_0) \\ \vdots \\ x_n(t_0) \end{bmatrix}$$

and knowledge of the input $r(t)$ for time $t \geq t_0$, the values of $\mathbf{x}(t)$ and $y(t)$ for $t \geq t_0$ are completely determined.

Since first-order differential equations are used in state models of continuous systems, by analogy, one would correctly anticipate that first-order difference equations assume this role for discrete systems. Thus the equations

$$\begin{bmatrix} x_1(k+1) \\ x_2(k+1) \\ \vdots \\ x_n(k+1) \end{bmatrix} = \begin{bmatrix} a_{11} & a_{12} & \cdots & a_{1n} \\ a_{21} & & & \\ \vdots & & & \\ a_{n1} & & \cdots & a_{nn} \end{bmatrix} \begin{bmatrix} x_1(k) \\ x_2(k) \\ \vdots \\ x_n(k) \end{bmatrix} + \begin{bmatrix} b_1 \\ b_2 \\ \vdots \\ b_n \end{bmatrix} r(k) \qquad (2.5\text{-}3)$$

and the output equation

$$y(k) = [c_1 \quad c_2 \quad \cdots \quad c_n] \begin{bmatrix} x_1(k) \\ x_2(k) \\ \vdots \\ x_n(k) \end{bmatrix} + dr(k) \qquad (2.5\text{-}4)$$

would constitute a state variable representation for a linear discrete time-invariant system with one input and one output as given by (2.2-3). Methods for generating relationships (2.5-3) and (2.5-4) from the basic nth-order difference equation (2.2-3) are developed in Chapter 5. In state vector notation, we write (2.5-3) as

$$\mathbf{x}(k+1) = \mathbf{A}\mathbf{x}(k) + \mathbf{B}r(k) \qquad (2.5\text{-}5)$$

where $\mathbf{A}$ and $\mathbf{B}$ are matrices of constants; (2.5-4) is written as

$$y(k) = \mathbf{C}\mathbf{x}(k) + dr(k) \qquad (2.5\text{-}6)$$

The $n \times n$ matrix $\mathbf{A}$ in (2.5-5) determines the dynamical behavior of the system in the absence of an input [i.e., $r(k) = 0$]. It provides the transition characteristics from the state at one instance of time to the state at the next instance of time and is appropriately called the *state transition matrix*. The $n \times 1$ matrix **B** is called the *input matrix*, while **C** is the $1 \times n$ *output matrix*, and d is the 1×1 *direct transmission* matrix.

More generally, when there exist m inputs and p outputs, the dimensions of the matrices **A**, **B**, **C**, and **D** will be $n \times n$, $n \times m$, $p \times n$ and $p \times m$, respectively.

It is now possible to determine the system's state at time k resulting from an initial state $\mathbf{x}(0)$ and the input sequence $\{r(0), r(1), \ldots, r(k-1)\}$. Evaluating (2.5-5) successively for $k = 0, 1, 2, \ldots$, we find

$$\mathbf{x}(1) = \mathbf{A}\mathbf{x}(0) + \mathbf{B}r(0)$$
$$\mathbf{x}(2) = \mathbf{A}\mathbf{x}(1) + \mathbf{B}r(1) = \mathbf{A}^2\mathbf{x}(0) + \mathbf{A}\mathbf{B}r(0) + \mathbf{B}r(1)$$
$$\mathbf{x}(3) = \mathbf{A}\mathbf{x}(2) + \mathbf{B}r(2) = \mathbf{A}^3\mathbf{x}(0) + \mathbf{A}^2\mathbf{B}r(0) + \mathbf{A}\mathbf{B}r(1) + \mathbf{B}r(2) \quad (2.5\text{-}7)$$
$$\cdot \quad \cdot \quad \cdot \quad \cdot \quad \cdot \quad \cdot \quad \cdot \quad \cdot \quad \cdot \quad \cdot \quad \cdot \quad \cdot \quad \cdot$$
$$\mathbf{x}(k) = \mathbf{A}^k \mathbf{x}(0) + \sum_{m=0}^{k-1} \mathbf{A}^m \mathbf{B} r(k - m - 1)$$

Expression (2.5-7) reveals that the state at discrete time k is dependent on two factors: (1) the initial state $\mathbf{x}(0)$, and (2) the control input sequence $r(0), r(1), \ldots, r(k-1)$. The fundamental matrix

$$\mathbf{\Phi}(k) = \mathbf{A}^k \quad (2.5\text{-}8)$$

relates the initial state contribution to $\mathbf{x}(k)$. It is possible to express the state and output vector in terms of the fundamental matrix; that is,

$$\mathbf{x}(k) = \mathbf{\Phi}(k)\mathbf{x}(0) + \sum_{m=0}^{k-1} \mathbf{\Phi}(m) \mathbf{B} r(k - m - 1) \quad (2.5\text{-}9)$$

and

$$y(k) = \mathbf{C}\mathbf{\Phi}(k)\mathbf{x}(0) + \sum_{m=0}^{k-1} \mathbf{C}\mathbf{\Phi}(m) \mathbf{B} r(k - m - 1) + dr(k) \quad (2.5\text{-}10)$$

EXAMPLE 2.5-1

Compute the first few states for the system described by the discrete state equations of Section 1.4.

$$\begin{bmatrix} x_1(k+1) \\ x_2(k+1) \end{bmatrix} = \begin{bmatrix} .368 & 0 \\ .632 & 1 \end{bmatrix} \begin{bmatrix} x_1(k) \\ x_2(k) \end{bmatrix} + \begin{bmatrix} .632 \\ .368 \end{bmatrix} r(k) \quad (1.4\text{-}9)$$
$$y(k) = c(k) = x_2(k)$$

The input sequence is given as a unit step function,

$$r(k) = \begin{cases} 1 & \text{for } k \geq 0 \\ 0 & \text{for } k < 0 \end{cases}$$

Since the matrices in (1.4-9) are computed for time periods of one second, the states that will be computed correspond to times

$$t = 0, 1, 2, 3, \ldots$$

The initial state will be assumed to be

$$\begin{bmatrix} x_1(0) \\ x_2(0) \end{bmatrix} = 0$$

At the end of the first period the state is

$$\begin{bmatrix} x_1(1) \\ x_2(1) \end{bmatrix} = \begin{bmatrix} .368 & 0 \\ .632 & 1 \end{bmatrix} \begin{bmatrix} 0 \\ 0 \end{bmatrix} + \begin{bmatrix} .632 \\ .368 \end{bmatrix} = \begin{bmatrix} .632 \\ .368 \end{bmatrix}$$

At the end of the second period,

$$\begin{bmatrix} x_1(2) \\ x_2(2) \end{bmatrix} = \begin{bmatrix} .368 & 0 \\ .632 & 1 \end{bmatrix} \begin{bmatrix} .632 \\ .368 \end{bmatrix} + \begin{bmatrix} .632 \\ .368 \end{bmatrix} = \begin{bmatrix} .864 \\ 1.136 \end{bmatrix}$$

At the end of the third period,

$$\begin{bmatrix} x_1(3) \\ x_2(3) \end{bmatrix} = \begin{bmatrix} .368 & 0 \\ .632 & 1 \end{bmatrix} \begin{bmatrix} .864 \\ 1.136 \end{bmatrix} + \begin{bmatrix} .632 \\ .368 \end{bmatrix} = \begin{bmatrix} .950 \\ 2.040 \end{bmatrix}$$

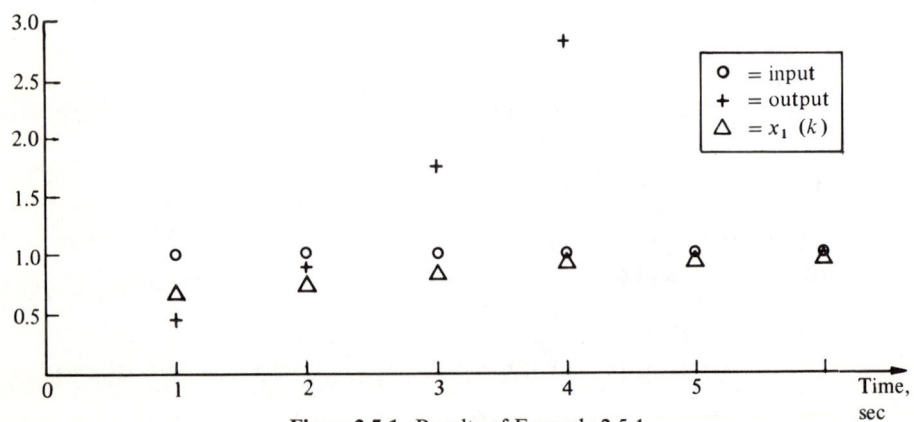

Figure 2.5-1. Results of Example 2.5-1.

and, similarly,

$$\begin{bmatrix} x_1(4) \\ x_2(4) \end{bmatrix} = \begin{bmatrix} .982 \\ 3.018 \end{bmatrix}, \quad \begin{bmatrix} x_1(5) \\ x_2(5) \end{bmatrix} = \begin{bmatrix} .993 \\ 3.996 \end{bmatrix}, \quad \begin{bmatrix} x_1(6) \\ x_2(6) \end{bmatrix} = \begin{bmatrix} 1.00 \\ 5.00 \end{bmatrix}$$

The output is given by the sequence

$$\{c(0), c(1), c(2), \ldots\} = \{0, .368, 1.136, 2.040, 3.008, 3.996, 5.000, \ldots\}$$

Figure 2.5-1 shows a plot of the results.

2.6 Continuous Systems with Piecewise Constant Inputs

As indicated in Chapter 1, there are many processes in which a continuous system has applied to it inputs that are constant over T-second intervals. Computer-controlled and sampled-data systems are such examples. We shall examine here what the discrete state model and the weighting sequence of such systems are.

2.6-1 Discrete State Model

The state model of a continuous system is given by

$$\dot{\mathbf{x}}(t) = \mathbf{F}\mathbf{x} + \mathbf{G}\mathbf{r}(t), \quad \mathbf{x}(t_0) = \mathbf{x}_0 \quad (2.6\text{-}1)$$
$$\mathbf{y}(t) = \mathbf{C}\mathbf{x} + \mathbf{D}\mathbf{r}(t) \quad (2.6\text{-}2)$$

The solution of (2.6-1) is

$$\mathbf{x}(t) = e^{\mathbf{F}(t-t_0)}\mathbf{x}_0(t_0) + \int_{t_0}^{t} e^{\mathbf{F}(t-\tau)}\mathbf{G}\mathbf{r}(\tau)d\tau \quad (2.6\text{-}3)$$

We are considering inputs that are constant over intervals of T seconds' duration; therefore, the input $r(t)$ is of the form

$$\mathbf{r}(t) = \mathbf{r}(kT), \quad kT \le t < (k+1)T$$

The initial condition at the beginning of the interval $[kT, (k+1)T]$ is

$$\mathbf{x}(t_0) = \mathbf{x}(kT)$$

To determine $\mathbf{x}(t)$ at the end of this interval we substitute into (2.6-3)

$$t_0 = kT$$
$$t = (k+1)T$$

and obtain

$$\mathbf{x}[(k+1)T] = e^{\mathbf{F}T}\mathbf{x}(kT) + \int_{kT}^{(k+1)T} e^{\mathbf{F}[(k+1)T-\tau]}\mathbf{Gr}(kT)d\tau \qquad (2.6\text{-}4)$$

The integral may be evaluated as follows:

$$\int_{kT}^{(k+1)T} e^{\mathbf{F}[(k+1)T-\tau]}\mathbf{Gr}(kT)d\tau = \left\{\int_{kT}^{(k+1)T} e^{\mathbf{F}[(k+1)T-\tau]}\mathbf{G}d\tau\right\}\mathbf{r}(kT)$$

since the input is constant for the integration interval. Furthermore, the integral is valid for all k. By making the change of variables $t = (k+1)T - \tau$, we have

$$\int_{kT}^{(k+1)T} e^{\mathbf{F}[(k+1)T-\tau]}\mathbf{G}d\tau = \int_{0}^{T} e^{\mathbf{F}t}\mathbf{G}dt$$

The last integral is a matrix of constants dependent on the parameter T. We write (2.5-4) now as

$$\mathbf{x}[(k+1)T] = e^{\mathbf{F}T}\mathbf{x}(kT) + \left\{\int_{0}^{T} e^{\mathbf{F}\tau}\mathbf{G}d\tau\right\}\mathbf{r}(kT) \qquad (2.6\text{-}5)$$

Equation (2.6-5) is a first-order vector difference equation, and its form is identical to (2.5-5). In fact, the two equations can be made to correspond to each other if we let

$$e^{\mathbf{F}T} = \mathbf{A}(T) = \mathbf{A} \qquad (2.6\text{-}6)$$

and

$$\int_{0}^{T} e^{\mathbf{F}\tau}\mathbf{G}d\tau = \mathbf{B}(T) = \mathbf{B} \qquad (2.6\text{-}7)$$

It is now clearly seen that both $\mathbf{A}(T)$ and $\mathbf{B}(T)$ are constant matrices whose elements are functions of T. Furthermore, two important properties of state transition matrices can be stated.

1. $\mathbf{A}^n(T) = \mathbf{A}(nT) = e^{\mathbf{F}nT}$ \qquad (2.6-8)
2. $\mathbf{A}(0) = \mathbf{I}$ \qquad (2.6-9)

Both relations are readily obtained by using (2.6-5).

In conclusion, the discrete state model for a continuous system with a piecewise constant input is given by

$$\mathbf{x}[(k+1)T] = \mathbf{A}(T)\mathbf{x}(kT) + \mathbf{B}(T)\mathbf{r}(kT) \qquad (2.6\text{-}10)$$

$$\mathbf{y}(kT) = \mathbf{C}\mathbf{x}(kT) + \mathbf{D}\mathbf{r}(kT) \qquad (2.6\text{-}11)$$

Since (2.6-10) and (2.6-11) are equivalent to (2.5-5) and (2.5-6) in form, it may be concluded that the mathematical model of a continuous system with piecewise constant inputs is equivalent to that of a linear discrete system with an input consisting of a number sequence. To distinguish these equivalences, the matrices $\mathbf{A}(T)$ and $\mathbf{B}(T)$ are written to show the explicit dependence on T in the former case, whereas in the latter case T does not appear [see (2.5-5)].

2.6-2 Weighting Sequence Model

If a continuous system with one input and one output has an impulse response $h(t)$, then the relationship existing between the input $r(t)$ and output $c(t)$ is

$$c(t) = \int_{-\infty}^{t} h(t - \tau) r(\tau) d\tau \qquad (2.6\text{-}12)$$

Again, the input $r(\tau)$ is taken to be piecewise constant; that is,

$$r(\tau) = r(kT), \qquad kT \leq \tau < (k+1)T \qquad (2.6\text{-}13)$$

or equivalently

$$r(\tau) = \sum_{k=0}^{\infty} r(kT)[u(\tau - kT) - u(\tau - kT - T)] \qquad (2.6\text{-}14)$$

where it is now assumed that $r(\tau) = 0$ for $\tau < 0$ and $u(t)$ is the unit step function. Inserting (2.6-14) into (2.6-12) and evaluating $c(t)$ at $t = nT$, we have

$$c(nT) = \sum_{k=0}^{\infty} \int_{0}^{nT} h(nT - \tau) r(kT)[u(\tau - kT) - u(\tau - kT - T)] d\tau \qquad (2.6\text{-}15)$$

By noting that the contributions of terms for $k \geq n$ are zero, (2.6-15) is rewritten as

$$c(nT) = \sum_{k=0}^{n-1} r(kT) \int_{kT}^{(k+1)T} h(nT - \tau) d\tau \qquad (2.6\text{-}16)$$

If we define

$$\hat{h}(nT - kT) = \int_{kT}^{(k+1)T} h(nT - \tau) d\tau = \int_{0}^{T} h(nT - kT - \tau) d\tau \qquad (2.6\text{-}17)$$

the expression for $c(nT)$ simplifies to

$$c(nT) = \sum_{k=0}^{n-1} r(kT) \hat{h}(nT - kT) \qquad (2.6\text{-}18)$$

But (2.6-18) is of the same form as (2.3-8), so that the properties developed in the previous sections of this chapter apply to this problem also. This is just another example that reinforces the fact that a continuous system with a piecewise constant input has dynamical characteristics similar to a discrete system driven by a sequence of numbers.

EXAMPLE 2.6-1

Determine the weighting sequence for the system with transfer function

$$\frac{C(s)}{R(s)} = \frac{1}{s+a} \qquad (2.6\text{-}19)$$

when piecewise constant inputs over T-second intervals are applied to it.
The impulse response for this system is

$$h(t) = e^{-at} \qquad t \geq 0$$

which, when inserted into (2.6-17), gives

$$\hat{h}(nT - kT) = \int_0^T e^{-a(nT-kT-\tau)} d\tau = \frac{(e^{aT} - 1)e^{-a(nT-kT)}}{a}$$

This is the weighting sequence for the system of (2.6-19). It allows us to determine the system's output at the sampling instants nT, which by (2.6-18) is

$$c(nT) = \frac{e^{aT} - 1}{a} e^{-anT} \sum_{k=0}^{n-1} r(kT) e^{akT} \qquad (2.6\text{-}20)$$

For example, if a unit step is applied to this system, that is, if

$$r(t) = u(t)$$

then

$$r(kT) = \begin{cases} 1 & \text{for } k \geq 0 \\ 0 & \text{for } k < 0 \end{cases}$$

Substituting $r(kT)$ into (2.6-20) gives

$$c(nT) = \frac{e^{aT} - 1}{a} e^{-anT} \sum_{k=0}^{n-1} e^{akT}$$

However,

$$\sum_{k=0}^{n-1} e^{akT} = \frac{1 - e^{-anT}}{1 - e^{aT}}$$

Therefore,

$$c(nT) = \frac{1 - e^{-anT}}{a} \quad \text{for } n \geq 0 \tag{2.6-21}$$

To see that (2.6-21) is in fact true, let $R(s) = 1/s$ in (2.6-19).

$$C(s) = \frac{1}{s(s + a)} = \frac{1}{a}\left[\frac{1}{s} - \frac{1}{s + a}\right]$$

Therefore,

$$c(t) = \frac{1}{a}[1 - e^{-at}] \quad \text{for } t \geq 0$$

which agrees with (2.6-21) for $t = nT$.

2.7 Relationships between Discrete System Representations

It is desirable to investigate several relationships that exist between the two methods of using linear difference equations in representing discrete systems. Obviously, both methods can be employed to describe the same system. The relationships to be derived will provide flexibility in converting from one form to another.

2.7-1 Discrete State Models from Digital Recursion Equations

Let the input and output number sequences of a discrete system be related by the general recursion formula

$$c(k) = b_0 r(k) + b_1 r(k - 1) + \cdots + b_n r(k - n)$$
$$- a_1 c(k - 1) - a_2 c(k - 2) - \cdots - a_n c(k - n) \tag{2.7-1}$$

There are many techniques for converting this nth-order difference equation to a discrete state model as given by (2.5-5) and (2.5-6). A detailed derivation of the individual methods is best carried out by using z-transform techniques and is, therefore, postponed until Chapter 5. The results of two popular methods, which have their counterparts in continuous state representations, are summarized here.

1. *Direct Programming*

A state model for the discrete system characterized by (2.7-1) is obtained by using a method widely known as *direct programming* and is given by

$$\begin{bmatrix} x_1(k+1) \\ x_2(k+1) \\ x_3(k+1) \\ \vdots \\ x_{n-1}(k+1) \\ x_n(k+1) \end{bmatrix} = \begin{bmatrix} -a_1 & -a_2 & -a_3 & \cdots & -a_{n-2} & -a_{n-1} & -a_n \\ 1 & 0 & 0 & & 0 & 0 & 0 \\ 0 & 1 & 0 & & 0 & 0 & 0 \\ \vdots & & & & & & \vdots \\ 0 & 0 & 0 & & 1 & 0 & 0 \\ 0 & 0 & 0 & \cdots & 0 & 1 & 0 \end{bmatrix} \times \begin{bmatrix} x_1(k) \\ x_2(k) \\ x_3(k) \\ \vdots \\ x_{n-1}(k) \\ x_n(k) \end{bmatrix} + \begin{bmatrix} 1 \\ 0 \\ 0 \\ \vdots \\ 0 \\ 0 \end{bmatrix} r(k) \qquad (2.7\text{-}2)$$

and

$$y(k) = c(k) = [\hat{b}_1 \quad \hat{b}_2 \quad \cdots \quad \hat{b}_{n-1} \quad \hat{b}_n] \begin{bmatrix} x_1(k) \\ x_2(k) \\ \vdots \\ x_n(k) \end{bmatrix} + b_0 r(k) \qquad (2.7\text{-}3)$$

where

$$\hat{b}_j = b_j - b_0 a_j \quad \text{for } j = 1, 2, \ldots, n$$

2. *Nested Programming*

Another method, *nested programming*, is frequently used for obtaining state space representations. It yields the representation

Sec. 2.7   Relationships between Discrete System Representations   57

$$\begin{bmatrix} x_1(k+1) \\ x_2(k+1) \\ \vdots \\ x_{n-1}(k+1) \\ x_n(k+1) \end{bmatrix} = \begin{bmatrix} -a_1 & 1 & 0 & \cdots & 0 & 0 \\ -a_2 & 0 & 1 & & 0 & 0 \\ \vdots & \vdots & \vdots & & \vdots & \vdots \\ -a_{n-1} & 0 & 0 & & 0 & 1 \\ -a_n & 0 & 0 & & 0 & 0 \end{bmatrix} \begin{bmatrix} x_1(k) \\ x_2(k) \\ \vdots \\ x_{n-1}(k) \\ x_n(k) \end{bmatrix}$$

$$+ \begin{bmatrix} b_1 - a_1 b_0 \\ b_2 - a_2 b_0 \\ \vdots \\ b_{n-1} - a_{n-1} b_0 \\ b_n - a_n b_0 \end{bmatrix} r(k) \qquad (2.7\text{-}4)$$

with the output given by

$$y(k) = c(k) = [1 \; 0 \; \cdots \; 0] \begin{bmatrix} x_1(k) \\ x_2(k) \\ \vdots \\ x_n(k) \end{bmatrix} + b_0 r(k) \qquad (2.7\text{-}5)$$

Note that each technique yields a form corresponding to (2.5-5) and (2.5-6).

EXAMPLE 2.7-1

Develop the direct programming discrete state model for the linear recursion formula

$$c(k) = -2c(k-1) + .5c(k-2) - .2c(k-3) \\ + .5r(k) - .4r(k-1) + .1r(k-2)$$

According to the notation used in (2.6-2) and (2.6-3), we have

$$a_1 = 2 \qquad\qquad b_0 = .5$$
$$a_2 = -.5 \quad \text{and} \quad b_1 = -.4$$
$$a_3 = .2 \qquad\qquad b_2 = .1$$

so that the state model is

$$\begin{bmatrix} x_1(k+1) \\ x_2(k+1) \\ x_3(k+1) \end{bmatrix} = \begin{bmatrix} -2 & .5 & -.2 \\ 1 & 0 & 0 \\ 0 & 1 & 0 \end{bmatrix} \begin{bmatrix} x_1(k) \\ x_2(k) \\ x_3(k) \end{bmatrix} + \begin{bmatrix} 1 \\ 0 \\ 0 \end{bmatrix} r(k)$$

$$c(k) = \begin{bmatrix} -1.4 & .35 & 0 \end{bmatrix} \begin{bmatrix} x_1(k) \\ x_2(k) \\ x_3(k) \end{bmatrix} + .5r(k)$$

EXAMPLE 2.7-2

Repeat Example 2.7-1 when the input sequence is delayed by one interval. The recursion formula is then

$$c(k) = -2c(k-1) + .5c(k-2) - .2c(k-3) \\ + .5r(k-1) - .4r(k-2) + .1r(k-3)$$

The corresponding state model is

$$\begin{bmatrix} x_1(k+1) \\ x_2(k+1) \\ x_3(k+1) \end{bmatrix} = \begin{bmatrix} -2 & .5 & -.2 \\ 1 & 0 & 0 \\ 2 & 1 & 0 \end{bmatrix} \begin{bmatrix} x_1(k) \\ x_2(k) \\ x_3(k) \end{bmatrix} + \begin{bmatrix} 1 \\ 0 \\ 0 \end{bmatrix} r(k)$$

$$c(k) = \begin{bmatrix} .5 & -.4 & .1 \end{bmatrix} \begin{bmatrix} x_1(k) \\ x_2(k) \\ x_3(k) \end{bmatrix}$$

2.7-2 Weighting Sequence from Discrete State Model

Let the discrete state model of a linear system be given as

$$\mathbf{x}(k+1) = \mathbf{A}\mathbf{x}(k) + \mathbf{B}r(k) \tag{2.7-6}$$
$$y(k) = c(k) = \mathbf{C}\mathbf{x}(k) + dr(k) \tag{2.7-7}$$

Since it is desired to find the weighting sequence, it is required that we find the system's response $c(k)$ when

1. The system is initially at rest; that is, $\mathbf{x}(0) = \mathbf{0}$.
2. The input

$$r(k) = \delta_0(k) = \begin{cases} 1 & \text{for } k = 0 \\ 0 & \text{for } k \neq 0 \end{cases}$$

is applied.

Under these restrictions, equations (2.7-6) and (2.7-7) are applied recursively, giving

$$c(0) = d \qquad \mathbf{x}(1) = \mathbf{B}$$
$$c(1) = \mathbf{CB} \qquad \mathbf{x}(2) = \mathbf{AB}$$
$$c(2) = \mathbf{CAB} \qquad \mathbf{x}(3) = \mathbf{A}^2\mathbf{B}$$
$$\cdot \quad \cdot \quad \cdot \quad \cdot \quad \cdot \qquad \cdot \quad \cdot \quad \cdot \quad \cdot \quad \cdot$$
$$c(k) = \mathbf{CA}^{k-1}\mathbf{B} \qquad \mathbf{x}(k) = \mathbf{A}^{k-1}\mathbf{B}$$

Therefore, the weighting sequence is given by

$$h(k) = \begin{cases} 0 & \text{for } k < 0 \\ d & \text{for } k = 0 \\ \mathbf{CA}^{k-1}\mathbf{B} & \text{for } k > 0 \end{cases} \qquad (2.7\text{-}8)$$

where the identity $\mathbf{A}^0 = \mathbf{I}$ has been incorporated for $h(1)$.

EXAMPLE 2.7-3

Determine the weighting sequence for the discrete process described by the equations

$$\begin{bmatrix} x_1(k+1) \\ x_2(k+2) \end{bmatrix} = \begin{bmatrix} -.3 & .4 \\ 1 & 0 \end{bmatrix} \begin{bmatrix} x_1(k) \\ x_2(k) \end{bmatrix} + \begin{bmatrix} 0 \\ 1 \end{bmatrix} r(k)$$

$$c(k) = \begin{bmatrix} 1 & -1 \end{bmatrix} \begin{bmatrix} x_1(k) \\ x_2(k) \end{bmatrix}$$

Applying (2.7-8), we obtain

$$h(k) = \begin{bmatrix} 1 & -1 \end{bmatrix} \begin{bmatrix} -.3 & .4 \\ 1 & 0 \end{bmatrix}^{k-1} \begin{bmatrix} 0 \\ 1 \end{bmatrix}, \qquad k \geq 1$$

To evaluate

$$\begin{bmatrix} -.3 & .4 \\ 1 & 0 \end{bmatrix}^{k-1}$$

we apply the Sylvester expansion formula as given in Appendix A2. The eigenvalues of the matrix are determined from the characteristic equation

$$\det \begin{bmatrix} \lambda + .3 & -.4 \\ -1 & \lambda \end{bmatrix} = (\lambda + .3)\lambda - .4 = (\lambda + .8)(\lambda - .5) = 0$$

so that $\lambda_1 = -.8$ and $\lambda_2 = .5$.

The constituent matrices are

$$\mathbf{A}_1 = \frac{\mathbf{A} - \lambda_2 \mathbf{I}}{\lambda_1 - \lambda_2} = \frac{1}{-1.3}\begin{bmatrix} -.3 - .5 & .4 \\ 1 & -.5 \end{bmatrix} = \frac{1}{1.3}\begin{bmatrix} .8 & -.4 \\ -1 & .5 \end{bmatrix}$$

$$\mathbf{A}_2 = \frac{\mathbf{A} - \lambda_1 \mathbf{I}}{\lambda_2 - \lambda_1} = \frac{1}{1.3}\begin{bmatrix} -.3 + .8 & .4 \\ 1 & .8 \end{bmatrix} = \frac{1}{1.3}\begin{bmatrix} .5 & .4 \\ 1 & .8 \end{bmatrix}$$

Note, as an arithmetic check, $\mathbf{A}_1 + \mathbf{A}_2 = \mathbf{I}$. Now

$$\begin{bmatrix} -.3 & .4 \\ 1 & 0 \end{bmatrix}^k = \frac{1}{1.3}\left\{ \begin{bmatrix} .8 & -.4 \\ -1 & .5 \end{bmatrix}(-.8)^k + \begin{bmatrix} .5 & .4 \\ 1 & .8 \end{bmatrix}(.5)^k \right\}$$

Substituting into $h(k)$, we obtain

$$h(k) = \frac{1}{1.3}\left\{ [1 \; -1]\begin{bmatrix} .8 & -.4 \\ 1 & .5 \end{bmatrix}\begin{bmatrix} 0 \\ 1 \end{bmatrix}(-.8)^{k-1} + [1 \; -1]\begin{bmatrix} .5 & .4 \\ 1 & .8 \end{bmatrix}\begin{bmatrix} 0 \\ 1 \end{bmatrix}(.5)^{k-1} \right\}$$

$$= \frac{1}{1.3}\{-.9(-.8)^{k-1} - .4(.5)^{k-1}\}$$

$$= -.692(-.8)^{k-1} - .308(.5)^{k-1}$$

2.8 Summary

Linear discrete systems may be represented in three different ways using time domain methods.
1. Difference equations or linear recursion formula.
2. Weighting sequences.
3. Discrete state equations.

All these methods are analogous to corresponding methods in the continuous time domain.

Procedures have been given for conversions from one representation to another. It is possible to convert a difference equation into a discrete state model, while a weighting sequence may be obtained from a discrete state model and vice versa.

A fourth method of representing linear discrete systems will be made available after an introduction to the z-transform. This method will offer a transfer function approach.

REFERENCES

1. Freeman, H., *Discrete-Time Systems*, Wiley, New York, 1965.
2. Schwarz, R. J. and B. Friedland, *Linear Systems*, McGraw-Hill, New York, 1965.

PROBLEMS

2.1 For a discrete system governed by the nth order difference equation (2.2-2) determine the first five terms of the weighting sequence.

2.2 Determine the weighting sequence for the system shown in Figure P2.2 in terms of the individual weighting sequences $h(k)$ and $c(k)$. The two systems are said to be connected in parallel.

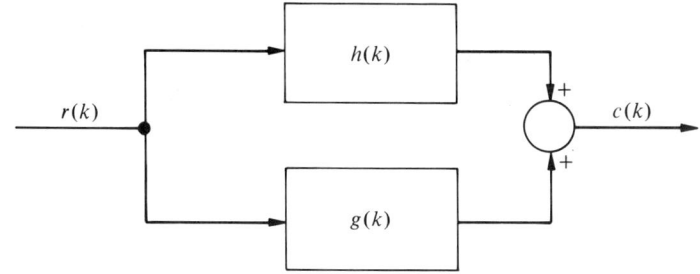

Figure P2.2.

2.3 Determine a recursive relationship that allows one to obtain the weighting sequence for the discrete feedback system shown in Figure P2.3.

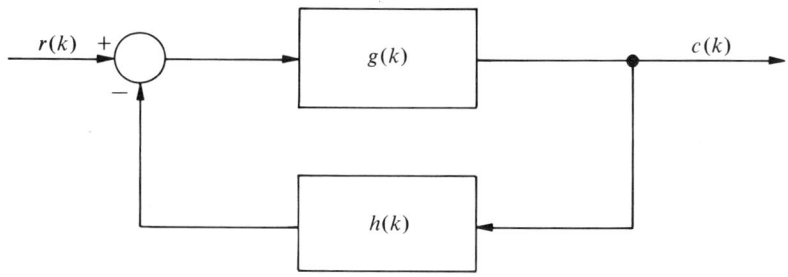

Figure P2.3.

2.4 Given two discrete systems in cascade with weighting sequences

$$h(k) = \begin{cases} \alpha^k & \text{for } k \geq 0 \\ 0 & \text{for } k < 0 \end{cases}$$

$$g(k) = \begin{cases} \beta^k & \text{for } k \geq 0 \\ 0 & \text{for } k < 0 \end{cases}$$

Find the equivalent weighting sequence.

2.5 Given the two discrete systems of problem 2.4 connected in the feedback configuration of Fig. P2.3, find the first five terms of the feedback systems weighting sequence.

2.6 For a discrete system characterized by the difference equation

$$c(k+1) - \beta c(k) = r(k-1)$$

find its response to the inputs

(a) $r(k) = u(k) = \begin{cases} 1 & k \geq 0 \\ 0 & k < 0 \end{cases}$ discrete step

(b) $r(k) = ku(k) = \begin{cases} k & k \geq 0 \\ 0 & k < 0 \end{cases}$ discrete ramp

Use the weighting sequence technique.

2.7 Obtain state variable representations for the systems indicated by the transfer functions

(a) $G(s) = \dfrac{s^2 + 3s + 1}{2s^2 + .15s + 1}$

(b) $G(s) = \dfrac{s + 5}{s(s^2 + 2s + 5)}$

2.8 Obtain a direct programming state model for the linear difference equations

(a) $2c(k) + .5c(k-1) = r(k) - .5r(k-1)$
(b) $c(k) = r(k) - .8r(k-1) - .6r(k-2) - .4r(k-3)$
(c) $c(k) = -2c(k-1) + c(k-2) + r(k) - .5r(k-2)$

2.9 Compute the weighting sequence corresponding to the discrete state equations by use of equation (2.7-8).

$$\begin{bmatrix} x_1(k+1) \\ x_2(k+1) \end{bmatrix} = \begin{bmatrix} .3 & -.2 \\ -.5 & .2 \end{bmatrix} \begin{bmatrix} x_1(k) \\ x_2(k) \end{bmatrix} + \begin{bmatrix} 0 \\ 1 \end{bmatrix} r(k)$$

$$c(k) = [.5 \quad -.5] \begin{bmatrix} x_1(k) \\ x_2(k) \end{bmatrix} + r(k)$$

2.10 Repeat Problem 2.9 by the method indicated by Example 2.7-3.

2.11 Develop a computer program to evaluate a general recursion equation. Apply this program to determine the sequence $c(k)$, $k = 0, 1, 2, \ldots$, where

$$c(k) = -.2c(k-1) + .1c(k-2) + .8r(k) - .6r(k-1)$$

and (a) $r(k) = 1, \quad k = 0$
$\phantom{\text{and (a) }}r(k) = 0, \quad k \neq 0$
(b) $r(k) = 1, \quad \text{all } k \geq 0$
$\phantom{\text{(b) }}r(k) = 0, \quad k < 0$

2.12 Difference equations may be efficiently exploited as generating algorithms for computer plots of certain geometrical shapes or waveforms. Determine a set of difference equations whose solution sequences represent the coordinates of a circle. Prepare a program to compute and plot such a circle. (*Hint:* Start with $\ddot{x} + x = 0$.)

2.13 Repeat Problem 2.12 for a triangular waveform.

2.14 Repeat Problem 2.12 for an *n*-sided polygon.

2.15 Compute the step response to a system with transfer function

$$G(s) = \frac{a}{s+a}$$

What linear discrete system would have a step response approximating that computed above?

Appendix 2A

COMPUTER PROGRAM FOR A(T) AND B(T)

The linear constant coefficient system

$$\dot{x}(t) = Fx(t) + Gu(t)$$

has the solution

$$x(t_1) = e^{F(t_1-t_0)}x(t_0) + \int_{t_0}^{t_1} e^{F(t_1-\tau)}Gu(\tau)d\tau$$

If $u(\tau)$ is piecewise constant over the interval (t_0, t_1) such that

$$u(\tau) = u(nT) \quad \text{for } nT \leq \tau < (n+1)T$$

and t_0 and t_1 are selected as

$$t_0 = nT$$
$$t_1 = (n+1)T$$

then the solution may be written as

$$x[(n+1)T] = e^{FT}x(nT) + \int_0^T e^{F\tau}Gu(n\tau)d\tau$$

or as the vector difference equation

$$x[(n+1)T] = A(T)x(nT) + B(T)u(nT)$$

The matrices $A(T)$ and $B(T)$ may be evaluated by a computer program according to the methods of Section 9.2.

Computer Program for A(T) and B(T)

The following is a program that implements the algorithms specified by equations (9.2-40) and (9.2-42). This program is written as a subroutine called MATEXP.

EXAMPLE

Consider the set of state equations shown below.

$$\frac{d}{dt}\begin{bmatrix} x_1 \\ x_2 \end{bmatrix} = \begin{bmatrix} 0 & 1 \\ 0 & -1 \end{bmatrix}\begin{bmatrix} x_1 \\ x_2 \end{bmatrix} + \begin{bmatrix} 0 \\ 1 \end{bmatrix}x_{in}(t)$$

It is desired to compute the EXP(AT) and the [∫ EXP(AT)] *B for $T = .01$. The FORTRAN statements to accomplish this are shown below:

```
       DIMENSION  F(2,2),G(2,1),A(2,2)INTEGA(2,2),B(2,1)
       REAL  INTEGA
       F(1,1) = 0.0
       F(1,2) = 1.0
       F(2,1) = 0.0
       F(2,2) = -1.0
       G(1,1) = 0.0
       G(2,1) = 1.0
       CALL  MATEXP  (F,G,A,B,2,1,.01)
```

Example Output:

The **F** matrix is

0.	1.0000E 00
0.	-1.0000E 00

The **G** matrix is

0.	1.0000E 00

The discrete transition matrix is

1.0000E 00	9.9500E-03
-0.	9.9500E-01

The discrete input matrix is

5.0000E-05	9.9500E-03

```
       SUBROUTINE  MATEXP(F,G,A,B,N,M,T)
       DIMENSION  F(5,5),G(5,5),A(5,5),INTEGA(5,5),B(5,5),ST(5,5)
       REAL  INTEGA
       INTEGER  POWER
       NORMFT = 0.0
       DO 1  I = 1,N
       DO 1  J = 1,N
       ST(I,J) = F(I,J)*T
    1  A(I,J) = ST(I,J)
```

```
          POWER = 100
          DO 7 I = 2, POWER
          FPOWR = POWER-I+2
          DO 5 J = 1,N
          DO 3 K = 1,N
    3   INTEGA(J,K) = A(J,K)/FPOWR
    5   INTEGA(J,J) = INTEGA(J,J)+1.0
    7   CALL MULTIO(ST,INTEGA,A,N,N,N)
          DO 9 J = 1,N
          A(J,J) = A(J,J)+1.
          DO 9 K = 1,N
    9   INTEGA(J,K) = T*INTEGA(J,K)
          CALL MULTIO(INTEGA,G,B,N,N,M)
          WRITE(6,21)
   21   FORMAT(16H THE F MATRIX IS ,//)
          DO 10 I = 1,N
          WRITE(6,20) (F(I,J),J = 1,N)
   20   FORMAT(1P10E13.4, ///)
   10   CONTINUE
          WRITE(6,11)
   11   FORMAT(/, 16H THE G MATRIX IS //)
          DO 22 I = 1,N
          WRITE(6,20) (G(I,J),J = 1,M)
   22   CONTINUE
          WRITE(6,16)
   16   FORMAT(/, 34H THE DISCRETE TRANSITION MATRIX IS,//)
          DO 17 I = 1,N
          WRITE(6,20) (A(I,J),J = 1,N)
   17   CONTINUE
          WRITE(6,19)
   19   FORMAT(/, 30H THE DISCRETE INPUT MATRIX IS //)
          DO 23 I = 1,N
          WRITE(6,20) (B(I,J),J = 1,M)
   23   CONTINUE
          RETURN
          END
```

Appendix 2B

THE TRANSITION MATRIX

The state variable equation for the time-varying linear system

$$\dot{\mathbf{x}}(t) = \mathbf{F}(t)\mathbf{x}(t) + \mathbf{G}(t)\mathbf{u}(t)$$
$$\mathbf{y}(t) = \mathbf{H}(t)\mathbf{x}(t) \quad (2\text{B-}1)$$

will have a unique solution of the form

$$\mathbf{x}(t) = \mathbf{\Phi}(t, t_0)\mathbf{x}(t_0) + \int_{t_0}^{t} \mathbf{\Phi}(t, \tau)\mathbf{G}(\tau)\mathbf{u}(\tau)\, d\tau \quad (2\text{B-}2)$$

The transition matrix $\mathbf{\Phi}(t, \tau)$ satisfies the matrix differential equation

$$\frac{\partial \mathbf{\Phi}(t, \tau)}{\partial t} = \mathbf{F}(t)\mathbf{\Phi}(t, \tau) \quad (2\text{B-}3)$$

for all t and τ with the initial condition

$$\mathbf{\Phi}(\tau, \tau) = \mathbf{I}$$

In the particular case in which the system is constant, i.e., $\mathbf{F}(t) = \mathbf{F}$ [usually also $\mathbf{G}(t) = \mathbf{G}$ and $\mathbf{H}(t) = \mathbf{H}$], the state variable equations of the system (2B-1) are written as

$$\dot{\mathbf{x}}(t) = \mathbf{F}\mathbf{x}(t) + \mathbf{G}\mathbf{u}(t)$$
$$\mathbf{y}(t) = \mathbf{H}\mathbf{x}(t) \quad (2\text{B-}4)$$

and the solution is

$$\mathbf{x}(t) = \mathbf{\Phi}(t - t_0)\mathbf{x}(t_0) + \int_{t_0}^{t} \mathbf{\Phi}(t - \tau)\mathbf{G}\mathbf{u}(\tau)d\tau \qquad (2\text{B-}5)$$

where the transition matrix may be determined by

$$\mathbf{\Phi}(t - t_0) = e^{\mathbf{F}(t - t_0)} \qquad (2\text{B-}6)$$

and where the exponential matrix is defined by

$$e^{\mathbf{F}t} = \mathbf{I} + \mathbf{F}t + \frac{\mathbf{F}^2 t^2}{2!} + \frac{\mathbf{F}^2 t^3}{3!} + \cdots \qquad (2\text{B-}7)$$

For general applications the exponential matrix may be evaluated by a computer program using an algorithm based upon (2B-7). See, for instance, Appendix 2A.

We consider here several additional methods of analytically determining the transition matrix.

Solution by Laplace Transform

When the entire system is constant it is possible to write the Laplace transform of (2B-5) as

$$\mathbf{X}(s) = [s\mathbf{I} - \mathbf{F}]^{-1}\{\mathbf{x}(t_0) + \mathbf{G}\mathbf{U}(s)\} \qquad (2\text{B-}8)$$

It is, therefore, seen that the transition matrix may be evaluated by the expression

$$\mathbf{\Phi}(t) = \mathscr{L}^{-1}\{[s\mathbf{I} - \mathbf{F}]^{-1}\} \qquad (2\text{B-}9)$$

This equation indicates a solution process that is entirely defined in terms of Laplace transform techniques. Required in this process are two steps: (1) the determination of $[s\mathbf{I} - \mathbf{F}]^{-1}$ and (2) the inverse Laplace transform $\mathscr{L}^{-1}\{[s\mathbf{I} - \mathbf{F}]^{-1}\}$. The inverse of the matrix $[s\mathbf{I} - \mathbf{F}]$ may be obtained by Leverrier's algorithm, which has been put into the following form by Frame.*

This algorithm makes use of the relation

$$[s\mathbf{I} - \mathbf{F}]^{-1} = \frac{1}{D(s)} \text{ adjoint } [s\mathbf{I} - \mathbf{F}] \qquad (2\text{B-}10)$$

where $D(s)$ is the determinant of $[s\mathbf{I} - \mathbf{F}]$. From the expansion into minors the determinant $[s\mathbf{I} - \mathbf{F}]$ may be evaluated by

$$D(s) = s^n - p_1 s^{n-1} - p_2 s^{n-2} - \cdots - p_n \qquad (2\text{B-}11)$$

*See Reference 1 at the end of this appendix. See also Section 5.4.

The Transition Matrix

where $p_1, p_2, \ldots, p_n$ are defined below. The adjoint may be evaluated by

$$\text{adjoint } [s\mathbf{I} - \mathbf{F}] = \mathbf{I}s^{n-1} + \mathbf{H}_1 s^{n-2} + \mathbf{H}_2 s^{n-3} + \cdots + \mathbf{H}_{n-1} \quad (2\text{B-}12)$$

where

$$\begin{aligned}
\mathbf{H}_1 &= \mathbf{F} - p_1 \mathbf{I}, & p_1 &= \text{trace } \mathbf{F} \\
\mathbf{H}_2 &= \mathbf{FH}_1 - p_2 \mathbf{I}, & p_2 &= \tfrac{1}{2} \text{ trace } (\mathbf{FH}_1) \\
\mathbf{H}_3 &= \mathbf{FH}_2 - p_3 \mathbf{I}, & p_3 &= \tfrac{1}{3} \text{ trace } (\mathbf{FH}_2) \\
&\cdots & &\cdots \\
\mathbf{H}_{n-1} &= \mathbf{FH}_{n-2} - p_{n-1} \mathbf{I}, & p_{n-1} &= \frac{1}{n-1} \text{ trace } (\mathbf{FH}_{n-2}) \\
\mathbf{H}_n &= \mathbf{FH}_{n-1} - p_n \mathbf{I} = 0, & p_n &= \frac{1}{n} \text{ trace } (\mathbf{FH}_{n-1})
\end{aligned} \quad (2\text{B-}13)$$

The trace of a matrix $\mathbf{F}$ is the sum of its diagonal elements, i.e.,

$$\text{trace } \mathbf{F} = \sum_{i=1}^{n} f_{ii}$$

EXAMPLE

Determine the inverse and the determinant of $[s\mathbf{I} - \mathbf{F}]$ when

$$\mathbf{F} = \begin{bmatrix} 2 & -1 & 0 \\ 1 & 1 & 2 \\ -1 & 0 & 1 \end{bmatrix}$$

From equation (2B-13) we have

$$p_1 = \text{trace } \mathbf{F} = 4$$

and

$$\mathbf{H}_1 = \mathbf{F} - p_1 \mathbf{I} = \begin{bmatrix} -2 & -1 & 0 \\ 1 & -3 & 2 \\ -1 & 0 & -3 \end{bmatrix}$$

Continuing, we evaluate

$$p_2 = \tfrac{1}{2} \text{ trace } (\mathbf{FH}_1) = \tfrac{1}{2} \text{ trace } \begin{bmatrix} -5 & 1 & -2 \\ -3 & -4 & -4 \\ 1 & 1 & -3 \end{bmatrix} = -6$$

$$\mathbf{H}_2 = \mathbf{FH}_1 + 6\mathbf{I} = \begin{bmatrix} 1 & 1 & -2 \\ -3 & 2 & -4 \\ 1 & 1 & 3 \end{bmatrix}$$

$$\tfrac{1}{3}\text{trace}\,(\mathbf{FH}_2) = \tfrac{1}{3}\text{trace}\begin{bmatrix} 5 & 0 & 0 \\ 0 & 5 & 0 \\ 0 & 0 & 5 \end{bmatrix} = 5$$

and

$$\mathbf{H}_3 = \mathbf{0}$$

Thus

$$\text{adjoint}\,[s\mathbf{I} - \mathbf{F}] = \mathbf{I}s^2 + \mathbf{H}_1 s + \mathbf{H}_2$$

$$= \begin{bmatrix} s^2 - 2s + s & -s + 1 & -2 \\ s - 3 & s^2 - 3s + 2 & 2s - 4 \\ -s + 1 & 1 & s^2 - 3s + 3 \end{bmatrix}$$

Also

$$D(s) = s^3 - 4s^2 + 6s - 5$$

This procedure for determining the inverse of $[s\mathbf{I} - \mathbf{A}]$ may be programmed as a numerical routine and made available as a standard library program. The inverse Laplace transform of $[s\mathbf{I} - \mathbf{A}]^{-1}$ is obtained by applying the procedures of partial fraction expansion.

Solution by Functions of a Matrix

The characteristic equation of a square matrix $\mathbf{F}$ is

$$D(\lambda) = |\mathbf{F} - \lambda\mathbf{I}| = 0 \qquad (2\text{B-}14)$$

This equation yields the polynomial

$$D(\lambda) = \lambda^n + a_{n-1}\lambda^{n-1} + \cdots + a_1\lambda + a_0 = 0$$

whose roots are the eigenvalues of $\mathbf{F}$. By the Cayley-Hamilton theorem the characteristic equation is also satisfied if λ is replaced by $\mathbf{F}$, i.e.,

$$D(\mathbf{F}) = \mathbf{0}$$

The Cayley-Hamilton theorem offers a means of evaluating the transition matrix as a function of a matrix.

We restrict our attention to those functions of matrices which can be represented as a series of powers of the matrix. This restriction is appropriate in the study of linear systems. Thus, for a function of a matrix we may write the expansion

$$f(\mathbf{F}) = c_0\mathbf{I} + c_1\mathbf{F} + c_2\mathbf{F}^2 + c_3\mathbf{F}^3 + \cdots + c_n\mathbf{F}^n + c_{n+1}\mathbf{F}^{n+1} + \cdots \tag{2B-15}$$

The Cayley-Hamilton theorem, however, says that

$$D(\mathbf{F}) = \mathbf{F}^n + a_{n-1}\mathbf{F}^{n-1} + \cdots + a_1\mathbf{F} + a_0\mathbf{I} = 0$$

or

$$\mathbf{F}^n = -a_{n-1}\mathbf{F}^{n-1} - \cdots - a_1\mathbf{F} - a_0\mathbf{I}$$

Multiplying both sides by $\mathbf{F}$, we find that the last equation becomes

$$\mathbf{F}^{n+1} = -a_{n-1}\mathbf{F}^n - \cdots - a_1\mathbf{F}^2 - a_0\mathbf{F}$$

Eliminating $\mathbf{F}^n$ from this equation, we get

$$\mathbf{F}^{n+1} = -a_{n-1}(-a_{n-1}\mathbf{F}^{n-1} - \cdots - a_1\mathbf{F} - a_0\mathbf{I}) \\ - a_{n-2}\mathbf{F}^{n-1} - \cdots - a_1\mathbf{F}^2 - a_0\mathbf{F}$$

This equation implies that $\mathbf{F}^n, \mathbf{F}^{n+1}, \ldots$, or, in general, all powers of $\mathbf{F}$ greater than $n - 1$ can be expressed as linear combinations of $\mathbf{F}^0, \mathbf{F}^1, \ldots, \mathbf{F}^{n-1}$. We use this result in equation (2B-15) to obtain a finite power series expansion for the function of a matrix,

$$f(\mathbf{F}) = \alpha_0\mathbf{I} + \alpha_1\mathbf{F} + \cdots + \alpha_{n-1}\mathbf{F}^{n-1} = \sum_{i=0}^{n-1} \alpha_i\mathbf{F}^i \tag{2B-16}$$

where the α_i's are combinations of the c's and a's.

EXAMPLE

Given

$$\mathbf{F} = \begin{bmatrix} -3 & -1 \\ 2 & 0 \end{bmatrix}$$

calculate $e^{\mathbf{F}t}$ by use of the Cayley-Hamilton theorem.
From equation (2B-16) we get

$$f(\mathbf{F}) = e^{\mathbf{F}t} = \alpha_0\mathbf{I} + \alpha_1\mathbf{F}$$

To determine the coefficients α_0 and α_1 we calculate the eigenvalues of $\mathbf{F}$. Thus

$$d(\lambda) = |\mathbf{F} - \lambda \mathbf{I}| = \begin{vmatrix} -3 - \lambda & -1 \\ 2 & -\lambda \end{vmatrix} = \lambda^2 + 3\lambda + 2 = 0$$

giving

$$\lambda_1 = -1 \quad \text{and} \quad \lambda_2 = -2$$

By the Cayley-Hamilton theorem $f(\mathbf{F})$ must be satisfied by λ_1 and λ_2. Hence

$$f(\lambda_1) = \alpha_0 + \alpha_1 \lambda_1 \quad \text{and} \quad f(\lambda_2) = \alpha_0 + \alpha_1 \lambda_2$$

or

$$e^{-t} = \alpha_0 - \alpha_1 \quad \text{and} \quad e^{-2t} = \alpha_0 - 2\alpha_1$$

from which we solve for α_0 and α_1:

$$\alpha_1 = e^{-t} - e^{-2t}$$
$$\alpha_0 = 2e^{-t} - e^{-2t}$$

Finally,

$$e^{\mathbf{F}t} = (2e^{-t} - e^{-2t})\mathbf{I} + (e^{-t} - e^{-2t})\mathbf{F}$$
$$= \begin{bmatrix} -e^{-t} + 2e^{-2t} & -e^{-t} + e^{-2t} \\ 2e^{-t} - 2e^{-2t} & 2e^{-t} - e^{-2t} \end{bmatrix}$$

For the particular case in which $\mathbf{F}$ has distinct eigenvalues, a function of matrix may also be evaluated by Sylvester's expansion theorem, which states that

$$f(\mathbf{F}) = \sum_{i=1}^{n} \mathbf{F}_i f(\lambda_i) \qquad (2\text{B-}17)$$

where

$$\mathbf{F}_i = \prod_{\substack{j=1 \\ j \neq i}}^{n} \frac{[\mathbf{F} - \lambda_j \mathbf{I}]}{\lambda_i - \lambda_j} \qquad (2\text{B-}18)$$

The matrices $\mathbf{F}_1, \mathbf{F}_2, \ldots, \mathbf{F}_n$ are called the *constituent matrices* of $\mathbf{F}$ and are dependent only on $\mathbf{F}$ and its n eigenvalues. This is a significant property, inasmuch as it permits the use of the same constituent matrices for the evaluation of any function of a matrix.*

*Sylvester's expansion theorem may be expanded to treat the case of multiple eigenvalues. See Reference 2.

EXAMPLE

For the matrix $\mathbf{F}$ of the previous example, evaluate $e^{\mathbf{F}t}$ and $\mathbf{F}^n$.

$$\mathbf{F} = \begin{bmatrix} -3 & -1 \\ 2 & 0 \end{bmatrix}, \quad \lambda_1 = -1, \quad \lambda_2 = -2$$

We determine

$$\mathbf{F}_1 = \frac{\begin{bmatrix} -3 & -1 \\ 2 & 0 \end{bmatrix} - \begin{bmatrix} -2 & 0 \\ 0 & -2 \end{bmatrix}}{-1 - (-2)} = \begin{bmatrix} -1 & -1 \\ 2 & 2 \end{bmatrix}$$

$$\mathbf{F}_2 = \frac{\begin{bmatrix} -3 & -1 \\ 2 & 0 \end{bmatrix} - \begin{bmatrix} -1 & 0 \\ 0 & -1 \end{bmatrix}}{-2 - (-1)} = \begin{bmatrix} 2 & 1 \\ -2 & -1 \end{bmatrix}$$

Therefore,

$$e^{\mathbf{F}t} = \mathbf{F}_1 e^{\lambda_1 t} + \mathbf{F}_2 e^{\lambda_2 t} = \begin{bmatrix} -1 & -1 \\ 2 & 2 \end{bmatrix} e^{-t} + \begin{bmatrix} 2 & 1 \\ -2 & -1 \end{bmatrix} e^{-2t}$$

This result agrees with the previous example. For $\mathbf{F}^n$ we use the same constituent matrices to obtain

$$\mathbf{F}^n = \mathbf{F}_1 \lambda_1^n + \mathbf{F}_2 \lambda_2^n = \begin{bmatrix} -1 & -1 \\ 2 & 2 \end{bmatrix}(-1)^n + \begin{bmatrix} 2 & 1 \\ -2 & -1 \end{bmatrix}(-2)^n$$

Constituent matrices satisfy the following properties:
1. Constituent matrices are mutually orthogonal.

$$\mathbf{F}_i \mathbf{F}_j = 0$$

2. A complete set of constituent matrices sums to the unit matrix.

$$\sum_{i=1}^n \mathbf{F}_i = \mathbf{I}$$

This property serves as a convenient numerical check on calculations.
3. A constituent matrix is an idempotent matrix.

$$\mathbf{F}_i^2 = \mathbf{F}_i$$

Properties of Transition Matrix

Using the above relations, we may show the following properties of the transition matrix:

1. $\Phi^{-1}(t, \tau) = \Phi(\tau, t)$
2. $\Phi(t_1, t_2)\Phi(t_2, t_3) = \Phi(t_1, t_3)$
3. $\dfrac{d}{dt} e^{\mathbf{F}t} = \mathbf{F} e^{\mathbf{F}t}$
4. $\mathbf{F} e^{\mathbf{F}t} = e^{\mathbf{F}t} \mathbf{F}$

In general,

$$f_1(\mathbf{F}) f_2(\mathbf{F}) = f_2(\mathbf{F}) f_1(\mathbf{F})$$

REFERENCES

1. Frame, J. S., "Matrix Functions and Applications," *IEEE Spectrum*, Vol. 1, No. 6, June 1964.
2. Gantmacher, F. R., *The Theory of Matrices*, Vol. 1, Chelsea Publishing Company, New York, 1959.
3. Bellman, R., *Introduction to Matrix Analysis*, McGraw-Hill, New York, 1960.
4. Pipes, L., *Matrix Methods for Engineering*, Prentice-Hall, Englewood Cliffs, N. J., 1963.
5. Timothy, L. K. and B. E. Bona, *State Space Analysis: An Introduction*, McGraw-Hill, New York, 1968.
6. Ogata, K., *State Space Analysis of Control Systems*, Prentice-Hall, Englewood Cliffs, N. J., 1967.

3

The Analysis of Discrete-time Systems: Time-domain Approach

3.1 Introduction

Several examples were given in Chapter 1 of typical systems operating in discrete time. The common characteristic of these systems is that they contain at least one discrete-time component. It is possible, for instance, for the feedback information to be available only at periodic instances when the feedback channel is time shared. Such a system requires a data-hold circuit, since it essentially operates on an open-loop basis between data transfers. An illustration of this type of system is given by Figure 3.1-1.

Another example of a system containing a discrete component is shown in Figure 3.1-2. Illustrated is a closed-loop system that operates in continuous time except for a digital computer inserted into the control loop as an active systems component. Again, a hold circuit is required to maintain control over the system during intervals between data transfers.

A third class of systems operating in discrete time may be distinguished as one in which all components operate in discrete time. Consider, for instance,

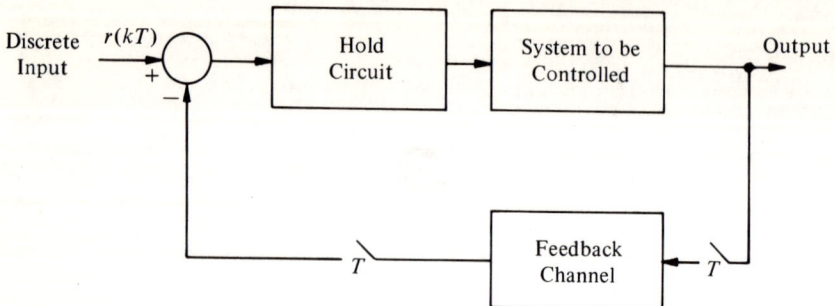

Figure 3.1-1. System with time-shared feedback channel.

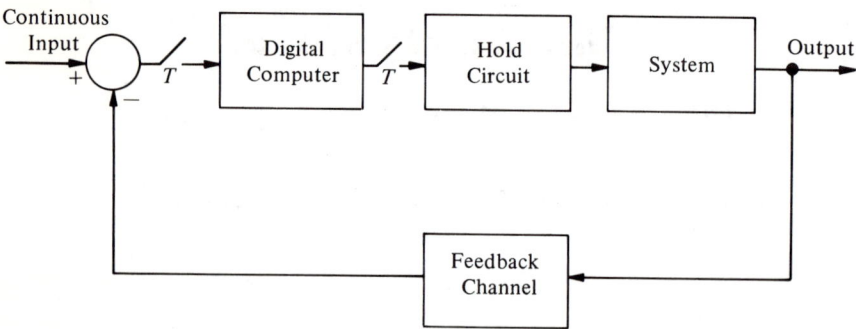

Figure 3.1-2. Computer control system.

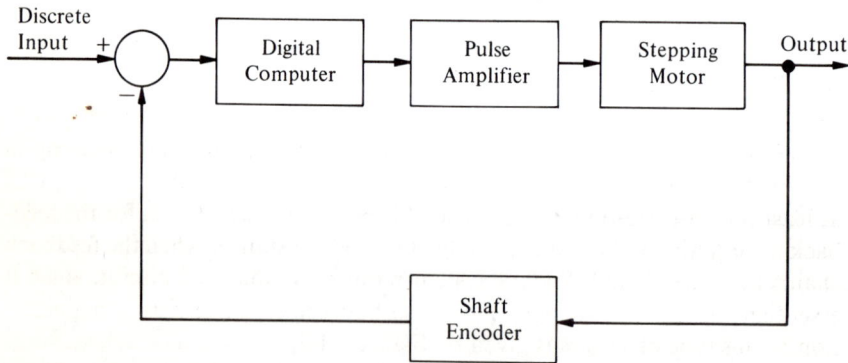

Figure 3.1-3. Digital control system, operating entirely in discrete time.

the system shown in Figure 3.1-3, which consists of a digital computer driving a digital stepping motor with the feedback being provided by a digital shaft encoder. The output of this system is continuous in time, although limited to quantized levels.

Discrete-time systems may be operated in an open-loop or closed-loop manner. The analysis of these systems is generally carried out by using one of two approaches: (1) by discrete state equations or (2) by the use of the z-transform. This chapter serves to demonstrate the application of discrete state techniques.

3.2 Data-hold Techniques

An essential ingredient in the satisfactory operation of a hybrid discrete-time system having components that operate both in discrete time and continuous time is a data-hold device. Its function is to convert a discrete-time function (sequence of numbers) into a continuous-time function in order to provide a suitable input to a continuous-time component.

When the input is a sampled analog signal, it is called a data-hold circuit. However, when the input is a discrete data signal in digital form, such as might originate from a digital computer, the data-hold device provides digital-to-analog conversion in addition to the hold action. Then it is simply called a digital-to-analog converter. As is shown in Chapter 7, a digital-to-analog converter automatically provides hold action. Since mathematically there is no distinction between the two cases, we can treat them alike and provide an identical analysis.

Figure 3.2-1 shows a block diagram representation of a digital-to-analog converter (DAC).

Figure 3.2-1. Block diagram representation of a DAC.

In general, the purpose of a DAC is that of generating a function of continuous time $h(t)$ from a sequence of numbers $g(nT)$ that are separated in time by T-second intervals. It is usually desirable to have the function $h(t)$ correspond roughly to an envelope of the input sequence $g(nT)$. Between sampling times [i.e., $NT \leq t < (N+1)T$], the DAC must extrapolate between the most recent sample and the next to follow.

In effect, the DAC has the properties normally ascribed to an extrapolator. An mth-order extrapolator will be defined as an extrapolator whose present output depends on $m+1$ past sample values. At each sampling time, a new member of the input sequence $g(nT)$ is available so that the extrapolating process must be reinitiated at the sampling instants.

A useful form of extrapolation is that of polynomial extrapolation. Here, it is assumed that the desired signal $h(t)$ may be adequately approximated by an mth-order polynomial; that is,

$$h(nT + \tau) = a_m \tau^m + a_{m-1} \tau^{m-1} + \cdots + a_0 \quad \text{for } 0 \leq \tau < T \quad (3.2\text{-}1)$$

Since it is desired that $h(t)$ be the envelope of the sequence $g(nT)$, it is natural to require that the output signal $h(t)$ have the value of the input sequence at the sampling times $t = kT$; that is,

$$h(t)|_{t=kT} = h(kT) = g(kT) \quad \text{for all values of } k$$

The coefficients $a_m, a_{m-1}, \ldots, a_0$ for any time interval $nT \leq t < (n+1)T$ may be evaluated by forcing $h(t)$ to satisfy the constraints

$$h(kT) = g(kT) \quad \text{for } k = n - m, n - m + 1, \ldots, n \quad (3.2\text{-}2)$$

That is, $h(nT + \tau)$ is a polynomial that passes through the immediate $m + 1$ past values of the input $g(kT)$. At each sampling time the coefficients $a_m, a_{m-1}, \ldots, a_0$ must be reevaluated, since a new data point is available.

Zero-order Hold $(m = 0)$

The simplest type of polynomial extrapolator arises when $h(t)$ is assumed to be a zero-order polynomial $(m = 0)$. In this case, we have, by (3.2-1),

$$h(nT + \tau) = g(nT) \quad \begin{array}{l} \text{for } 0 \leq \tau < T \\ n = 0, \pm 1, \pm 2, \ldots \end{array} \quad (3.2\text{-}3)$$

Figure 3.2-2(a) illustrates a typical response of a zero-order hold.

First-order Hold $(m = 1)$

If $h(t)$ is assumed to be a first-order polynomial, then

$$h(nT + \tau) = a_1 \tau + a_0 \quad \text{for } 0 \leq \tau < T \quad (3.2\text{-}4)$$

with the requirements corresponding to (3.2-2) being

$$h(nT) = g(nT)$$
$$h([n-1]T) = g([n-1]T)$$

which, when τ is set equal to 0 and $-T$ in (3.2-4), gives us

$$a_0 = g(nT), \quad a_1 = \frac{g(nT) - g([n-1]T)}{T}$$

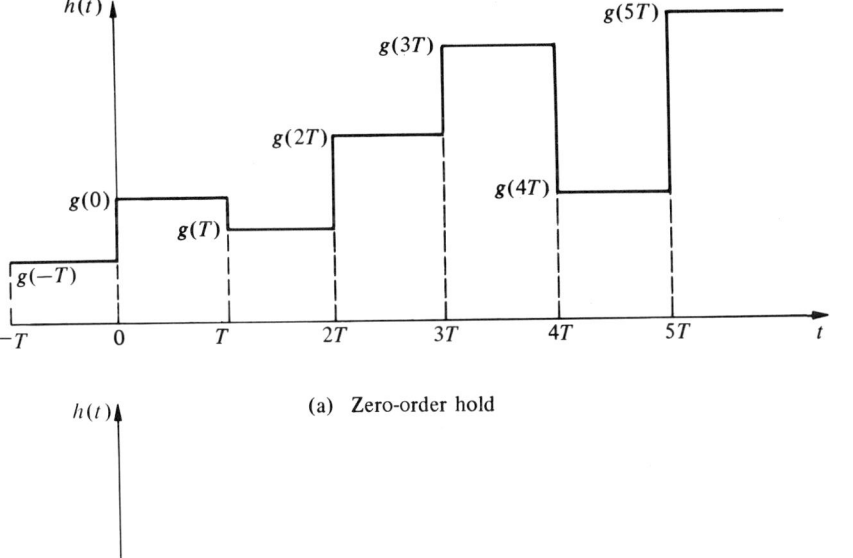

Figure 3.2-2. Typical responses of zero- and first-order holds.

Therefore, the first-order extrapolator is characterized by

$$h(nT + \tau) = \frac{g(nT) - g([n-1]T)}{T}\tau + g(nT) \quad \text{for } 0 \leq \tau < T \quad (3.2\text{-}5)$$
$$n = 0, \pm 1, \pm 2, \ldots$$

A typical response of the first-order hold is shown in Figure 3.2-2(b).

The higher-order holds ($m \geq 2$) may be generated in a like manner. In general, most modern systems do not use higher-order holds because of the inherent delay which they introduce into the system's response. Also, they are highly noise-sensitive and much more difficult to implement.

3.3 Open-loop Sampled-data Systems

An example of a basic discrete-time system consists of a sampling element, a hold circuit, and a continuous-time system, as shown schematically in Figure 3.3-1. The input $r(t)$ is sampled periodically at intervals T seconds apart to

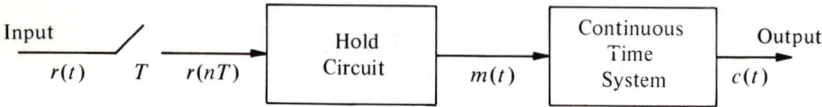

Figure 3.3-1. Open-loop sampled-data system.

generate the sequence of numbers $r(nT)$. The hold circuit changes this discrete-time function to a piecewise continuous function. If it is a zero-order hold, then a function of the form shown by Figure 3.2-2(a) is produced. For higher-order holds, more complicated forms will be generated. The output of the hold circuit represents the input to the continuous-time system. The analysis of such systems is directed at determining the response of the continuous-time system. We shall, therefore, derive a mathematical model suitable to carry out this objective.

The zero-order hold is the most commonly employed hold device. In view of the graph of Figure 3.2-2(a), the output of a zero-order hold is a periodically piecewise constant function. That is,

$$m(t_k + \tau) = r(t_k), \qquad 0 \leq \tau < t_{k+1} - t_k$$

where $r(t_k)$ is the value of the sampled function $r(t)$, at time $t = t_k$. If the sampling occurs at constant intervals then $t_k = kT$ and $t_{k+1} - t_k = T$. In this case, the hold-circuit output can be written as

$$m(kT + \tau) = r(kT), \qquad 0 \leq \tau < T \qquad (3.3\text{-}1)$$

In Chapter 2 we developed the state equations for a linear, time-invariant, continuous system governed by the continuous state equations

$$\dot{\mathbf{x}}(t) = \mathbf{F}\mathbf{x}(t) + \mathbf{G}m(t)$$
$$c(t) = \mathbf{C}\mathbf{x}(t) + dm(t)$$

when its input $m(t)$ is of the form (3.3-1). Namely,

$$\mathbf{x}[(k+1)T] = \mathbf{A}(T)\mathbf{x}(kT) + \mathbf{B}r(kT) \qquad (3.3\text{-}2)$$
$$c(kT) = \mathbf{C}\mathbf{x}(kT) + dr(kT) \qquad (3.3\text{-}3)$$

where $\mathbf{x}(kT)$ is the continuous system's state at time kT, $r(kT)$ is the value of the input signal $r(t)$ at time kT, and

$$\mathbf{A}(T) = e^{\mathbf{F}T}, \qquad \mathbf{B}(T) = \int_0^T e^{\mathbf{F}\tau} \mathbf{G}\, d\tau$$

The matrices $\mathbf{F}$ and $\mathbf{G}$ are the system matrix and the input matrix of the continuous-time system, respectively. Equations (3.3-2) and (3.3-3) may be used to calculate the open-loop system response to the input $r(t)$ at the sampling times kT. It must be understood that the sampling operation permits the passage of the function $r(t)$ only at the sampling instants nT. We consider an example to demonstrate the use of (3.3-2) and (3.3-3).

EXAMPLE

Compute the response of the system shown in Figure 3.3-2 to a square wave input and a sinusoidal wave input for various sampling rates.

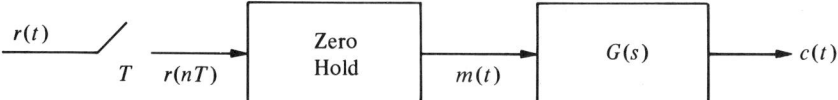

Figure 3.3-2. Open-loop sampled system with zero-order hold.

Let the transfer function of the continuous time system be

$$G(s) = \frac{s+1}{(s+2)(s+10)} \tag{3.3-4}$$

Using nested programming to develop a state model, we obtain

$$\frac{d}{dt}\begin{bmatrix} x_1 \\ x_2 \end{bmatrix} = \begin{bmatrix} -12 & 1 \\ -20 & 0 \end{bmatrix}\begin{bmatrix} x_1 \\ x_2 \end{bmatrix} + \begin{bmatrix} 1 \\ 1 \end{bmatrix} m(t) \tag{3.3-5}$$

$$c(t) = x_1$$

By any one of the methods of Appendix 2 we determine that

$$e^{\mathbf{F}T} = \mathbf{A}(T) = \begin{bmatrix} -\tfrac{1}{4}e^{-2T} + \tfrac{5}{4}e^{-10T} & \tfrac{1}{8}e^{-2T} - \tfrac{1}{8}e^{-10T} \\ -\tfrac{5}{2}e^{-2T} + \tfrac{5}{2}e^{-10T} & \tfrac{5}{4}e^{-2T} - \tfrac{1}{4}e^{-10T} \end{bmatrix} \tag{3.3-6}$$

and

$$\begin{aligned}
\mathbf{B}(T) &= \int_0^T \mathbf{A}(\tau)\mathbf{G}\,d\tau \\
&= \int_0^T \begin{bmatrix} -\tfrac{1}{8}e^{-2\tau} + \tfrac{9}{8}e^{-10\tau} \\ -\tfrac{5}{4}e^{-2\tau} + \tfrac{9}{4}e^{-10\tau} \end{bmatrix} d\tau \\
&= \begin{bmatrix} \tfrac{1}{20} + \tfrac{1}{16}e^{-2T} - \tfrac{9}{80}e^{-10T} \\ -\tfrac{2}{5} + \tfrac{5}{8}e^{-2T} - \tfrac{9}{40}e^{-10T} \end{bmatrix}
\end{aligned} \tag{3.3-7}$$

For the specific case, where $T = 1$ for instance, we obtain

$$\mathbf{A}(1) = \begin{bmatrix} -.0338 & .0169 \\ -.338 & .169 \end{bmatrix}$$

and

$$\mathbf{B}(1) = \begin{bmatrix} .0584 \\ -.315 \end{bmatrix}$$

Thus, for this example, the discrete state equations for a piecewise constant input are

$$\begin{bmatrix} x_1[(k+1)T] \\ x_2[(k+1)T] \end{bmatrix} = \begin{bmatrix} -.0338 & .0169 \\ -.338 & .169 \end{bmatrix} \begin{bmatrix} x_1(kT) \\ x_2(kT) \end{bmatrix} + \begin{bmatrix} .0584 \\ -.315 \end{bmatrix} r(kT) \quad (3.3\text{-}8)$$

$$c(kT) = x_2(kT)$$

Equations (3.3-8) serve as the mathematical model for this problem.

It is interesting to study the effect of the sample and hold operations on the input signal. Figure 3.3-3 and 3.3-4 show waveforms for $m(t)$ and $c(t)$ for the square wave and sinusoidal wave inputs at various sampling frequencies. A study of these waveforms reveals two important aspects of sampling a signal and then passing it through a zero-order hold circuit. First, we note that the higher the sampling frequency, the better the zero-order hold is capable of generating a time function that represents a good reproduction of the input.

The second observation we wish to make is that the output of the zero-order hold is an approximate reproduction of the input but appears with a time lag relative to the unsampled signal. An exact reproduction is possible in the case of the square wave. However, the sinusoidal wave may suffer greatly when passed through the zero-order hold.

Both figures show the output $c(t)$ for each of the sampling periods used. For the square wave input, the output $c(t)$ is identical to the output for the equivalent unsampled system ($T = 0$) except with time lag equal to the sampling period as long as T is less than one-half the period of the square wave input. For the sine wave input, the output becomes seriously affected as the sampling period increases. For $T = .1$ the output is very nearly equal to that for the unsampled system ($T = 0$). However, when the sampling period is increased to $T = .5$ and $T = 1.0$, it is very apparent that the output cannot be regarded as the sinusoidal response of a linear system.

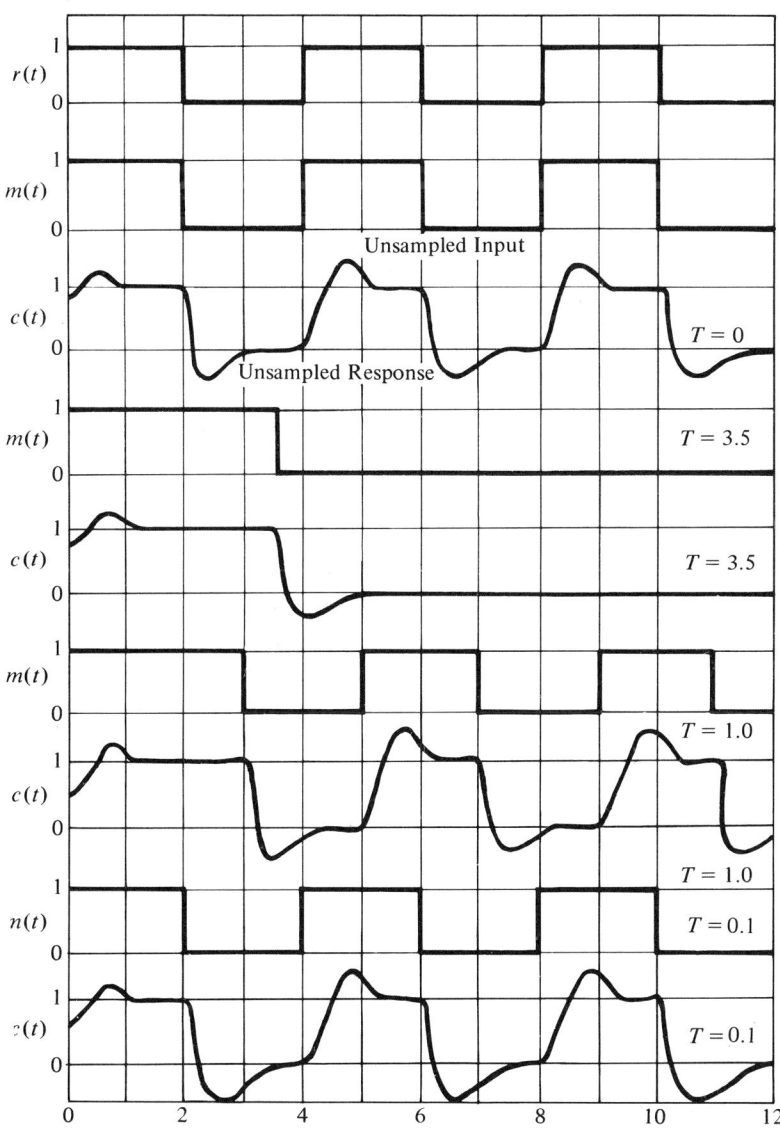

Figure 3.3-3. Response of a sampled-data system to a squarewave input for various sampling rates.

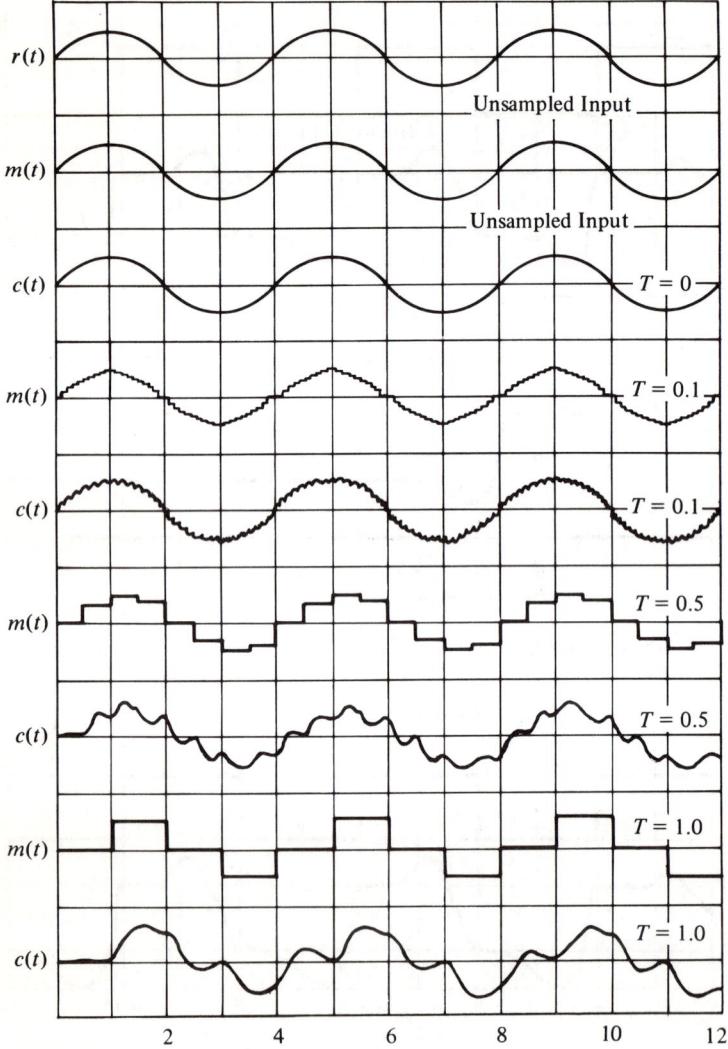

Figure 3.3-4. Response of a sampled system to a sinusoidal input for various sampling rates.

The sampled-data system described above was simulated on an analog computer at a speed 10 times slower than real time. A diagram of the analog computer circuit used is shown in Figure 3.3-5. The sample and hold circuit is implemented by use of an integrator, which is switched between the OPERATE and RESET modes in synchronism with the sampling rate. The

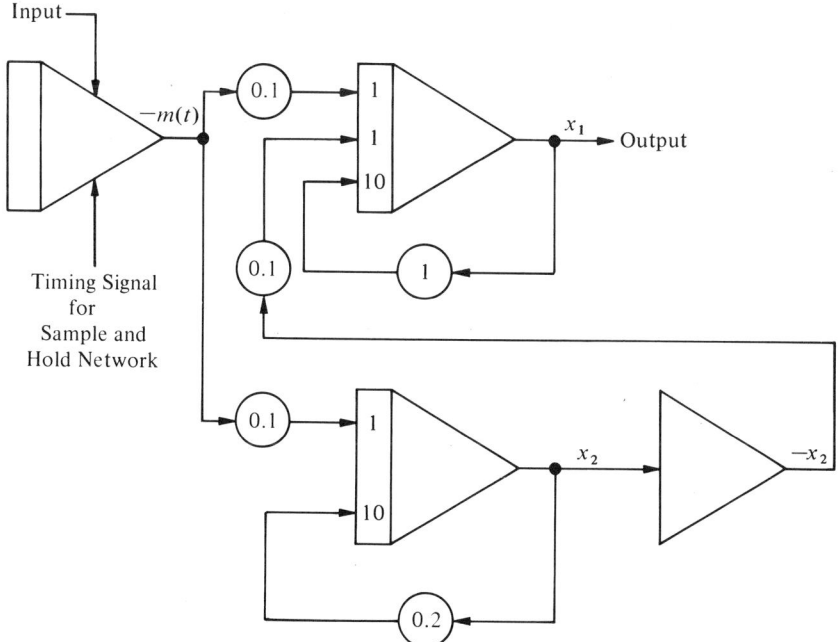

Figure 3.3-5. Analog simulation of sampled-data system.

input is supplied through the initial condition terminal. When the integrator is in RESET mode, it tracks the input function; when it is in OPERATE mode, it stores the input function. If the RESET time is kept very short and the integrator's time constant is very small, the integrator effectively samples the input during RESET and holds it during OPERATE. An integrator operating in this fashion is called a *track-store unit*, which is frequently utilized in hybrid computations (see Chapter 9, Section 9.5).

The timing signal may be generated by means of the circuit shown in Figure 3.3-6(a). A triangular wave generator output is fed into a comparator network, which is biased with a fixed voltage slightly less than the peak output of the triangular wave generator. The comparator produces a pulse signal of the same frequency as the wave generator. The wave shapes are shown in Figure 3.3-6(b).

The above example effectively demonstrates several important characteristics of sampled systems with zero-order holds. Most outstanding are the facts that the sampling process introduces data loss and that the holding process introduces time lag.

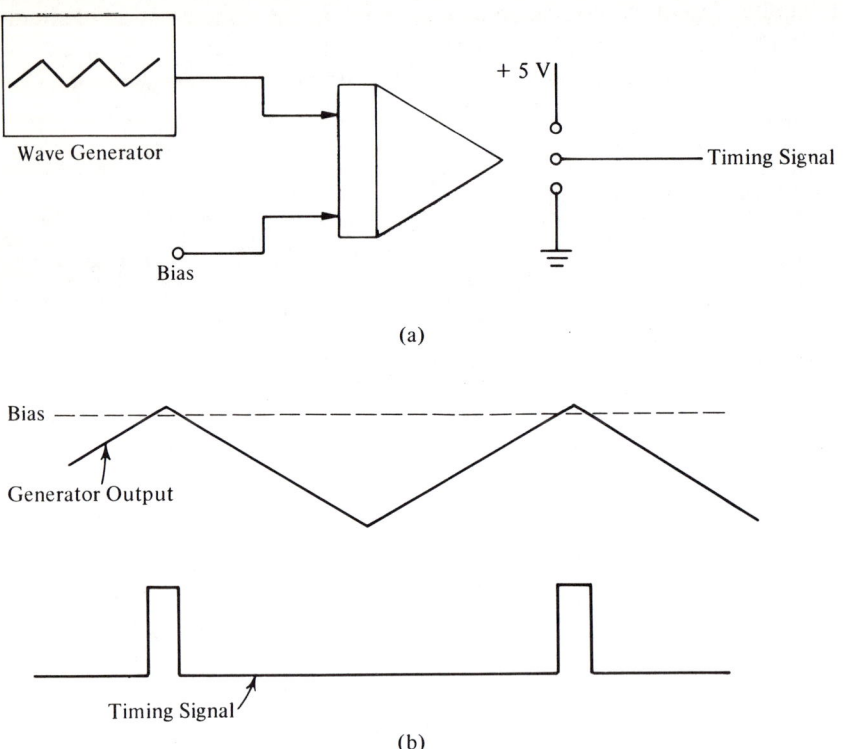

Figure 3.3-6. Generation of timing signal.

3.4 Discrete State Equations of Closed-loop Sampled-data Systems

When certain signals in a conventional closed-loop feedback system are used only at discrete times, such a system may be viewed as a sampled-data system. A symbolic representation of a typical sampled-data system is shown in Figure 3.4-1. Here the sampling operation is applied to the error signal. Frequently, it is required to sample the feedback signal. To illustrate procedures applicable to the analysis of sampled-data closed-loop systems, let us consider the following example.

EXAMPLE 3.4-1

For the system shown in Figure 3.4-2 calculate the response to a step input for sampling periods of .1, 1, and 4 seconds.

Sec. 3.4 Discrete State Equations of Closed-loop Sampled-data Systems

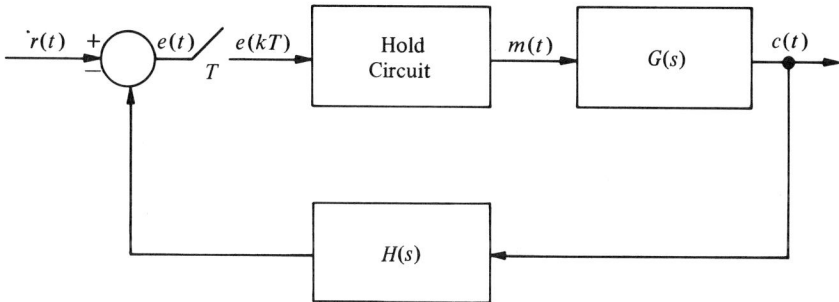

Figure 3.4-1. Closed-loop sampled-data system.

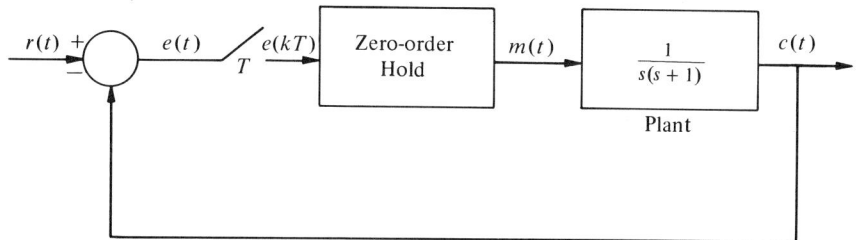

Figure 3.4-2. Block diagram of a sampled system.

Since $e(t) = r(t) - c(t)$, we have

$$e(kT) = r(kT) - c(kT) \tag{3.4-1}$$

The state equations for the plant under control (see Section 1.5) are

$$\frac{d}{dt}\begin{bmatrix} x_1 \\ x_2 \end{bmatrix} = \begin{bmatrix} -1 & 0 \\ 1 & 0 \end{bmatrix}\begin{bmatrix} x_1 \\ x_2 \end{bmatrix} + \begin{bmatrix} 1 \\ 0 \end{bmatrix} m(t) \tag{3.4-2}$$

Since $m(t)$ is the output of a zero-order hold, it is a piecewise constant input. Consequently, we may develop discrete state equations relating $c(kT)$ to $e(kT)$.

The transition matrix corresponding to (3.4-2) is obtained from Section 1.5.

$$\mathbf{A}(T) = e^{\mathbf{F}T} = \begin{bmatrix} e^{-T} & 0 \\ 1 - e^{-T} & 1 \end{bmatrix}$$

and the discrete input matrix is

$$\mathbf{B}(T) = \begin{bmatrix} 1 - e^{-T} \\ T - 1 + e^{-T} \end{bmatrix}$$

The discrete state equations are given by

$$\begin{bmatrix} x_1[(k+1)T] \\ x_2[(k+1)T] \end{bmatrix} = \begin{bmatrix} e^{-T} & 0 \\ 1 - e^{-T} & 1 \end{bmatrix} \begin{bmatrix} x_1(kT) \\ x_2(kT) \end{bmatrix} + \begin{bmatrix} 1 - e^{-T} \\ T - 1 + e^{-T} \end{bmatrix} e(kT)$$

(3.4-3)

$$c(kT) = x_2(kT)$$

Substituting (3.4-1) into (3.4-3) yields

$$\begin{bmatrix} x_1[(k+1)T] \\ x_2[(k+1)T] \end{bmatrix} = \begin{bmatrix} e^{-T} & 0 \\ 1 - e^{-T} & 1 \end{bmatrix} \begin{bmatrix} x_1(kT) \\ x_2(kT) \end{bmatrix}$$
$$+ \begin{bmatrix} 1 - e^{-T} \\ T - 1 + e^{-T} \end{bmatrix} [r(kT) - x_2(kT)]$$

$$c(kT) = x_2(kT)$$

which simplifies to

$$\begin{bmatrix} x_1[(k+1)T] \\ x_2[(k+1)T] \end{bmatrix} = \begin{bmatrix} e^{-T} & e^{-T} - 1 \\ 1 - e^{-T} & 2 - T - e^{-T} \end{bmatrix} \begin{bmatrix} x_1(kT) \\ x_2(kT) \end{bmatrix}$$
$$+ \begin{bmatrix} 1 - e^{-T} \\ T - 1 + e^{-T} \end{bmatrix} r(kT) \qquad (3.4\text{-}4)$$

$$c(kT) = x_2(kT)$$

These relationships represent the closed-loop discrete state equations. We consider now the response of this system for three sampling periods ($T = .1, 1$, and 4 seconds) subject to the step input; therefore,

$$r(kT) = 1, \quad k = 0, 1, 2, \ldots$$

and we make the assumption that the system is initially at rest; that is,

$$x_1(0) = x_2(0) = 0$$

For $T = .1$, expression (3.4-4) becomes

$$\begin{bmatrix} x_1[.1(k+1)] \\ x_2[.1(k+1)] \end{bmatrix} = \begin{bmatrix} .905 & -.095 \\ .095 & .995 \end{bmatrix} \begin{bmatrix} x_1(.1k) \\ x_2(.1k) \end{bmatrix} + \begin{bmatrix} .095 \\ .005 \end{bmatrix}$$

$$c(.1k) = x_2(.1k)$$

By repeated application of these equations for $k = 0, 1, \ldots$, we obtain

$$\{c(0), \ c(.1), \ c(.2), \ \cdots \}$$
$$= \{0, \ .005, \ .019, \ .041, \ .071, \ .106, \ .146, \ \ldots\}$$

Sec. 3.4   Discrete State Equations of Closed-loop Sampled-data Systems

For $T = 1$ the equations are

$$\begin{bmatrix} x_1(k+1) \\ x_2(k+1) \end{bmatrix} = \begin{bmatrix} .368 & -.632 \\ .632 & .632 \end{bmatrix} \begin{bmatrix} x_1(k) \\ x_2(k) \end{bmatrix} + \begin{bmatrix} .632 \\ .368 \end{bmatrix}$$

and the first few numbers in the output sequence are

$$\{c(0), \ c(1), \ c(2), \ \ldots\}$$
$$= \{.0, \ .368, \ 1.000, \ 1.399, \ 1.399, \ 1.147, \ .894, \ \ldots\}$$

For $T = 4$ we have

$$\begin{bmatrix} x_1[4(k+1)] \\ x_2[4(k+1)] \end{bmatrix} = \begin{bmatrix} .0183 & -.98 \\ .98 & 2.02 \end{bmatrix} \begin{bmatrix} x_1(4k) \\ x_2(4k) \end{bmatrix} + \begin{bmatrix} .98 \\ 3.02 \end{bmatrix}$$

The output is

$$\{c(0), \ c(4), \ c(8), \ \ldots\}$$
$$= \{.0, \ 3.02, \ -2.11, \ 5.34, \ -4.82, \ 8.6, \ -8.8, \ \ldots\}$$

The responses for the three cases are plotted in Figure 3.4-3. Clearly, for $T = .1$ and $T = 1$, the response is underdamped. For both cases the response is stable and follows the input. On the other hand, when the sampling period is increased to $T = 4$, the response shows that the sampled system has become unstable, since the oscillation amplitude grows with time.

It is interesting to compare these three responses with the equivalent unsampled system. The closed-loop transfer function is

$$C(s) = \frac{1}{s^2 + s + 1} R(s)$$

and for $R(s) = 1/s$ the output is

$$C(s) = \frac{.1}{s^2 + 2(\frac{1}{2})s + 1} \frac{1}{s}$$

From the transfer function it is indicated that the damping factor is $\zeta = .5$. The response can be easily computed; when compared with the response of the .1 second sampled system, it is found to be almost identical to it. The difference is so small that the plot over the time scale selected in Figure 3.4-3 would not reveal it.

The above example serves to demonstrate a procedure to be followed in the analysis of a sampled system using discrete state techniques. It furthermore alerts the designer to the fact that the selection of the sampling rate is no

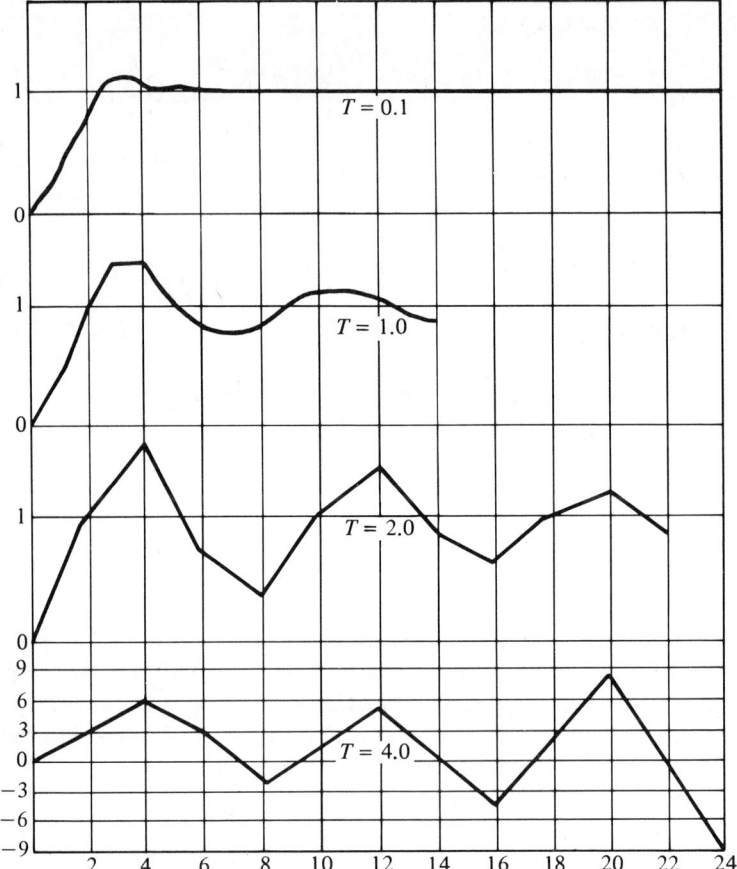

Figure 3.4-3. Response of system for various sampling rates. Input is unit step. (Note scale change on bottom curve.)

arbitrary matter. For reasons of design simplicity a low sampling rate is advisable; this, however, works against the requirement of good data passage through the sample and hold devices. Furthermore, the stability of the system is threatened if the sampling period is selected too large. More will be said about the stability of a sampled system in a subsequent section.

In order to generalize the concepts developed by this example, let us consider the configuration shown in Figure 3.4-4. Since $m(t)$ is a piecewise constant signal, it follows by the discussion of Chapter 2, Section 2.5, that a state variable representation for the plant will be of the form

$$\mathbf{x}[(k+1)T] = \mathbf{A}(T)\mathbf{x}(kT) + \mathbf{B}(T)m(kT)$$
$$y(kT) = c(kT) = \mathbf{C}\mathbf{x}(kT)$$

(3.4-5)

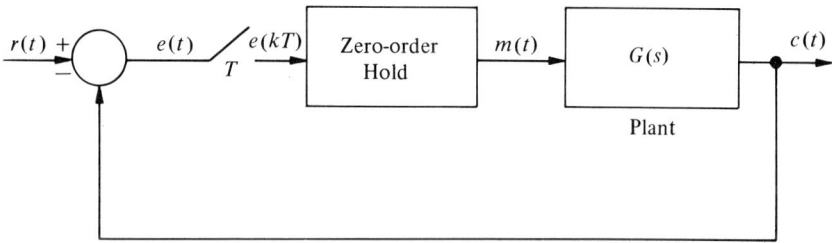

Figure 3.4-4. Generalized sampled-data system.

where the direct transmission matrix is assumed zero. Now

$$m(kT) = e(kT) = r(kT) - c(kT) = r(kT) - \mathbf{C}\mathbf{x}(kT)$$

so that (3.4-5) becomes

$$\mathbf{x}[(k+1)T] = \{\mathbf{A}(T) - \mathbf{B}(T)\mathbf{C}\}\mathbf{x}(kT) + \mathbf{B}(T)r(kT)$$
$$= \bar{\mathbf{A}}(T)\mathbf{x}(kT) + \mathbf{B}(T)r(kT) \tag{3.4-6}$$
$$c(kT) = \mathbf{C}\mathbf{x}(kT) \tag{3.4-7}$$

A comparison of the open-loop dynamics of the plant under control (3.4-5) with its closed-loop form as given by (3.4-6) indicates that an essential difference lies in the makeup of the transition matrix.* The open-loop transition matrix $\mathbf{A}(T)$ has been transformed to $\mathbf{A}(T) - \mathbf{B}(T)\mathbf{C}$ by the feedback process. This feedback property may generate many desirable characteristics, such as a more rapidly responding system, a more stable system, etc. However, it is important that one investigate these characteristics carefully, as feedback can also have a destabilizing effect and can generate other undesirable characteristics.

This generalization approach may be extended to more complex systems (e.g., the system shown in Figure 3.4-1) in a straightforward manner.

3.5 The Discrete-state Analysis of Computer Control Systems

A computer control system is inherently a sampled-data system. Typically, it is of the form shown in Figure 3.5-1, where a digital computer is included in the forward loop of the system. As will be seen in later chapters, the presence of a digital computer opens up a wealth of considerably more sophis-

---

*The closed-loop terminology arises because the output signal $c(t)$ is fed back and subtracted from the input signal $r(t)$ to generate $e(t)$. With no feedback (open-loop) we would have $e(t) = r(t)$.

ticated design approaches than are possible from a more conventional approach. For this reason, it will be desirable to have computer control systems requiring the feedback of more variables than just the output. Although Figure 3.5-1 represents the basic structure of a computer control system, it is

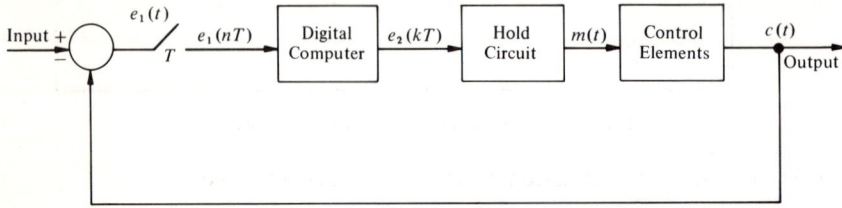

Figure 3.5-1. Schematic of a computer control system.

insufficient to indicate the sophisticated details of a computer control system in which the capabilities of the computer are fully exploited. Let the system shown above suffice as an introduction to the analysis of a system containing a digital computer as an active system element.

In the simplest application of a computer control system, the digital computer is employed to implement the linear recursion equation

$$e_2(kT) = b_0 e_1(kT) + b_1 e_1[(k-1)T] + \cdots + b_n e_1[(k-n)T] \\ - a_1 e_2[(k-1)T] - a_2 e_2[(k-2)T] - \cdots - a_n e_2[(k-n)T] \tag{3.5-1}$$

where $e_1(kT)$ and $e_2(kT)$ represent the input and output sequence of the digital computer at the discrete time kT.

It was shown in Chapter 2 how a linear recursion formula can be changed into a discrete state equation. If we follow those procedures, the discrete state equations in direct programming form corresponding to (3.5-1) are

$$\begin{bmatrix} x_1[(k+1)T] \\ x_2[(k+1)T] \\ \vdots \\ x_n[(k+1)T] \end{bmatrix} = \begin{bmatrix} -a_1 & -a_2 & \cdots & -a_n \\ 1 & 0 & \cdots & 0 \\ 0 & 1 & \cdots & 0 \\ \vdots & \vdots & & \vdots \\ 0 & \cdots & 0 & 0 \end{bmatrix} \begin{bmatrix} x_1(kT) \\ x_2(kT) \\ \vdots \\ x_n(kT) \end{bmatrix} + \begin{bmatrix} 1 \\ 0 \\ 0 \\ \vdots \\ 0 \end{bmatrix} e_1(kT) \tag{3.5-2}$$

$$e_2(kT) = [\hat{b}_1 \ \hat{b}_2 \ \cdots \ \hat{b}_n] \begin{bmatrix} x_1(kT) \\ x_2(kT) \\ \vdots \\ x_n(kT) \end{bmatrix} + b_0 e_1(kT) \tag{3.5-3}$$

where $\hat{b}_j = b_j - b_0 a_j$ for $j = 1, 2, \ldots, n$.

Sec. 3.5 The Discrete-state Analysis of Computer Control Systems

With the recursion equation of the digital computer converted into state variable form, it is now an easy matter to proceed with the analysis of a typical computer control system.

EXAMPLE 3.5-1

Derive the closed-loop state equations for the system shown in Figure 3.5-2.

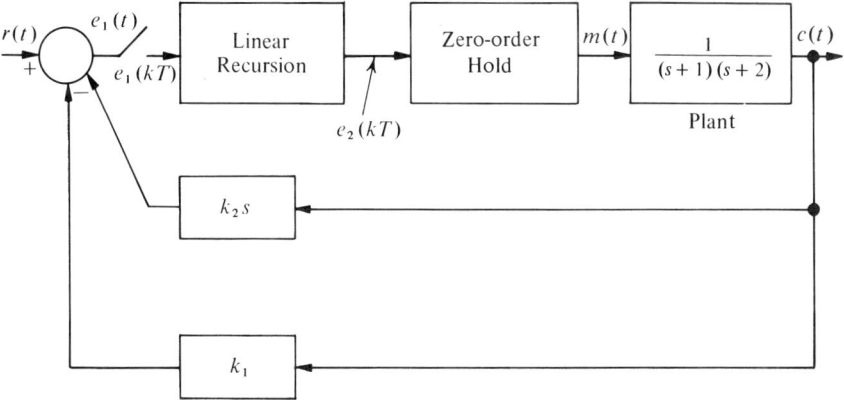

Figure 3.5-2. Computer control system.

Let the linear recursion equation be given as

$$e_2(kT) = 1.2e_1(kT) - .4e_1[(k-1)T] - .25e_2[(k-1)T] \quad (3.5\text{-}4)$$

The corresponding state equations are

$$\begin{aligned} x_3[(k+1)T] &= -.25x_3(kT) + e_1(kT) \\ e_2(kT) &= -.1x_3(kT) + 1.2e_1(kT) \end{aligned} \quad (3.5\text{-}5)$$

The continuous state equations of the plant obtained by using the direct programming method are

$$\begin{bmatrix} \dot{x}_1 \\ \dot{x}_2 \end{bmatrix} = \begin{bmatrix} -3 & -2 \\ 1 & 0 \end{bmatrix} \begin{bmatrix} x_1 \\ x_2 \end{bmatrix} + \begin{bmatrix} 1 \\ 0 \end{bmatrix} m(t) \quad (3.5\text{-}6)$$

$$c(t) = x_2$$

Since $m(t)$ is a piecewise constant input, we can easily derive the discrete state equations of (3.5-6) to be

$$\begin{bmatrix} x_1[(k+1)T] \\ x_2[(k+1)T] \end{bmatrix} = \begin{bmatrix} -e^{-T} + 2e^{-2T} & -2e^{-T} + 2e^{-2T} \\ e^{-T} - e^{-2T} & 2e^{-T} - e^{-2T} \end{bmatrix} \begin{bmatrix} x_1(kT) \\ x_2(kT) \end{bmatrix}$$

$$+ \begin{bmatrix} e^{-T} - e^{-2T} \\ \tfrac{1}{2}e^{-T} + \tfrac{1}{2}e^{-2T} \end{bmatrix} m(kT) \quad (3.5\text{-}7)$$

Two feedbacks paths are included in the system such that

$$e_1(t) = r(t) - k_1 c(t) - k_2 \dot{c}(t) \tag{3.5-8}$$

Since the state model adopted for the plant is a direct programming version, we have

$$\begin{aligned} c(t) &= x_2 \\ \dot{c}(t) &= x_1 \end{aligned} \tag{3.5-9}$$

At the sampling instants (3.5-8) yields

$$e_1(kT) = r(kT) - k_1 c(kT) - k_2 \dot{c}(kT)$$

or, using (3.5-9), we obtain

$$e_1(kT) = r(kT) - k_1 x_2(kT) - k_2 x_1(kT) \tag{3.5-10}$$

Also, we note that

$$m(kT) = e_2(kT) \tag{3.5-11}$$

Equations (3.5-5), (3.5-7), (3.5-10), and (3.5-11) may now be combined to establish the closed-loop discrete state equations. For convenience we shall write (3.5-7) in the form

$$\begin{bmatrix} x_1[(k+1)T] \\ x_2[(k+1)T] \end{bmatrix} = \begin{bmatrix} a_{11} & a_{12} \\ a_{21} & a_{22} \end{bmatrix} \begin{bmatrix} x_1(k) \\ x_2(k) \end{bmatrix} + \begin{bmatrix} b_1 \\ b_2 \end{bmatrix} m(k) \tag{3.5-12}$$

Using this notation, we find that the combined state equations are

$$\begin{bmatrix} x_1[(k+1)T] \\ x_2[(k+1)T] \\ x_3[(k+1)T] \end{bmatrix} = \begin{bmatrix} a_{11} - 1.2 b_1 k_2 & a_{12} - 1.2 b_1 k_2 & -0.1 b_1 \\ a_{21} - 1.2 b_2 k_2 & a_{22} - 1.2 b_2 k_1 & -0.1 b_2 \\ -k_2 & -k_1 & -0.25 \end{bmatrix} \begin{bmatrix} x_1(kT) \\ x_2(kT) \\ x_3(kT) \end{bmatrix}$$

$$+ \begin{bmatrix} 1.2 b_1 \\ 1.2 b_2 \\ 1 \end{bmatrix} r(kT) \tag{3.5-13}$$

Equations (3.5-13) may be solved iteratively to calculate the response of the computer control system to any input $r(t)$. One should observe that the resultant controlled system is of third order, as indicated by (3.5-13). This is due to the fact that the system being controlled is second order and the digital computer is programmed to implement a first-order difference equation. In general, for the configuration shown in Figure 3.5-1, the order of the overall control system is equal to the sum of the orders of the system being controlled and the difference equation implemented by the digital computer.

Sec. 3.5   The Discrete-state Analysis of Computer Control Systems   95

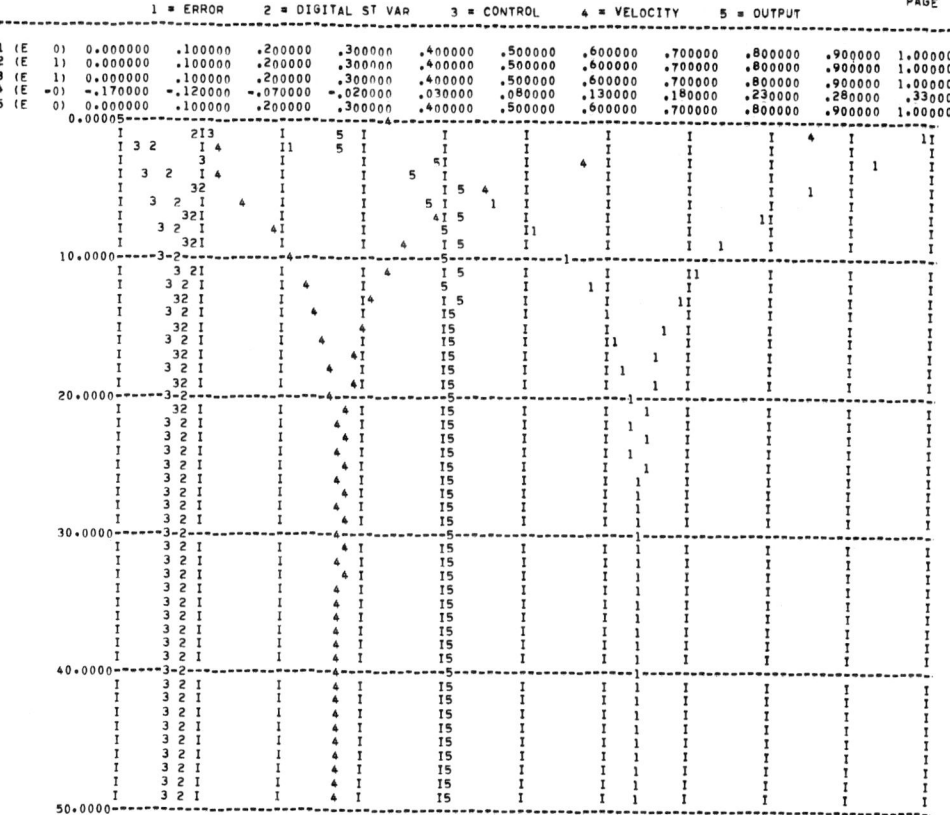

Figure 3.5-3.

A computer program has been prepared to compute the response of this system to a step input. Solved iteratively are equations (3.5-5), (3.5-7), and (3.5-10). The results are plotted out in Figure 3.5-3 on the computer-generated printer plot on which the variables are identified as follows:

Plot 1: Error $= e_1(k)$
Plot 2: Digital state variable $= x_3(k)$
Plot 3: Control $= e_2(k)$
Plot 4: Velocity $= x_1(k)$
Plot 5: Position $= x_2(k)$

The program used to generate the printer plot is explained in Appendix 3A.

```
      DIMENSION  X(5),  Y(2)
      READ  (5,14)  T,TLIMIT,  NPT,  NV
      READ  (5,13)  (HEAD(K),  K = 1,8)
   14 FORMAT  (2F10.4,2I10)
```

```
   13 FORMAT  (8A10)
      R  =  1.
      E1  =  EXP(-T)
      E2  =  EXP(-2.*T)
      A11  =  E1+2.*E2
      A12  =  -2.*E1+2.*E2
      A21  =  E1-E2
      A22  =  2.*E1-E2
      B1  =  E1-E2
      B2  =  .5*(E1+E2)
      TIME  =  0.
      DO 23  K  =  1,NV
   23 X(K)  =  0.
      REWIND  1
      WRITE  (6,15)
   15 FORMAT  (1H1,8X,4HTIME,9X,5HERROR,8X,10HDIG ST VAR,  7X,7HCONTROL,
     X  7X,X7X,8HVELOCITY,8X,6HOUTPUT)
   49 WRITE  (6,11)  TIME,(X(I),I  =  1,NV)
   11 FORMAT  (7(5X,F10.6))
      WRITE  (1,12)  TIME,  (X(I),I  =  1,NV)
   12 FORMAT  (7O20)
      IF  (TIME.GT.TLIMIT)  GO  TO  51
      X(1)  =  R-X(5)-2.*X(4)
      X(2)  =  .25*X(2)+X(1)
      X(3)  =  -.1*X(2)+1.2*X(1)
      Y(1)  =  A11*X(4)+A12*X(5)+B1*X(3)
      Y(2)  =  A21*X(4)+A22*X(5)+B2*X(3)
      X(4)  =  Y(1)
      X(5)  =  Y(2)
      TIME  =  TIME+T
      GO  TO  49
   51 CALL  PLOTX
      CALL  EXIT
      END
```

By incorporating a digital computer in a control system, the design engineer has the capability of generating control characteristics not usually obtainable using standard control elements. Even if the iterative processes that the digital computer carries out are restricted to be linear, a wide design choice is available. Linear iterative processes may be programmed to perform such operations as differentiation, integration, filtering, prediction, etc.

To give a simple demonstration of this, assume that the digital computer in Figure 3.5-1 has been programmed to numerically differentiate the function $e_1(t)$. We shall use the iteration process developed in Section 1.4, namely

$$e_2(kT) = \frac{1}{T}[e_1(kT) - e_1(kT - T)] \quad (3.5\text{-}14)$$

To illustrate that this iteration actually has properties ascribed to the differentiation operation, let $e_1(t) = tu(t)$ so that

$$e_1(kT) = \begin{cases} kT & \text{for } k \geq 0 \\ 0 & \text{for } k < 0 \end{cases}$$

Assume that $e_2(-T) = 0$; that is, the discrete system represented by (3.5-14) is initially at rest. Evaluating (3.5-14) for $k = 0, 1, 2, \ldots$, we have

$$e_2(0) = 0$$

$$e_2(T) = \frac{1}{T}[T - 0] = 1$$

$$e_2(2T) = \frac{1}{T}[2T - T] = 1$$

$$\cdot \quad \cdot \quad \cdot \quad \cdot \quad \cdot \quad \cdot \quad \cdot$$

$$e(kT) = \frac{1}{T}[kT - T] = 1 \quad \text{for } k \geq 1$$

Thus, the output of the digital computer will be a sequence of ones, which, when fed into the zero-order hold circuit shown in Figure 3.5-1, results in

$$m(t) = u(t - T) = \begin{cases} 1 & \text{for } t \geq T \\ 0 & \text{for } t < T \end{cases}$$

An ideal differentiating network would have produced the output $m(t) = (d/dt) e_1(t) = u(t)$. Thus, for the input $e_1(t) = tu(t)$, the given iteration process in conjunction with a zero-order hold has characteristics normally associated with the differential operator. Again, this was a very simple example to illustrate the fact that a digital computer may be used to implement characteristics usually associated with analog elements. In addition, the digital computer may be programmed to perform extremely complex nonlinear operations that are difficult, if not impossible, to implement by analog elements. In summary, the inclusion of a digital computer in a control system opens up new avenues of design not available with standard analog components.

3.6 Stability of Discrete Systems

The stability of linear feedback control systems depends predominantly on the gain of the control loop, on the poles and zeros of the controlled system, on the magnitude of transportation lags, and perhaps on several other less important physical characteristics. A criterion for the stability of continuous-time systems consists of testing whether the eigenvalues of the system matrix or the closed-loop poles all have negative real parts. In the analysis of discrete-time systems one other important design parameter enters into the consideration of stability; this is the sampling period T. In Example 3.4-1, it was demonstrated that a stable sampled-data system can be made unstable by increasing the sampling period beyond a certain point.

Here we wish to consider a stability criterion for discrete-time systems. To this effect we consider the state equations of typical linear discrete systems.

$$\mathbf{x}[(k+1)T] = \underset{\phi}{\mathbf{A}}\mathbf{x}(kT) + \underset{\Delta}{\mathbf{B}}\mathbf{r}(kT) \qquad (3.6\text{-}1)$$

By an iterative approach we have shown that the solution is given by

$$\mathbf{x}(kT) = \mathbf{A}^k \mathbf{x}(0) + \sum_{n=0}^{k-1} \mathbf{A}^{k-1-n} \mathbf{B}\mathbf{r}(nT) \qquad (3.6\text{-}2)$$

For the purpose of a stability analysis it is necessary to consider only the homogeneous solution to (3.6-1); that is, for $\mathbf{r}(nT) = \mathbf{0}$,

$$\mathbf{x}(k) = \mathbf{A}^k \mathbf{x}(0) \qquad (3.6\text{-}3)$$

By the use of (3.6-3) we can establish a stability criterion for a discrete system. The matrix $\mathbf{A}^k$ is the state transition matrix $\mathbf{A}$ raised to the kth power.

If we assume that the eigenvalues of the state transition matrix are all distinct, then $\mathbf{A}^k$ may be expressed as a series by means of the Sylvester expansion theorem (see Appendix 2). Thus, if $\mathbf{A}$ is an $n \times n$ matrix with eigenvalues $\lambda_1, \lambda_2, \ldots, \lambda_n$, then

$$\mathbf{A}^k = \sum_{i=1}^{n} \mathbf{A}_i \lambda_i^k \qquad (3.6\text{-}4)$$

where the matrices $\mathbf{A}_i$ are the constituent matrices of $\mathbf{A}$. Substituting (3.6-4) into (3.6-3) yields

$$\mathbf{x}(k) = \sum_{i=1}^{n} \mathbf{A}_i \lambda_i^k \mathbf{x}(0) \qquad (3.6\text{-}5)$$

From this equation it is apparent that the sequence of vectors

$$\{\mathbf{x}(0),\ \mathbf{x}(1),\ \ldots,\ \mathbf{x}(k),\ \ldots\}$$

can converge to zero, for arbitrary $\mathbf{x}(0)$, only if the terms λ_i^k converge individually to zero. Therefore, for a discrete system to be stable, the eigenvalues of the state transition matrix must satisfy the following condition

$$|\lambda_i| < 1 \quad \text{for } i = 1, 2, \ldots, n \qquad (3.6\text{-}6)$$

Although we have considered here only the case where all eigenvalues are distinct, it can be shown that the stability criterion (3.6-6) is general and applies to systems with any degree of multiple eigenvalues.

EXAMPLE 3.6-1

Test the stability of the system of Example 3.4-1 for $T = 1$ and $T = 4$. The state transition matrices for the two cases were found to be

(a) $T = 1$

$$\mathbf{A} = \begin{bmatrix} .368 & -.632 \\ .632 & .632 \end{bmatrix}$$

The eigenvalues are determined from the relation

$$|\lambda \mathbf{I} - \mathbf{A}| = \begin{vmatrix} \lambda - .368 & .632 \\ -.632 & \lambda - .632 \end{vmatrix} = \lambda^2 - \lambda + .632 = 0$$

which yields

$$\lambda_{1,2} = .5 \pm j\sqrt{.382} = .5 \pm j.625$$
$$|\lambda_1| = |\lambda_2| = \sqrt{(.5)^2 + (.625)^2} = .796$$

Since the absolute value of each eigenvalue is less than unity, the system is stable.

(b) $T = 4$

$$\mathbf{A} = \begin{bmatrix} .0183 & -.98 \\ .98 & -2.02 \end{bmatrix}$$

$$\begin{vmatrix} \lambda - .0183 & -.98 \\ .98 & \lambda + 2.02 \end{vmatrix} = \lambda^2 + 2.002\lambda + .95$$

Clearly, $|\lambda_1|$ and $|\lambda_2|$ are greater than one, although only by a small margin; thus, the system is unstable. This example effectively demonstrates that the selection of the sampling period is a critical element in determining a sampled system's stability. It also verifies the conclusion that we reached in Example 3.4-1, namely, that this system was stable for $T = 1$ but unstable for $T = 4$.

3.6-1 Regions of Stability

We have established that the location of the roots of the characteristic equation of a system can be used as a criterion of stability. For the sake of comparison we summarize here the results as they apply to continuous- and discrete-time systems.

Continuous Time	Discrete Time

System equation:

$$\dot{\mathbf{x}}(t) = \mathbf{F}\mathbf{x}(t) + \mathbf{G}\mathbf{r}(t)$$
$$\mathbf{y}(t) = \mathbf{C}\mathbf{x}(t) + \mathbf{D}\mathbf{r}(t)$$

$$\mathbf{x}[(k+1)T] = \mathbf{A}\mathbf{x}(kT) + \mathbf{B}\mathbf{r}(kT)$$
$$\mathbf{y}(kT) = \mathbf{C}\mathbf{x}(kT) + \mathbf{D}\mathbf{r}(kT)$$

System matrix:

$$\mathbf{F} \qquad\qquad \mathbf{A}$$

Characteristic equation:

$$|\lambda\mathbf{I} - \mathbf{F}| = 0 \qquad\qquad |\lambda\mathbf{I} - \mathbf{A}| = 0$$

Stability criterion:

$$\operatorname{Re}\{\lambda_i\} < 0 \qquad\qquad |\lambda_i| < 1$$

Region of stability:

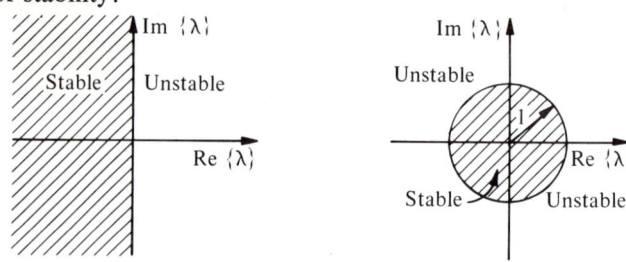

Figure 3.6-1. Regions of stability.

3.7 Analysis of a Digital Process Controller

The process control industry has a potentially large market for computer control technology. It is not unusual to find in a single process a substantial number of control loops, each one of which involves a single controlled variable such as temperature, flow, heat, etc. The process control industry has been using for many years the so-called PID controller, which provides proportional, integral, and derivative control action for a given loop. The configuration for this controller is illustrated in Figure 3.7-1, where it is shown as part of a typical control loop. Ideally, the controller is intended to provide the three control functions exactly as the transfer functions indicate. However, because of common physical limitations, neither the derivative nor the integral operation can be perfectly achieved.

Normally a PID controller is realized by analog means. We demonstrate now how one may replace the PID controller by a digital computer programmed to generate the functions of differentiation and integration numerically. The proposed digital control loop is shown in Figure 3.7-2.

The digital computer is programmed so that the output sequence is given by

$$e_2(k) = K_p e_{21}(k) + K_i e_{22}(k) + K_d e_{23}(k) \qquad (3.7\text{-}1)$$

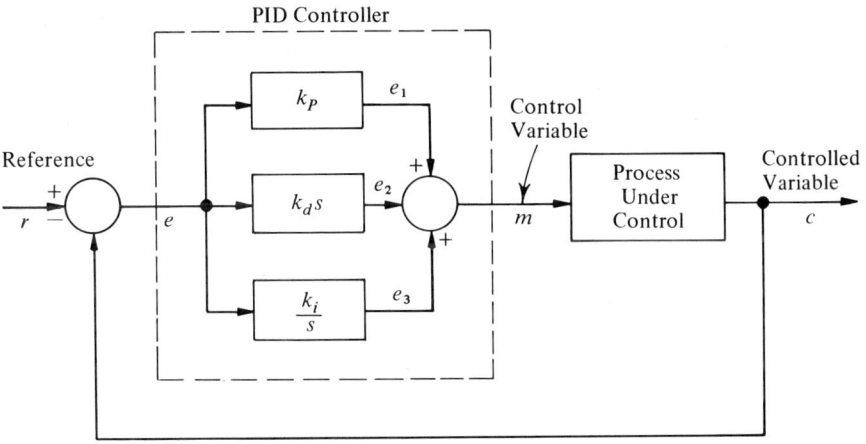

Figure 3.7-1. Typical PID control loop.

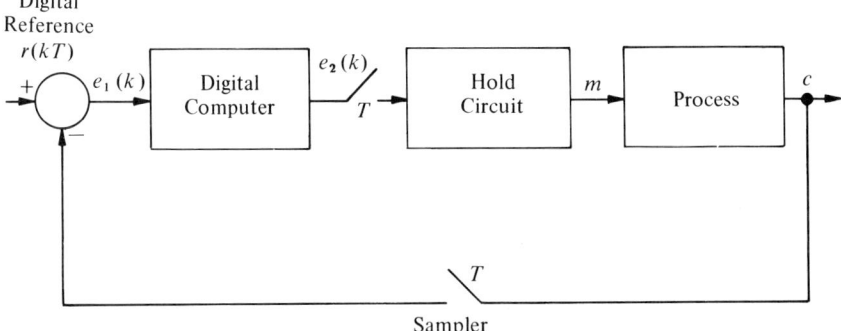

Figure 3.7-2. System with digital computer replacing PID controller.

The three components of the sum are generated as follows:

Proportional control

$$e_{21}(k) = e_1(k) \tag{3.7-2a}$$

Integral control

$$e_{22}(k) = e_{22}(k-1) + Te_1(k) \tag{3.7-2b}$$

Derivative control

$$e_{23}(k) = \frac{1}{T}[e_1(k) - e_1(k-1)] \tag{3.7-2c}$$

Other forms of numerical integration and differentiation could have been selected (e.g., see Chapter 9).

Although equations (3.7-2a) through (3.7-2c) may be readily programmed in their present form, it is desirable to obtain a discrete state model of the discrete PID controller in order to facilitate other analytical studies.

We recognize that the integral and derivative control are described by first-order linear difference equations of the general type

$$x_{\text{out}}(k) = b_0 x_{\text{in}}(k) + b_1 x_{\text{in}}(k-1) - a_1 x_{\text{out}}(k-1) \qquad (3.7\text{-}3)$$

The state model according to the direct programming method corresponding to this equation is given by

$$\begin{aligned} x(k+1) &= -a_1 x(k) + r(k) \\ y(k) &= (b_1 - b_0 a_1) x(k) + b_0 r(k) \end{aligned} \qquad (3.7\text{-}4)$$

Applying equation (3.7-4) to the integrator, we have $a_1 = -1$, $b_0 = T$, $b_1 = 0$, so

$$\begin{aligned} x_i(k+1) &= x_i(k) + e_1(k) \\ e_{22}(k) &= T x_i(k) + T e_1(k) \end{aligned} \qquad (3.7\text{-}5)$$

where the subscript i denotes integration. Similarly, for the differentiator we have $a_1 = 0$, $b_0 = 1/T$, $b_1 = -1/T$, so

$$\begin{aligned} x_d(k+1) &= e_1(k) \\ e_{22}(k) &= -\frac{1}{T} x_d(k) + \frac{1}{T} e_1(k) \end{aligned} \qquad (3.7\text{-}6)$$

where the subscript d denotes differentiation.

The complete state model can now be assembled from equations (3.7-2a), (3.7-5), and (3.7-6).

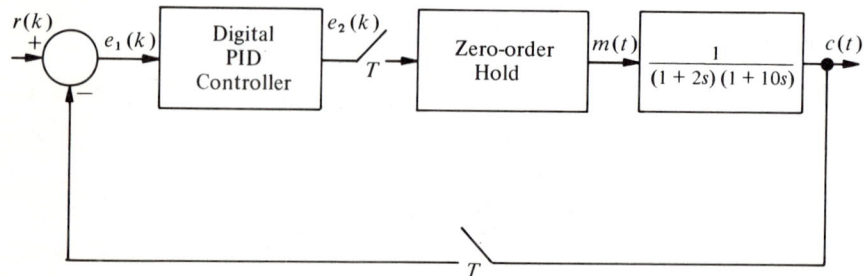

Figure 3.7-3. Digital control loop.

$$\begin{bmatrix} x_i(k+1) \\ x_d(k+1) \end{bmatrix} = \begin{bmatrix} 1 & 0 \\ 0 & 0 \end{bmatrix} \begin{bmatrix} x_i(k) \\ x_d(k) \end{bmatrix} + \begin{bmatrix} 1 \\ 1 \end{bmatrix} e_1(k)$$

$$e_2(k) = \begin{bmatrix} T & -\dfrac{1}{T} \end{bmatrix} \begin{bmatrix} x_i(k) \\ x_d(k) \end{bmatrix} + \left(K_p + K_i T + K_d \dfrac{1}{T} \right) e_1(k)$$

(3.7-7)

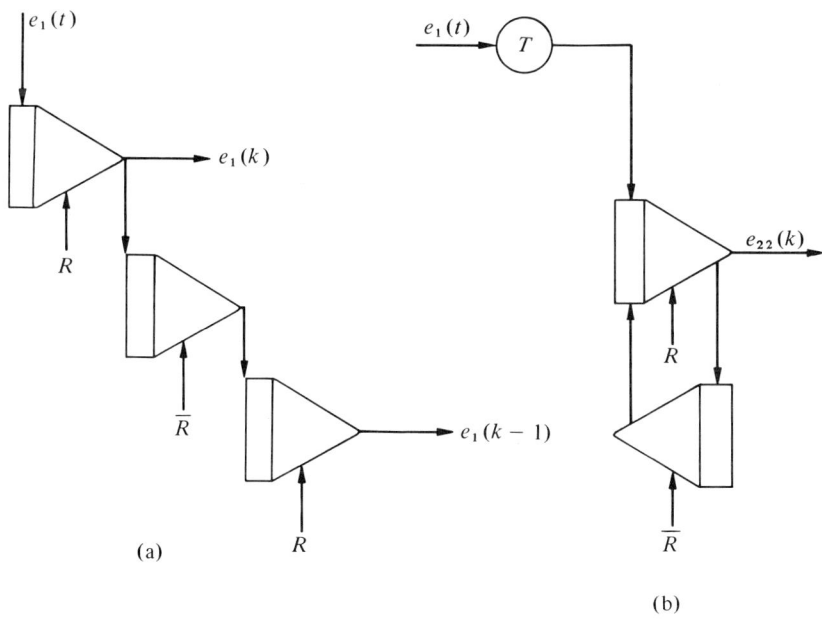

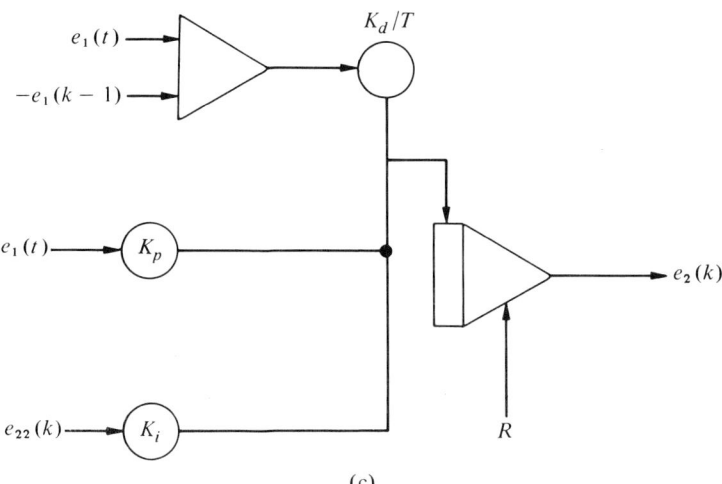

Figure 3.7-4. Analog-hybrid implementation of discrete PID controller.

Consider now the application of this digital controller to the control loop shown in Figure 3.7-3. In order economically to justify the employment of a digital computer for the control of a process, it would seem reasonable that the computer control a great number of loops on a time-shared basis. This is possible if the computer is switched through a multiplexor periodically to each of the control loops. The computer's high speed will permit the execution of the program for this control loop in less than 100 microseconds. It is apparent that, for example, if the sampling period is about one second, easily 1000 such loops can be handled by a single computer without encountering time problems.

In the application of a digital computer as a PID controller, the quantities K_p, K_i, K_d, and T must be determined. It has been the practice in the process control industry to adjust the gain constants on the job until a desirable response is obtained. Alternatively, a simulation of the entire system may be performed which permits parameter adjustment.

A hybrid computer is ideally suited for this task. The plant may be simulated on the analog computer, while the general-purpose digital computer is

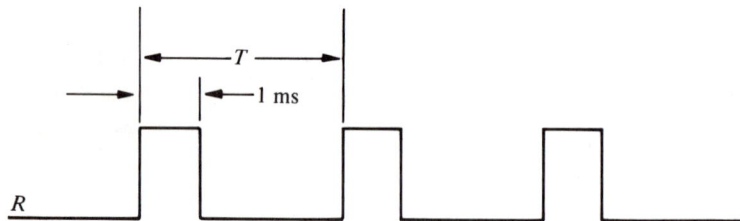

Figure 3.7-5. Timing signals for R.

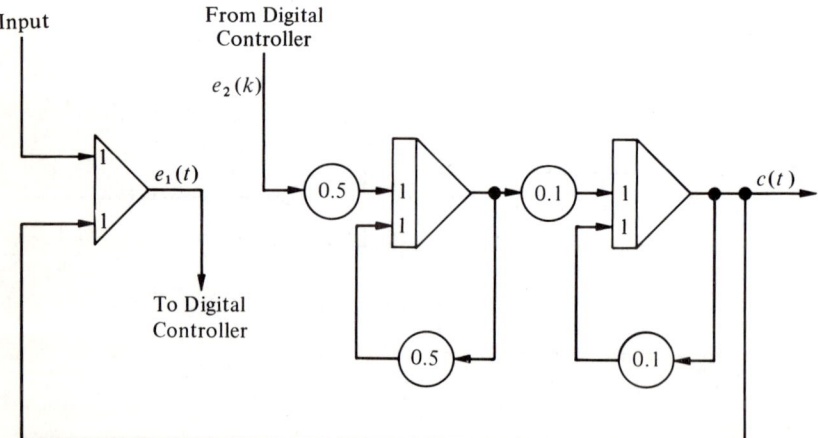

Figure 3.7-6. Simulation of process.

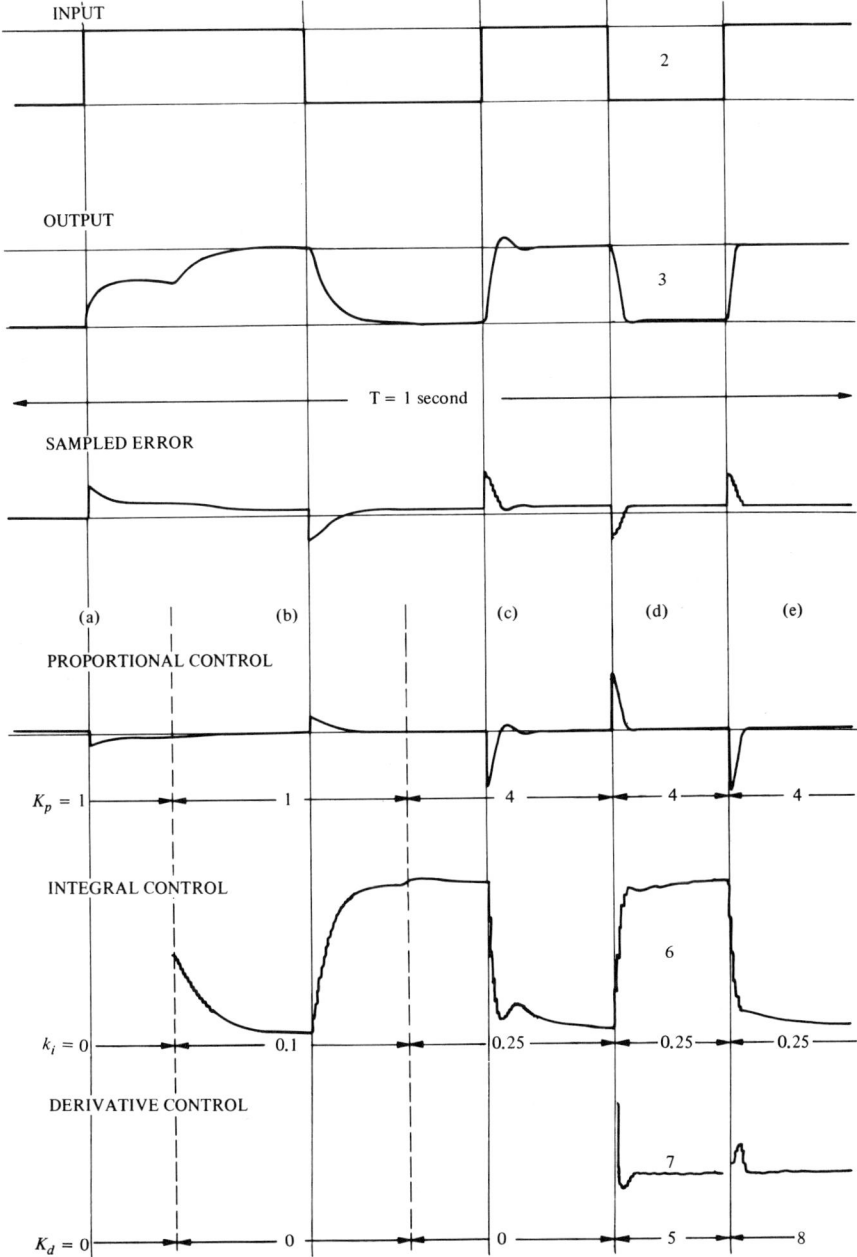

Figure 3.7-7.

available to program the PID controller. The hybrid computer also has the necessary data converter to transform signals from analog to digital form and vice versa. The programming of this problem by a hybrid computer is explained in Chapter 9.

One may also use an analog computer with discrete operation capability, called an analog/hybrid computer. This type of computer is equipped with digital logic timing networks with whose support integrators may be operated as discrete memory devices. Such a computer is used here to simulate the system.

The digital PID controller is simulated as shown in Figure 3.7-4. Figure 3.7-4(a) contains the diagram for the generation of $e_1(k)$ and $e_1(k-1)$ by use of two track-store units that are timed in a complementary fashion. Figure 3.7-4(b) shows the circuit that implements the integrator. By means of an analog accumulator, the integrator difference equation is generated. Finally, Figure 3.7-4(c) shows the generation of the output by summing the various signals making up equation (3.7-1).

The timing signal R is of the shape shown in Figure 3.7-5. It controls the RESET mode. It is understood that the OPERATE mode is controlled by the complement of RESET. The process to be controlled is programmed according to the diagram of Figure 3.7-6.

Step responses for various control settings are shown in Figure 3.7-7 in the form of a six-channel recording. Part (a) shows the response of the system for a sampling rate of one second and only proportional control. Since the process is a type 0 process and $K_p = 1$, the steady-state error is equal to 50 percent of the input. In part (b) the result of introducing integral control is an elimination of the steady-state error; a complete response for the settings ($T = 1$, $K_p = 1$, $K_i = .1$) is shown. In part (c) the gain settings for K_p and K_i have been increased to 4 and .25, respectively; the response now exhibits a definite overshoot. This overshoot is eliminated by the introduction of derivative control for which two runs are shown with gain settings of $K_d = 5$ and $K_d = 8$.

3.8 Response of Sampled Systems between Sampling Instants

It is evident by now that the analytical techniques developed in the preceding section for the analysis of sampled-data systems provide information on the system variables only at the sampling instants. Thus, even though the system to be analyzed has a continuous-time output such as a computer control system, the output is known only at discrete times synchronous with the sampling period. Quite frequently, however, it is important to know the time history of the continuous-time variables for more than the sampling instants. This is actually quite easily accomplished. Consider the response of a linear system via state variable techniques. Given the arbitrary initial conditions $x(t_0)$ at t_0, we may write

$$\mathbf{x}(t) = \mathbf{A}(t - t_0)\mathbf{x}(t_0) + \int_{t_0}^{t} \mathbf{A}(t - t_1)\mathbf{B}\mathbf{x}_{\text{in}}(t_1)dt_1 \qquad (3.8\text{-}1)$$

where $\mathbf{x}_{\text{in}}(t)$ is the input to the system.

If this linear system represents the continuous time part of a sampled system such as the one shown in Figure 3.3-2, then we identify $\mathbf{x}_{\text{in}}(t)$ with $m(t)$ and

$$m(kT + \tau) = r(kT), \qquad 0 \leq \tau < T \qquad (3.8\text{-}2)$$

where $r(kT)$ is the discrete input to the zero-order hold.

Letting $t_0 = kT$ and $t = kT + \tau$, $0 \leq \tau < T$, we write (3.8-1) as

$$\mathbf{x}(kT + \tau) = \mathbf{A}(\tau)\mathbf{x}(kT) + \int_{kT}^{kT+\tau} \mathbf{A}(kT + \tau - t_1)\mathbf{B}r(kT)dt_1 \qquad (3.8\text{-}3)$$

The integral was shown in Chapter 2 to simplify to

$$\int_{kT}^{kT+\tau} \mathbf{A}(kT + \tau - t_1)\mathbf{B}r(kT)dt_1 = \left\{\int_{0}^{\tau} \mathbf{A}(t_1)\mathbf{B}dt_1\right\}r(kT) = \mathbf{B}(\tau)r(kT)$$

Therefore, (3.8-3) reduces to

$$\mathbf{x}(kT + \tau) = \mathbf{A}(\tau)\mathbf{x}(kT) + \mathbf{B}(\tau)r(kT), \qquad 0 \leq \tau < T \qquad (3.8\text{-}4)$$

This equation may be used to calculate the response of any continuous time part in a sampled system during the sampling interval, provided that the state variables are known at the beginning of the interval and that the input is held constant during the interval. This equation may be best put to work by using computer techniques to evaluate it. This is shown in Chapter 9, Section 9.2.

REFERENCES

1. Dorf, R. C., *Time-Domain Analysis and Design of Control Systems*, Addison-Wesley, Reading, Mass., 1964.
2. Kuo, B. C., *Linear Networks and Systems*, McGraw-Hill, New York, 1967.
3. Monroe, A. J., *Digital Processes for Sampled-Data Systems*, Wiley, New York, 1963.

4. Koenig, H. E., Y. Tokad, and H. K. Kesavan, *Discrete Physical Systems*, McGraw-Hill, New York, 1967.
5. Timothy, L. K. and B. E. Bona, *State Space Analysis—An Introduction*, McGraw-Hill, New York, 1967.
6. Chen, C. F. and I. J. Haas, *Elements of Control Systems Analysis*, Prentice-Hall, Englewood Cliffs, New Jersey, 1968.

PROBLEMS

3.1 Find the discrete state variable representation for the system shown in Figure P3.1. Use the technique of Section 3.3.

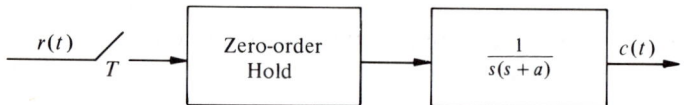

Figure P3.1. Block diagram of sampled system.

3.2 Referring to Figure 3.4-1, let $G(s) = 1/s^2$ and $H(s) = 1$; find the discrete state variable representation for this system. Calculate the first four terms of the response $c(nT)$ [i.e., $c(0)$, $c(1)$, $c(2)$, $c(3)$] when the input $r(t) = tu(t)$.

3.3 Again referring to Figure 3.4-1, let $G(s) = 1/s^2$, $H(s) = s$; find the discrete state variable representation for this system.

3.4 Determine the discrete state variable representation of the system shown in Figure P3.4, where the digital computer implements the numerical differential operation

$$e_2(kT) = \frac{1}{T}[e_1(kT) - e_1(kT - T)]$$

Find the first four terms of the response when $r(t) = u(t)$.

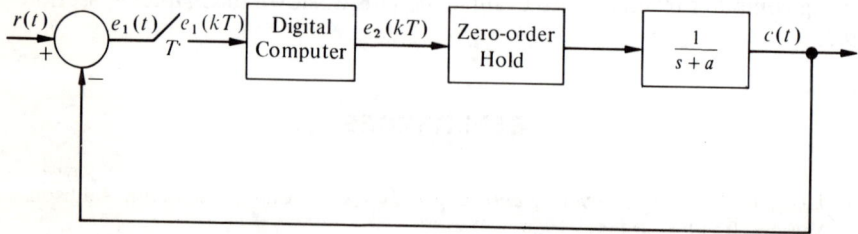

Figure P3.4. Computer control system.

3.5 Examine the stability of the system given in Problem 3.2.

3.6 Find the closed-loop state transition matrix for the sampled-data system shown in Figure P3.6. Determine the eigenvalues of this matrix. The digital computer implements the following difference equations:

$$e_2(kT) = 2e_1(kT) + .2e_{22}(kT)$$
$$e_{22}(kT) = e_{22}[(k-1)T] + e_1(kT)$$

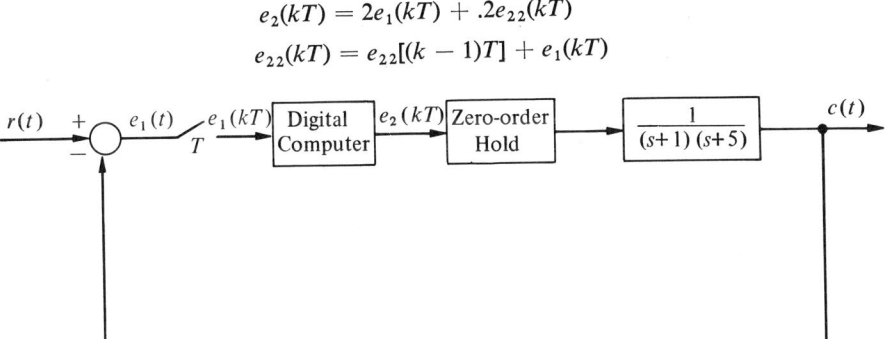

Figure P3.6. Computer control system.

3.7 Investigate the steady-state error of the system shown in Figure P3.6 when the input $r(t)$ is a step input.

3.8 Determine the state transition equations for the system shown in Figure P3.6 according to equations (2.5-9) and (2.5-10). Find the first few terms of the output sequence $c(k)$. Evaluate the matrix $\mathbf{A}^k$ by functions of a matrix.

3.9 Perform a digital computer simulation of the systems shown in Figure P3.6.

3.10 Perform an analog/hybrid computer simulation of the system shown in Figure P3.6.

3.11 Investigate the stability, with T as a parameter, of the system shown in Figure P3.11.

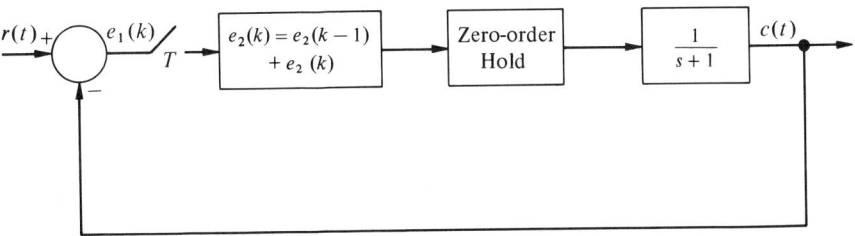

Figure P3.11. Computer control system.

Appendix 3A

COMPUTER-GENERATED PRINTER PLOT

The high-speed line printer is utilized to generate a plot of time-dependent variables. The plot may contain up to five simultaneous variables. A typical example is presented by Example 3.5-2, as shown in Figure 3.5-3. The plot is portioned into pages consistent with the width of computer paper. Each page has a heading made up of a field of 80 alphanumeric characters as supplied by a data card, identifying the variables. Scaling factors and vertical scales for each variable are shown. Each page has room for 50 time points.

A program utilizing this computer plot must contain a number of statements necessary to write the data to be plotted on tape and call the plotting subroutine.

The statements performing the storing of the data on tape involve the following

```
      REWIND  INUNIT
      WRITE  (INUNIT,2)  TIME,(X(I),I = 1,NV)
   2  FORMAT  (7O20)
```

The calling statement is

```
      CALL  PLOTX  (NPT,NV,INUNIT)
```

where   TIME = time variable
            X(I) = array of variables to be plotted
            NPT = number of points along time axis
             NV = number of variables to be plotted.
         INUNIT = tape unit used for storage of data

The plotting routine and its various subroutines are listed below.

```
          SUBROUTINE  PLOTX  (NPT,NVAR,INUNIT)
          LOGICAL  READ,SKIP,TRAN
          REAL  MIN(6),MAX(6),INCR(6),LABEL(11)
          REAL  PT(6)
          INTEGER  LINE(11),LINCNT,BLANK,PGCNT,PAGE,IDENT(6),SCALF(6),PTS
          DATA  IDENT,MINUS,IIJ/1H1,1H2,1H3,1H4,1H5,1H6,10H---------,1HI/
          DATA  BLANK/1H /
          DIMENSION  HEAD(8)
          READ  (5,1)  (HEAD(K),K = 1,8)
        1 FORMAT  (8A10)
          READ  =  .TRUE.
          SKIP  =  .FALSE.
          PTS  =  1
          REWIND  INUNIT
          DO 23  K  =  1,NVAR
          MAX(K)  =  0.
       23 MIN(K)  =  0.
          PAGE  =  1
          DO 21  I  =  1,NPT
          READ  (INUNIT,300)TIME,  (PT(J),J  =  1,NVAR)
          DO 22  K  =  1,NVAR
          MAX(K)  =  AMAX1(MAX(K),PT(K))
       22 MIN(K)  =  AMIN1(MIN(K),PT(K))
       21 CONTINUE
C         MIN ARRAY CONTAINS MIN. VALUES FOR EACH CORRESPONDING
C         SET OF POINTS TO BE PLOTTED
C         MAX CONTAINS MAX. VALUES FOR EACH CORRESPONDING
C         SET OF POINTS TO BE PLOTTED
          REWIND  INUNIT
C         CHECK RANGE OF NVAR (NO. OF VARIABLES TO BE PLOTTED)
          IF(NVAR.GT.0.AND.NVAR.LE.6)GO TO 10
C         OUTPUT DIAGNOSTIC IF NVAR OUT OF RANGE AND RETURN
C         TO CALLING PROGRAM WITH NO PLOTTING PERFORMED.............
          WRITE(6,200)NVAR
      200 FORMAT(6H NVAR  =  I10/
        1 35H NO. VARIABLES TO BE PLOTTED. LE. 5)
          RETURN
C         CALCULATE SCALING CONSTANTS INCR
       10 DO 20  I  =  1,NVAR
          CALL  SSCALE(MIN(I),MAX(I),100.,INCR(I))
       20 CONTINUE
C         OUTPUT HEADING FOR PAGE
       30 WRITE(6,100)PAGE,(HEAD(I),I = 1,8)
          IF(NVAR.EQ.6)GO TO 35
          ISKIP  =  6-NVAR
          DO 31  I  =  1,ISKIP
C         THIS LOOP SKIPS LINES TO FUDGE THE HEADING UP A BIT
          WRITE(6,101)
       31 CONTINUE
```

```
35         DO 36 I = 1,NVAR
C          CALCULATE SCALE FACTORS FOR OUTPUT LABELS
C          IF SCALE FACTORS HAVE BEEN CALCULATED ALREADY SKIP
C          THIS PART......
           IF(SKIP)GO TO 37
           SCALF(I) = IFIX(ALOG10(AMAX1(ABS(MIN(I)),ABS(MAX(I)))))
38         CONTINUE
           TRAN = .TRUE.
           FACTOR = 10.0**SCALF(I)
           IF(FACTOR.LE.1.0E-20)PRINT 105,FACTOR,MIN(I),MAX(I),I
105        FORMAT(1X,3E15.8,I5)
           IF(ABS(MIN(I)/FACTOR).GE.10.0)TRAN = .FALSE.
           IF(ABS(MAX(I)/FACTOR).GE.10.0)TRAN = .FALSE.
           IF(TRAN)GO TO 37
           SCALF(I) = SCALF(I)+1
           GO TO 38
37         LABEL(1) = MIN(I)
C          GENERATE Y-AXIS LABELS
           DO 39 J = 2,11
39         LABEL(J) = LABEL(J-1)*INCR(I)*10.0
           DO 40 J = 1,11
40         LABEL(J) = LABEL(J)/(10.0**SCALF(I))
           WRITE(6,102)I,SCALF(I),LABEL
36         CONTINUE
           SKIP = .TRUE.
C          INITIALIZE LINE AND PAGE COUNTS
           LINCNT = 1
           PGCNT = 1
C          INITIALIZE OUTPUT LINE TO---------------------------------
41         DO 42 I = 1,11
42         LINE(I) = MINUS
C          CHECK TO SEE IF THERE ARE ANY POINTS TO BE PLOTTED.......
           IF(PTS.GT.NPT)GO TO 60
           IF(READ)PTS = PTS+1
C          READ POINTS TO BE PLOTTED FROM INUNIT
           IF(READ)READ(INUNIT,300)TIME,(PT(I),I = 1,NVAR)
           READ = .TRUE.
           DO 43 I = 1,NVAR
C          THIS LOOP INSERTS CHARACTERS IN THE OUTPUT LINE
           INDEX = IFIX((PT(I)-MIN(I))/INCR(I))+1
           CALL PUTCHR(LINE,IDENT(I),INDEX)
43         CONTINUE
C          OUTPUT THE LINE
60         WRITE(6,103)TIME,(LINE(I),I = 1,11)
           IF(PGCNT.LT.6)GO TO 45
           PAGE = PAGE+1
           READ = .FALSE.
           IF(PTS.GT.NPT)RETURN
           GO TO 30
45         PGCNT = PGCNT+1
C          INITIALIZE PLOT LINE
```

```
48              DO 46 I = 1,11
46              LINE(I) = BLANK
C               INSERT VERTICAL MARKERS
                DO 47 I = 1,101,10
                CALL PUTCHR(LINE,IIJ,I)
47              CONTINUE
C               CHECK TO SEE IF MORE POINTS TO BE PLOTTED
C               IF NO MORE POINTS TO BE PLOTTED
C               THEN FINISH PAGE AND RETURN
                IF(PTS.GT.NPT)GO TO 70
                PTS = PTS+1
C               GET SET OF POINTS TO BE PLOTTED FROM INUNIT
                READ(INUNIT,300)TIME,(PT(I),I = 1,NVAR)
300             FORMAT(7O20)
                DO 49 I = 1,NVAR
C               LOOP FOR PLACEMENT OF POINTS ON OUTPUT TIME.......
                INDEX = IFIX((PT(I) - MIN(I))/INCR(I))+1
                CALL PUTCHR(LINE,IDENT(I),INDEX)
49              CONTINUE
C               OUTPUT LINE
70              WRITE(6,104)(LINE(I),I = 1,11)
                IF(LINCNT-9)71,72,72
72              LINCNT = 1
                GO TO 41
71              LINCNT = LINCNT+1
                GO TO 48
100             FORMAT(1H1,112X,4HPAGE,I3/20X,8A10/1X,119(1H-))
101             FORMAT(1H)
102             FORMAT(I2,3H,  (E,I4,1H),11(1X,F9.6))
103             FORMAT(5X,F11.4,10A10,A1)
104             FORMAT(16X,10A10,A1)
                END
                SUBROUTINE SELZER(MIN,MAX,INCR,LNG)
C               SUBROUTINE TO INSERT ZERO AS AN AXIS VALUE WHEN
C               DATA TO BE PLOTTED SWINGS FROM A NEG. MIN. VALUE
C               TO A POS. MAX. VALUE
                REAL MAX,MIN,INCR,LENGTH,LNG,SCAL(3)
                DATA SCAL/1.0,2.0,5.0/
                AIRA = MAX/ABS(MIN)
                IF(AIRA.LT.1.0)GO TO 10
                IRA = AIRA
                IF(IRA.GT.9)IRA = 9
                LENGTH = LNG*FLOAT(IRA)/FLOAT(IRA+1)
                LENGTH = FLOAT(IFIX(LENGTH))
                INCR = MAX/LENGTH
                MAG = ALOG10(INCR)
                MAG = MAG-1
5               DO 1 I = 1,3
                IF(INCR.LT.SCAL(I)*10.0**MAG)GO TO 2
1               CONTINUE
                MAG = MAG+1
```

```
            GO TO 5
2           INCR = SCAL(I)*10.0**MAG
            MAX = INCR*LENGTH
            MIN = -(LNG-LENGTH)*INCR
            RETURN
10          IRA = 1.0/AIRA
            IF(IRA.GT.9)IRA = 9
            LENGTH = LNG*FLOAT(IRA)/FLOAT(IRA+1)
            LENGTH = FLOAT(IFIX(LENGTH))
            INCR = ABS(MIN)/LENGTH
            MAG = ALOG10(INCR)
            MAG = MAG-1
7           DO 3 I = 1,3
            IF(INCR.LT.SCAL(I)*10.0**MAG)GO TO 4
3           CONTINUE
            MAG = MAG+1
            GO TO 7
4           INCR = SCAL(I)*10.0**MAG
            MIN = -INCR*LENGTH
            MAX = (LNG-LENGTH)*INCR
            RETURN
            END

            SUBROUTINE SSCALE(MIN,MAX,LENGTH,INCR)
            DIMENSION SCAL(3)
            REAL MIN,MAX,LENGTH,INCR
            DATA SCAL/1.0,2.0,5.0/
C           GET POWER OF TEN OF MIN VALUE
            MAGMIN = 0.0
            IF(MIN.EQ.0.0)GO TO 12
            IF(MIN.LT.0.0.AND.MAX.GT.0.0)GO TO 14
            MAGMIN = IFIX(ALOG10(ABS(MIN)))
C           FIND NICE NEW VALUE FOR MIN
1           DO 10 I = 1,3,2
            J = 4-I
            IF(MIN.LT.0.0)GO TO 100
            XXMIN = SCAL(J)*10.0**MAGMIN
            WRITE(6,900)MAGMIN,XXMIN,MIN
            IF(XXMIN.LE.ABS(MIN))GO TO 11
            GO TO 10
100         CONTINUE
            XXMIN = SCAL(I)*10.0**MAGMIN
            WRITE(6,900)MAGMIN,XXMIN,MIN
            IF(XXMIN.GE.ABS(MIN))GO TO 11
900         FORMAT(8H MAGMIN = I4,9H NEW MIN = E15.8,5H MIN =
           1E15.8)
10          CONTINUE
            IF(MIN.LE.0.0)GO TO 101
            MAGMIN = MAGMIN-1
            GO TO 1
101         CONTINUE
            MAGMIN = MAGMIN+1
```

```
              GO TO 1
11            SIGN = 1.0
              IF(MIN.LT.0)SIGN = -1.0
C             CALCULATE NEW VALUE FOR MIN
              MIN = XXMIN*SIGN
12            CONTINUE
              DIFF = (MAX-MIN)/LENGTH
              MAGDIF = IFIX(ALOG10(ABS(DIFF)))
2             DO 20 I = 1,3
C             THIS LOOP SELECTS NICE VALUE FOR SCALE INCREMENT
              IF(SCAL(I)*10.0**MAGDIF.GE.DIFF)GO TO 21
20            CONTINUE
              MAGDIF = MAGDIF+1
              GO TO 2
C             CALCULATE SCALE INCREMENT
21            INCR = SCAL(I)*10.0**MAGDIF
C             CALCULATE NEW MAXIMUM
              MAX = MIN+INCR*LENGTH
              RETURN
14            CALL SELZER(MIN,MAX,INCR,LENGTH)
              RETURN
              END

              SUBROUTINE PUTCHR(LINE,CHAR,INDEX)
              INTEGER LINE(11),CHAR,WORD,POS,DUMY(10)
              WORD = (INDEX-1)/10+1
              IF(WORD.GT.11)GO TO 1
              POS = INDEX-(WORD-1)*10
              DECODE(10,200,LINE(WORD))DUMY
              DUMY(POS) = CHAR
              ENCODE(10,200,LINE(WORD))DUMY
              RETURN
1             PRINT 100,WORD,INDEX
              RETURN
200           FORMAT(10A1)
100           FORMAT(21H ERROR WORD TO LARGE = ,I5,7H INDEX = ,I5)
              RETURN
              END
```

Appendix 3B

EIGENVALUES OF A MATRIX

This program will compute the eigenvalues of an arbitrary square matrix using the $[s\mathbf{I} - \mathbf{A}]$ inversion algorithm of Appendix 2B.

The program is called by the calling statement

$$\text{CALL EIGVAL (A, B, C, N, P, EIGEN)}$$

where A = the square matrix of order N whose eigenvalues are to be determined,
B and C = dummy matrices of order N,
N = the order of the matrix,
P = a real N-vector containing the coefficients of the characteristic polynomial

$$s^n + p_1 s^{n-1} + p_2 s^{n-2} + \cdots + p_{n-1} s + p_n$$

and EIGEN = a complex N-vector containing the N eigenvalues of the matrix A.

The calling program must provide appropriate dimension statements for A, B, C, P, and EIGEN, since the subroutine uses variable dimensioning.

The eigenvalue program makes use of a subroutine for matrix multiplication, SUBROUTINE MULTIQ, and a subroutine for polynomial root solving, SUBROUTINE POLRT of the IBM Scientific Subroutine Package.

Subroutine EIGVAL provides as output the coefficients of the characteristic polynomial and the eigenvalues of the matrix.

```
      SUBROUTINE EIGVAL (A,B,C,N,P,EIGEN)
      DIMENSION A(N,N),B(N,N),P(N),C(N,N)
      DIMENSION POL(15),ROOTR(15),ROOTI(15),PX(15)
      COMPLEX EIGEN(N)
      DO 1 I = 1,N
      DO 1 J = 1,N
   1  B(I,J) = A(I,J)
      DO 2 M = 1,N
      P(M) = 0.0
      FM = M
      DO 21 I = 1,N
  21  P(M) = P(M)+B(I,I)/FM
      DO 3 I = 1,N
   3  B(I,I) = B(I,I)-P(M)
      CALL MULTIQ (A,B,C,N,N,N)
      DO 6 K = 1,N
      DO 6 L = 1,N
   6  B(K,L) = C(K,L)
   2  CONTINUE
      DO 4 I = 1,N
      J = N-I+1
   4  POL(J) = -P(I)
      N1 = N+1
      POL(N1) = 1.0
      WRITE (6,10)(POL(I),I = 1,N1)
  10  FORMAT(1H1,*THE COEFFICIENTS OF CHAR. POL. IN INCRG POW*
     X *OF LAMBDA ARE*,//, 15(F12.4/))
      CALL POLRT (POL,PX,N,ROOTR,ROOTI,IER)
      DO 15 I = 1,N
  15  EIGEN(I) = CMPLX(ROOTR(I),ROOTI(I))
      WRITE (6,11)(EIGEN(I),I = 1,N)
  11  FORMAT(2X,*THE EIGENVALUES ARE*,//15(F12.4,5X,F12.4//))
      IF(IER.NE.0)WRITE(6,100)
 100  FORMAT(1H1,*ROOT SOLVING ROUTINE WAS UNABLE TO*
     1 *DETERMINE ALL EIGENVALUES.*/
     2 50H RESULTS FOR THIS DATA SET ARE PROBABLY
       INCORRECT. ,///)
      RETURN
      END


      SUBROUTINE MULTIQ(A,B,C,N1,N2,N3)
      DIMENSION A(N1,N2),B(N2,N3),C(N1,N2)
      DO 1 I = 1,N1
      DO 1 K = 1,N3
      C(I,K) = 0.
      DO 1 J = 1,N2
   1  C(I,K) = C(I,K)+A(I,J)*B(J,K)
      RETURN
      END
```

4

z-Transformation
and Linear Discrete Systems

4.1 Introduction

Transformation techniques have been extensively used by engineers over the years. In conventional linear control and communication applications, the Laplace and Fourier transforms have been particularly valuable. These transforms can be used to obtain solutions to the linear differential equations that characterize such control or communication systems. The resultant algebraic problem replaces the more difficult original integration problem.

In linear discrete systems, the linear difference equation characterizes the dynamical relationship that exists between the system's input and output signals. One must be able to solve these difference equations in order to determine the system's response to given inputs. By using the z-transform, the solutions to such difference equations become essentially algebraic in nature. Just as the Laplace transformation transforms a function of continuous time into a function of a complex variable s, the z-transformation transforms a function of discrete time (sequence of numbers) into a function of a complex variable z.

The z-transform has a history dating back to the 18th century, when it was used by DeMoivre in his studies on probability theory. More recently, the z-transform has been widely used in the analysis of sampled-data systems. It has a history that parallels the Laplace transform in this regard. Developed by the mathematician centuries ago, it has been reapplied by the engineer to serve as a powerful tool in system analysis.

Application of the z-transform to linear discrete systems allows one to introduce the concept of *transfer function* to such systems. In words, the z-transform of the output signal is equal to the product of the system's transfer function with the z-transform of the input signal. The transfer functions of linear continuous- and discrete-time systems have analogous properties.

This chapter introduces the z-transform and develops some of its more important properties. Emphasis of the z-transform to applications in linear discrete systems will be made.

4.2 z-Transform

The z-transform is a transformation that operates on a sequence of numbers producing a function of the complex variable z. In engineering applications, this sequence of numbers generally arises either (1) in an iterative process carried out by a digital computer, or (2) in the sampling of a function of continuous time. This sequence of numbers is formally written as

$$f(k): \quad \{\ldots, f(-1), f(0), f(1), f(2), \ldots\}$$

where the argument is used to indicate the order in which the number occurs in the sequence; that is, $f(5)$ immediately follows $f(4)$ and precedes $f(6)$. In the sampling of a function of continuous time, the sampling period T is an important parameter, and, ideally, the sequence arising from this process should properly be written as $f(kT)$. However, we shall drop the explicit appearance of T in the initial development of the z-transform.

If the sequence of numbers $f(k)$ is defined only for positive integers k, then the one-sided z-transform of $f(k)$ is defined by

$$\mathscr{Z}[f(k)] = F(z) = \sum_{k=0}^{\infty} f(k) z^{-k} \qquad (4.2\text{-}1)$$

When $f(k)$ is given both for positive and negative integers, the two-sided z-transform of $f(k)$ is defined by

$$\mathscr{Z}[f(k)] = F(z) = \sum_{k=-\infty}^{\infty} f(k) z^{-k} \qquad (4.2\text{-}2)$$

It is noted that both the one-sided and two-sided z-transforms are series in powers of z^{-1}. The one-sided z-transform is a negative power series, and therefore it converges for all z outside a circle of radius R, centered about the origin in the complex z plane. The region in the complex z plane for which the z-transform $F(z)$ converges plays a prominent role in characterizing the sequence $f(k)$, which generates $F(z)$. To put this into terms with which the reader is familiar, its importance is equivalent to the significance attached to the left and right half-planes of the complex s plane associated with the Laplace transform. Appendix 4A treats such topics as region of convergence for series as given by (4.2-1) and (4.2-2). After an initial reading of this chapter, it is recommended that the reader become familiar with the contents of this appendix.

We shall be concerned mainly with the one-sided z-transform as defined by (4.2-1). It is a power series in z^{-1} and classically is called a *negative power series* because of the exponent of z. In general, if the representation (4.2-1) has a finite value for a given value of z, this z is said to lie in the so-called convergence region of the z plane. Otherwise, it lies in the divergence region, where the representation (4.2-1) is unbounded (plus or minus infinity). For the one-sided z-transform, the convergence and divergence regions will be of the shape shown in Figure 4.2-1. This divergence region is simply a circle

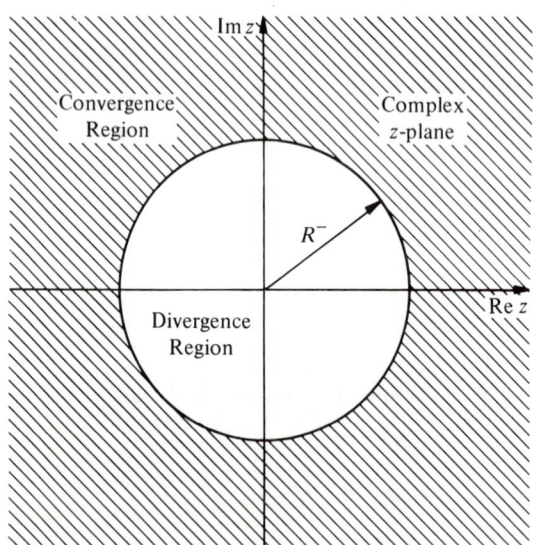

Figure 4.2-1. Regions of convergence and divergence for power series.

of radius R^- centered about the origin. On the boundary $|z| = R^-$ the one-sided z-transform (4.2-1) may either converge or diverge. Depending on the sequence $f(k)$, the one-sided z-transform may diverge everywhere ($R^- = \infty$), in some finite region (finite R^-), or almost nowhere ($R^- = 0$). For most engineering applications, the one-sided z-transform will have a convenient closed-form solution in its region of convergence which will generally be a ratio of polynomials in z; that is,

$$F(z) = \frac{b_0 z^m + b_1 z^{m-1} + \cdots + b_m}{z^n + a_1 z^{n-1} + \cdots + a_n} \qquad (4.2\text{-}3)$$

After factoring the numerator and denominator polynomials of $F(z)$, we have

$$F(z) = b_0 \frac{(z - z_1)(z - z_2) \cdots (z - z_m)}{(z - p_1)(z - p_2) \cdots (z - p_n)} \qquad (4.2\text{-}4)$$

The numbers z_i are referred to as the *zeros of $F(z)$*, for the obvious reason that $F(z_i) = 0$, whereas the numbers p_i are called the *poles of $F(z)$*. At a pole, $F(z)$ is unbounded [i.e., $F(p_i) = \pm\infty$]. The location of the poles and zeros of $F(z)$ determines the characteristics of the sequence of numbers $f(k)$ that generate it.

In applications involving Laplace transforms, one generally obtains a transform that is a ratio of polynomials in s. Such polynomials are of the form (4.2-3) with z replaced by s. The zeros and poles of Laplace transforms play a dominant role in determining the pertinent characteristics of con-

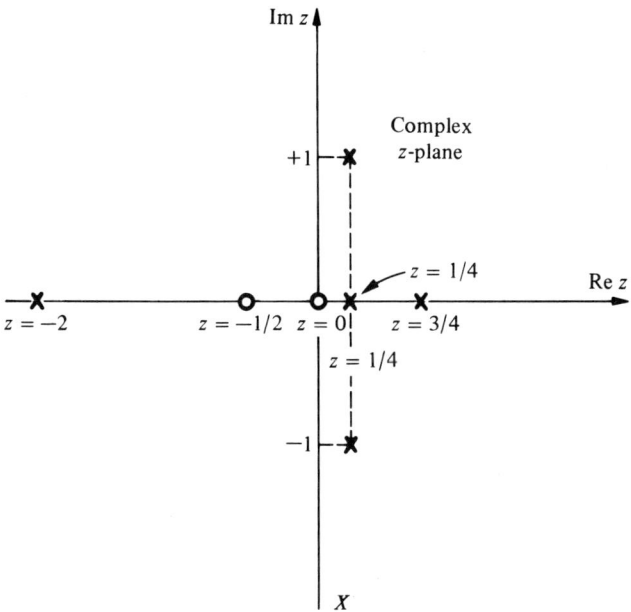

Figure 4.2-2. Notation used for zero-pole locations.

tinuous-time systems. This will also be the case for discrete-time systems. Since the location of the zeros and poles of $F(z)$ is of such significance, a graphical method for displaying their placements is essential. Figure 4.2-2 shows a plot of the pole-zero pattern for the function

$$F(z) = \frac{z(z + \frac{1}{2})}{(z - \frac{3}{4})(z + 2)(z - \frac{1}{4} + j)(z - \frac{1}{4} - j)}$$

A zero is indicated by a circle, while a pole is denoted by a cross. In general, a zero or pole may be a complex number.

4.3 Linearity of z-Transform

The z-transform is an operator that transforms a sequence of numbers $f(k)$ into a function of the complex variable z [i.e., $F(z)$]. It possesses an important transformation property, namely linearity. Linearity implies that, if any two sequences of numbers $f(k)$ and $g(k)$ and scalars a and b are given, the sequence $h(k)$ formed by the linear combination

$$h(k) = af(k) + bg(k)$$

has the z-transform

$$\begin{aligned} H(z) = \mathscr{Z}[h(k)] &= \mathscr{Z}[af(k) + bg(k)] = a\mathscr{Z}[f(k)] + b\mathscr{Z}[g(k)] \\ &= aF(z) + bG(z) \end{aligned}$$

where $F(z)$ and $G(z)$ are the z-transforms of the number sequences $f(k)$ and $g(k)$, respectively.

For a proof we turn to the definition of the z-transform.

$$\begin{aligned} \mathscr{Z}[af(k) + bg(k)] &= \sum_{k=-\infty}^{\infty} [af(k) + bg(k)]z^{-k} \\ &= a \sum_{k=-\infty}^{\infty} f(k)z^{-k} + b \sum_{k=0}^{\infty} g(k)z^{-k} \\ &= a\mathscr{Z}[f(k)] + b\mathscr{Z}[g(k)] \\ &= aF(z) + bG(z) \end{aligned} \qquad (4.3\text{-}1)$$

4.4 Determination of z-Transform Pairs

We shall now develop techniques for determining closed-form expressions for the one-sided z-transform of typical sequences. A sequence of numbers that occurs frequently is*

---

*The sequence (4.4-1) for discrete-time systems plays a role analogous to the time function $f(t) = e^{at}$ for continuous-time systems.

Sec. 4.4 Determination of z-Transform Pairs

$$f(k) = \begin{cases} a^k & \text{for } k \geq 0 \\ 0 & \text{for } k < 0 \end{cases} \quad (4.4\text{-}1)$$

Therefore,

$$F(z) = \sum_{k=0}^{\infty} a^k z^{-k} = \sum_{k=0}^{\infty} (az^{-1})^k \quad (4.4\text{-}2)$$

This is a geometric series in the argument az^{-1}. It is a well-known fact (see Appendix 4A) that such a geometric series converges only when its argument has absolute value less than one, that is, for

$$|az^{-1}| < 1$$

or equivalently

$$|z| > |a| \quad (4.4\text{-}3)$$

This region is the set of all z which lie outside a circle of radius $|a|$ centered at the origin in the complex z plane (see Figure 4.4-1). To get an appreciation for this fact, let us investigate the three points indicated in Figure 4.4-1.

(i) $z = a/2$: A point in the divergence region. In this case (4.4-2) becomes

$$F\!\left(\frac{a}{2}\right) = \lim_{z \to a/2} F(z) = \lim_{z \to a/2} \sum_{k=0}^{\infty} az^{-k} = a + \sum_{k=1}^{\infty} 2^k$$

which is obviously unbounded.

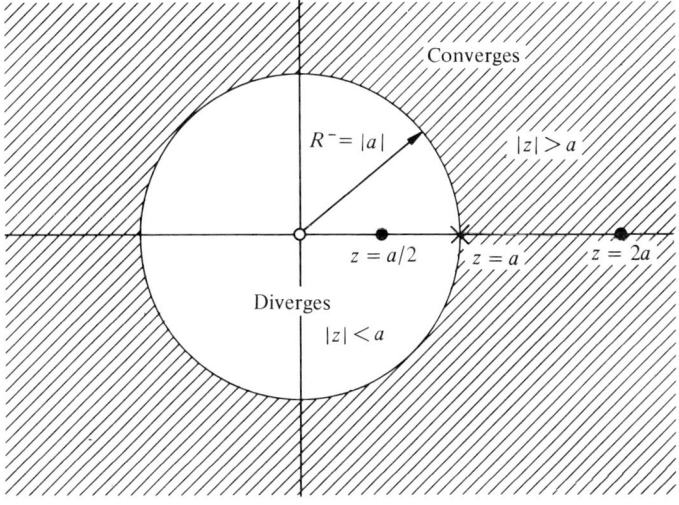

Figure 4.4-1. Region of convergence for the one-sided sequence $f(k) = a^k$.

(ii) $z = a$: A point on the boundary. Evaluating (4.4-2) at $z = a$ gives

$$F(a) = \lim_{z \to a} F(z) = \lim_{z \to a} \sum_{k=0}^{\infty} az^{-k} = a + \sum_{k=1}^{\infty} 1$$

which again is unbounded. The student is cautioned not to generalize this result; not all boundary points yield an unbounded $F(z)$.

(iii) $z = 2a$: A point in the convergence region. Letting $z = 2a$ in (4.4-2) yields

$$F(2a) = \lim_{z \to 2a} F(z) = \lim_{z \to 2a} \sum_{k=0}^{\infty} az^{-k} = \sum_{k=0}^{\infty} (\tfrac{1}{2})^k = 2$$

and, as expected, $F(z)$ has a finite value in its region of convergence.

A further demonstration that (4.4-3) is the region of convergence is obtained when we determine the closed-form expression for (4.4-2). Define

$$F_N(z) = \sum_{k=0}^{N} (az^{-1})^k$$

Clearly, as $N \to \infty$, $F_N(z) \to F(z)$. Let us now subtract $az^{-1} F_N(z)$ from $F_N(z)$; that is,

$$F_N(z) = 1 + (az^{-1}) + (az^{-1})^2 + \cdots + (az^{-1})^N$$
$$az^{-1} F_N(z) = (az^{-1}) + (az^{-1})^2 + \cdots + (az^{-1})^N + (az^{-1})^{N+1}$$

$$F_N(z) - az^{-1} F_N(z) = 1 - (az^{-1})^{N+1}$$

which, after we have solved for $F_N(z)$, results in

$$F_N(z) = \frac{1 - (az^{-1})^{N+1}}{1 - az^{-1}} = \frac{1}{1 - az^{-1}} - \frac{(az^{-1})^{N+1}}{1 - (az^{-1})}$$

We now let $N \to \infty$. If $|az^{-1}| < 1$, the second term on the right side goes to zero. Conversely, if $|az^{-1}| > 1$, the second term is unbounded. We have established the following relationship:

$$\mathscr{Z}[a^k] = F(z) = \begin{cases} \dfrac{1}{1 - az^{-1}} & \text{for } |z| > |a| \\ \text{unbounded} & \text{for } |z| < |a| \end{cases} \quad (4.4\text{-}4)$$

The alert student will observe that $F(z)$ has a pole (at $z = a$) on the boundary separating the convergence and divergence regions. In general, it will always be true that the convergence boundary will pass through one or more poles of $F(z)$.

Fortunately for the user of the one-sided z-transform (the transform whose index of summation ranges from 0 to ∞), the region of convergence may be selected so that convergence of the power series is never a problem. This is most certainly true for the z-transformation of those time sequences encountered in the study of discrete systems. Expression (4.4-4) forms a z-transform pair; that is, the sequence $f(k) = a^k$ for $k \geq 0$ generates the transform $F(z) = 1/(1 - az^{-1})$, and vice versa. It is not uncommon in many texts to multiply the numerator and denominator of (4.4-4) by z to give

$$F(z) = \frac{z}{z - a}$$

This is, of course, equivalent to (4.4-4), but we prefer to use expressions involving z^{-1} and not z because of their more natural usage in making a transition to iterative equations.

EXAMPLE 4.4-1

To serve as motivation for what is to follow, let us examine the simple discrete system studied in Example 2.3-1. It was characterized by the first-order difference equation

$$c(k) - \alpha c(k-1) = r(k) \tag{4.4-5}$$

We determined its weighting sequence by applying the Kronecker delta input and setting the corresponding response $c(k)$ equal to the weighting sequence. This resulted in

$$h(k) = \begin{cases} \alpha^k & \text{for } k \geq 0 \\ 0 & \text{for } k < 0 \end{cases}$$

Let us now find the z-transforms of the Kronecker delta input, the weighting sequence, and rely heavily upon intuition obtained from Laplace transform theory to determine the transfer function of system (4.4-5). The z-transform of the Kronecker delta function is

$$R(z) = \mathscr{Z}[r(k)] = \mathscr{Z}[\delta_0(k)] = 1 \cdot z^0 + 0 \cdot z^{-1} + 0 \cdot z^{-2} + \cdots = 1$$

since $\delta_0 k = 0$ for all k except $k = 0$, where it equals 1.

Similarly, the z-transform of the weighting sequence is

$$H(z) = \mathscr{Z}[h(k)] = \sum_{k=0}^{\infty} \alpha^k z^{-k} = \frac{1}{1 - \alpha z^{-1}} \quad \text{for } z > \alpha \tag{4.4-6}$$

This follows from (4.4-4).

By direct analogy to the relationship between the transfer function of a continuous-time system and its impulse response, we stipulate that $H(z)$, as given by equation (4.4-6), is the transfer function of the system given by equation (4.4-5). In fact, we write

$$C(z) = H(z) \cdot R(z) \qquad (4.4\text{-}7)$$

$\quad\;\;$ z-transform $\qquad$ z-transform $\qquad$ z-transform
$\quad\;\;$ of response $\qquad$ of weighting $\;\;$ of input
$\qquad\qquad\qquad\;\;$ sequence

This relationship, although derived for the special case for which $R(z) = 1$, is valid for an arbitrary input.

This result will now be obtained by an alternative method, to be more fully developed in Section 4.8. Taking the z-transform of (4.4-5) results in

$$C(z) - \alpha z^{-1} C(z) = R(z)$$

and, solving for $C(z)/R(z)$, we have

$$H(z) = \frac{C(z)}{R(z)} = \frac{1}{1 - \alpha z^{-1}} \qquad (4.4\text{-}8)$$

which is exactly what we obtained in (4.4-6) by a more indirect route. This example illustrates the importance of z-transforms and its application to the study of linear discrete systems.

EXAMPLE 4.4-2

Further to demonstrate the utility of z-transformation theory, let us again use intuition to determine the response of the system considered in Example 4.3-1 to a discrete step input

$$r(k) = \begin{cases} 1 & \text{for } k \geq 0 \\ 0 & \text{for } k < 0 \end{cases}$$

This input $r(k)$ is of the form (4.4-1) with $a = 1$, so that its z-transform is given by

$$R(z) = \frac{1}{1 - z^{-1}} \quad \text{for } |z| > 1$$

Inserting this relationship into (4.4-7), we have

$$C(z) = H(z) R(z) = \frac{1}{1 - \alpha z^{-1}} \frac{1}{1 - z^{-1}}$$

$$= \frac{1}{1 - \alpha} \left[\frac{1}{1 - z^{-1}} - \frac{\alpha}{1 - \alpha z^{-1}} \right]$$

Using the z-transform pair relationship developed in (4.4-4), we have

$$c(k) = \frac{1 - \alpha^{k+1}}{1 - \alpha} \qquad \text{for } k \geq 0$$

which is in agreement with the results obtained by using the lengthier time-domain approach taken in Example 2.3-2.

EXAMPLE 4.4-3

To demonstrate how the sequence of numbers as given by (4.4-1) could be generated by the sampling of a continuous-time function, consider the

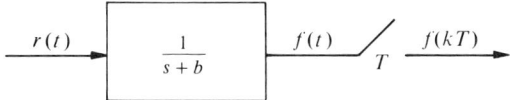

Figure 4.4-2. Sampling process.

system shown in Figure 4.4-2. Let the input $r(t)$ be the Dirac delta function $r(t) = \delta(t)$. The continuous system response is given by

$$f(t) = e^{bt}u(t) \qquad t \geq 0$$

so that after $f(t)$ has been sampled, the output $f(kT)$ is given by

$$f(kT) = e^{bkT}u(kT) = \begin{cases} (e^{bT})^k & \text{for } k \geq 0 \\ 0 & \text{for } k < 0 \end{cases}$$

By letting $a = e^{bT}$ and $f(k) = f(kT)$, we have generated the desired sequence of numbers spaced T seconds apart.

By setting $a = e^{j\omega T}$ in (4.4-4), we may conveniently obtain the z-transform of the sampled $\cos \omega t$ and $\sin \omega t$ functions. Thus

$$\mathscr{Z}[e^{j\omega kT}] = \frac{1}{1 - e^{j\omega T}z^{-1}} \qquad \text{for } |z| > 1$$

which, by Euler's identity, allows us to determine $\mathscr{Z}[\cos \omega kT]$; that is,

$$\mathscr{Z}[\cos \omega kT] = \mathscr{Z}\left[\frac{e^{j\omega kT} + e^{-j\omega kT}}{2}\right]$$

Using the fact that z is a linear transform, we have, by (4.3-1),

$$\mathscr{Z}[\cos \omega kT] = \frac{1}{2}\left[\frac{1}{1 - e^{j\omega T}z^{-1}} + \frac{1}{1 - e^{-j\omega T}z^{-1}}\right] \qquad \text{for } |z| > 1$$

$$= \frac{1}{2}\left[\frac{2 - z^{-1}[e^{j\omega T} + e^{-j\omega T}]}{1 - z^{-1}[e^{j\omega T} + e^{-j\omega T}] + z^{-2}}\right]$$

and finally

$$\mathscr{Z}[\cos \omega k T] = \frac{1 - z^{-1} \cos \omega T}{1 - 2z^{-1} \cos \omega T + z^{-2}} \quad \text{for } |z| > 1 \quad (4.4\text{-}9)$$

Similarly,

$$\mathscr{Z}[\sin \omega k T] = \frac{z^{-1} \sin \omega T}{1 - 2z^{-1} \cos \omega T + z^{-2}} \quad \text{for } |z| > 1 \quad (4.4\text{-}10)$$

4.4-1 Complex Differentiation

The process of differentiating known z-transforms offers another method for generating z-transform pairs. In its region of convergence, a power series may be differentiated any number of times and still remain convergent in this region. Therefore, any order derivative of a given z-transform $F(z)$ converges in the same region as $F(z)$. Consider the arbitrary sequence $f(k)$ and its one-sided z-transform.

$$F(z) = \sum_{k=0}^{\infty} f(k) z^{-k} \quad \text{which converges for } |z| > R^-$$

Differentiating with respect to z, we obtain

$$\frac{dF(z)}{dz} = \sum_{k=0}^{\infty} -k f(k) z^{-k-1}$$

Multiplying both sides by $-z$, we have

$$\sum_{k=0}^{\infty} k f(k) z^{-k} = -z \frac{dF(z)}{dz}$$

Therefore,

$$\mathscr{Z}[k f(k)] = -z \frac{dF(z)}{dz} \quad \text{for } |z| > R^- \quad (4.4\text{-}11)$$

EXAMPLE 4.4-4

Determine the z-transform of the discrete ramp function; that is,

$$g(k) = \begin{cases} k & \text{for } k \geq 0 \\ 0 & \text{for } k < 0 \end{cases}$$

We rewrite $g(k)$ as $k f(k)$, where

$$f(k) = \begin{cases} 1 & \text{for } k \geq 0 \\ 0 & \text{for } k < 0 \end{cases}$$

We may now apply (4.4-11)

$$\mathscr{Z}[g(k)] = \mathscr{Z}[kf(k)]$$
$$= -z\frac{dF(z)}{dz}$$

where $F(z)$ is the z-transform of $f(k)$ and is given by

$$F(z) = \frac{1}{1-z^{-1}} \quad \text{for } |z| > 1$$

Therefore,

$$\mathscr{Z}[k] = -z\frac{d}{dz}\left[\frac{1}{1-z^{-1}}\right]$$

or

$$\mathscr{Z}[k] = \frac{z^{-1}}{(1-z^{-1})^2} \quad \text{for } |z| > 1 \tag{4.4-12}$$

A more general differentiating theorem may be used; it is

$$\mathscr{Z}[k^m f(k)] = \left(-z\frac{d}{dz}\right)^m F(z) \quad \text{for } |z| > R^- \tag{4.4-13}$$

where the operation $(-z\, d/dz)^m$ means—apply the operation $(-z\, d/dz)$ m times in succession.

4.4-2 Complex Integration

The differentiating process may be reversed; for example, suppose it is desired to determine

$$F(z) = \mathscr{Z}\left[\frac{1}{k}\right] = \sum_{k=1}^{\infty} \frac{z^{-k}}{k}$$

Differentiating with respect to z gives

$$\frac{dF(z)}{dz} = \sum_{k=1}^{\infty} z^{-k-1} = -z^{-2}\sum_{k=0}^{\infty} z^{-k} = \frac{-z^{-2}}{1-z^{-1}} \quad \text{for } |z| > 1$$

Solving this differential equation results in

$$F(z) = \mathscr{Z}\left[\frac{1}{k}\right] = -\log(1-z^{-1}) \quad \text{for } |z| > 1$$

Table 4.4-1 Common One-sided z-Transform Pairs

	$f(k)$ for $k \geq 0$	$F(z) = \sum_{k=0}^{\infty} f(k)z^{-k}$	Region of Convergence $\|z\| > R^-$
1	1	$\dfrac{1}{1 - z^{-1}}$	1
2	a^k	$\dfrac{1}{1 - az^{-1}}$	$\|a\|$
3	k	$\dfrac{z^{-1}}{(1 - z^{-1})^2}$	1
4	k^2	$\dfrac{z^{-1}(1 + z^{-1})}{(1 - z^{-1})^3}$	1
5	k^3	$\dfrac{z^{-1}(1 + 4z^{-1} + z^{-2})}{(1 - z^{-1})^4}$	1
6	$\dfrac{a^k}{k!}$	$e^{az^{-1}}$	0
7	$\sin \omega Tk$	$\dfrac{z^{-1} \sin \omega T}{1 - 2z^{-1} \cos \omega T + z^{-2}}$	1
8	$\cos \omega Tk$	$\dfrac{1 - z^{-1} \cos \omega T}{1 - 2z^{-1} \cos \omega T + z^{-2}}$	1
9	ka^{k-1}	$\dfrac{z^{-1}}{(1 - az^{-1})^2}$	$\|a\|$
10	$k^2 a^{k-1}$	$\dfrac{z^{-1}(1 + az^{-1})}{(1 - az^{-1})^3}$	$\|a\|$
11	$k^3 a^{k-1}$	$\dfrac{z^{-1}(1 + 4az^{-1} + a^2 z^{-2})}{(1 - az^{-1})^4}$	$\|a\|$

Table 4.4-2 Common Sampled z-Transform Pairs

	$f(t)$ $t \geq 0$	$f(kT)$ $k \geq 0$	$F(z)$ $F(z) = \sum_{k=0}^{\infty} f(kT) z^{-k}$	Region of Convergence R^-
1	1	1	$\dfrac{1}{1 - z^{-1}}$	1
2	t	kT	$\dfrac{Tz^{-1}}{(1 - z^{-1})^2}$	1
3	t^2	$(kT)^2$	$\dfrac{T^2 z^{-1}(1 + z^{-1})}{(1 - z^{-1})^2}$	1
4	t^3	$(kT)^3$	$T^3 \left[\dfrac{3z^{-1}(1 + z)^{-1}}{(1 - z^{-1})^4} + \dfrac{z^{-1}(1 + 2z^{-1})}{(1 - z^{-1})^3} \right]$	1
5	e^{-at}	e^{-akT}	$\dfrac{1}{1 - e^{-aT} z^{-1}}$	e^{-aT}
6	te^{-at}	kTe^{-akT}	$\dfrac{Te^{-aT} z^{-1}}{(1 - e^{-aT} z^{-1})^2}$	e^{-aT}
7	$t^2 e^{-at}$	$(kT)^2 e^{-akT}$	$\dfrac{T^2 e^{-aT} z^{-1}(1 + e^{-aT} z^{-1})}{(1 - e^{-aT} z^{-1})^3}$	e^{-aT}
8	$\sin \omega t$	$\sin k\omega T$	$\dfrac{\sin \omega T z^{-1}}{1 - 2z^{-1} \cos \omega T + z^{-2}}$	1
9	$\cos \omega t$	$\cos k\omega T$	$\dfrac{1 - z^{-1} \cos \omega T}{1 - 2z^{-1} \cos \omega T + z^{-2}}$	1

Thus, the process of integration is introduced. For the general sequence $f(k)$, we have

$$\mathscr{Z}\left[\frac{f(k)}{k}\right] = G(z) = \sum_{k=0}^{\infty} \frac{f(k)}{k} z^{-k}$$

$$\frac{dG(z)}{dz} = -\sum_{k=0}^{\infty} f(k) z^{-k-1} = -z^{-1} F(z) \qquad (4.4\text{-}14)$$

The integral expression that satisfies (4.4-14) is

$$G(z) = \int_{z}^{\infty} \frac{F(\omega)}{\omega} d\omega + \alpha$$

As $z \to \infty$,

$$G(z) \longrightarrow \alpha = \lim_{k \to 0} \frac{f(k)}{k}$$

We therefore have

$$\mathscr{Z}\left[\frac{f(k)}{k}\right] = \int_{z}^{\infty} \frac{F(\omega)}{\omega} d\omega + \lim_{k \to 0} \frac{f(k)}{k} \qquad (4.4\text{-}15)$$

Expression (4.4-15) may be used to derive further transform pairs. Table 4.4-1 gives some of the more commonly used z-transform pairs.

The process of sampling a function of continuous time occurs so frequently in computer-controlled systems that a separate table listing important z-transform pairs of such time functions is desirable. Table 4.4-2 gives such a listing.

4.5 Properties of the One-sided z-Transform

Some of the more important properties of the one-sided z-transform are listed in Table 4.5-1. The linearity and multiplication by k^m properties have already been treated. Proofs of the remaining properties will now be given with the exception of the convolution property, which is reserved for Section 4.6.

4.5-1 Shifting Property

The z-transform of the sequence shifted to the left one time interval is obtained by specifying the identity $g(k) = f(k+1)$, which gives

Table 4.5-1 Properties of the One-sided z-Transform

Property	Discrete Sequence	z-Transform
1. Linearity	$af(k) + bg(k)$	$aF(z) + bG(z)$
2. Shifting $m \geq 0$	$f(k+m)$	$z^m F(z) - \sum_{j=0}^{m-1} f(j) z^{m-j}$
	$f(k-m)$	$z^{-m} F(z)$
3. Summation	$\sum_{j=0}^{k} f(j)$	$\dfrac{z}{z-1} F(z)$
4. Multiplication by a^k	$a^k f(k)$	$F(a^{-1} z)$
5. Convolution	$\sum_{m=0}^{k} h(k-m) r(m)$	$H(z) R(z)$
6. Multiplication by k^m	$k^m f(k)$	$\left(-z \dfrac{d}{dz}\right)^m F(z)$
7. Forward Difference	$\Delta^m f(k)$	$(z-1)^m F(z) - z \sum_{j=0}^{m-1} (z-1)^{m-j-1} \Delta^j f(0)$
8. Backward Difference	$\nabla^m f(k)$	$(1 - z^{-1})^m F(z)$
9. Initial Value Theorem:	$f(0) = \lim\limits_{z \to \infty} F(z)$	
10. Final Value Theorem:	$f(\infty) = \lim\limits_{z \to 1} (1 - z^{-1}) F(z) = \lim\limits_{z \to 1} (z-1) F(z)$ if $(1 - z^{-1}) F(z)$ is analytic for $|z| \geq 1$.	

$$\mathcal{Z}[g(k)] = \mathcal{Z}[f(k+1)]$$
$$= \sum_{k=0}^{\infty} f(k+1)z^{-k}$$
$$= z \sum_{k=0}^{\infty} f(k+1)z^{-k-1}$$
$$= z \sum_{k=0}^{\infty} f(k)z^{-k} - f(0)$$
$$= zF(z) - zf(0) \quad \text{for } |z| > R^-$$

For the more general case, we let $g(k) = f(k+m)$ so that

$$\mathcal{Z}[g(k)] = \mathcal{Z}[f(k+m)]$$
$$= \sum_{k=0}^{\infty} f(k+m)z^{-k}$$
$$= z^m \sum_{k=0}^{\infty} f(k+m)z^{-k-m}$$
$$= z^m \left[\sum_{k=0}^{\infty} f(k)z^{-k} - \sum_{j=0}^{m-1} f(j)z^{-j} \right]$$
$$= z^m F(z) - \sum_{j=0}^{m-1} f(j)z^{m-j} \quad \text{for } |z| > R^-$$

When the sequence $f(k)$ is shifted to the right m discrete times, we have

$$\mathcal{Z}[f(k-m)] = \sum_{k=0}^{\infty} f(k-m)z^{-m} = z^{-m} \sum_{k=0}^{\infty} f(k-m)z^{-k+m}$$

and since $f(k) = 0$ for $k < 0$

$$\mathcal{Z}[f(k-m)] = z^{-m} \sum_{k=m}^{\infty} f(k-m)z^{-k+m}$$
$$= z^{-m} F(z) \quad \text{for } |z| > R^-$$

4.5-2 Summation

For summation, we form the sequence

$$g(k) = \sum_{j=0}^{k} f(j) \quad \text{for } k = 0, 1, 2, \ldots$$

and observe the identity

$$g(k) - g(k-1) = f(k) \quad k = 0, 1, \ldots$$

where $g(k) = 0$ for $k < 0$. Therefore,

$$\mathcal{Z}[f(k)] = \mathcal{Z}[g(k) - g(k-1)]$$
$$F(z) = G(z) - z^{-1}G(z)$$

which gives

$$G(z) = \frac{z}{z-1} F(z) \quad \text{for } |z| > R^-$$

4.5-3 Multiplication by a^k

$$\mathscr{Z}[a^k f(k)] = \sum_{k=0}^{\infty} a^k f(k) z^{-k} = \sum_{k=0}^{\infty} f(k)(a^{-1}z)^{-k}$$
$$= F(a^{-1}z) \quad \text{for } |a^{-1}z| > R^-$$

4.5-4 Backward Difference

The definition of the backward difference is

$$\nabla f(k) = f(k) - f(k-1)$$

Therefore,

$$\begin{aligned}\mathscr{Z}[\nabla f(k)] &= \mathscr{Z}[f(k)] - \mathscr{Z}[f(k-1)] \\ &= F(z) - z^{-1} F(z) \quad \text{for } |z| > R^- \\ &= (1 - z^{-1}) F(z)\end{aligned} \quad (4.5\text{-}1)$$

A proof by induction will be used to demonstrate the validity of Property 8 of Table 4.5-1. Assume that

$$\mathscr{Z}[\nabla^m f(k)] = (1 - z^{-1})^m F(z) \quad (4.5\text{-}2)$$

is true. Then

$$\mathscr{Z}[\nabla^{m+1} f(k)] = \mathscr{Z}[\nabla(\nabla^m f(k))]$$

which by (4.5-1) is given by

$$\begin{aligned}\mathscr{Z}[\nabla^{m+1} f(k)] &= (1 - z^{-1}) \mathscr{Z}[\nabla^m f(k)] \\ &= (1 - z^{-1})^{m+1} F(z)\end{aligned} \quad (4.5\text{-}3)$$

Since (4.5-2) is true for $m = 1$, and since for general m (4.5-3) follows from (4.5-2), the induction proof is completed.

4.5-5 Forward Difference

The forward difference is defined by

$$\Delta f(k) = f(k+1) - f(k)$$

so that

$$\mathscr{Z}[\Delta f(k)] = \mathscr{Z}[f(k+1) - f(k)]$$

which by Properties 1 and 2 (Table 4.5-1) becomes

$$\mathscr{Z}[\Delta f(k)] = (z-1)\mathscr{Z}[f(k)] - zf(0)$$

The z-transform for the second forward difference is therefore

$$\mathscr{Z}[\Delta^2 f(k)] = (z-1)\mathscr{Z}[\Delta f(k)] - z\Delta f(0)$$
$$= (z-1)^2 F(z) - z(z-1)f(0) - z\Delta f(0)$$

The general relation of Table 4.5-1, Property 7, may be proved by induction.

4.5-6 Initial Value Theorem

Frequently, one is given the one-sided z-transform $F(z)$ and it is desired to determine the initial value $f(0)$ of the sequence that generated $F(z)$. One possibility is to find the inverse of $F(z)$ and evaluate $f(0)$ directly. However, a much more straightforward method exists. Recall that $F(z)$ is given by

$$F(z) = f(0) + f(1)z^{-1} + f(2)z^{-2} + \cdots$$

which converges uniformly for all $|z| > R^-$; therefore, as $z \to \infty$ the terms multiplied by powers of z^{-1} disappear. Thus

$$f(0) = \lim_{z \to \infty} F(z) \tag{4.5-4}$$

4.5-7 Final Value Theorem

In order to determine the behavior of $f(k)$ as $k \to \infty$ from its z-transform $F(z)$, the final value theorem is very useful. If $F(z)$ has poles outside the unit circle, this implies that $f(k)$ will have terms that approach infinity as $k \to \infty$; that is, terms of the form $k^m a_i^k$ for $m = 0, 1, \ldots$ with $|a_i| > 1$. It is, therefore, necessary for $F(z)$ to be analytic for $|z| > 1$ in order that such terms not exist.*
If this is the case, then we may proceed as follows. We have

$$\mathscr{Z}[f(k+1) - f(k)] = \lim_{m \to \infty} \sum_{k=0}^{m} [f(k+1) - f(k)]z^{-k}$$

---

*A function $F(z)$, which is a ratio of polynomials in z, is said to be analytic in a region if it has no poles in this region.

The transform of the left side by the shifting property is

$$(z - 1)F(z) - f(0) = \lim_{m \to \infty} \sum_{k=0}^{m} [f(k+1)] - f(k)]z^{-k} \quad (4.5\text{-}5)$$

Letting $z \to 1$ on both sides of (4.5-5) and assuming that the limits with respect to m and z may be interchanged, we have

$$\lim_{z \to 1} [(z - 1)F(z) - f(0)] = \lim_{m \to \infty} [f(m+1) - f(0)]$$

which simplifies to

$$f(\infty) = \lim_{m \to \infty} f(m+1) = \lim_{z \to 1} (z - 1)F(z) \quad (4.5\text{-}6)$$

if this limit exists.

EXAMPLE 4.5-1

To demonstrate the initial and final value theorems, let us consider the system treated in Examples 4.3-1 and 4.3-2. We have previously determined that the z-transform of the system's response to a discrete step input was given by

$$C(z) = \frac{1}{(1 - \alpha z^{-1})(1 - z^{-1})} \quad (4.5\text{-}7)$$

where for this particular application we shall assume $|\alpha| < 1$. To find the initial value of the time response $c(0)$, we may apply the initial value theorem (4.5-4).

$$c(0) = \lim_{z \to \infty} C(z) = \lim_{z \to \infty} \frac{1}{(1 - \alpha z^{-1})(1 - z^{-1})} = 1$$

The value of $c(k)$ as k approaches infinity is obtained through (4.5-6); that is,

$$c(\infty) = \lim_{z \to 1} (z - 1)C(z)$$
$$= \lim_{z \to 1} (z - 1) \frac{1}{(1 - \alpha z^{-1})(1 - z^{-1})}$$
$$= \lim_{z \to 1} \frac{z}{1 - \alpha z^{-1}} = \frac{1}{1 - \alpha}$$

In Example 4.3-2, it was shown that the closed-form solution for $c(k)$ was

$$c(k) = \frac{1 - \alpha^{k+1}}{1 - \alpha} \quad \text{for } k \geq 0 \qquad (4.5\text{-}8)$$

from which it follows that $c(0) = 1$ and $c(\infty) = 1/(1 - \alpha)$ (if it is assumed that $|\alpha| < 1$), which is in agreement with the results obtained when the initial and final value theorems are used. The advantage accrued by using the initial and final value theorems is that the design engineer may directly obtain values of $c(0)$ and $c(\infty)$ without recourse to inverting $C(z)$, sometimes a difficult task. Expression (4.5-8) gives an indication of why the function $C(z)$ must be analytic outside the unit circle (i.e., $|z| > 1$) in order for the final value theorem to be applied. If $|\alpha| > 1$, then (4.5-7) has a pole at $z = \alpha$, which is outside the unit circle. Blindly applying the final value theorem, one would obtain the false result

$$c(\infty) = \frac{1}{1 - \alpha}$$

However, for $|\alpha| > 1$, the closed-form expression for $c(k)$ as given by (4.5-8) indicates that

$$c(k) \approx \frac{-\alpha^{k+1}}{1 - \alpha} \quad \text{for large } k$$

so that $c(\infty)$ is unbounded. Therefore, one must be careful when applying the final value theorem—*all poles of $C(z)$ must lie inside the unit circle* (with the possible exception of a singular pole at $z = 1$).

EXAMPLE 4.5-2

In using discrete approximations for the numerical integration of differential equations, it is important to determine whether the steady-state values of the approximation and the differential equations agree. To do this we apply the final value theorems of the z-transforms to determine the steady-state value of the Euler approximation for the digital simulation of the transfer function

$$C(s) = \frac{s + 2}{(s + 1)(s + 5)} R(s) \qquad (4.5\text{-}9)$$

when $R(s) = 1/s$, a step input.

Applying the final value theorem of the Laplace transform, we have

$$c(\infty) = \lim_{s \to 0} sC(s) = \lim_{s \to 0} s \left[\frac{s + 2}{(s + 1)(s + 5)} \cdot \frac{1}{s} \right] = \frac{2}{5}$$

To apply the Euler approximation as outlined in Chapter 9, Section 9.2, we first convert the transfer function (4.5-9) into a state model. Using direct programming, we obtain

$$\frac{d}{dt}\begin{bmatrix} x_1(t) \\ x_2(t) \end{bmatrix} = \begin{bmatrix} -6 & -5 \\ 1 & 0 \end{bmatrix}\begin{bmatrix} x_1(t) \\ x_2(t) \end{bmatrix} + \begin{bmatrix} 1 \\ 0 \end{bmatrix} r(t) \qquad (4.5\text{-}10)$$

$$c(t) = [1 \quad 2]\begin{bmatrix} x_1(t) \\ x_2(t) \end{bmatrix} \qquad (4.5\text{-}11)$$

The Euler approximation to equation (4.5-10) is

$$\begin{bmatrix} x_1(k+1) \\ x_2(k+1) \end{bmatrix} = \begin{bmatrix} 1-6T & -5T \\ T & 1 \end{bmatrix}\begin{bmatrix} x_1(k) \\ x_2(k) \end{bmatrix} + \begin{bmatrix} T \\ 0 \end{bmatrix} r(k) \qquad (4.5\text{-}12)$$

We now take the z-transform of equation (4.5-12) and obtain after collecting terms (we assume that the initial conditions are zero)

$$\begin{bmatrix} z-1+6T & 5T \\ -T & z-1 \end{bmatrix}\begin{bmatrix} X_1(z) \\ X_2(z) \end{bmatrix} = \begin{bmatrix} T \\ 0 \end{bmatrix} R(z)$$

Solving for $X_1(z)$ and $X_2(z)$, we obtain

$$\begin{bmatrix} X_1(z) \\ X_2(z) \end{bmatrix} = \frac{1}{(z-1)(z-1+6T)+5T^2} \begin{bmatrix} z-1 & -5T \\ T & z-1+6T \end{bmatrix}\begin{bmatrix} T \\ 0 \end{bmatrix} R(z)$$

Let $R(z)$ be a step input so that $R(z) = z/(z-1)$. Applying now the final value theorem of the z-transform, we have

$$x_1(\infty) = \lim_{z \to 1}(z-1)X_1(z) = 0$$

$$x_2(\infty) = \tfrac{1}{5}$$

Finally, using equation (4.5-11), we obtain the final value of the output

$$c(\infty) = [1 \quad 2]\begin{bmatrix} x_1(\infty) \\ x_2(\infty) \end{bmatrix} = \tfrac{2}{5}$$

It is thereby demonstrated that the steady-state value of the Euler approximation is identical to that of the system to be simulated.

4.6 Transfer Function of a Linear Discrete System

As was demonstrated in Chapter 2, for a discrete system governed by a linear, time-invariant difference equation, the relationship between the output sequence $c(k)$ and the input sequence $r(k)$ for a system initially at rest is

$$c(k) = \sum_{m=-\infty}^{\infty} h(k-m)r(m) \qquad (4.6\text{-}1)$$

where $h(k)$ is the system's weighting sequence. This is a convolution summation which, when the z-transform of $c(k)$ is taken, gives

$$\mathscr{Z}[c(k)] = C(z) = \sum_{k=-\infty}^{\infty} c(k)z^{-k} \qquad (4.6\text{-}2)$$

or

$$C(z) = \sum_{k=-\infty}^{\infty} \left[\sum_{m=-\infty}^{\infty} h(k-m)r(m) \right] z^{-k} \qquad (4.6\text{-}3)$$

The order of summation in (4.6-3) may be reversed, since the power series involved are uniformly convergent in a segment of the complex plane. Therefore,

$$C(z) = \sum_{m=-\infty}^{\infty} r(m) \sum_{k=-\infty}^{\infty} h(k-m)z^{-k}$$

Making the change of variables $n = k - m$ gives

$$C(z) = \sum_{m=-\infty}^{\infty} r(m) \sum_{n=-\infty}^{\infty} h(n)z^{-n-m}$$

$$C(z) = \sum_{m=-\infty}^{\infty} r(m)z^{-m} \sum_{n=-\infty}^{\infty} h(n)z^{-n}$$

Letting $R(z) = \mathscr{Z}[r(k)]$ and $H(z) = \mathscr{Z}[h(k)]$ results in

$$C(z) = H(z)R(z) \qquad (4.6\text{-}4)$$

Equation (4.6-4) gives a very convenient method for determining the z-transform of the output sequence for any input sequence with z-transform $R(z)$. It will become particularly valuable when we develop methods for determining the sequence $c(k)$, which generates $C(z)$. The region of convergence of (4.6-4) is that common region of the complex z plane where $H(z)$ and $R(z)$ both converge. Dividing both sides of (4.6-4) by $R(z)$ gives

$$\frac{C(z)}{R(z)} = H(z)$$

$H(z)$ is called the *transfer function* of the given discrete system. It will now be shown that $H(z)$ is the z-transform of the system's response to the Kronecker delta input

$$r(k) = \delta_0(k) \begin{cases} 1 & \text{for } k = 0 \\ 0 & \text{for } k \neq 0 \end{cases}$$

This is proved by applying (4.6-4) and observing that

$$R(z) = \sum_{k=-\infty}^{\infty} r(k)z^{-k} = 1$$

Therefore,

$$C(z) = H(z)$$

But, by definition, the system's response to this input is the weighting sequence (see Chapter 2).

In Chapter 2, it was shown that if two linear systems with weighting sequence $h(k)$ and $g(k)$ are connected in cascade, then they may be replaced by an equivalent linear system. The equivalent system's weighting sequence $f(k)$ is related to $h(k)$ and $g(k)$ by

$$f(k) = \sum_{j=-\infty}^{\infty} g(k-j)h(j) \tag{4.6-5}$$

Taking the z-transform of both sides of (4.6-5) and noting that the right side is a convolution summation, we have

$$F(z) = G(z)H(z) \tag{4.6-6}$$

which is the equivalent transfer function of two linear systems connected in cascade. Equation (4.6-6) has a form that is analogous to cascaded continuous systems and Laplace transform transfer functions. Namely, the transfer

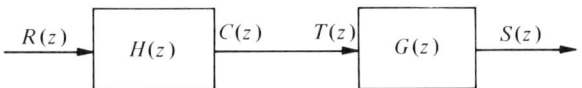

Figure 4.6-1. Two systems in cascade.

function of two systems connected in cascade is simply the product of the two individual systems' transfer functions. A different derivation of (4.6-6) will now be presented. Consider two systems connected in cascade so that the output of the first system serves as the input to the second system. This is illustrated in Figure 4.6-1. The relationships existing between the various transforms are

$$C(z) = H(z)R(z)$$
$$S(z) = G(z)T(z)$$

But $T(z) = C(z)$; therefore,

$$S(z) = G(z)H(z)R(z)$$

for which

$$\frac{S(z)}{R(z)} = G(z)H(z) = F(z)$$

which demonstrates the validity of (4.6-6) by an alternate approach.

EXAMPLE 4.6-1

Determine the transfer function of the system governed by the difference equation

$$c(k) - \alpha c(k-1) = r(k) \qquad (4.6\text{-}7)$$

From Example 2.2-1, $h(k)$ is given by

$$h(k) = \alpha^k \quad k \geq 0$$
$$h(k) = 0 \quad k < 0$$

Therefore,

$$H(z) = \mathscr{Z}[\alpha^k] = \frac{1}{1 - \alpha z^{-1}}$$

Suppose two systems governed by (4.6-7) are connected in cascade; what is the overall system transfer function? In this case, using (4.4-6), we have

$$F(z) = H(z)H(z) = \frac{1}{(1 - \alpha z^{-1})^2}$$

so that from entry 9 of Table 4.4-1 and Property 2 of Table 4.5-1, the weighting sequence of the cascaded systems is

$$f(k) = (k+1)\alpha^k \quad \text{for } k \geq 0$$

4.7 Inverse z-Transform of One-sided z-Transforms

Obviously, if the z-transform is to be of much use, methods for obtaining the sequence of numbers that generate a given z-transform are essential.

There exist primarily three methods for determining the sequence $f(k)$ that generates any given $F(z)$: the direct division method, the partial-fraction expansion method, and the inversion integral method.

4.7-1 Direct Division Method

The simplest method will be considered first. If $F(z)$, which is a ratio of polynomials in z, converges for all $|z| > R^-$, then the generating sequence $f(k)$ is obtained as follows: Divide the denominator of $F(z)$ into the numerator in such a manner that a power series in increasing powers of z^{-1} is obtained. Since

$$F(z) = \sum_{k=0}^{\infty} f(k) z^{-k} \quad (4.7\text{-}1)$$

is the z-transform for such a convergent series, equating the coefficients of like powers in z^{-1} from the result of the division process with (4.7-1) gives the desired sequence $f(k)$.

EXAMPLE 4.7-1

Determine the first five terms of the sequence $f(k)$ when

$$F(z) = \frac{z^{-1} - 3}{z^{-2} - 2z^{-1} + 1} \quad \text{for } |z| > 1 \quad (4.7\text{-}2)$$

Rewriting $F(z)$

$$F(z) = \frac{-3z^2 + z}{z^2 - 2z + 1}$$

and dividing the numerator by the denominator, we obtain

$$F(z) = -3 - 5z^{-1} - 7z^{-2} - 9z^{-3} - 11z^{-4} - \cdots$$

Therefore,

$$f(0) = -3,\ f(1) = -5,\ f(2) = -7,\ f(3) = -9,\ f(4) = -11, \ldots$$

This method is useful only in determining the first few terms of a sequence. If the general expression for $f(k)$ is desired, the following two methods serve this purpose.

4.7-2 Partial-fraction Expansion Method

Perhaps the most popular and direct method for finding the inverse of a z-transform is the partial-fraction expansion method. Let the z-transform $F(z)$ be written in terms of a ratio of polynomials in z; that is,

$$F(z) = \frac{N(z)}{D(z)} = \frac{a_m z^m + a_{m-1} z^{m-1} + \cdots + a_0}{b_p z^p + b_{p-1} z^{p-1} + \cdots + b_0} \qquad (4.7\text{-}3)$$

We are restricting $F(z)$ to be a rational function of z, since this is the form that usually appears in applications. If the integers are such that $m > p$, *it is necessary* to divide $D(z)$ into $N(z)$ until the order of the remainder polynomial is of at least one degree less than $D(z)$; that is,

$$F(z) = \sum_{k=0}^{m-p} c_k z^k + \frac{N_1(z)}{D(z)} \qquad (4.7\text{-}4)$$

The polynomial $D(z)$ is now factored to determine its p roots, some of which may repeat. These roots are the poles of $F(z)$ and determine the characteristic behavior of $f(k)$. A partial fraction expansion of $\hat{F}(z)$ is made, where $\hat{F}(z) = N_1(z)/D(z)$. This results in

$$\hat{F}(z) = F_1(z) + F_2(z) + \cdots + F_q(z)$$

where each $F_k(z)$ is an elementary function that has a pole located at $z = a_k$ of degree one, two, etc. These elementary functions are written so that their composite terms are in the form shown in Table 4.7-1. This gives a particularly convenient technique for obtaining a closed-form expression for multiple poles. For example, let $F(z)$ have a pole of order three at $z = a_1$. The

Table 4.7-1 Transform Pairs Used in Partial-fraction Expansion Method

Fundamental z-Transform Term $F_j(z)$	Convergence Region $\|z\| > R^- = a$ $f(k)$ for $k \geq 1$ $f(0) = 0$
$\dfrac{1}{z-a}$	a^{k-1}
$\dfrac{z}{(z-a)^2}$	$k a^{k-1}$
$\dfrac{z(z+a)}{(z-a)^3}$	$k^2 a^{k-1}$
$\dfrac{z(z^2 + 4az + a^2)}{(z-a)^4}$	$k^3 a^{k-1}$
$\dfrac{z(z^3 + 11az^2 + 11a^2 z + a^3)}{(z-a)^5}$	$k^4 a^{k-1}$

partial-fraction expansion of $F(z)$ will give the elementary function $F_1(z)$ and others. $F_1(z)$ is written as

$$F_1(z) = b_3 \frac{z(z + a_1)}{(z - a_1)^3} + b_2 \frac{z}{(z - a_1)^2} + b_1 \frac{1}{z - a_1}$$

where the coefficients b_1, b_2, and b_3 are evaluated by standard methods; for example

$$b_3 = \frac{(z - a_1)^3}{z(z + a_1)} \hat{F}(z) \bigg|_{z = a_1, \text{ etc.}}$$

Since $F_1(z)$ converges for $|z| > a_1$, the corresponding generating sequence is

$$f_1(k) = b_1 a_1^{k-1} + b_2 k a_1^{k-1} + b_3 k^2 a_1^{k-1} \quad \text{for } k \geq 1$$
$$= 0 \quad \text{for } k < 1$$

The terms $c_k z^k$ in the summation portion of (4.7-5) are generated by the positive integer segment of the sequence, since

$$F(z) = f(m - p)z^{m-p} + \cdots + f(2)z^2 + f(1)z + f(0) + \frac{N_1(z)}{D(z)}$$

Therefore, the elements $c_k z^k$ are generated by

$$c_k z^k \longrightarrow f(k) = c_k \quad \text{for } k \leq 0$$

EXAMPLE 4.7-2

Determine $f(k)$ for

$$F(z) = \frac{-3z^3 + z^2}{z^3 - 4z^2 + 5z - 2} \quad \text{for } |z| > 2$$

Since the numerator is of the same degree as the denominator, a division is required to put it in the form of (4.7-5); this results in

$$F(z) = -3 + \frac{-11z^2 + 15z - 6}{z^3 - 4z^2 + 5z - 2}$$

$F(z)$ has a single pole at $z = 2$ and a double pole at $z = 1$. Making a partial-fraction expansion of $F(z)$ gives

$$F(z) = -3 + b_2 \frac{z}{(z - 1)^2} + b_1 \frac{1}{z - 1} + c_1 \frac{1}{z - 2}$$

from which one finds that $b_1 = 7$, $b_2 = 2$, $c_1 = -20$. Therefore, by Table 4.7-1, the generating sequence is

$$f(0) = -3$$
$$f(k) = 2k(1)^{k-1} + 7(1)^{k-1} - 20(2)^{k-1}$$
$$ = 2k + 7 - 20(2)^{k-1} \quad \text{for } k \geq 1$$
$$ = 0 \quad \text{for } k < 0$$

4.7-3 Inversion Integral Method

The most general technique for obtaining the inverse of a z-transform is the inversion integral method. It utilizes some basic concepts from complex variable theory which are beyond the scope of this text. It will suffice to identify the inversion formula as

$$f(k) = \frac{1}{2\pi j} \oint_\Gamma F(z) z^{k-1} \, dz \quad (4.7\text{-}5)$$

For a detailed explanation of the meaning of this integral and its derivation the reader is referred to a text on complex variable theory and its application to the theory of transformations. Briefly, (4.7-5) is a closed-line integral in the z plane along the closed curve Γ. In this case, the closed curve Γ is taken to be the unit circle (i.e., circle of radius one centered at the origin of the z plane).

4.8 Linear Difference Equations

Consider a linear, time-invariant discrete system characterized by the linear difference equation

$$c(k) + a_1 c(k-1) + \cdots + a_n c(k-n)$$
$$= b_0 r(k) + b_1 r(k-1) + \cdots + b_n r(k-n) \quad (4.8\text{-}1)$$

where $r(k)$ and $c(k)$ are the system's input and output, respectively, at the kth iteration. For all values of k, the right and left sides of (4.8-1) are equal. Treating each side as a function of discrete time k, we see that the sequence $g(k)$ defined by

$$g(k) = c(k) + a_1 c(k-1) + \cdots + a_n c(k-n)$$
$$= b_0 r(k) + b_1 r(k-1) + \cdots + b_n r(k-n)$$

is generated. Taking the z-transform of this sequence, we have

$$\mathscr{Z}[g(k)] = \mathscr{Z}[c(k) + a_1 c(k-1) + \cdots + a_n c(k-n)]$$
$$= \mathscr{Z}[b_0 r(k) + b_1 r(k-1) + \cdots + b_n r(k-n)]$$

Using the linearity property results in

$$\mathscr{Z}[c(k)] + a_1 \mathscr{Z}[c(k-1)] + \cdots + a_n \mathscr{Z}[c(k-n)]$$
$$= b_0 \mathscr{Z}[r(k)] + b_1 \mathscr{Z}[r(k-1)] + \cdots + b_n \mathscr{Z}[r(k-n)]$$

Finally, applying the shifting property gives

$$C(z) + a_1 z^{-1} C(z) + \cdots + a_n z^{-n} C(z)$$
$$= b_0 R(z) + b_1 z^{-1} R(z) + \cdots + b_n z^{-n} R(z) \quad (4.8\text{-}2)$$

or equivalently

$$C(z) = \frac{b_0 + b_1 z^{-1} + \cdots + b_n z^{-n}}{1 + a_1 z^{-1} + \cdots + a_n z^{-n}} R(z) \quad (4.8\text{-}3)$$

Comparison of (4.8-3) with (4.6-4) reveals that

$$H(z) = \frac{b_0 + b_1 z^{-1} + \cdots + b_n z^{-n}}{1 + a_1 z^{-1} + \cdots + a_n z^{-n}} \quad (4.8\text{-}4)$$

is the transfer function of the discrete system governed by the linear difference equation (4.8-1). It is also the z-transform of the weighting sequence, as was shown in Section 4.6. At the risk of being repetitious, let us demonstrate this fact again. It will be recalled that the weighting sequence is the response of a linear system, initially at rest, to the Kronecker delta input; that is,

$$r(k) = \delta_0(k) = \begin{cases} 1 & \text{for } k = 0 \\ 0 & \text{for } k \neq 0 \end{cases}$$

Thus

$$R(z) = \mathscr{Z}[r(k)] = \mathscr{Z}[\delta_0(k)] = 1$$

which when inserted into (4.8-3) gives

$$C(z) = \mathscr{Z}[h(k)] = H(z) = \frac{b_0 + b_1 z^{-1} + \cdots + b_n z^{-n}}{1 + a_1 z^{-1} + \cdots + a_n z^{-n}}$$

To obtain the weighting sequence of the linear discrete system (4.8-1), one simply forms (4.8-4) and takes its inverse z-transform.

EXAMPLE 4.8-1

Determine the weighting sequence for the linear discrete system

$$c(k) - \alpha c(k-1) = r(k)$$

For this system, (4.8-4) is

$$H(z) = \frac{1}{1 - \alpha z^{-1}}$$

which has the inverse z-transform

$$h(k) = \begin{cases} \alpha^k & \text{for } k \geq 0 \\ 0 & \text{for } k < 0 \end{cases}$$

This result agrees with that of Example 2.2-1 and the heuristic approach taken in Example 4.4-1.

Some texts prefer to write the linear difference equation relating the system's input sequence $r(k)$ to its output sequence as

$$c(n+k) + a_1 c(n+k-1) + \cdots + a_n c(k)$$
$$= b_0 r(n+k) + b_1 r(n+k-1) + \cdots + b_n r(k) \quad (4.8\text{-}5)$$

To determine the relationship between $C(z)$ and $R(z)$, we take the z-transform of (4.8-5) and use Properties 1 and 2 of Table 4.5-1. This is demonstrated by the following example.

EXAMPLE 4.8-2

Find the system's response $C(z)$ to the general input $R(z)$ when

$$c(k+2) + 3c(k+1) + 2c(k) = r(k+2) + 4(k+1) + r(k)$$

Taking the z-transform gives

$$C(z)[z^2 + 3z + 2] - c(0)z^2 - c(1)z - 3c(0)z$$
$$= R(z)[z^2 + 4z + 1] - r(0)z^2 - r(1)z - 4r(0)z$$

Dividing by $z^2 + 3z + 2$ and rearranging, we obtain

$$C(z) = \frac{z^2 + 4z + 1}{z^2 + 3z + 2} R(z)$$
$$+ \frac{z^2[c(0) - r(0)] + z[c(1) - r(1) + 3c(0) - 4r(0)]}{z^2 + 3z + 2}$$

The expression for $C(z)$ may be directly evaluated for any $r(k)$. For example, if $r(k)$ is the discrete step input, that is, if

$$r(k) = \begin{cases} 1 & \text{for } k \geq 0 \\ 0 & \text{for } k < 0 \end{cases}$$

we have

$$R(z) = \frac{1}{1 - z^{-1}} = \frac{z}{z - 1}$$

Thus

$$C(z) = \frac{z(z^2 + 4z + 1)}{(z + 1)(z^2 + 3z + 2)} + \frac{z^2[c(0) - 1] + z[c(1) + 3c(0) - 5]}{z^2 + 3z + 2}$$

$c(k)$ may be obtained by taking the inverse z-transform of this expression. It should be observed that a portion of the response $c(k)$ depends on the initial conditions $c(0)$ and $c(1)$.

An interesting and extremely important property follows if it is assumed that the system of (4.8-5) is initially at rest for $k < 0$ [i.e., $c(k) = 0$ for $k < 0$] and $r(k) = 0$ for $k < 0$. Consider the general second-order linear discrete system that is treated in the following example.

EXAMPLE 4.8-3

Determine the system's response $C(z)$ to the general input $R(z)$ if the system is initially at rest, $r(k) = 0$ for $k < 0$, and

$$c(k + 2) + a_1 c(k + 1) + a_2 c(k) = b_0 r(k + 2) + b_1 r(k + 1) + b_2 r(k)$$
(4.8-6)

z-transforming (4.8-6), we obtain

$$C(z)[z^2 + a_1 z + a_2] - c(0)z^2 - c(1)z - a_1 c(0)z$$
$$= R(z)[b_0 z^2 + b_1 z + b_2] - b_0 r(0)z^2 - b_0 r(1)z - b_1 r(0)z$$

Therefore,

$$C(z) = \frac{b_0 z^2 + b_1 z + b_2}{z^2 + a_1 z + a_2} R(z)$$
$$+ \frac{z^2[c(0) - b_0 r(0)] + z[c(1) + a_1 c(0) - b_1 r(0) - b_0 r(1)]}{z^2 + a_1 z + a_2}$$

(4.8-7)

Evaluating (4.8-2) at $k = -2, -1$ and recalling that $c(k) = r(k) = 0$ for $k < 0$, we find

$$k = -2: \quad c(0) = b_0 r(0)$$
$$k = -1: \quad c(1) + a_1 c(0) = b_0 r(1) + b_0 r(0)$$

which when incorporated into (4.8-7) results in

$$C(z) = \frac{b_0 z^2 + b_1 z + b_2}{z^2 + a_1 z + a_2} R(z)$$

In general, when $r(k) = c(k) = 0$ for $k < 0$, the relationship existing between the system's input and output for a system characterized by (4.8-5) is

$$C(z) = \frac{b_0 z^n + b_1 z^{n-1} + \cdots + b_n}{z^n + a_1 z^{n-1} + \cdots + a_n} R(z) \qquad (4.8\text{-}8)$$

Multiplying the numerator and denominator of (4.8-8) by z^{-n}, we arrive at the result given by (4.8-3). This indicates that systems (4.8-1) and (4.8-5) are equivalent, from an input-output dynamical viewpoint.

4.9 Response of a Linear Discrete System

The transfer function concept can greatly simplify the calculations required in determining the response of a linear discrete system to a given input sequence. For example, let us examine system (4.8-1), which has the transfer function (4.8-4). To determine the z-transform of the system's response $c(k)$ to a given input $r(k)$, we use the relationship

$$C(z) = \frac{b_0 z^n + b_1 z^{n-1} + \cdots + b_n}{z^n + a_1 z^{n-1} + \cdots + a_n} R(z) = H(z) R(z) \qquad (4.9\text{-}1)$$

where $R(z) = \mathscr{Z}[r(k)]$. It is assumed that the poles p_i of the transfer function $H(z)$ are simple and different from those of $R(z)$, so that a partial-fraction expansion of (4.9-1) has the form

$$C(z) = e_0 + \frac{e_1}{z - p_1} + \frac{e_2}{z - p_2} + \cdots + \frac{e_n}{z - p_n}$$
$$+ \{\text{expansion poles of } R(z)\} \qquad (4.9\text{-}2)$$

Using Table 4.7-1, we find that the inverse z-transform is given by

$$c(k) = e_1 (p_1)^{k-1} + e_2 (p_2)^{k-1} + \cdots + e_n (p_n)^{k-1}$$
$$+ \mathscr{Z}^{-1} \{\text{expansion poles of } R(z)\} \text{ for } k \geq 1 \qquad (4.9\text{-}3)$$

The $e_m(p_m)^{k-1}$ terms of (4.9-3) are the *transient response* portion of $c(k)$ and are due primarily to poles of the transfer function in the partial-fraction expansion process. The

$$\mathscr{Z}^{-1}\{\text{expansion poles of } R(z)\} \qquad (4.9\text{-}4)$$

term is the *steady-state* response of the system and arises solely because of the poles of $R(z)$. It is apparent that the parameters $a_1, a_2, \ldots, a_n, b_0, b_1, \ldots, b_n$, characterizing the transfer function $H(z)$, greatly influence the resultant response $c(k)$. The set $a_1, a_2, \ldots, a_n$ directly determines the p_i, while the set $a_1, a_2, \ldots, a_n, b_1, \ldots, b_n$ indirectly determines the e_i and the magnitudes in expression (4.9-4). To demonstrate this, let us find the response of the system with transfer function

$$H(z) = \frac{b_1 z + b_2}{z^2 - (p_1 + p_2)z + p_1 p_2}$$

to the discrete step input. Here $a_1 = -(p_1 + p_2)$ and $a_2 = p_1 p_2$ with p_1 and p_2 assumed to be not equal to 1. The z-transform of the input is

$$R(z) = \frac{1}{1 - z^{-1}} = \frac{z}{z - 1}$$

so that

$$C(z) = \frac{b_1 z^2 + b_2 z}{(z - p_1)(z - p_2)(z - 1)} = \frac{e_1}{z - p_1} + \frac{e_2}{z - p_2}$$
$$+ \left\{ \frac{\frac{b_1 + b_2}{(1 - p_1)(1 - p_2)}}{z - 1} \right\} \qquad (4.9\text{-}5)$$

with

$$e_1 = \frac{p_1(b_1 p_1 + b_2)}{(p_1 - p_2)(p_1 - 1)}, \qquad e_2 = \frac{p_2(b_1 p_2 + b_2)}{(p_2 - p_1)(p_2 - 1)}$$

The response is obtained by inverse-transforming (4.9-5), which gives

$$c(k) = e_1(p_1)^{k-1} + e_2(p_2)^{k-1} + \frac{b_1 + b_2}{(1 - p_1)(1 - p_2)} \quad \text{for } k \geq 1 \quad (4.9\text{-}6)$$

It is apparent from (4.9-6) that the *steady-state* response of this system to a discrete step input is a discrete step of magnitude

$$\frac{b_1 + b_2}{(1 - p_1)(1 - p_2)}$$

Thus, in this example, the steady-state response is of the same form as the input signal. This will also be true in the more general case, as one can easily see, since the steady-state response is given by (4.9-4).

From (4.9-3), we observe that the transient response goes to zero for increasing discrete time k only if all the poles of the system's transfer function have absolute value less than one. The reader may wish to compare this conclusion with the discussion of Section 3.6.

In most applications, it is desired that the linear discrete system operate on the input signal $r(k)$ in some desirable manner. We are, therefore, interested in influencing the steady-state response in some manner and correspondingly in making the transient response negligible by comparison. This, in effect, restricts our attention to stable systems.

To treat the more general case, let us assume that the input to a system with transfer function

$$H(z) = \frac{b_0 z^m + b_1 z^{m-1} + \cdots + b_m}{(z - p_1)(z - p_2) \cdots (z - p_n)} \quad \text{with } m \leq n \quad (4.9\text{-}7)$$

is

$$r(0) = \beta_0$$
$$r(k) = \beta_1(\alpha_1)^{k-1} + \beta_2(\alpha_2)^{k-1} + \cdots + \beta_q(\alpha_q)^{k-1} \quad \text{for } k \geq 1 \quad (4.9\text{-}8)$$

where for simplicity of presentation we have assumed that the numbers $\alpha_1, \alpha_2, \ldots, \alpha_p, p_1, p_2, \ldots, p_q$ are distinct. This in no way restricts the generality of the concepts to be presented.

The z-transform of the input $r(k)$ is

$$\begin{aligned} R(z) &= \beta_0 + \frac{\beta_1}{z - \alpha_1} + \frac{\beta_2}{z - \alpha_2} + \cdots + \frac{\beta_q}{z - \alpha_q} \\ &= \frac{\hat{\alpha}_0 z^q + \hat{\alpha}_1 z^{q-1} + \cdots + \hat{\alpha}_q}{(z - \alpha_1)(z - \alpha_2) \cdots (z - \alpha_q)} \end{aligned} \quad (4.9\text{-}9)$$

so that $C(z)$ is given by

$$\begin{aligned} C(z) = H(z)R(z) &= \frac{\sigma_0 z^{q+m} + \sigma_1 z^{q+m-1} + \cdots + \sigma_{q+m}}{(z - p_1) \cdots (z - p_n)(z - \alpha_1) \cdots (z - \alpha_q)} \\ &= \gamma + \frac{e_1}{z - p_1} + \frac{e_2}{z - p_2} + \cdots + \frac{e_n}{z - p_n} \\ &\quad + \left\{ \frac{\hat{\beta}_1}{z - \alpha_1} + \frac{\hat{\beta}_2}{z - \alpha_2} + \cdots + \frac{\hat{\beta}_q}{z - \alpha_q} \right\} \end{aligned}$$

where for $m < n$, the coefficients e_j and $\hat{\beta}_j$ are evaluated by

$$e_j = \lim_{z \to p_j} (z - p_j) H(z) R(z)$$
$$\hat{\beta}_j = \lim_{z \to \alpha_j} (z - \alpha_j) H(z) R(z)$$

The steady-state response $c_{ss}(k)$ is

$$c_{ss}(k) = \hat{\beta}_1(\alpha_1)^{k-1} + \hat{\beta}_2(\alpha_2)^{k-1} + \cdots + \hat{\beta}_q(\alpha_q)^{k-1} \quad \text{for } k \geq 1 \quad (4.9\text{-}10)$$

A comparison of this steady-state response with the input signal (4.9-8) indicates that they are of the same form, the only distinction being in the coefficients β_j and $\hat{\beta}_j$. Clearly, the effect of the discrete system on the steady-state response is through the relation

$$\hat{\beta}_j = \lim_{z \to \alpha_j} (z - \alpha_j) H(z) R(z) = \beta_j H(\alpha_j) \quad (4.9\text{-}11)$$

If it is desired that the discrete system remove (filter out) the $\beta_1 \alpha_1^k$ term of the steady-state response, this can be accomplished by making $\hat{\beta}_1$ as small as possible. Ideally, in making $\hat{\beta}_1 = 0$, we totally remove the α_1^k term in the steady-state response. This is accomplished by having

$$\lim (z - \alpha_1) H(z) R(z) = \beta_1 H(\alpha_1) = 0$$

which requires $H(\alpha_1) = 0$. In other words, the α_1^k contribution is completely filtered out if the transfer function has a zero at $z = \alpha_1$.

The transient response is

$$c_t(k) = e_1(p_1)^{k-1} + e_2(p_2)^{k-1} + \cdots + e_n(p_n)^{k-1} \quad \text{for } k \geq 1 \quad (4.9\text{-}12)$$

In order that the transient response be negligible in comparison to the steady-state response for moderately small k, it is necessary that the magnitude of the poles of the transfer function be small relative to those of the input $R(z)$; that is,

$$|p_i| \ll |\alpha_j| \quad \text{for } i = 1, 2, \ldots, n$$
$$j = 1, 2, \ldots, q$$

This is appreciated if one compares (4.9-10) with (4.9-12). In terms of a z-plane plot of the poles of $H(z)$ and $R(z)$, the poles of $H(z)$ must lie much closer to the origin than those of $R(z)$. Figure 4.9-1(a) shows such a proper pole distribution, whereas Figure 4.9-1(b) illustrates an undesirable distribution of poles. The transient response of the system-input pole plot of Figure 4.9-1(b) will not decay at a faster rate than the steady-state response, usually an undesirable feature. The plots are restricted to the unit circle, as we are concerned only with stable systems and bounded inputs.

By properly selecting the poles and zeros of the transfer function, it is possible to pass or reject different types of input sequences. In effect, the linear discrete system has characteristics normally ascribed to a filter. This should not be totally surprising, since, if we rely on the analogies that have been shown to exist between continuous and discrete systems, it is common

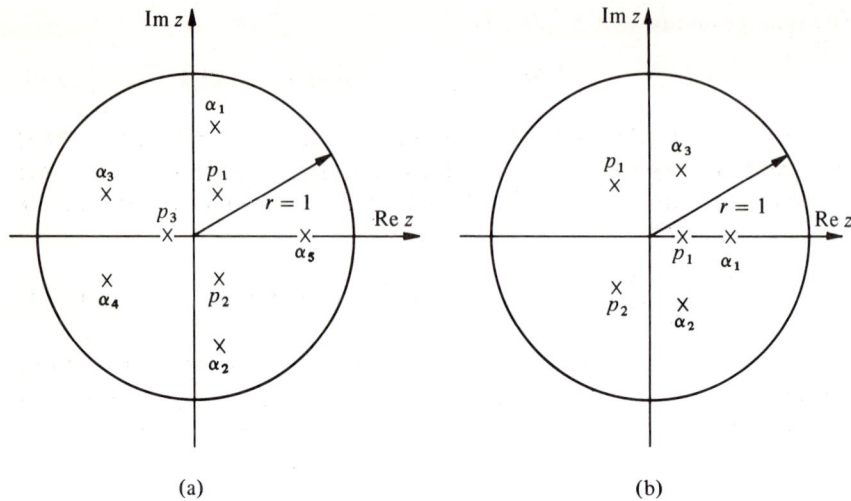

Figure 4.9-1. (a) Proper pole pattern; (b) improper pole pattern.

to refer to a linear continuous system with transfer function $H(s)$ as a filter when it is used in continuous-time communication applications. The point to be made is that a linear discrete system may be used to perform data-processing functions (numerical filtering) as well as to carry out control-oriented iterations.

One might question the seemingly restricted type of input (4.9-8) that has been postulated. However, it can be shown that a very large class of input sequences $r(k)$ may be approximated by such an expression for sufficiently large values of q. We might properly compare this type of discrete input to its counterpart in continuous systems

$$r(t) = \beta_0 + \beta_1 e^{-\alpha_1 t} + \beta_2 e^{-\alpha_2 t} + \cdots + \beta_q e^{-\alpha_q t}$$

This class of inputs and the steady-state response resulting from their application to continuous systems receive much attention in classical system theory. It should be expected (and correctly so) that inputs and their corresponding responses of the form (4.9-8) play a prominent role in the analysis and synthesis of linear discrete systems.

In summary, a properly designed discrete system will have a transfer function $H(z)$ which

(i) Has poles that lie relatively close to the origin in comparison to the poles of the input signal $R(z)$, which the system is to pass.

(ii) Has zeros that cancel the poles of that portion of $R(z)$ which the system is to filter out (reject).

Of course, the synthesis of a desirable discrete system requires a more complex process than merely the satisfaction of steps (i) and (ii). They are

Sec. 4.9 Response of a Linear Discrete System

listed only to indicate the importance attached to the zeros and poles of the discrete system transfer function.

EXAMPLE 4.9-1

Determine the steady-state response characteristics of the discrete system characterized by the second-order difference equation

$$c(k) = r(k-1) + \tfrac{1}{4}r(k-2) - \tfrac{1}{4}c(k-1) + \tfrac{3}{8}c(k-2)$$

This system has the transfer function

$$H(z) = \frac{z + \tfrac{1}{4}}{(z - \tfrac{1}{2})(z + \tfrac{3}{4})} \qquad (4.9\text{-}13)$$

In order to find the steady-state response characteristics to the input

$$r(k) = \beta(\alpha)^k \quad \text{with } \alpha \text{ real} \qquad (4.9\text{-}14)$$

we shall plot $|H(z)|$ versus z. Figure 4.9-2 shows the results of such a plot, where we allow z to take on only real values, as α is assumed real. This plot demonstrates that the steady-state response to (4.9-14) is small for α close to $-\tfrac{1}{4}$ [a zero of $H(z)$] and becomes unbounded near $\alpha = \tfrac{1}{2}, -\tfrac{3}{4}$ [poles of $H(z)$].

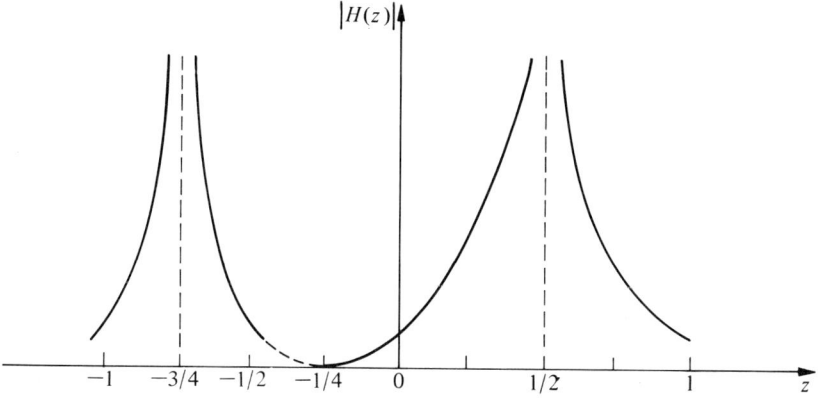

Figure 4.9-2. Plot of $|H(z)|$ vs. z for Example 4.9-1.

EXAMPLE 4.9-2

Suppose that an input of the form

$$r(k) = \begin{cases} \alpha + \beta a^k & \text{for } k \geq 0 \\ 0 & \text{for } k < 0 \end{cases}$$

is applied to a system with transfer function of the form

$$H(z) = b\frac{z - z_1}{z - p_1}$$

We do not know the values of α or β, but we wish to select the parameters b, p_1, and z_1 of $H(z)$ so that the discrete system will filter out the a^k portion of $f(k)$ but pass the constant α. In this mode, we are using the discrete system as a filter.

The z-transform of $r(k)$ is

$$R(z) = \frac{\alpha z}{z - 1} + \frac{\beta z}{z - a} = \frac{z[(\alpha + \beta)z - (a\alpha + \beta)]}{(z - 1)(z - a)}$$

so that

$$C(z) = b\frac{z(z - z_1)[(\alpha + \beta)z - (a\alpha + \beta)]}{(z - 1)(z - a)(z - p_1)}$$

Without going into a more elaborate presentation, it is desirable to select the zero of $H(z)$ so as to cancel out the pole of the undesirable portion of $R(z)$. That is, let $z_1 = a$; therefore,

$$\begin{aligned}
C(z) &= b\frac{z[(\alpha + \beta)z - (a\alpha + \beta)]}{(z - 1)(z - p_1)} \\
&= b(\alpha + \beta) + b\frac{(\alpha + \alpha p_1 + \beta p_1 - a\alpha)z - p_1(\alpha + \beta)}{(z - 1)(z - p_1)} \\
&= b(\alpha + \beta) + \frac{\frac{\alpha b(1 - a)}{1 - p_1}}{z - 1} + \frac{\frac{bp_1(\alpha p_1 + \beta p_1 - a\alpha - \beta)}{p_1 - 1}}{z - p_1}
\end{aligned}$$

so that the response $c(k)$ is given by

$$c(0) = (\alpha + \beta)b$$

$$c(k) = \frac{b}{1 - p_1}[\alpha(1 - a) + \{\beta + a\alpha - p_1\alpha - \beta p_1\}p_1^k] \quad \text{for } k \geq 1$$

Since we desire the transient response to decay rapidly, we set $p_1 = 0$. In addition, we let $b = 1/(1 - a)$, so that

$$c(0) = \frac{\alpha + \beta}{1 - a}$$

$$c(k) = \alpha \quad \text{for } k \geq 1$$

Thus, the system with transfer function

$$H(z) = \frac{1}{1-a}\frac{z-a}{z}$$

or the corresponding difference equation

$$c(k) = \frac{1}{1-a}[r(k) - ar(k-1)]$$

will effectively filter the input $r(k)$ and generate the desired response after two iterations when the input hypothesized is applied.

4.10 z-Transform of a Product of Two Functions*

Let the sequence $h(k) = f(k)g(k)$ be given; the z-transform of $h(k)$ is, by definition,

$$\mathscr{Z}[h(k)] = \mathscr{Z}[f(k)g(k)] = \sum_{k=0}^{\infty} f(k)g(k)z^{-k} \quad \text{for } |z| > R \quad (4.10\text{-}1)$$

It is assumed that the sequences have the z-transforms

$$F(z) = \mathscr{Z}[f(k)] \quad \text{for } |z| > R_1$$
$$G(z) = \mathscr{Z}[g(k)] \quad \text{for } |z| > R_2$$

so that by the inversion integral

$$g(k) = \frac{1}{2\pi j}\oint_C G(p)p^{k-1}dp \quad (4.10\text{-}2)$$

where the closed curve C lies within the region $|z| > R_2$. Incorporating (4.10-2) into (4.10-1) gives

$$\mathscr{Z}[f(k)g(k)] = \sum_{k=0}^{\infty} f(k)z^{-k}\frac{1}{2\pi j}\oint_C G(p)p^{k-1}dp \quad (4.10\text{-}3)$$

Since $\mathscr{Z}[f(k)g(k)]$ converges uniformly for $|z| > R$, the operations of summation and integration may be interchanged. Therefore,

$$\mathscr{Z}[f(k)g(k)] = \frac{1}{2\pi j}\oint_C G(p)p^{-1}\sum_{k=0}^{\infty} f(k)(p^{-1}z)^{-k}dp \quad (4.10\text{-}4)$$

*This section relies heavily on residue theory and is intended only for students having a background in complex variable theory.

The summation

$$\sum_{k=0}^{\infty} f(k)(p^{-1}z)^{-k}$$

converges uniformly to $F(p^{-1}z)$ for $|p^{-1}z| > R_1$ or $|p| < |z|/R_1$. Therefore,

$$\mathcal{Z}[f(k)g(k)] = \frac{1}{2\pi j}\oint_C G(p)F(p^{-1}z)p^{-1}dp \qquad (4.10\text{-}5)$$

and since $G(p)$ converges for $|p| > R_2$, this integral converges for those values of p over which $G(p)$ and $F(p^{-1}z)$ both converge. This region is given by

$$\frac{|z|}{R_1} > |p| > R_2$$

EXAMPLE 4.10-1

Determine $\mathcal{Z}[ke^{ak}]$.
Let

$$f(k) = k, \qquad g(k) = e^{ak}$$

We have, by Table 4.4-1,

$$F(z) = \frac{z^{-1}}{(1-z^{-1})^2} \quad \text{for } |z| > 1$$

$$G(z) = \frac{1}{1-e^a z^{-1}} \quad \text{for } |z| > e^a$$

Therefore, by (4.10-5),

$$\mathcal{Z}[ke^{ak}] = \frac{1}{2\pi j}\oint_C \frac{z^{-1}}{(1-e^a p^{-1})(1-pz^{-1})^2}dp \quad \text{for } |z| > |p| > e^a$$

where the closed curve C is in the region $|z| > e^a$; using residue theory, we have

$$\mathcal{Z}[ke^{ak}] = \text{residue } \frac{pz^{-1}}{(p-e^a)(1-pz^{-1})^2} \quad \text{at } p = e^a$$

$$= \frac{e^{-a}z^{-1}}{(1-e^a z^{-1})^2}$$

4.10-1 Parseval's Theorem

A very useful special case exists when $g(k) = f(k)$, namely,

$$\mathcal{Z}[f^2(k)] = \frac{1}{2\pi j}\oint_C F(p)F(p^{-1}z)p^{-1}dp \qquad (4.10\text{-}6)$$

where

$$\mathscr{Z}[f(k)] = F(z) \quad \text{for } |z| > R_1$$

The integral (4.10-6) converges for

$$\frac{|z|}{R_1} > |p| > R_1$$

and if $R_1 < 1$, it will converge for $z = 1$. Letting $z = 1$ in (4.10-6) results in

$$\sum_{k=0}^{\infty} f^2(k) = \frac{1}{2\pi j} \oint_C F(z)F(z^{-1})z^{-1} dz \qquad (4.10\text{-}7)$$

Equation (4.10-7) is the discrete form of Parseval's theorem. It is particularly useful in the study of linear discrete systems that are driven by stationary random processes. Since $R_1 < 1$, this implies that (4.10-7) holds only for $f(k)$ that approach zero as $k \to \infty$.

4.11 Conclusion

The z-transform has been shown to possess properties that demonstrate its usefulness in analyzing linear, time-invariant discrete systems. Again, the analogy theme of linear discrete and continuous systems is evident. The similarity of the Laplace and z-transforms and their various properties is clear. This being the case, it is possible for the engineer to draw heavily on his previous experience in continuous systems.

The application of the z-transform to state space representations of linear discrete systems and sampled-data systems is treated in the next two chapters.

REFERENCES

1. Aseltine, J. A., *Transform Method in Linear System Analysis*, McGraw-Hill, New York, 1958.

2. Churchill, R. V., *Introduction to Complex Variables and Applications*, McGraw-Hill, New York, 1948.

3. DeRusso, P. M., R. J. Roy and C. M. Close, *State Variables for Engineers*, Wiley, New York, 1965.

4. Freeman, H., *Discrete-Time Systems*, Wiley, New York, 1965.

5. Jury, E. I., *Theory and Application of the z-Transform Method*, Wiley, New York, 1964.

6. Monroe, A. J., *Digital Processes for Sampled-Data Systems*, Wiley, New York, 1962.

7. Schwarz, R. J. and B. Friedland, *Linear Systems*, McGraw-Hill, New York, 1965.

8. Ragazzini, J. R. and G. F. Franklin, *Sampled-Data Control Systems*, McGraw-Hill, New York, 1958.

9. Tou, J. T., *Digital and Sampled-Data Control Systems*, McGraw-Hill, New York, 1959.

PROBLEMS

4.1 Determine the z-transform of the following sequences and find their region of convergence.

(a) $f(0) = 1, f(1) = 3, f(4) = 0,$
$f(k) = 0$ for $k \neq 0, 1, 4$

(b) $f(k) = \begin{cases} 1/k & \text{for } k > 0 \\ 0 & \text{for } k < 0 \end{cases}$

4.2 Demonstrate that the sequence in Problem 4.1-b converges and diverges for different values of z on the boundary separating its convergence and divergence regions (i.e., $|z| = 1$).

4.3 Determine the z-transform of the following sequences and the corresponding regions of convergence. The first term corresponds to $k = 0$ in all cases.

(a) 1, 1, 2, 3, 5, 8, 13, 21, 34, ...
(b) 0, 2, 2(2)², 3(2)³, 4(2)⁴, 5(2)⁵, ...
(c) 0, 1 × 3, 2 × 4, 3 × 5, 4 × 6, ...

4.4 Determine the z-transform and the region of convergence for the following sequences.

(a) $f(k) = a^{k-1}$ for $k \geq 1$
$ f(k) = 0$ for $k < 1$
(b) $f(k) = ka^{k-1}$ for $k \geq 1$
$ f(k) = 0$ for $k < 1$
(c) $f(k) = k^3 a^{k-1}$ for $k \geq 1$
$ f(k) = 0$ for $k < 1$

4.5 Demonstrate the validity of entries 2, 3, and 6 of Table 4.4-2.

4.6 Find the inverse z-transforms of the following complex functions.

(a) $F(z) = z^2 + 3z + 5 + \dfrac{2}{z^2 + 5z + 6}$ for $|z| > 3$

(b) $F(z) = \dfrac{z^2 + 5z + 3}{z^2 + \frac{1}{4}z - \frac{1}{8}}$ for $|z| > \frac{1}{2}$

4.7 A system is characterized by the difference equation

$$c(k+2) - c(k+1) - 2c(k) = 6r(k+2) - 2r(k+1)$$

(a) Determine its transfer function.
(b) Determine the z-transform of the output $c(k)$ sequence generated by the input sequence

$r(0) = 1, \quad r(1) = -1, \quad r(2) = 1, \quad r(3) = -1, \quad r(k) = 0 \quad \text{for } k \geq 4$

4.8 Apply the initial and final value thereoms to Problem 4.6.

4.9 Obtain the results displayed by Figure 2.3-4 by applying the z-transform.

4.10 Determine the transfer function of two cascaded systems, each described by the difference equation

$$c(k) = .5c(k-1) + r(k)$$

4.11 Repeat Problem 4.10 when the difference equation is

$$c(k+1) = .5c(k) + r(k)$$

4.12 Verify the results obtained by Example 2.3-1 by using the z-transform.

4.13 Repeat Example 2.7-1, using the z-transform.

4.14 Given the step response of a discrete system, devise a formula by which the weighting sequence and the transfer function of the system can be obtained.

4.15 Determine the weighting sequence for the system whose transfer function is

$$G(z) = \frac{5z^2 + 2z + 1}{z^2 + 3z + 2}$$

Obtain this result by two different methods.

Appendix 4A

POWER SERIES

Given the sequence of numbers $f(0), f(1), \ldots$, the z-transform of this sequence is defined to be

$$\mathscr{Z}[f(k)] = F(z) = \sum_{k=0}^{\infty} f(k) z^{-k} \qquad (4\text{A-}1)$$

where z is taken to be a complex variable. This expression is called the *one-sided z-transform*, since it transforms the sequence $f(k)$ only for positive k. The one-sided z-transform is actually a power series in the variable z^{-1}, so it has all the properties associated with power series. A detailed understanding of the z-transform is possible only with an intimate knowledge of complex variable theory. Such a knowledge is not possessed by the normal undergraduate engineering student. As a compromise, we shall briefly outline the salient features of power series.

The sequence of complex numbers $z_0, z_1, z_2, \ldots$ is said to *converge* to the limit ω:

$$\lim_{k \to \infty} z_k = \omega$$

if for every positive ϵ, there exists an integer K such that

$$|z_k - \omega| < \epsilon \quad \text{for all } k > K$$

If a sequence does not converge, it is said to *diverge*. Thus a convergent sequence is one whose terms approach arbitrarily close to its limit ω as k increases.

EXAMPLE 4A-1

Show that the sequence with the general term

$$z_k = \frac{1 - a^k}{1 - a}$$

converges to $1/(1 - a)$ for all complex numbers a such that $|a| < 1$.

We have to show that there exists an integer K such that for any positive ϵ

$$\left|\frac{1 - a^k}{1 - a} - \frac{1}{1 - a}\right| = \left|\frac{-a^k}{1 - a}\right| < \epsilon \quad \text{for } k > K$$

or

$$|a|^k < \epsilon |1 - a|$$

and since $|a| < 1$, after taking the logarithm of both sides, we have

$$k \log(|a|) > \log \epsilon + \log(|1 - a|)$$

A value of K that guarantees the required property for convergence is then

$$K = \frac{\log \epsilon + \log(|1 - a|)}{\log(|a|)}$$

or if this expression is not an integer, then K is set equal to the next largest integer.

If the complex number a is greater than one in absolute value, it may be shown that this sequence diverges.

The infinite sum of a sequence $z_0, z_1, z_2, \ldots$ of complex numbers is denoted by

$$z_0 + z_1 + z_2 + \cdots = \sum_{k=0}^{\infty} z_k \qquad (4A\text{-}2)$$

This infinite series value is said to converge to the (sum) S if the partial sum sequence S_k converges to S; that is,

$$\lim_{k \to \infty} S_k = S \qquad (4A\text{-}3)$$

where

$$S_k = \sum_{j=0}^{k} z_j$$

When the sequence S_k diverges, the series (4A-2) is said to diverge. This series is said to be absolutely convergent if the series

$$\sum_{k=0}^{\infty} |z_k|$$

converges. Any series that is absolutely convergent also converges in the sense of (4A-3).

Two important theorems relative to series convergence will be given without proof. A proof of these thereoms may be found in most standard advanced calculus texts.

Theorem 4A-1. Ratio Test: If the ratio

$$\lim_{k \to \infty} \left| \frac{z_{k+1}}{z_k} \right|$$

converges to L, then the series (4A-2) is absolutely convergent if $L < 1$ and is divergent if $L > 1$. More generally, if this ratio never exceeds a number r for sufficiently large k, then the series is absolutely convergent for $r < 1$ and it diverges if $r > 1$.

Theorem 4A-2. Root Test: If the sequence

$$\omega_k = \sqrt{|z_k|}$$

converges to L, as k approaches infinity then the series (4A-2) converges absolutely if $L < 1$ and diverges if $L > 1$. More generally, if for sufficiently large k, ω_k never exceeds a number r, then the series converges absolutely if $r < 1$ and diverges if $r > 1$.

In both the ratio and root tests, nothing can be determined when $L = 1$. Other techniques will be required to indicate convergence or divergence of such series.

EXAMPLE 4A-2

Determine whether the series

$$\sum_{k=0}^{\infty} a^k \qquad (4A\text{-}4)$$

converges or diverges; a is a complex number.

(i) Applying the ratio test with $z_k = a^k$, we have

$$\left|\frac{z_{k+1}}{z_k}\right| = \left|\frac{a^{k+1}}{a^k}\right| = |a|$$

so that $L = |a|$. The series (4A-4) is absolutely convergent if the complex number a is less than one in absolute value and it diverges for $|a| > 1$.

(ii) Application of the root test gives

$$\sqrt[k]{|z_k|} = \sqrt[k]{|a^k|} = |a|$$

and the same conclusion reached in part (i) is made. In fact, for $|a| < 1$, we have

$$\sum_{k=0}^{\infty} a^k = \frac{1}{1-a}$$

which may be shown by letting the partial sum S_k be

$$S_k = \sum_{j=0}^{k} a^j = \frac{1 - a^{k+1}}{1 - a}$$

which converges to $1/(1-a)$ if $|a| < 1$, as $k \to \infty$.

A series of the form

$$\sum_{k=0}^{\infty} c_k(z - z_0)^k = c_0 + c_1(z - z_0) + c_2(z - z_0)^2 + \cdots \qquad (4\text{A-}5)$$

is called a positive power series in powers of $(z - z_0)$. A power series of negative powers of $(z - z_0)$

$$\sum_{k=0}^{\infty} c_k(z - z_0)^{-k} = c_0 + c_1(z - z_0)^{-1} + c_2(z - z_0)^{-2} + \cdots \qquad (4\text{A-}6)$$

is called a negative power series.

The power series (4A-5) converges at $z = z_0$ to the value c_0. It may or may not converge for other values of z. The region in the complex z plane for which (4A-5) converges is called the *region of convergence*. This region of convergence may be determined by use of the following theorem.

Theorem 4A-3. Every power series

$$\sum_{k=0}^{\infty} c_k(z - z_0)^k$$

converges absolutely in a circle of radius R^+ centered at z_0, i.e., $|z - z_0| < R^+$, and diverges outside this circle, i.e., $|z - z_0| > R^+$.

The value of R^+ may be zero, a positive number, or ∞. If $R^+ = 0$, the series converges only at $z = z_0$, whereas if $R^+ = \infty$, the series converges everywhere.

The radius R^+ is evaluated by

$$R^+ = \lim_{k \to \infty} \left| \frac{c_k}{c_{k+1}} \right| \quad \text{if this limit exists} \qquad (4A\text{-}7)$$

or by

$$R^+ = \lim_{k \to \infty} \frac{1}{\sqrt[k]{|c_k|}} \quad \text{if this limit exists} \qquad (4A\text{-}8)$$

Proof. For any fixed z, apply the ratio test where

$$z_k = c_k(z - z_0)^k$$

$$\left| \frac{z_{k+1}}{z_k} \right| = \left| \frac{c_{k+1}(z - z_0)^{k+1}}{c_k(z - z_0)^k} \right| = \left| \frac{c_{k+1}}{c_k} \right| |z - z_0|$$

For the power series to converge, we must have, by Theorem 4A-1,

$$\lim_{k \to \infty} \left| \frac{c_{k+1}}{c_k} \right| |z - z_0| < 1$$

or

$$|z - z_0| < \lim \left| \frac{c_k}{c_{k+1}} \right| = R^+$$

so that the power series converge absolutely for all z that satisfy this inequality and diverge for all z such that

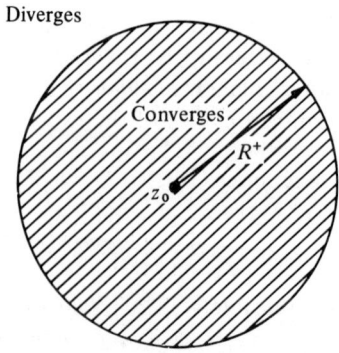

Figure 4A-1. Region of convergence of positive power series.

$$|z - z_0| > R^+$$

The value of R^+ as given in (4A-8) is proved by applying the root test.

The region of convergence for a positive power series (4A-5) is shown in Figure 4A-1.

EXAMPLE 4A-3

Determine the region of convergence for the power series

$$\sum_{k=0}^{\infty} a^k z^k$$

We have $c_k = a^k$, from which

$$R^+ = \lim_{k \to \infty} \left| \frac{a^k}{a^{k+1}} \right| = \left| \frac{1}{a} \right|$$

The series converges for all z such that

$$|z| < \frac{1}{|a|}$$

The power series in negative powers of $(z - z_0)$ has convergence properties that are the inverse of those of (4A-5).

Theorem 4A-4. Every negative power series

$$\sum_{k=0}^{\infty} c_k (z - z_0)^{-k}$$

converges absolutely outside a circle of radius R^- centered at z_0; i.e., $|z - z_0| > R^-$ and diverges inside this circle, i.e., $|z - z_0| < R^-$.

The radius R^- is evaluated as

$$R^- = \lim_{k \to \infty} \left| \frac{c_{k+1}}{c_k} \right| \quad \text{if this limit exists}$$

or by

$$R^- = \lim_{k \to \infty} \sqrt[k]{|c_k|} \quad \text{if this limit exists}$$

The proof of this theorem exactly parallels that of Theorem 4A-3. The region of convergence for a negative power series (4A-6) is shown in Figure 4A-2.

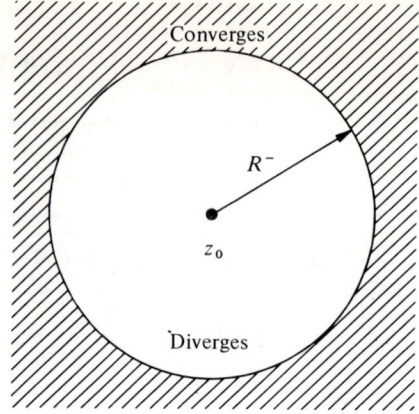

Figure 4A-2. Convergence region of a negative power series.

EXAMPLE 4A-4

Determine the radius of convergence for the power series

$$\sum_{k=0}^{\infty} a^k z^{-k}$$

Since this is a negative power series, we apply Theorem 4A-4 with $c_k = a^k$, obtaining

$$R^- = \lim_{k \to \infty} \left| \frac{a^{k+1}}{a^k} \right| = |a|$$

or

$$R^- = \lim_{k \to \infty} \sqrt[k]{|a^k|} = |a|$$

This series converges for all z such that

$$|az^{-1}| < 1 \quad \text{or} \quad |z| > |a|$$

and, as was shown in Section 4.4, a closed-form expression for this series is

$$\frac{1}{1 - az^{-1}} \quad \text{for } |z| > |a|$$

There exist Laurent series that are defined over the integers $-\infty \leq k \leq \infty$; that is,

$$\sum_{k=-\infty}^{\infty} c_k (z - z_0)^k \qquad (4A\text{-}9)$$

In order to determine the region of convergence, it is best to break this series into two parts.

$$\sum_{k=-\infty}^{\infty} c_k(z-z_0)^k = \sum_{k=0}^{\infty} c_k(z-z_0)^k + \sum_{k=-1}^{-\infty} c_k(z-z_0)^k$$

The region of convergence for the first part

$$|z-z_0| < R^+ \qquad (4A\text{-}10)$$

and for the second part

$$|z-z_0| > R^- \qquad (4A\text{-}11)$$

may be obtained by methods already demonstrated. The region (values of z) for which this series (4A-9) converges is then given by those points that are common to both (4A-10) and (4A-11); that is,

$$R^- < |z-z_0| < R^+$$

In some cases, $R^- > R^+$, so that the series diverges everywhere. Figure 4A-3 illustrates the region of convergence for a typical Laurent series.

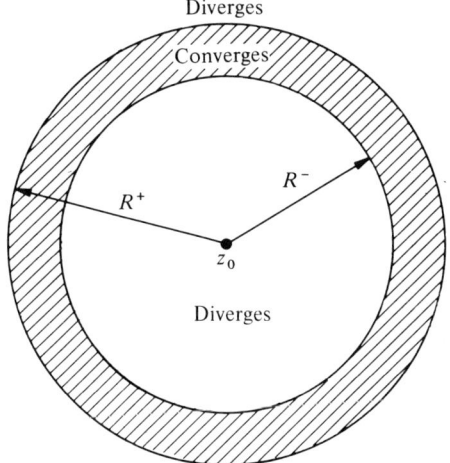

Figure 4A-3. Region of convergence for Laurent series (4A-9).

EXAMPLE 4A-5

Determine the region of convergence for the Laurent series

$$\sum_{k=-\infty}^{\infty} c_k z^k$$

where

$$c_k = (\tfrac{1}{2})^k \quad \text{for } k \geq 0$$
$$ = 4^k \quad \text{for } k < 0$$

We have $R^+ = 2$, $R^- = 4$, so that this series converges for all z such that

$$4 < |z| < 2$$

However, this is an empty region. The series therefore diverges for all z.

EXAMPLE 4A-6

Determine the region of convergence for the Laurent series

$$\sum_{k=-\infty}^{\infty} c_k z^k$$

where

$$c_k = (\tfrac{1}{4})^k \quad \text{for } k \geq 0$$
$$ = 2^k \quad \text{for } k < 0$$

It follows that $R^+ = 4$, $R^- = 2$, so that this series converges for all z such that

$$2 < |z| < 4$$

PROBLEMS

1 Determine whether the following sequences converge, and if so determine their limits.

(a) $\dfrac{k^2 + 3k + 1}{k^3 + 5}$

(b) $k \log\left(1 + \dfrac{1}{k}\right)$

(c) $\dfrac{k}{2^k}$

2 Determine the convergence or divergence properties of the following series.

(a) $\displaystyle\sum_{k=1}^{\infty} \dfrac{1}{k^2 + 2}$

(b) $\displaystyle\sum_{k=1}^{\infty} \dfrac{2^k}{n^k}$

(c) $\sum_{k=1}^{\infty} \dfrac{1!}{3 \cdot 5 \cdot \ \cdots \ \cdot (2k+3)}$

3 Find the z-transform of the sequence

$$\begin{aligned} f(k) &= -k3^k & \text{for } k < 0 \\ &= 7 & \text{for } k = 0 \\ &= 2^{-k} & \text{for } k > 0 \end{aligned}$$

and determine its region of convergence.

5

State Variable Representation

5.1 Introduction

Around the end of the 1950's, the concept of representing a dynamical system by a set of first-order differential or difference equations became a standard tool of the research engineer. These techniques have since become generally known as state variable representations. Such representations have become an important tool of the control and communication engineer during the intervening years. These concepts often allow one to carry out a meaningful system design entirely in the time domain as opposed to popular transformation methods. That this is important follows from basically two facts:

1. The system may be nonlinear, so transformation methods (e.g., Laplace, Fourier, z-transform) may not be directly applicable.
2. Time-domain concepts give one better insight into the analysis or synthesis of the system.

A state variable representation of a system differs from the more conventional representations. For instance, in a transform representation, only the relationships between the input and output variables must be known. A state variable representation gives a total description of a system, both the internal as well as the external dynamics.

5.2 Concept of State

We define a *system* as a mathematical abstraction that utilizes three types of variables to represent or model the dynamics of some process. The three variables are called the *input*, the *output* and the *state* variables.

The input variables u_i serve as external forces that influence the dynamics or motion of the system. The output variables y_j are the characteristic variables that are directly observable (measurable) by an external observer.* The state variables x_q characterize the internal dynamics of the system. These variables are formulated in such a manner that if one knows the values of the state variables x_q at a given instant of time together with the values of the input variables u_i for that time and all future time, then the output variables y_j and the state variables x_q are completely determined for the present and all future time. By the word "time" is meant either continuous time t or discrete time

$$\{\ldots, t_{-1}, t_0, t_1, t_2, \ldots\}$$

If by the concept of time for a given system we imply continuous time, the system is said to be *continuous*; it is said to be *discrete* if discrete time is implied.

A discrete system is a system in which the state variables u_i, y_j, x_q are sequences that depend on discrete time [i.e., $u_i(t_k)$, $y_j(t_k)$, $x_q(t_k)$], whereas a continuous system is characterized by these variables being functions of continuous time [i.e., $u_i(t)$, $y_j(t)$, $x_q(t)$].

In many cases, a system is made up of both discrete and continuous subsystems. Whenever a digital computer is utilized as an element in conjunction with continuous systems, such a mixed system is generated. These systems will be termed *hybrid systems*. They are extremely important in modern control and communication applications. Frequently, they may be treated as discrete systems by formulating the problem properly. This formulation is one of the objectives of this text.

A block diagram of a state variable representation is illustrated in Figure 5.2-1. We note that the input and output variables appear to be external to the system, whereas the state variables are internal.

---

*We shall depart from our practice of using r and c to denote the input and output variables so as to conform to the most widely accepted state variable notation.

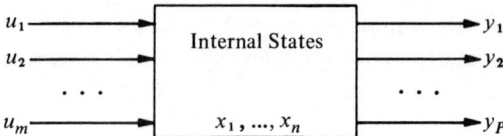

Figure 5.2-1. State variable representation of a system.

For notational economy, the variables will be represented by the input vector **u**, where

$$\mathbf{u} = \begin{bmatrix} u_1 \\ u_2 \\ \vdots \\ u_m \end{bmatrix} \qquad (5.2\text{-}1)$$

the output vector **y**, where

$$\mathbf{y} = \begin{bmatrix} y_1 \\ y_2 \\ \vdots \\ y_p \end{bmatrix} \qquad (5.2\text{-}2)$$

and the state vector **x**, where

$$\mathbf{x} = \begin{bmatrix} x_1 \\ x_2 \\ \vdots \\ x_n \end{bmatrix} \qquad (5.2\text{-}3)$$

For a given process, the state variable representation is not unique. However, all such representations have one characteristic in common for a given system; namely, the number of elements in the state vector (5.2-3) is equal and minimal. This number of elements n is referred to as the *order* of the system.

For a continuous system, the state variable $x_q(t)$ depends on an initial state $\mathbf{x}(t_0)$ of the system and the input $\mathbf{u}(t)$ over the time interval $[t_0, t)$.*

For continuous systems whose dynamics are governed by differential equations, the state variables are usually expressed in the form

$$\dot{x}_q(t) = f_q[\mathbf{x}(t_0), \mathbf{u}[t_0, t), t] \quad \text{for } q = 1, 2, \ldots, n \text{ and } t \geq t_0 \qquad (5.2\text{-}4)$$

*The notation $[t_0, t_1)$ means all time t in the interval $t_0 \leq t < t_1$.

where the $f_q(\ )$ are scalar functions that depend on the variables $x_q(t_0)$, $u_i[t_0, t)$. Equation (5.2-4) may be put into the vector form

$$\dot{\mathbf{x}}(t) = \mathbf{f}[\mathbf{x}(t_0), \mathbf{u}[t_0, t), t] \qquad (5.2\text{-}5)$$

where $\dot{\mathbf{x}}(t)$ is the $n \times 1$ vector with elements $dx_q(t)/dt$; that is,

$$\dot{\mathbf{x}}(t) = \begin{bmatrix} \dfrac{dx_1(t)}{dt} \\ \dfrac{dx_2(t)}{dt} \\ \cdot \\ \cdot \\ \cdot \\ \dfrac{dx_n(t)}{dt} \end{bmatrix}$$

and the $n \times 1$ vector $\mathbf{f}$ is given by

$$\mathbf{f}[\mathbf{x}(t_0), \mathbf{u}[t_0, t), t] = \begin{bmatrix} f_1[\mathbf{x}(t_0), \mathbf{u}[t_0, t), t] \\ f_2[\mathbf{x}(t_0), \mathbf{u}[t_0, t), t] \\ \cdot \quad \cdot \quad \cdot \quad \cdot \\ \cdot \quad \cdot \quad \cdot \quad \cdot \\ \cdot \quad \cdot \quad \cdot \quad \cdot \\ f_n[\mathbf{x}(t_0), \mathbf{u}[t_0, t), t] \end{bmatrix}$$

The output variables were said to be determined uniquely by the initial state vector $\mathbf{x}(t_0)$ and the input vector $\mathbf{u}(t)$, which implies that

$$y_j(t) = g_j[\mathbf{x}(t), \mathbf{u}(t), t] \qquad j = 1, 2, \ldots, p \qquad (5.2\text{-}6)$$

where g_j is a scalar function of the vectors $\mathbf{x}$ and $\mathbf{u}$. Putting (5.2-6) in vector form, we obtain

$$\mathbf{y}(t) = \mathbf{g}[\mathbf{x}(t), \mathbf{u}(t), t] \qquad (5.2\text{-}7)$$

where $\mathbf{g}$ is a $p \times 1$ vector given by

$$\mathbf{g}[\mathbf{x}(t), \mathbf{u}(t), t] = \begin{bmatrix} g_1[\mathbf{x}(t), \mathbf{u}(t), t] \\ g_2[\mathbf{x}(t), \mathbf{u}(t), t] \\ \cdot \quad \cdot \quad \cdot \quad \cdot \\ \cdot \quad \cdot \quad \cdot \quad \cdot \\ \cdot \quad \cdot \quad \cdot \quad \cdot \\ g_p[\mathbf{x}(t), \mathbf{u}(t), t] \end{bmatrix}$$

The appearance of the variable t in (5.2-4) and (5.2-6) is inserted for the case when the system is time-varying. If t does not appear explicitly in either (5.2-4) or (5.2-6), the system is said to be time invariant. Heuristically, this implies that none of the parameters characterizing the system's dynamics vary with time.

In summary, for systems whose dynamics are governed by differential equations, the state equations are of the form

$$\dot{\mathbf{x}}(t) = \mathbf{f}[\mathbf{x}(t), \mathbf{u}(t), t] \tag{5.2-8}$$

$$\mathbf{y}(t) = \mathbf{g}[\mathbf{x}(t), \mathbf{u}(t), t] \tag{5.2-9}$$

Equation (5.2-8) represents a system of n first-order differential equations.

For discrete systems, the state variables $x_q(t_k)$ depend on the initial state vector $\mathbf{x}(t_0)$ of the system and the input vectors $\{\mathbf{u}(t_0), \mathbf{u}(t_0), \ldots, \mathbf{u}(t_{k-1})\}$. If the state vector $\mathbf{x}(t_k)$ is known, then $x_q(t_{k+1})$ is completely determined by a knowledge of the input vector $\mathbf{u}(t_k)$; that is,

$$x_q(t_{k+1}) = f_q[\mathbf{x}(t_k), \mathbf{u}(t_k), t_k] \quad \text{for } q = 1, 2, \ldots, n \tag{5.2-10}$$

or, in vector form,

$$\mathbf{x}(t_{k+1}) = \mathbf{f}[\mathbf{x}(t_k), \mathbf{u}(t_k), t_k] \tag{5.2-11}$$

Similarly, the output vector $\mathbf{y}(t_k)$ becomes

$$\mathbf{y}(t_k) = \mathbf{g}[\mathbf{x}(t_k), \mathbf{u}(t_k), t_k] \tag{5.2-12}$$

If, as is normally the case, the discrete times t_k are equidistant from each other,

$$t_k - t_{k-1} = T \quad \text{for any integer } k \tag{5.2-13}$$

Then (5.2-11) and (5.2-12) can be notationally simplified to

$$\mathbf{x}(k+1) = \mathbf{f}[\mathbf{x}(k), \mathbf{u}(k), kT] \tag{5.2-14}$$

$$\mathbf{y}(k) = \mathbf{g}[\mathbf{x}(k), \mathbf{u}(k), kT] \tag{5.2-15}$$

Equation (5.2-14) represents a system of n first-order difference equations.

Linear continuous and discrete systems are characterized by equations (5.2-8), (5.2-9), and (5.2-14), (5.2-15), being of the form

$$\dot{\mathbf{x}}(t) = \mathbf{F}(t)\mathbf{x}(t) + \mathbf{G}(t)\mathbf{u}(t) \tag{5.2-16}$$

$$\mathbf{y}(t) = \mathbf{C}(t)\mathbf{x}(t) + \mathbf{D}(t)\mathbf{u}(t) \tag{5.2-17}$$

and

$$\mathbf{x}(k+1) = \mathbf{A}(k)\mathbf{x}(k) + \mathbf{B}(k)\mathbf{u}(k) \tag{5.2-18}$$

$$\mathbf{y}(k) = \mathbf{C}(k)\mathbf{x}(k) + \mathbf{D}(k)\mathbf{u}(k) \tag{5.2-19}$$

respectively, where $\mathbf{F}(t)$ and $\mathbf{A}(k)$ are called the $n \times n$ system matrices, $\mathbf{G}(t)$ and $\mathbf{B}(k)$ are the $n \times m$ driving matrices, $\mathbf{C}(t)$ and $\mathbf{C}(k)$ are the $p \times n$

output matrices, and $\mathbf{D}(t)$ and $\mathbf{D}(k)$ are the $p \times m$ transmission matrices, respectively.

These systems are said to be linear, since all the variables $\mathbf{x}$, $\mathbf{y}$, and $\mathbf{u}$ occur in a linear form.

Figure (5.2-2) depicts a vector diagram representation for the linear continuous and discrete systems characterized by (5.2-16) through (5.2-19).

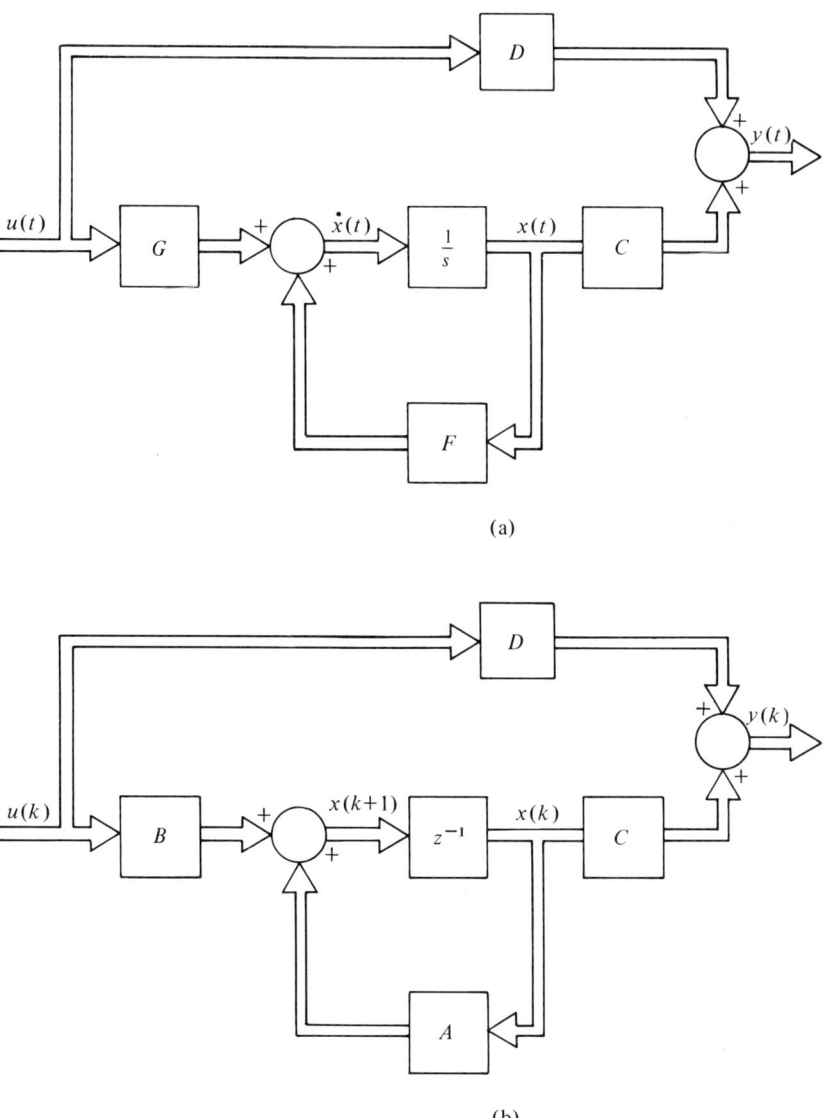

Figure 5.2-2. Vector diagram representation of (a) linear continuous system; (b) linear discrete system.

It is observed from Figure 5.2-2 that the basic structures of linear continuous and discrete systems are quite similar.

The set of all possible input vectors **u** to a system forms the input space **U**. For example, the elements of **u** may be restricted in magnitude as $|u_i| \leq a_i$ for $i = 1, 2, \ldots, m$. The set of all possible output vectors **y** forms the output space **Y**, whereas all the possible state vectors **x** form the state space **X**.

5.3 State Variable Representations of Linear Discrete Systems

With the concepts of state as presented in Section 5.2 serving as a background, we shall now develop four methods for representing a linear discrete system in a state variable representation. Each method has its counterpart in continuous state variable representations, a fact which by now should be expected.

We shall restrict our discussion to discrete systems whose dynamics are governed by a difference equation of the form

$$y(k) = b_0 u(k) + b_1 u(k-1) + \cdots + b_m u(k-m)$$
$$- a_1 y(k-1) - a_2 y(k-2) - \cdots - a_n y(k-n)$$
$$\text{with } m \leq n \quad (5.3\text{-}1)$$

or its equivalent transfer function

$$G(z) = \frac{Y(z)}{U(z)} = \frac{b_0 + b_1 z^{-1} + \cdots + b_m z^{-m}}{1 + a_1 z^{-1} + \cdots + a_n z^{-n}} \quad (5.3\text{-}2)$$

$$= \frac{b_0 z^n + b_1 z^{n-1} + \cdots + b_m z^{n-m}}{z^n + a_1 z^{n-1} + \cdots + a_n} \quad (5.3\text{-}3)$$

where $u(k)$ and $y(k)$ are the system's input and output, respectively, at discrete time k. Many linear discrete systems have such a dynamical representation. For those systems that do not exactly fit this form, obvious extensions of the four methods to be presented may be made. For example, the system with transfer function $z^{-1}/(z-1)$ is not in the desired form, but may be treated by slight modifications.

5.3-1 Partial-fraction Expansion Method

The system to be investigated is characterized by the transfer function

$$G(z) = \frac{Y(z)}{U(z)} = \frac{b_0 z^m + b_1 z^{m-1} + \cdots + b_m}{(z+z_1)(z+z_2)\cdots(z+z_n)} \quad \text{with } m \leq n \quad (5.3\text{-}4)$$

where the denominator polynomial of (5.3-3) is in factored form. If $m = n$, it is necessary to put (5.3-4) into proper form; that is,

$$G(z) = b_0 + \frac{\hat{b}_1 z^{n-1} + \hat{b}_2 z^{n-2} + \cdots + \hat{b}_n}{(z+z_1)(z+z_2)\cdots(z+z_n)}$$
$$= b_0 + \hat{G}(z) \tag{5.3-5}$$

This is accomplished by performing a division of denominator into numerator. It is now possible to expand (5.3-5) by a partial-fraction expansion. We shall treat the case where $G(z)$ has distinct poles first.

(i) *$G(z)$ Has Distinct Poles*

In this case, $G(z)$, as given by (5.3-5), becomes

$$\frac{Y(z)}{R(z)} = G(z) = b_0 + \frac{d_1}{z+z_1} + \cdots + \frac{d_n}{z+z_n} \tag{5.3-6}$$

where

$$d_j = \lim_{z \to -z_j} [(z+z_j)G(z)]$$

This form of the transfer function may be represented by the state variable diagram shown in Figure 5.3-1.

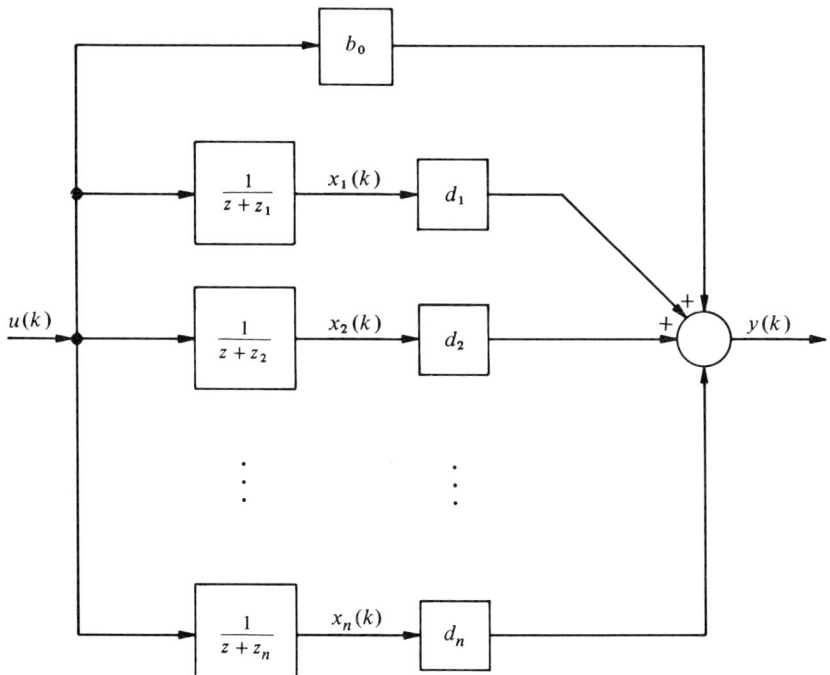

Figure 5.3-1. State variable diagram of the transfer function by the partial-fraction method (distinct poles).

The output signal $Y(z)$ and its input $U(z)$ are related by

$$Y(z) = \left[b_0 + \sum_{j=1}^{n} \frac{d_j}{z + z_j}\right] U(z) \qquad (5.3\text{-}7)$$

Upon selecting the state variables as indicated in the state variable diagram, we have

$$X_m(z) = \frac{1}{z + z_m} U(z) \quad \text{for } m = 1, 2, \ldots, n$$

or its discrete-time equivalent

$$x_m(k+1) = -z_m x_m(k) + u(k) \quad \text{for } m = 1, 2, \ldots, n \qquad (5.3\text{-}8)$$

Putting (5.3-8) into matrix form gives the state variable vector difference equation

$$\mathbf{x}(k+1) = \begin{bmatrix} -z_1 & 0 & \cdots & & 0 \\ 0 & -z_2 & 0 & & 0 \\ \vdots & & \ddots & & \vdots \\ 0 & \cdots & & 0 & -z_n \end{bmatrix} \mathbf{x}(k) + \begin{bmatrix} 1 \\ 1 \\ \vdots \\ 1 \end{bmatrix} u(k) \qquad (5.3\text{-}9)$$

and the output signal is, by (5.3-7),

$$y(k) = [d_1 \quad d_2 \quad \cdots \quad d_n]\mathbf{x}(k) + b_0 u(k) \qquad (5.3\text{-}10)$$

where

$$\mathbf{x}(k) = \begin{bmatrix} x_1(k) \\ x_2(k) \\ \vdots \\ x_n(k) \end{bmatrix}$$

defines the state vector by this representation method. We have represented the nth-order system governed by (5.3-4) by the system of matrix difference equations

$$\mathbf{x}(k+1) = \mathbf{A}\mathbf{x}(k) + \mathbf{B}u(k)$$
$$y(k) = \mathbf{C}\mathbf{x}(k) + du(k)$$

where the matrices $\mathbf{A}$, $\mathbf{B}$, $\mathbf{C}$, and d are given in (5.3-9) and (5.3-10). In using the partial-fraction expansion method to obtain a state variable representation of a system that has distinct poles, the system matrix $\mathbf{A}$ is seen to be of a diagonal form. The diagonal elements are the poles of the transfer function $G(z)$.

EXAMPLE 5.3-1

Give a state variable representation of the discrete system whose transfer function is

$$G(z) = \frac{z^2 + 2z + 1}{(z+2)(z+3)}$$

Putting $G(z)$ into proper form and then making a partial-fraction expansion, we obtain

$$G(z) = 1 - \frac{3z + 5}{(z+2)(z+3)} = 1 + \frac{1}{z+2} - \frac{4}{z+3}$$

Letting

$$x_1(k+1) = -2x_1(k) + u(k)$$
$$x_2(k+1) = -3x_2(k) + u(k)$$

be our choice of state variables as indicated in (5.3-8), we then have

$$\begin{bmatrix} x_1(k+1) \\ x_2(k+1) \end{bmatrix} = \begin{bmatrix} -2 & 0 \\ 0 & -3 \end{bmatrix} \begin{bmatrix} x_1(k) \\ x_2(k) \end{bmatrix} + \begin{bmatrix} 1 \\ 1 \end{bmatrix} u(k)$$

whereas the output signal is given by

$$y(k) = x_1(k) - 4x_2(k) + u(k) = \begin{bmatrix} 1 & -4 \end{bmatrix} \begin{bmatrix} x_1(k) \\ x_2(k) \end{bmatrix} + u(k)$$

This state variable representation could have been obtained directly from (5.3-9) and (5.3-10) by observing that $z_1 = -2$, $z_2 = -3$, $d_1 = 1$, and $d_2 = -4$.

(ii) *G(z) Has Multiple Poles*

An extension of the preceding technique to the case when $G(z)$ has multiple poles follows directly. To illustrate the procedure, let it be assumed that $G(z)$ has a multiple pole of order p at $z = -z_n$ with all its other poles being distinct. $G(z)$ has the transfer function

$$G(z) = \frac{b_0 z^m + b_1 z^{m-1} + \cdots + b_{m-1} z + b_m}{(z+z_1)(z+z_2)\cdots(z+z_{n-p})(z+z_n)^p} \quad \text{with } m \leq n \quad (5.3\text{-}11)$$

If $m = n$, we first put $G(z)$ into proper form, as in (5.3-5). Without loss of generality, we can then assume $m < n$. Expanding (5.3-11) by partial-fraction expansion, we obtain

$$G(z) = \frac{d_1}{z+z_1} + \frac{d_2}{z+z_2} + \cdots + \frac{d_{n-p}}{z+z_{n-p}} + \frac{e_1}{z+z_n} + \frac{e_2}{(z+z_n)^2}$$
$$+ \cdots + \frac{e_p}{(z+z_p)^p} \quad (5.3\text{-}12)$$

where the coefficients $e_1, e_2, \ldots, e_p$ are obtained by standard partial-fraction expansion techniques. The system as given by (5.3-12) may be represented by the state variable diagram of Figure 5.3-2.

Making the following choice of state variables with the aid of (5.3-12) and Figure 5.3-2,

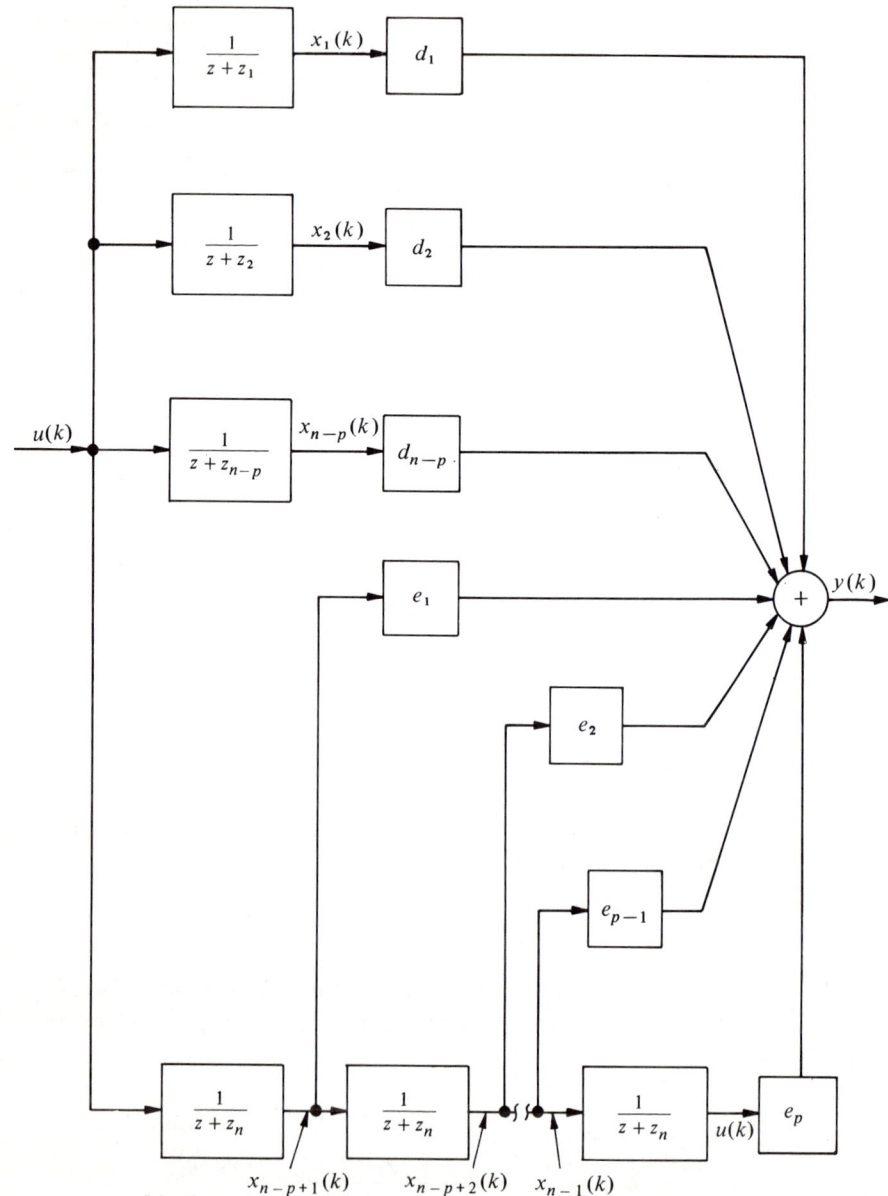

Figure 5.3-2. State variable diagram for a system whose transfer function has multiple poles by partial-fraction expansion.

$$x_1(k+1) = -z_1 x_1(k) + u(k)$$
$$x_2(k+1) = -z_2 x_2(k) + u(k)$$
$$\vdots$$
$$x_{n-p+1}(k+1) = -z_n x_{n-p+1}(k) + u(k) \qquad (5.3\text{-}13)$$
$$x_{n-p+2}(k+1) = -z_n x_{n-p+2}(k) + x_{n-p+1}(k)$$
$$\vdots$$
$$x_n(k+1) = -z_n x_n(k) + x_{n-1}(k)$$

we find that the output signal $y(k)$ may be written in terms of the state variables as

$$y(k) = d_1 x_1(k) + d_2 x_2(k) + \cdots + d_{n-p} x_{n-p}(k)$$
$$+ e_1 x_{n-p+1}(k) + \cdots + e_p x_n(k) \qquad (5.3\text{-}14)$$

The matrix representation of (5.3-13) and (5.3-14) is then

$$\begin{bmatrix} x_1(k+1) \\ x_2(k+1) \\ \vdots \\ x_{n-p}(k+1) \\ \vdots \\ x_n(k+1) \end{bmatrix} = \begin{bmatrix} -z_1 & 0 & \cdots & & & & & 0 \\ 0 & -z_2 & & & & & & 0 \\ \vdots & & \ddots & & & & & \vdots \\ & & & -z_{n-p} & 0 & \cdots & & 0 \\ & & & 0 & z_n & 0 & \cdots & 0 \\ & & & & 1 & z_n & 0 & \cdots & 0 \\ & & & & 0 & 1 & \ddots & & \vdots \\ & & & & \vdots & & & & 0 \\ 0 & \cdots & & & 0 & 0 & \cdots & 0 & 1 & z_n \end{bmatrix}$$
$$\times \begin{bmatrix} x_1(k) \\ x_2(k) \\ \vdots \\ \mathbf{x}_{n-p}(k) \\ \vdots \\ x_n(k) \end{bmatrix} + \begin{bmatrix} 1 \\ 1 \\ \vdots \\ 1 \\ 0 \\ 0 \\ \vdots \\ 0 \end{bmatrix} u(k) \qquad (5.3\text{-}15)$$

and

$$y(k) = c(k) = [d_1 \quad d_2 \quad \ldots \quad d_{n-p} \quad e_1 \quad e_2 \quad \ldots \quad e_p]\mathbf{x}(k) \quad (5.3\text{-}16)$$

where the state vector $\mathbf{x}(k)$ is given by

$$\mathbf{x}(k) = \begin{bmatrix} x_1(k) \\ x_2(k) \\ \vdots \\ x_n(k) \end{bmatrix}$$

The system matrix $\mathbf{A}$, which is obtained by a partial-fraction expansion of $G(z)$ when it has multiple poles, is said to be in a Jordan canonical form.

EXAMPLE 5-3.2

Determine a Jordan canonical state variable representation for the system with transfer function

$$G(z) = \frac{1}{(z+1)^2(z+2)}$$

Making a partial-fraction expansion of $G(z)$ results in

$$G(z) = \frac{1}{(z+1)^2} - \frac{1}{z+1} + \frac{1}{z+2}$$

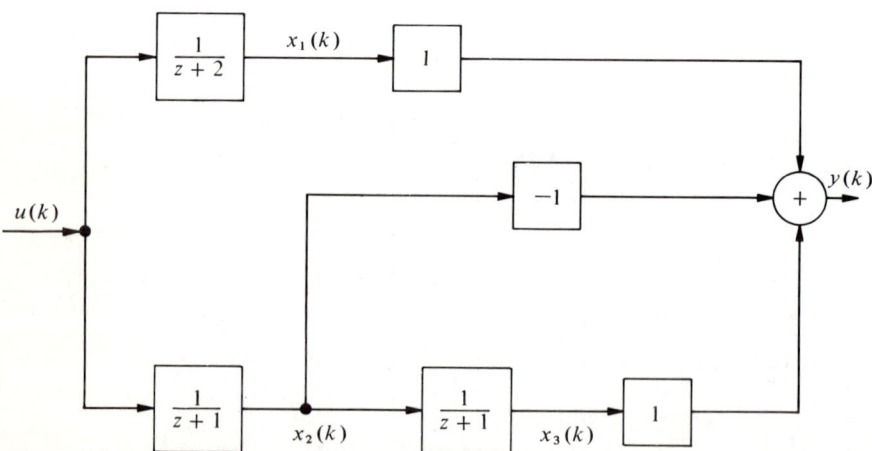

Figure 5.3-3. State variable diagram for Example 5.3-2.

This form of the transfer function is shown as a state variable diagram in Figure 5.3-3.

The corresponding state space equations become

$$\mathbf{x}(k+1) = \begin{bmatrix} -2 & 0 & 0 \\ 0 & -1 & 0 \\ 0 & 1 & -1 \end{bmatrix} \mathbf{x}(k) + \begin{bmatrix} 1 \\ 1 \\ 0 \end{bmatrix} u(k)$$

$$y(k) = \begin{bmatrix} 1 & -1 & 1 \end{bmatrix} \mathbf{x}(k)$$

5.3-2 Iterative Programming

An alternate approach for obtaining a state variable representation of a system whose transfer function is in factored form is available. Let the transfer function have the form

$$G(z) = \frac{Y(z)}{U(z)} = \frac{\alpha(z + z_1)(z + z_2) \cdots (z + z_m)}{(z + p_0)(z + p_1) \cdots (z + p_{n-1})} \quad (5.3\text{-}17)$$

We shall be concerned only with transfer functions where $m \leq n$. Since the maximum order of the numerator and denominator polynomials is taken to be n, this system is characterized by an nth-order difference equation. We therefore require that a state vector be an $n \times 1$ vector.

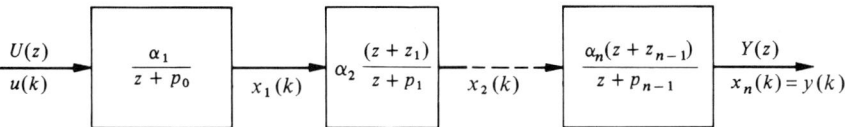

Figure 5.3-4. Iterative programming representation of $G(z)$, where $m = n - 1$.

Without loss of generality, assume $m = n - 1$. A representation of $G(z)$ may be made as shown in Figure 5.3-4. The identity

$$\alpha = \alpha_0 \alpha_1 \ldots \alpha_{n-1}$$

is used. The order in which the zeros and poles of $G(z)$ appear is obviously not important. As state variables, the outputs of the various individual blocks serve the requirements of a state variable representation. This follows since if we know the values of $x_1(k), x_2(k), \ldots, x_n(k)$ at time k and $u(k)$ for $k \geq k_0$, then the values of these state variables and the output $y(k)$ may be determined for all $k \geq k_0$.

If $m < n - 1$, then $n - m - 1$ of the blocks in Figure 5.3-4 will have only denominator dynamics; that is, they will be of the form

$$\frac{\alpha_j}{z + p_j}$$

EXAMPLE 5.3-3

Put the system considered in Example 5.3-1 into a state variable representation by the method of iterative programming.

Rewriting $G(z)$ as

$$G(z) = 1 - 3\frac{z + \frac{5}{3}}{(z + 2)(z + 3)}$$

we may represent $G(z)$ as shown in Figure 5.3-5.

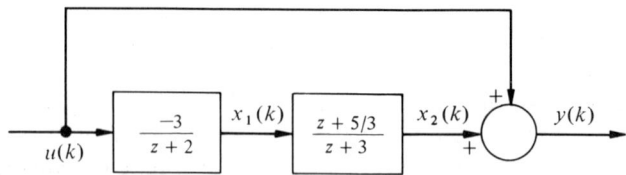

Figure 5.3-5. Representation of the $G(z)$ in Example 5.3-3.

Making the indicated selection of state variables, we obtain the first-order difference equations

$$x_1(k + 1) + 2x_1(k) = -3u(k)$$
$$x_2(k + 1) + 3x_2(k) = x_1(k + 1) + \tfrac{5}{3}x_1(k)$$

After substituting the expression for $x_1(k + 1)$ of the first equation into the second, we have

$$x_1(k + 1) = -2x_1(k) - 3u(k)$$
$$x_2(k + 1) = -\tfrac{1}{3}x_1(k) - 3x_2(k) - 3u(k)$$

The output signal $y(k)$ is given by

$$y(k) = x_2(k) + u(k)$$

Putting these equations into vector form, we find that the iterative programming technique gives the following state variable representation.

$$\begin{bmatrix} x_1(k+1) \\ x_2(k+1) \end{bmatrix} = \begin{bmatrix} -2 & 0 \\ -\frac{1}{3} & -3 \end{bmatrix} \begin{bmatrix} x_1(k) \\ x_2(k) \end{bmatrix} + \begin{bmatrix} -3 \\ -3 \end{bmatrix} u(k)$$

$$y(k) = \begin{bmatrix} 0 & 1 \end{bmatrix} \begin{bmatrix} x_1(k) \\ x_2(k) \end{bmatrix} + u(k)$$

5.3-3 Direct Programming

In many cases, the factored form of the system's transfer function $G(z)$ may not be given. Instead of factoring $G(z)$ to determine its poles and zeros in order to use partial-fraction expansion or iterative programming to obtain a state variable representation, a more direct method may be applied, called appropriately *direct programming*.

Let $G(z)$ be written as

$$G(z) = \frac{Y(z)}{U(z)} = \frac{b_0 z^n + b_1 z^{n-1} + \cdots + b_{n-1} z + b_n}{z^n + a_1 z^{n-1} + \cdots + a_{n-1} z + a_n} \quad (5.3\text{-}18)$$

If the order of the numerator is less than n, then some of the coefficients (i.e., b_0, etc.) will be zero. If $b_0 \neq 0$, we may put $G(z)$ into proper form as in (5.3-5). So without loss of generality, we may assume $b_0 = 0$. Multiplying the numerator and denominator of (5.3-18) by z^{-n} and crossmultiplying gives

$$\frac{Y(z)}{b_1 z^{-1} + b_2 z^{-2} + \cdots + b_{n-1} z^{-n+1} + b_n z^{-n}}$$
$$= \frac{U(z)}{1 + a_1 z^{-1} + \cdots + a_{n-1} z^{-n+1} + a_n z^{-n}}$$
$$= Q(z)$$

Therefore,

$$Q(z) = -a_1 z^{-1} Q(z) - \cdots - a_{n-1} z^{-n+1} Q(z) - a_n z^{-n} Q(z) + U(z) \quad (5.3\text{-}19)$$

and

$$Y(z) = b_1 z^{-1} Q(z) + b_2 z^{-2} Q(z) + \cdots + b_{n-1} z^{-n+1} Q(z) + b_n z^{-n} Q(z) \quad (5.3\text{-}20)$$

Defining the state variables by

$$X_1(z) = z^{-1} Q(z)$$
$$X_2(z) = z^{-2} Q(z)$$
$$\cdot \quad \cdot \quad \cdot \quad \cdot \quad \cdot$$
$$X_n(z) = z^{-n} Q(z)$$

we observe that

$$zX_2(z) = X_1(z)$$
$$zX_3(z) = X_2(z)$$
$$\cdot \quad \cdot \quad \cdot \quad \cdot \quad \cdot$$
$$zX_n(z) = X_{n-1}(z)$$

so that

$$\begin{aligned} x_2(k+1) &= x_1(k) \\ x_3(k+1) &= x_2(k) \\ \cdot \quad \cdot \quad &\cdot \quad \cdot \quad \cdot \\ x_n(k+1) &= x_{n-1}(k) \end{aligned} \quad (5.3\text{-}21)$$

From the relationship $zX_1(z) = Q(z)$, equation (5.3-19) is generated by the first-order difference equation

$$x_1(k+1) = -a_1 x_1(k) - a_2 x_2(k) - \ldots - a_{n-1} x_{n-1}(k) \\ - a_n x_n(k) + u(k) \quad (5.3\text{-}22)$$

and equation (5.3-20) is equivalent to

$$y(k) = b_1 x_1(k) + b_2 x_2(k) + \cdots + b_{n-1} x_{n-1}(k) + b_n x_n(k) \quad (5.3\text{-}23)$$

Putting (5.3-21) through (5.3-23) into vector form, we have the direct programming state variable representation of (5.3-18) with $b_0 = 0$.

$$\begin{bmatrix} x_1(k+1) \\ x_2(k+1) \\ \vdots \\ x_n(k+1) \end{bmatrix} = \begin{bmatrix} -a_1 & -a_2 & \cdots & -a_{n-1} & -a_n \\ 1 & 0 & \cdots & 0 & 0 \\ \cdot & \cdot & \cdots & \cdot & \cdot \\ 0 & 0 & \cdots & 1 & 0 \end{bmatrix} \begin{bmatrix} x_1(k) \\ x_2(k) \\ \vdots \\ x_n(k) \end{bmatrix} + \begin{bmatrix} 1 \\ 0 \\ \vdots \\ 0 \end{bmatrix} u(k)$$
(5.3-24)

and

$$y(k) = c(k) = [b_1 \quad b_2 \quad \cdots \quad b_n] \begin{bmatrix} x_1(k) \\ x_2(k) \\ \vdots \\ x_n(k) \end{bmatrix} \quad (5.3\text{-}25)$$

The state variable diagram that we obtain by using the direct programming method is shown in Figure 5.3-6.

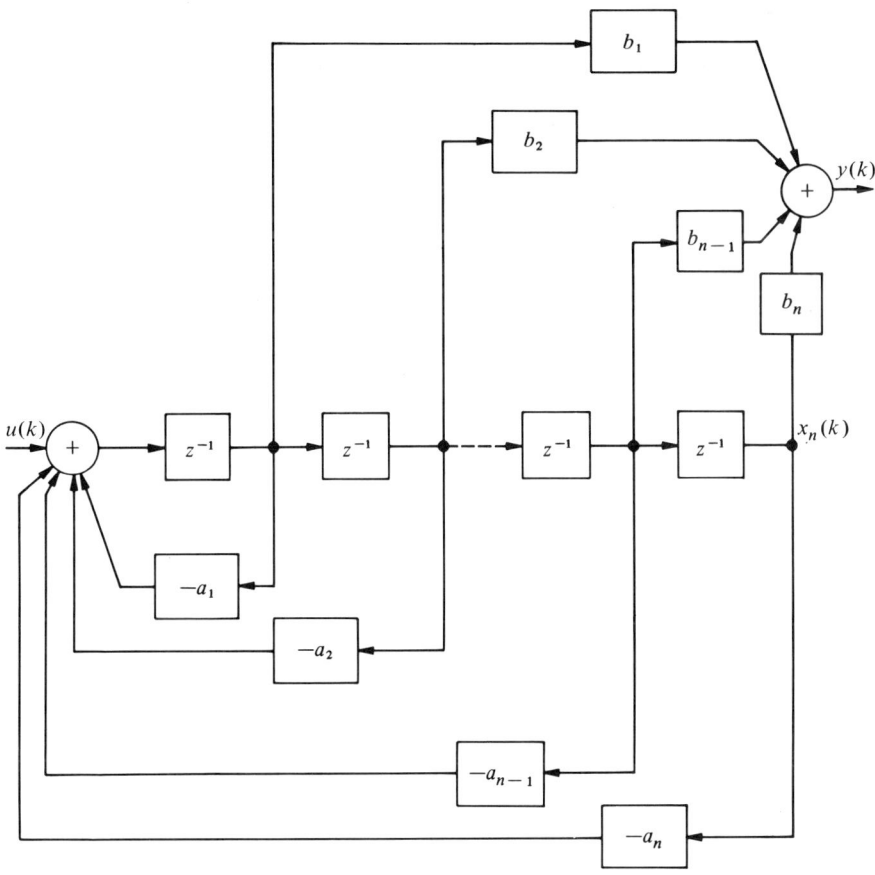

Figure 5.3-6. State variable via direct programming procedure ($m < n$).

If $b_0 \neq 0$, then we put $G(z)$ into proper form, as in (5.3-5); that is,

$$G(z) = b_0 + \hat{G}(z)$$

The state variable representation as given by (5.3-24) remains the same; however, (5.3-25) becomes

$$y(k) = c(k) = [\hat{b}_1 \quad \hat{b}_2 \quad \ldots \quad \hat{b}_{n-1} \quad \hat{b}_n] \begin{bmatrix} x_1(k) \\ x_2(k) \\ \vdots \\ x_n(k) \end{bmatrix} + b_0 u(k) \quad (5.3\text{-}26)$$

where the coefficients $\hat{b}_j$ are the numerator coefficients of $\hat{G}(z)$.

EXAMPLE 5.3-4

Using the direct programming method, represent the system examined in Example 5.3-1 by state variable techniques.

Noting that $n = 2$ and $b_0 \neq 0$, we first put $G(z)$ into proper form.

$$G(z) = 1 + \frac{-3z - 5}{z^2 + 5z + 6}$$

Therefore, $a_1 = 5$, $a_2 = 6$, $\hat{b}_0 = -3$, $\hat{b}_1 = -5$, so that we have, by (5.3-24) and (5.3-26),

$$\begin{bmatrix} x_1(k+1) \\ x_2(k+1) \end{bmatrix} = \begin{bmatrix} -5 & -6 \\ 1 & 0 \end{bmatrix} \begin{bmatrix} x_1(k) \\ x_2(k) \end{bmatrix} + \begin{bmatrix} 1 \\ 0 \end{bmatrix} u(k)$$

$$c(k) = [-3 \quad -5] \begin{bmatrix} x_1(k) \\ x_2(k) \end{bmatrix} + u(k)$$

The state variable diagram for this system is shown in Figure 5.3-7.

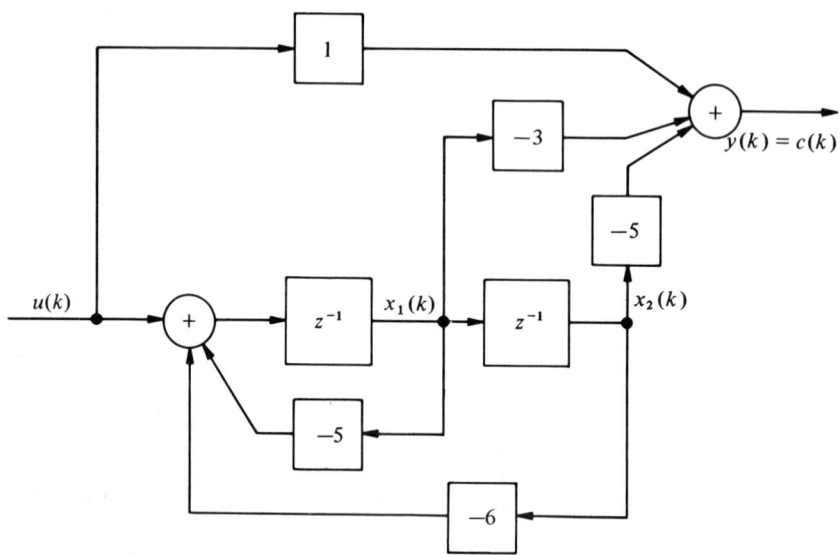

Figure 5.3-7. State variable diagram by direct programming techniques.

5.3-4 Nested Programming

When the system's transfer function is in an unfactored form, such as that given by (5.3-18), the method of nested programming may also be used to give a state variable representation. Upon multiplying numerator and denominator of (5.3-18) by z^{-n} and collecting terms of equal powers of z^{-1}, we obtain

$$Y(z) - b_0 U(z) + z^{-1}[a_1 Y(z) - b_1 U(z)] + z^{-2}[a_2 Y(z) - b_2 U(z)]$$
$$+ \cdots + z^{-n+1}[a_{n-1} Y(z) - b_{n-1} U(z)] + z^{-n}[a_n Y(z) - b_n U(z)] = 0$$

Rearranging terms gives

$$Y(z) = b_0 U(z) + z^{-1}\{b_1 U(z) - a_1 Y(z) + z^{-1}[b_2 U(z) - a_2 Y(z) \\ + z^{-1}(b_3 U(z) - a_3 Y(z) + \cdots)]\} \quad (5.3\text{-}27)$$

By defining the state variables to be

$$X_n(z) = z^{-1}[b_n U(z) - a_n Y(z)]$$
$$X_{n-1}(z) = z^{-1}[b_{n-1} U(z) - a_{n-1} Y(z) + X_n(z)]$$
$$X_{n-2}(z) = z^{-1}[b_{n-2} U(z) - a_{n-2} Y(z) + X_{n-1}(z)] \quad (5.3\text{-}28)$$
$$\cdots\cdots\cdots\cdots\cdots\cdots\cdots\cdots\cdots$$
$$X_2(z) = z^{-1}[b_2 U(z) - a_2 Y(z) + X_3(z)]$$
$$X_1(z) = z^{-1}[b_1 U(z) - a_1 Y(z) + X_2(z)]$$

we find that equation (5.3-27) simplifies to

$$Y(z) = b_0 U(z) + X_1(z) \quad (5.3\text{-}29)$$

With the aid of (5.3-29), the difference equation equivalents of (5.3-28) and (5.3-29) are seen to be

$$x_n(k+1) = -a_n x_1(k) + (b_n - a_n b_0) u(k)$$
$$x_{n-1}(k+1) = -a_{n-1} x_1(k) + x_n(k) + (b_{n-1} - a_{n-1} b_0) u(k)$$
$$x_{n-2}(k+1) = -a_{n-2} x_1(k) + x_{n-1}(k) + (b_{n-2} - a_{n-2} b_0) u(k) \quad (5.3\text{-}30)$$
$$\cdots\cdots\cdots\cdots\cdots\cdots\cdots\cdots\cdots$$
$$x_2(k+1) = -a_2 x_1(k) + x_3(k) + (b_2 - a_2 b_0) u(k)$$
$$x_1(k+1) = -a_1 x_1(k) + x_2(k) + (b_1 - a_1 b_0) u(k)$$

and

$$y(k) = b_0 u(k) + x_1(k) \quad (5.3\text{-}31)$$

Putting (5.3-30) and (5.3-31) into vector form, we obtain

$$\begin{bmatrix} x_1(k+1) \\ x_2(k+1) \\ \vdots \\ x_{n-1}(k+1) \\ x_n(k+1) \end{bmatrix} = \begin{bmatrix} -a_1 & 1 & 0 & \cdots & 0 & 0 \\ -a_2 & 0 & 1 & 0 & \cdots & 0 \\ \vdots & \vdots & \vdots & & \vdots & \vdots \\ -a_{n-1} & 0 & 0 & \cdots & 0 & 1 \\ -a_n & 0 & 0 & \cdots & 0 & 0 \end{bmatrix} \begin{bmatrix} x_1(k) \\ x_2(k) \\ \vdots \\ x_{n-1}(k) \\ x_n(k) \end{bmatrix}$$

$$+ \begin{bmatrix} b_1 - a_1 b_0 \\ b_2 - a_2 b_0 \\ \vdots \\ b_{n-1} - a_{n-1} b_0 \\ b_n - a_n b_0 \end{bmatrix} u(k) \qquad (5.3\text{-}32)$$

$$y(k) = [1 \ 0 \ \cdots \ 0] \begin{bmatrix} x_1(k) \\ x_2(k) \\ \vdots \\ x_n(k) \end{bmatrix} + b_0 u(k) \qquad (5.3\text{-}33)$$

An interesting relationship exists between the system matrices obtained by using direct and nested programming techniques. It will be observed from (5.3-24) and (5.3-32) that they are transposes of one another.

The characterizing first-order difference equations of (5.3-30) and the output signal given by (5.3-31) can be represented by the state variable diagram shown in Figure 5.3-8.

EXAMPLE 5.3-5

Apply the method of nested programming to obtain a state variable representation for the system studied in Example 5.3-1.

We have

$$G(z) = \frac{z^2 + 2z + 1}{z^2 + 5z + 6}$$

so that

$$b_0 = 1, \quad b_1 = 2, \quad b_2 = 1$$
$$a_0 = 1, \quad a_1 = 5, \quad a_2 = 6$$

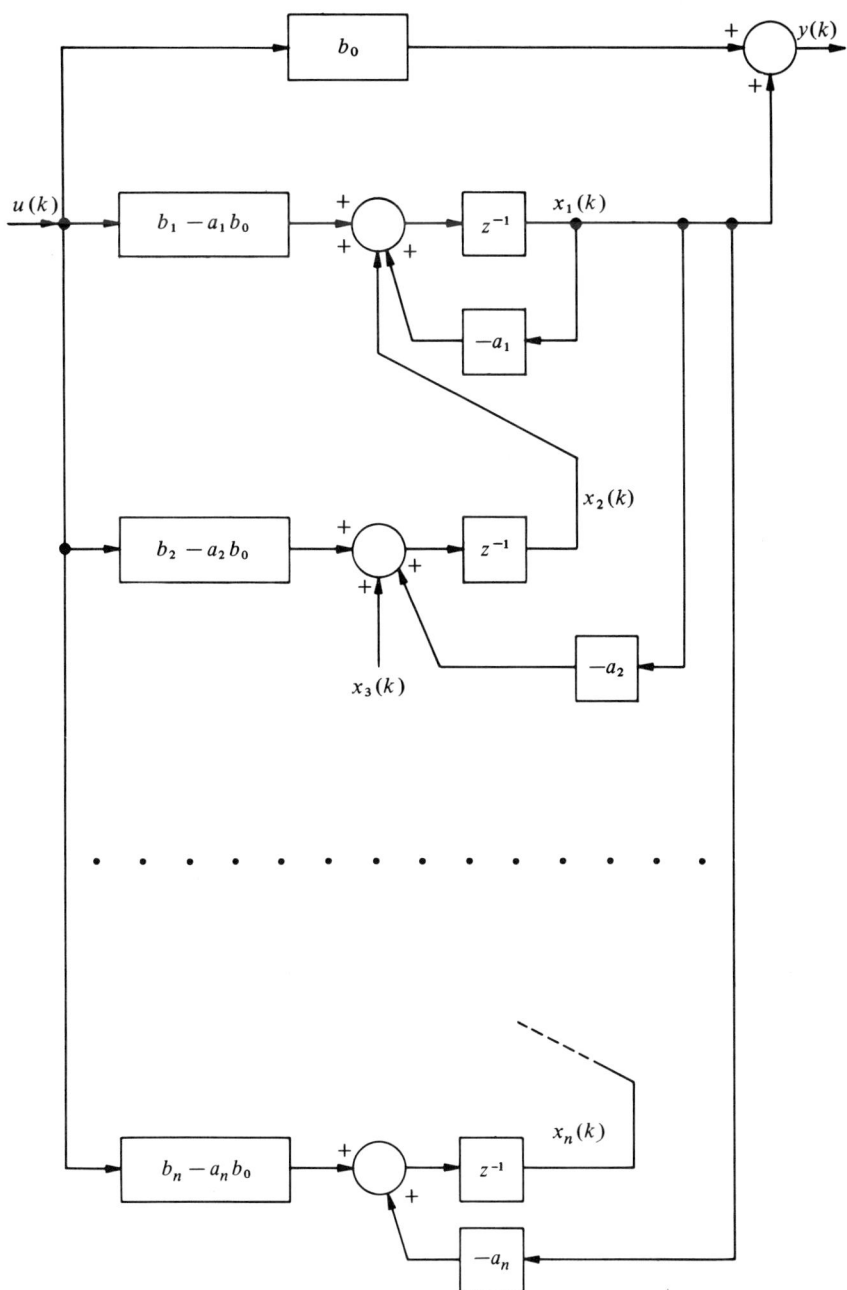

Figure 5.3-8. State variable diagram obtained by nested programming methods.

Making the state variable selection as given in (5.3-30) with $n = 2$, we obtain from (5.3-32) and (5.3-33)

$$\begin{bmatrix} x_1(k+1) \\ x_2(k+1) \end{bmatrix} = \begin{bmatrix} -5 & 1 \\ -6 & 0 \end{bmatrix} \begin{bmatrix} x_1(k) \\ x_2(k) \end{bmatrix} + \begin{bmatrix} -3 \\ -5 \end{bmatrix} u(k)$$

$$y(k) = \begin{bmatrix} 1 & 0 \end{bmatrix} \begin{bmatrix} x_1(k) \\ x_2(k) \end{bmatrix} + u(k)$$

The corresponding state variable diagram is shown in Figure 5.3-9.

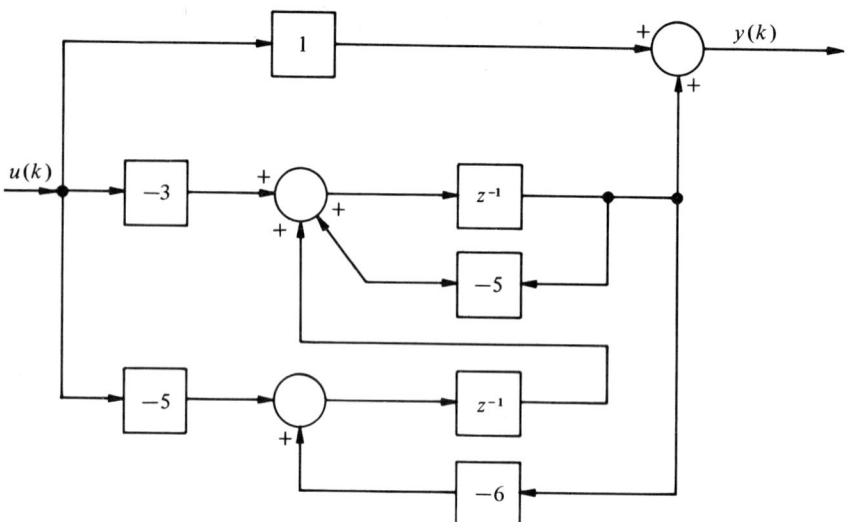

Figure 5.3-9. State variable diagram for Example 5.3-5.

5.4 Discrete State Equations

The z-transform may be conveniently applied to the solution of discrete state equations. Let the discrete state model of a linear system with m independent inputs $u_i(k)$ and p outputs $y_j(k)$ be given as

$$\mathbf{x}(k+1) = \mathbf{Ax}(k) + \mathbf{Bu}(k) \tag{5.4-1}$$

$$\mathbf{y}(k) = \mathbf{Cx}(k) + \mathbf{Du}(k) \tag{5.4-2}$$

where $\mathbf{x}(k)$ is an $n \times 1$ state vector, $\mathbf{u}(k)$ is the $m \times 1$ control vector with components $u_i(k)$, and $\mathbf{y}(k)$ is the $p \times 1$ output vector with components $y_j(k)$. The z-transform of the vector sequence $\mathbf{x}(k)$ is defined by

$$\mathscr{L}[\mathbf{x}(k)] = \mathscr{L}\begin{bmatrix} x_1(k) \\ x_2(k) \\ \vdots \\ x_n(k) \end{bmatrix} = \begin{bmatrix} \mathscr{L}[x_1(k)] \\ \mathscr{L}[x_2(k)] \\ \vdots \\ \mathscr{L}[x_n(k)] \end{bmatrix} = \begin{bmatrix} X_1(z) \\ X_2(z) \\ \vdots \\ X_n(z) \end{bmatrix} \quad (5.4\text{-}3)$$

with $X_q(z)$ being the z-transform of the qth component of the vector $\mathbf{x}(k)$. It is not difficult to show that

$$\mathscr{L}[\mathbf{Ax}(k)] = \mathbf{A}\mathbf{X}(z)$$

where $\mathbf{X}(z) = \mathscr{L}[\mathbf{x}(k)]$ is an $n \times 1$ vector.

With these facts in mind, we proceed to take the one-sided z-transform of (5.4-1) and (5.4-2) to obtain

$$z\mathbf{X}(z) - z\mathbf{x}(0) = \mathbf{A}\mathbf{X}(z) + \mathbf{B}\mathbf{U}(z) \quad (5.4\text{-}4)$$
$$\mathbf{Y}(z) = \mathbf{C}\mathbf{X}(z) + \mathbf{D}\mathbf{U}(z) \quad (5.4\text{-}5)$$

where $\mathbf{U}(z)$ is the $m \times 1$ vector that is the z-transform of $\mathbf{u}(k)$ and $\mathbf{x}(0)$ is the value of the state vector at $k = 0$. $\mathbf{Y}(z)$ is the $p \times 1$ vector that is the z-transform of $\mathbf{y}(k)$.

Solving (5.4-4) for $\mathbf{X}(z)$

$$[z\mathbf{I} - \mathbf{A}]\mathbf{X}(z) = z\mathbf{x}(0) + \mathbf{B}\mathbf{U}(z)$$

and premultiplying this expression by $[z\mathbf{I} - \mathbf{A}]^{-1}$, we conclude that

$$\mathbf{X}(z) = [z\mathbf{I} - \mathbf{A}]^{-1} z\mathbf{x}(0) + [z\mathbf{I} - \mathbf{A}]^{-1} \mathbf{B}\mathbf{U}(z) \quad (5.4\text{-}6)$$

Equation (5.4-6) may be used to determine $\mathbf{x}(kT) = \mathbf{x}(k)$, provided that $\mathbf{x}(0)$ and $\mathbf{U}(z)$ are known. The solution to (5.4-6) requires the inversion of a matrix, which may be accomplished either analytically or by use of a computer routine. Second, the inverse z-transform of the resulting expression is required. The solution of discrete state equations by z-transform methods is completely analogous to the solution of continuous-time state equations by Laplace transform methods, requiring the same two steps, namely, the inversion of a matrix and the inverse transformation of a z-transform.

EXAMPLE 5.4-1

For a discrete system with state variable representation given by

$$\begin{bmatrix} x_1(k+1) \\ x_2(k+1) \end{bmatrix} = \begin{bmatrix} -.5 & -.3 \\ .2 & 0 \end{bmatrix} \begin{bmatrix} x_1(k) \\ x_2(k) \end{bmatrix} + \begin{bmatrix} 1 \\ 0 \end{bmatrix} u(k), \quad \begin{bmatrix} x_1(0) \\ x_2(0) \end{bmatrix} = 0$$

determine $x(k)$ when the input $u(k)$ is the discrete step, that is,

$$u(k) = \begin{cases} 1 & \text{for } k \geq 0 \\ 0 & \text{for } k < 0 \end{cases}$$

Using (5.4-6), we obtain

$$\begin{bmatrix} X_1(z) \\ X_2(z) \end{bmatrix} = \begin{bmatrix} z + .5 & .3 \\ -.2 & z \end{bmatrix}^{-1} \left\{ \begin{bmatrix} 1 \\ 0 \end{bmatrix} \frac{z}{z-1} \right\}$$

Proceeding with the inverse, we have

$$\begin{bmatrix} X_1(z) \\ X_2(z) \end{bmatrix} = \frac{1}{z^2 + .5z + .6} \begin{bmatrix} z & -.3 \\ .2 & z + .5 \end{bmatrix} \begin{bmatrix} 1 \\ 0 \end{bmatrix} \frac{z}{z-1}$$

Upon multiplication

$$\begin{bmatrix} X_1(z) \\ X_2(z) \end{bmatrix} = \frac{z}{(z-1)(z+.2)(z+.3)} \begin{bmatrix} z \\ .2 \end{bmatrix}$$

Thus, to determine $x_2(k)$, for example, we have to find the inverse z-transform for the expression

$$X_2(z) = \frac{.2z}{(z-1)(z+.2)(z+.3)}$$

A partial-fraction expansion is first made and results in

$$X_2(z) = \frac{a}{z-1} + \frac{b}{z+.2} + \frac{c}{z+.3}$$

where

$$a = \left. \frac{.2z}{(z+.2)(z+.3)} \right|_{z=1} = .128$$

and similarly

$$b = .333$$
$$c = -.461$$

so that

$$X_2(z) = \frac{.128}{z-1} + \frac{.333}{z+.2} - \frac{-.461}{z+.3}$$

After inverting, we have

$$x_2(k) = .128 + .333(-.2)^{k-1} - .461(-.3)^{k-1} \quad \text{for } k \geq 1$$

Note that both $x_2(0)$ and $x_2(1)$ are zero.

In order to determine the relationship between the time-domain and z-transform approaches, let us postulate the following problem. Suppose that we are given an initial state $\mathbf{x}(0)$, and the vector inputs $\mathbf{u}(0), \mathbf{u}(1), \ldots, \mathbf{u}(N-1)$; what is the state at discrete time N? Successively applying (5.4-1), we have

$$\mathbf{x}(1) = \mathbf{A}\mathbf{x}(0) + \mathbf{B}\mathbf{u}(0)$$
$$\mathbf{x}(2) = \mathbf{A}\mathbf{x}(1) + \mathbf{B}\mathbf{u}(1) = \mathbf{A}^2\mathbf{x}(0) + \mathbf{A}\mathbf{B}\mathbf{u}(0) + \mathbf{B}\mathbf{u}(1)$$
$$\mathbf{x}(3) = \mathbf{A}\mathbf{x}(2) + \mathbf{B}\mathbf{u}(2) = \mathbf{A}^3\mathbf{x}(0) + \mathbf{A}^2\mathbf{B}\mathbf{u}(0) + \mathbf{A}\mathbf{B}\mathbf{u}(1) + \mathbf{B}\mathbf{u}(2)$$

or in general

$$\mathbf{x}(N) = \mathbf{A}^N \mathbf{x}(0) + \sum_{i=0}^{N-1} \mathbf{A}^{N-i-1} \mathbf{B}\mathbf{u}(i) \quad \text{for } N = 1, 2, 3, \ldots \quad (5.4\text{-}7)$$

Denoting the matrix $\mathbf{A}^N$ by $\mathbf{\Phi}(N)$, called the fundamental matrix, we find that equation (5.4-7) becomes

$$\mathbf{X}(N) = \mathbf{\Phi}(N)\mathbf{x}(0) + \sum_{i=0}^{N-1} \mathbf{\Phi}(N - i - 1)\mathbf{B}\mathbf{u}(i) \quad (5.4\text{-}8)$$

The state vector $\mathbf{x}(N)$ is seen to be composed of a factor dependent only on the initial state $\mathbf{x}(0)$ and another factor dependent solely on the control input vectors $\mathbf{u}(0), \mathbf{u}(1), \ldots, \mathbf{u}(N-1)$. It is this second factor that the control engineer has the ability to specify and thereby to influence the resultant state vectors in some manner. The first factor $\mathbf{\Phi}(N)\mathbf{x}(0)$ gives the system's response if no control is applied. The output vector is

$$\mathbf{y}(N) = \mathbf{C}\mathbf{\Phi}(N)\mathbf{x}(0) + \sum_{i=0}^{N-1} \mathbf{C}\mathbf{\Phi}(N - i - 1)\mathbf{B}\mathbf{u}(i) + \mathbf{D}\mathbf{u}(N) \quad (5.4\text{-}9)$$

A comparison of (5.4-6) with (5.4-8) indicates that

$$\mathbf{\Phi}(k) = \mathscr{Z}^{-1}\{z[z\mathbf{I} - \mathbf{A}]^{-1}\} \quad (5.4\text{-}10)$$

so that the fundamental matrix is the inverse z-transform of the $n \times n$ matrix $z[z\mathbf{I} - \mathbf{A}]^{-1}$. Now,

$$[z\mathbf{I} - \mathbf{A}]^{-1} = \frac{1}{D(z)} \text{ adjoint } [z\mathbf{I} - \mathbf{A}] \qquad (5.4\text{-}11)$$

where

$$D(z) = \text{determinant of the } n \times n \text{ matrix } [z\mathbf{I} - \mathbf{A}]$$

An algorithm for determining the inverse of $[z\mathbf{I} - \mathbf{A}]$ will now be outlined.* First, express $D(z)$ as

$$D(z) = z^n - p_1 z^{n-1} - p_2 z^{n-2} - \cdots - p_n \qquad (5.4\text{-}12)$$

The adjoint of $[z\mathbf{I} - \mathbf{A}]$ may be written as

$$\text{Adjoint } [z\mathbf{I} - \mathbf{A}] = \mathbf{I} z^{n-1} + \mathbf{H}_1 z^{n-2} + \mathbf{H}_2 z^{n-3} + \cdots + \mathbf{H}_{n-1}$$
$$(5.4\text{-}13)$$

where the coefficients p_i and matrices $\mathbf{H}_i$ are obtained from the sequential relationships

$$\begin{aligned}
\mathbf{H}_1 &= \mathbf{A} - p_1 \mathbf{I}, & \text{where } p_1 &= \text{trace } \mathbf{A} \\
\mathbf{H}_2 &= \mathbf{A}\mathbf{H}_1 - p_2 \mathbf{I}, & \text{where } p_2 &= \tfrac{1}{2} \text{ trace } (\mathbf{A}\mathbf{H}_1) \\
\mathbf{H}_3 &= \mathbf{A}\mathbf{H}_2 - p_3 \mathbf{I}, & \text{where } p_3 &= \tfrac{1}{3} \text{ trace } (\mathbf{A}\mathbf{H}_2) \qquad (5.4\text{-}14) \\
&\cdots & &\cdots \\
\mathbf{H}_n &= \mathbf{A}\mathbf{H}_{n-1} - p_n \mathbf{I}, & \text{where } p_n &= \frac{1}{n} \text{ trace } (\mathbf{A}\mathbf{H}_{n-1})
\end{aligned}$$

The trace of any $n \times n$ matrix is defined to be the sum of its diagonal elements. By means of (5.4-11) through (5.4-14), it is possible systematically to evaluate $[z\mathbf{I} - \mathbf{A}]^{-1}$ from which we in turn may find $\Phi(k)$ by (5.4-10).

EXAMPLE 5.4-2

Determine the state vector representation of the discrete system characterized by

$$c(k) + ac(k-1) = u(k-2)$$

using the method of direct programming. Evaluate the resultant fundamental matrix.

*See also Appendix 2B, in which the same algorithm is outlined for the inversion of $[s\mathbf{I} - \mathbf{A}]$.

The transfer function of this system is given by

$$H(z) = \frac{C(z)}{U(z)} = \frac{1}{z(z+a)}$$

From (5.3-18), we have that $b_0 = b_1 = a_2 = 0$, $a_0 = 1$, $a_1 = a$, $b_2 = 1$. Therefore, the direct programming representation is, by (5.3-24) and (5.3-25),

$$\begin{bmatrix} x_1(k+1) \\ x_2(k+1) \end{bmatrix} = \begin{bmatrix} -a & 0 \\ 1 & 0 \end{bmatrix} \begin{bmatrix} x_1(k) \\ x_2(k) \end{bmatrix} + \begin{bmatrix} 1 \\ 0 \end{bmatrix} u(k)$$

$$y(k) = c(k) = \begin{bmatrix} 0 & 1 \end{bmatrix} \begin{bmatrix} x_1(k) \\ x_2(k) \end{bmatrix}$$

Therefore,

$$[z\mathbf{I} - \mathbf{A}] = \begin{bmatrix} z+a & 0 \\ -1 & z \end{bmatrix}$$

$$z[z\mathbf{I} - \mathbf{A}]^{-1} = \begin{bmatrix} \dfrac{z}{z+a} & 0 \\ \dfrac{1}{z+a} & 1 \end{bmatrix}$$

Thus, by (5.4-10),

$$\boldsymbol{\Phi}(k) = \mathscr{L}^{-1} \begin{bmatrix} \dfrac{z}{z+a} & 0 \\ \dfrac{1}{z+a} & 1 \end{bmatrix}$$

or

$$\boldsymbol{\Phi}(0) = \begin{bmatrix} 1 & 0 \\ 0 & 1 \end{bmatrix}$$

$$\boldsymbol{\Phi}(k) = \begin{bmatrix} a^k & 0 \\ a^{k-1} & 0 \end{bmatrix} \quad \text{for } k = 1, 2, \ldots$$

5.5 Transfer Matrix for Discrete Systems

An nth-order linear, time-invariant discrete system with m inputs, $u_i(k)$, and p outputs, $y_j(k)$ has been shown to have a state variable representation of the form

$$\mathbf{x}(k+1) = \mathbf{A}\mathbf{x}(k) + \mathbf{B}\mathbf{u}(k) \qquad (5.5\text{-}1)$$

$$\mathbf{y}(k) = \mathbf{C}\mathbf{x}(k) + \mathbf{D}\mathbf{u}(k) \qquad (5.5\text{-}2)$$

Taking the z-transform of (5.5-1) and (5.5-2) gives us

$$X(z) = [zI - A]^{-1}zx(0) + [zI - A]^{-1}BU(z)$$
$$Y(z) = CX(z) + DU(z)$$

Assuming that the initial state x(0) is zero, we find that the resultant output vector is

$$Y(z) = \{C[zI - A]^{-1}B + D\}U(z)$$
$$= H(z)U(z) \tag{5.5-3}$$

where

$$H(z) = C[zI - A]^{-1}B + D = [H_{ij}(z)] \tag{5.5-4}$$

is the $p \times m$ transfer matrix that characterizes the input-output dynamics of the given discrete system. The response of the ith output signal to the set of m inputs is, by (5.4-3),

$$Y_i(z) = \sum_{j=1}^{m} H_{ij}(z)U_j(z) \quad \text{for } i = 1, 2, \ldots, p$$

The element $H_{ij}(z)$ of the transfer matrix is the z-transform of the ith output in response to a Kronecker delta sequence applied to the jth input.

Making use of identity (5.5-3), we have

$$H(z) = C\{[zI - A]^{-1}z\}z^{-1}B + D$$

so that, by (5.4-10), the inverse z-transform of the $p \times m$ transfer matrix $H(z)$ is

$$H(k) = \mathscr{Z}^{-1}[H(z)] = C\Phi(k - 1)B + D\delta_0(k) \tag{5.5-5}$$

Since $\Phi(k) = A^k$, (5.5-5) simplifies to

$$H(0) = D \tag{5.5-6}$$
$$H(k) = CA^{k-1}B \quad \text{for } k \geq 1 \tag{5.5-7}$$

For the case when the system has one input and one output, then $H(k)$ is a 1×1 matrix, which has been called the *weighting sequence* in Chapter 2. Since $Y(z)$ is equal to the product of a transfer matrix $H(z)$ with the input vector transform $U(z)$, it may be shown that its inverse is a convolution summation, given by

$$\mathbf{y}(k) = \sum_{j=0}^{k-1} \mathbf{H}(k-j)\mathbf{u}(j) \quad \text{for } k \geq 1$$

$$\mathbf{y}(0) = \mathbf{D}\mathbf{u}(0)$$

For the purpose of comparison we identify analogous relations which exist between continuous and discrete systems in Table 5.5-1.

Table 5.5-1 Analogies between Continuous and Discrete Systems

	Continuous System	Discrete System
Dynamics	$\dot{\mathbf{x}}(t) = \mathbf{F}\mathbf{x}(t) + \mathbf{G}\mathbf{u}(t)$ $\mathbf{y}(t) = \mathbf{C}\mathbf{x}(t) + \mathbf{D}\mathbf{u}(t)$	$\mathbf{x}(k+1) = \mathbf{A}\mathbf{x}(k) + \mathbf{B}\mathbf{u}(k)$ $\mathbf{y}(k) = \mathbf{C}\mathbf{x}(k) + \mathbf{D}\mathbf{u}(k)$
Transition matrix	$\mathbf{\Phi}(t) = e^{\mathbf{F}t}$	$\mathbf{\Phi}(k) = \mathbf{A}^k$
Transform of transition matrix	$\mathbf{\Phi}(s) = [s\mathbf{I} - \mathbf{F}]^{-1}$	$\mathbf{\Phi}(z) = z[z\mathbf{I} - \mathbf{A}]^{-1}$
Transfer matrix	$\mathbf{H}(s) = \mathbf{C}\mathbf{\Phi}(s)\mathbf{G} + \mathbf{D}$	$\mathbf{H}(z) = z^{-1}\mathbf{C}\mathbf{\Phi}(z)\mathbf{B} + \mathbf{D}$
Inverse of transfer matrix	$\mathbf{H}(t) = \mathbf{C}\mathbf{\Phi}(t)\mathbf{G} + \mathbf{D}\delta(t)$	$\mathbf{H}(k) = \mathbf{C}\mathbf{\Phi}(k-1)\mathbf{B} \quad k \geq 1$ $\mathbf{H}(k) = \mathbf{D} \quad k = 0$

REFERENCES

1. DeRusso, P. M., R. J. Roy, and C. M. Close, *State Variables for Engineers*, Wiley, New York, 1967.
2. Freeman, H., *Discrete-Time Systems*, Wiley, New York, 1965.
3. Gupta, S. C., *Transform and State Variable Methods in Linear Systems*, Wiley, New York, 1966.
4. Tou, J. T., *Modern Control Theory*, McGraw-Hill, New York, 1964.
5. Timothy, L. K. and B. E. Bona, *State Space Analysis: An Introduction*, McGraw-Hill, New York, 1968.

PROBLEMS

5.1 Under what conditions are the matrices in equations (5.2-17) and (5.2-19) identical?

5.2 Determine discrete state variable representations for the transfer frunctions

(a) $G(z) = \dfrac{2 + z^{-1}}{1 + z^{-1}}$

(b) $G(z) = \dfrac{5z}{z^2 + 2z + 2}$

Use all four techniques.

5.3 Determine the inverse of the matrix $[z\mathbf{I} - \mathbf{A}]$ by the inversion algorithm when

$$\mathbf{A} = \begin{bmatrix} .2 & -.3 \\ 1 & 0 \end{bmatrix}$$

5.4 Verify that

$$\mathbf{\Phi}(k) = \mathscr{Z}^{-1}\{z[z\mathbf{I} - \mathbf{A}]^{-1}\}$$

for the matrix $\mathbf{A}$ given in Problem 5.3.

5.5 By use of the z-transform determine the step response of the system described by the state equations

$$\begin{bmatrix} x_1(k+1) \\ x_2(k+1) \end{bmatrix} = \begin{bmatrix} -.7 & 1 \\ -.12 & 0 \end{bmatrix} \begin{bmatrix} x_1(k) \\ x_2(k) \end{bmatrix} + \begin{bmatrix} 1 \\ 0 \end{bmatrix} u(k)$$

$$y(k) = \begin{bmatrix} 1 & -1 \end{bmatrix} \begin{bmatrix} x_1(k) \\ x_2(k) \end{bmatrix}$$

5.6 Which one of the four methods of state variable representation lends itself to represent a multiple input-multiple output system as described by

(a) $c(k) + a_1 c(k-1) + a_2 c(k-2) = b_0 u_1(k) + b_1 u_1(k-1)$
$\qquad + b_2 u_1(k-2) + e_0 u_2(k) + e_1 u_2(k-1) + e_2 u_2(k-2)$

(b) $\begin{bmatrix} y_1(z) \\ y_2(z) \end{bmatrix} \begin{bmatrix} \dfrac{1}{1 - z^{-1}} & \dfrac{2 + z^{-1}}{1 - z - 1} \\ \dfrac{.5}{1 + .5z^{-1}} & \dfrac{.2 + z^{-1}}{1 + .5z^{-1}} \end{bmatrix} \begin{bmatrix} U_1(z) \\ U_2(z) \end{bmatrix}$

5.7 Demonstrate the derivation for nested and direct programming by using the following transfer function:

$$G(z) = \dfrac{2z^2 + z + 1}{z^2 + 4z + 3}$$

6

Analysis of Linear Discrete-time Systems: z-Domain Approach

6.1 Introduction

There are basically two approaches to the analysis of linear discrete-time systems. One is the direct use of difference equations in some form, commonly known as the time-domain approach. You were introduced to the principal time-domain methods in Chapter 3. An alternate approach through the application of the z-transform will be the objective of the present chapter.

Sampled-data theory was developed primarily for the purpose of studying those systems operating partially in discrete and partially in continuous time but whose overall dynamics may be essentially characterized by discrete-time methods. It will be recalled from Chapters 1 and 3 that a sampled-data system is one in which some, but not all, of the signals appear at one or more points in the form of a sequence of numbers. Figure 6.1-1 illustrates a typical sampled-data control system. We briefly review the operation of such a system. The function of the sampler and analog-to-digital converter (ADC) is to change the continuous input signal $e(t)$ into a sequence of numbers

$f(nT)$. Unless otherwise specified, it is assumed that

$$f(nT) = e(nT) = e(t)|_{t=nT}$$

where T is the sampling period of the ADC. We may visualize such a sampler and ADC as a switch that closes instantaneously every T seconds to generate the sequence $e(nT)$. The digital computer, operating on the sequence of numbers $f(nT)$, generates a sequence of numbers $g(nT)$. It is the function of the digital computer to operate on the $f(nT)$ so as to create some desirable effect on the control system. The hold circuit and the digital-to-analog converter (DAC) convert the sequence of numbers $g(nT)$ into a piecewise continuous function of time $h(t)$, which serves to drive the continuous plant being controlled.

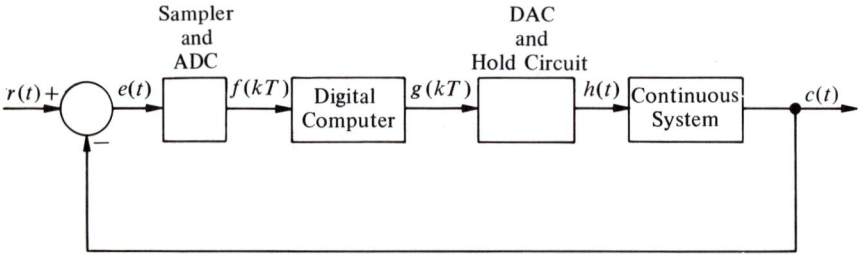

Figure 6.1-1. Typical sampled-data system.

The goal of this chapter is to develop techniques for analyzing sampled-data systems such as that shown in Figure 6.1-1 by the use of the z-transform.

To permit the z-transform analysis of sampled-data systems, we shall introduce the concept of impulse sampling, which, in conjunction with continuous systems, may be used mathematically to replace a class of samplers and digital-to-analog converters. This simplifies the resultant analysis, since we have effectively removed the hybrid DAC subsystem and replaced it with a continuous subsystem. The z-transform may then be readily applied. A study of systems that include the impulse sampler is then made. After development of these concepts, it is shown how they may be used to analyze many sampled-data systems.

6.2 Artifice of Impulse Sampling

A sampled-data system has both continuous and discrete signals present. It is advantageous to use special methods because of the existence of these basically different signals in the same system. For example, consider the hybrid subsystem shown in Figure 6.2-1. It was shown in Section 3.2 that the relationship existing between $h(t)$ and $g(nT)$ for a zero-order hold is given by

$$h(kT + \tau) = g(kT) \quad \text{for } 0 \leq \tau < T \tag{6.2-1}$$

where k takes on integer values.

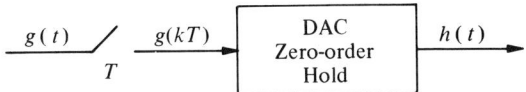

Figure 6.2-1. Hybrid subsystem.

Let us now determine the Laplace transform of the output signal $h(t)$ to a general input signal $g(t)$, which for convenience is assumed to be zero for $t < 0$. From (6.2-1), $h(t)$ is related to $g(t)$ by

$$h(t) = \sum_{n=0}^{\infty} g(kT)[u(t - kT) - u(t - kT - T)] \tag{6.2-2}$$

Taking the Laplace transform of (6.2-2) and using the property that

$$\mathscr{L}[u(t - t_0)] = \frac{e^{-st_0}}{s}$$

we arrive at the result

$$\mathscr{L}[h(t)] \equiv H(s) = \sum_{k=0}^{\infty} g(kT) \left[\frac{e^{-skT} - e^{-s(k+1)T}}{s} \right] \tag{6.2-3}$$

This expression can be written in the form

$$H(s) = \left(\frac{1 - e^{-sT}}{s} \right) \sum_{k=0}^{\infty} g(kT) e^{-skT} \tag{6.2-4}$$

by factoring out those terms that are not a function of the summation index k. It is noted that (6.2-4) is the product of two transform functions; that is,

$$H(s) = T(s)G^*(s) \tag{6.2-5}$$

where

$$T(s) = \frac{1 - e^{-sT}}{s} \tag{6.2-6}$$

$$G^*(s) = \sum_{k=0}^{\infty} g(kT) e^{-skT} \tag{6.2-7}$$

This product form is suggestive of a transfer function relationship. Since $G^*(s)$ is a function of the input signal $g(t)$, it is natural to think of $T(s)$ as the transfer function relating the transform of the output signal $H(s)$ to the input signal $G^*(s)$. $G^*(s)$ as given by (6.2-7) can be shown to be generated from the time function

$$g^*(t) = \sum_{k=0}^{\infty} g(t)\delta(t - kT) \qquad (6.2\text{-}8)$$

This follows since

$$\mathscr{L}[g^*(t)] = G^*(s) = \int_0^\infty \sum_{k=0}^{\infty} g(t)\delta(t - kT)e^{-st} dt$$

$$G^*(s) = \sum_{k=0}^{\infty} \int_0^\infty g(t)e^{-st}\delta(t - kT) dt$$

$$= \sum_{k=0}^{\infty} g(kT)e^{-skT}$$

where it is assumed that the operations of integration and summation may be interchanged. Thus the result given in (6.2-7) is obtained. The function $g^*(t)$ given in (6.2-8) is a sequence of weighted Dirac delta (impulse) functions separated by T-second intervals. One may visualize (6.2-8) as being generated by the fictitious system shown in Figure 6.2-2. The switch shown

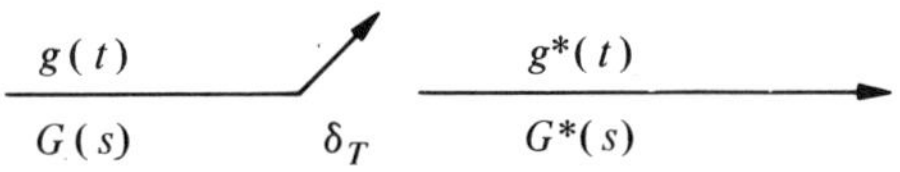

Figure 6.2-2. Impulse sampler.

is to be thought of as closing instantaneously every T seconds and generating the impulses $g(kT)\delta(t - kT)$. It is impossible to implement a device that generates impulses.[†] However, the impulse sampler is introduced for strictly mathematical conveniences. This is an extremely important concept to bear in mind. Many works on sampled-data theory do not stress the fact that the impulse sampler is utilized only for analysis purposes, thereby leaving the student with the false impression that such samplers are subsystems of actual sampled-data systems.

It has been demonstrated that the system shown in Figure 6.2-1 has the same output response given by (6.2-4) as does a system with the transfer function $T(s)$ of (6.2-6) being driven by the input $g^*(t)$ of (6.2-8). These two

---

[†]Recall the simulation of a sampler by an analog computer, as shown in Chapter 3, Figures 3.3-5 and 3.3-6. It illustrates well the principle upon which the design of commercially available sampler and sample-and-hold amplifiers is based.

systems are therefore mathematically equivalent with regard to their input-output relationship. This equivalence is shown in Figure 6.2-3. In Figure 6.2-3(a), the switch closes and opens instantaneously every T seconds to generate the sequence of numbers $g(nT)$. The switch shown in Figure 6.2-3(b) has the input-output relationship

$$g^*(t) = \sum_{k=-\infty}^{\infty} g(t)\delta(t - kT) \qquad (6.2\text{-}9)$$

which is a time-weighted sequence of impulses. The sampling process of (6.2-9) is called *impulse sampling* for obvious reasons. It is important that one understand the basic difference between these two sampling processes. For illustrative purposes, the two sampling processes are distinguished by the manner in which the sampling switches are drawn, as shown in Figure 6.2-3.

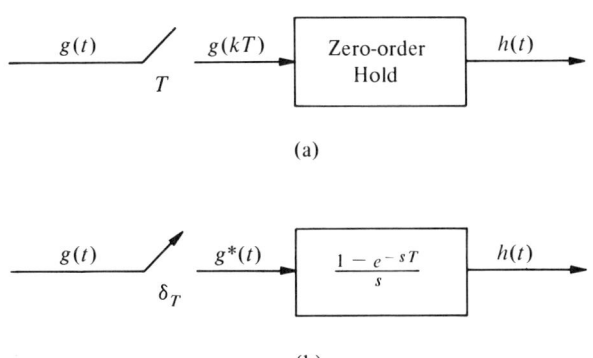

Figure 6.2-3. Equivalent systems: (a) actual system; (b) mathematical equivalent system.

Since the first-order hold circuit is used in many systems, it would be desirable to replace it by a mathematically equivalent continuous system. The characteristic of such a hold circuit is, by (3.2-5),

$$h(kT + \tau) = \frac{g(kT) - g([k-1]T)}{T}\tau + g(kT) \quad \text{for } 0 \leq \tau < T \qquad (6.2\text{-}10)$$

A first-order hold has a response in the time interval $kT \leq t < (k+1)T$, which is seen to depend on the samples $g(kT)$ and $g([k-1]T)$. Let the input

$$r(t) = g(kT)\delta(t - kT) + g([k-1]T)\delta(t - [k-1]T)$$

be fed into a system with transfer function

$$\frac{C(s)}{R(s)} = G_1(s) = \frac{Ts+1}{T}\left(\frac{1-e^{-sT}}{s}\right)^2 \qquad (6.2\text{-}11)$$

The system is assumed to be at rest before $r(t)$ is applied. The Laplace transform of $r(t)$ is easily shown to be

$$R(s) = g(kT)e^{-ksT} + g([k-1]T)e^{-(k-1)sT}$$

Therefore,

$$C(s) = \frac{Ts+1}{Ts^2}[1 - 2e^{-sT} + e^{-2sT}][g(kT)e^{-ksT} + g([k-1]T)e^{-(k-1)sT}]$$

$$= \frac{Ts+1}{Ts^2}[g(kT)\{e^{-ksT} - 2e^{-(k+1)sT} + e^{-(k+2)sT}\}$$

$$+ g([k-1]T)\{e^{-(k-1)sT} - 2e^{-ksT} + e^{-(k-1)sT}\}]$$

Using the basic property from Laplace transform theory, which states that

$$\mathscr{L}^{-1}[F(s)e^{-sT}] = f(t-T)u(t-T)$$

where

$$\mathscr{L}[f(t)] = F(s)$$

we find the inverse Laplace transform of $C(s)$.

$$c(t) = \frac{g(kT) - g([k-1]T)}{T}(t - kT) + g(kT) \quad \text{for } kT \le t < (k+1)T$$

$c(t) = 0$ elsewhere

This is in agreement with the response of the first-order data hold of (6.2-10). Therefore, the system of (6.2-11) in conjunction with an impulse sampler is

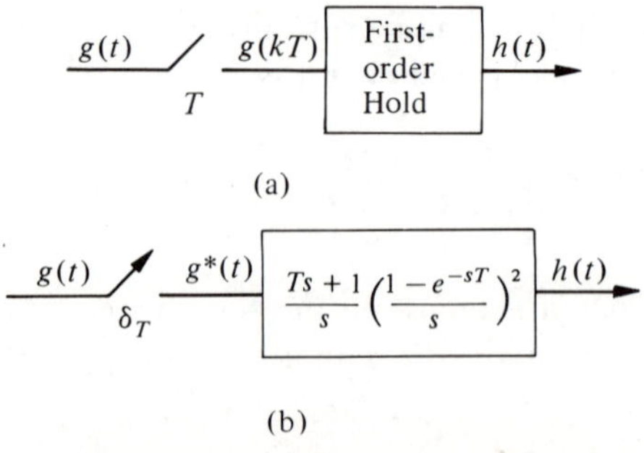

$$\xrightarrow{g(t)} \underset{T}{\diagup} \xrightarrow{g(kT)} \boxed{\begin{array}{c}\text{First-}\\\text{order}\\\text{Hold}\end{array}} \xrightarrow{h(t)}$$

(a)

$$\xrightarrow{g(t)} \underset{\delta_T}{\diagup} \xrightarrow{g^*(t)} \boxed{\frac{Ts+1}{s}\left(\frac{1-e^{-sT}}{s}\right)^2} \xrightarrow{h(t)}$$

(b)

Figure 6.2-4. Equivalent systems.

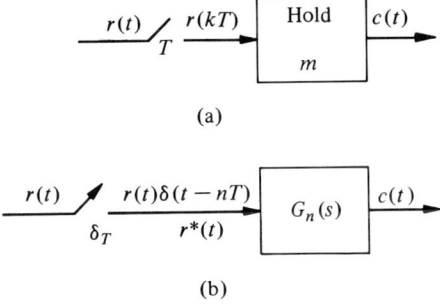

Figure 6.2-5. (a) Actual hold circuit; (b) equivalent subsystem.

seen to be equivalent to a first-order hold. This equivalence is shown in Figure 6.2-4. For higher-order hold circuits it will be found that equivalences similar to those shown in Figures 6.2-3 and 6.2-4 may be found. In general, if the hold circuit is an mth-order polynomial extrapolator, then its equivalent representation will be depicted as shown in Figure 6.2-5.

6.3 Systems with Impulse Sampling

By the artifice of impulse sampling, we may replace a hybrid component such as a digital-to-analog converter with an equivalent continuous component in conjunction with an impulse sampler. Equivalent continuous components are more amenable to classical analysis methods such as Laplace and z-transforms. It is, therefore, essential that one understand the basic characteristics of continuous systems driven by a train of impulses.

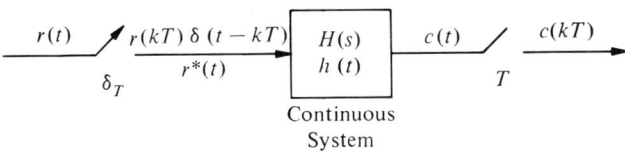

Figure 6.3-1. System with impulse sampling.

Consider the system shown in Figure 6.3-1. It is assumed that the impulse-generating switch and the output ADC switch are synchronized. The continuous system is defined to have the physically realizable impulse response $h(t)$ [i.e., $h(t) = 0$ for $t < 0$]. When the input $r^*(t)$ is applied to this continuous system, where

$$r^*(t) = \sum_{k=0}^{\infty} r(t)\delta(t - kT) \tag{6.3-1}$$

the corresponding output is obtained by utilizing the convolution integral relating the input and output signals for linear continuous systems. This results in

$$c(t) = \sum_{k=0}^{\infty} r(kT)h(t - kT)$$

It has been assumed for convenience in (6.3-1) that $r(t) = 0$ for negative t. At the sampling times $t = nT$, $c(t)$ is given by

$$c(nT) = \sum_{k=0}^{\infty} r(kT)h(nT - kT)$$

which is in the form of a convolution summation. The z-transform of this output sequence $c(nT)$ has been previously shown to be

$$C(z) = R(z)H(z) \qquad (6.3\text{-}2)$$

where $R(z)$ and $H(z)$ are the z-transforms of the sequences $r(kT)$ and $h(kT)$, respectively.

To obtain $H(z)$, given $H(s)$, one can use either of two techniques, which will be called the *direct* and *indirect methods*. Since the direct method requires essentially only a background in Laplace transform theory, its use is recommended for the undergraduate as well as for the graduate student.

6.3-1 Direct Method

The direct method is a very logical and straightforward process. Let $h(t)$ be the inverse Laplace transform of $H(s)$. To determine $H(z)$, we evaluate $h(t)$ at the sampling times and take the z-transform of this resultant sequence; that is,†

$$H(z) = \sum_{k=0}^{\infty} h(kT)z^{-k}$$

where

$$h(t) = \mathscr{L}^{-1}[H(s)]$$

An example will illustrate the steps required in using the direct method.

EXAMPLE 6.3-1

Determine $H(z)$ corresponding to

$$H(s) = \frac{a}{s(s + a)}$$

†An analogous development may be made for two-sided Laplace transform functions.

Sec. 6.3 Systems with Impulse Sampling

By taking the inverse Laplace transform, we obtain

$$h(t) = \mathscr{L}^{-1}[H(s)] = \mathscr{L}^{-1}\left[\frac{1}{s} - \frac{1}{s+a}\right]$$
$$= (1 - e^{-at})u(t)$$

so that

$$h(kT) = 1 - e^{-akT} \quad \text{for } k \geq 0$$
$$h(kT) = 0 \quad \text{for } k < 0$$

The z-transform of this sequence is given by

$$H(z) = \sum_{k=0}^{\infty}(1 - e^{-akT})z^{-k}$$
$$= \frac{1}{1-z^{-1}} - \frac{1}{1-e^{-aT}z^{-1}}$$
$$= \frac{z^{-1}(1-e^{-aT})}{(1-z^{-1})(1-e^{-aT}z^{-1})} \quad \text{for } |z| > 1$$

The use of the direct method is considerably enhanced by the use of tables that match up Laplace transforms and z-transforms of corresponding sampled-time functions. Table 6.3-1 gives such relationships for frequently used Laplace transforms.

Table 6.3-1 Laplace z-Transform Pairs

$F_s(s)$	$F(z)$
$\dfrac{1}{s}$	$\dfrac{1}{1-z^{-1}}$
$\dfrac{1}{s^2}$	$\dfrac{T}{(1-z^{-1})^2}$
$\dfrac{1}{s^3}$	$\dfrac{T^2 z^{-1}(1+z^{-1})}{(1-z^{-1})^3}$
$\dfrac{1}{s+a}$	$\dfrac{1}{1-z^{-1}e^{-aT}}$
$\dfrac{1}{(s+a)^2}$	$\dfrac{Tz^{-1}e^{-aT}}{(1-z^{-1}e^{-aT})^2}$
$\dfrac{a}{s(s+a)}$	$\dfrac{(1-e^{-aT})z^{-1}}{(1-z^{-1})(1-z^{-1}e^{-aT})}$
$\dfrac{a}{s^2(s+a)}$	$\dfrac{Tz^{-1}}{(1-z^{-1})^2} \dfrac{(1-e^{-aT})z^{-1}}{a(1-z^{-1})(1-z^{-1}e^{-aT})}$

6.3-2 Indirect Method

This method is not recommended for students who do not have a basic understanding of complex variable theory. Such students should rely solely on the direct method.

Theorem 6.3-1. The transform of the sequence $h(0), h(T), h(2T), \ldots$, where

$$h(kT) = h(t)|_{t=kT}$$

and $h(t) = \mathcal{L}^{-1}[H(s)]$ for rational $H(s)$ is given by

$$\mathcal{L}[h(kT)] = \left[\text{sum of residues of } \frac{H(p)}{1 - z^{-1}e^{pT}} \text{ at poles of } H(p)\right] + \alpha \quad (6.3\text{-}3)$$

where α is selected so that

$$\lim_{s \to \infty} sH(s) = \lim_{z \to \infty} \mathcal{L}[h(kT)]$$

This relationship guarantees that the initial values $h(0)$ of $H(s)$ and $H(z)$ agree.

Proof. The proof is contained in Appendix 6A.

EXAMPLE 6.3-2

Determine $H(z)$ for $H(s) = 1/(s + a)$ by the direct and the indirect methods.

(i) $h(t) = \mathcal{L}^{-1}[H(s)] = e^{-at}$

Then

$$h(kT) = e^{-akT} \quad \text{for } k = 0, 1, 2, \ldots$$

Therefore,

$$H(z) = \frac{1}{1 - e^{-aT}z^{-1}}$$

(ii) $\hat{H}(z) = \left[\text{residue of } \frac{1}{(p+a)[1 - z^{-1}e^{pT}]} \text{ at pole } p = -a\right]$

$$= \frac{1}{1 - z^{-1}e^{-aT}}$$

Now,

$$h(0) = \lim_{t \to \infty} h(t) = \lim_{s \to \infty} sH(s) = 1$$

Similarly,

$$\hat{h}(0) = h(nT)|_{n=0} = \lim_{z \to \infty} H(z) = 1$$

and the initial value checks. Therefore,

$$H(z) = \frac{1}{1 - z^{-1}e^{-aT}}$$

By the introduction of the fictitious impulse sampler it is possible to analyze some relatively general sampled-data systems. This is perhaps best illustrated by considering the system shown in Figure 6.3-2. Let us determine

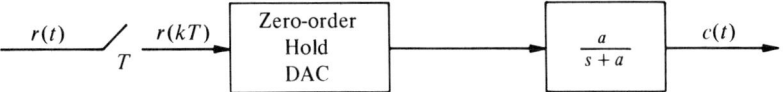

Figure 6.3-2. Simple sampled-data system.

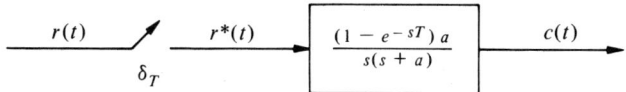

Figure 6.3-3. Equivalent system.

the value of the output signal $c(t)$ at the sampling times, which results from a general input $r(t)$. Because of the equivalences established in Section 6.2, the system shown in Figure 6.3-3 is mathematically equivalent to the actual system of Figure 6.3-2. But the system of Figure 6.3-3 is of the form given in Figure 6.3-1, for which it was shown that

$$C(z) = H(z)R(z)$$

where

$$R(z) = \mathscr{Z}[r(kT)], \qquad C(z) = \mathscr{Z}[c(kT)]$$

$$H(s) = \frac{(1 - e^{-sT})a}{s(s + a)} \tag{6.3-4}$$

We can determine the $H(z)$ that corresponds to (6.3-4) by using either the direct or indirect methods. Both methods will be used to point up the basic difference in complexities of each.

(a) *Direct Method*

$$h(t) = \mathscr{L}^{-1}\left[\frac{(1 - e^{-sT})a}{s(s + a)}\right] = \mathscr{L}^{-1}\left[\left(\frac{1}{s} - \frac{1}{s + a}\right)(1 - e^{-sT})\right]$$
$$= (1 - e^{-aT})u(t) - (1 - e^{-a(t-T)})u(t - T)$$

so that

$$h(kT) = (1 - e^{-akT})u(kT) - (1 - e^{-a(k-1)T})u(kT - T)$$

Therefore,

$$H(z) = \frac{1}{1 - z^{-1}} - \frac{1}{1 - e^{-aT}z^{-1}} - z^{-1}\left(\frac{1}{1 - z^{-1}}\right) - z^{-1}\left(\frac{1}{1 - e^{-aT}z^{-1}}\right)$$

$$= \frac{z^{-1}(1 - e^{-aT})}{(1 - e^{-aT}z^{-1})}$$

(b) *Indirect Method*

By (6A-8), we have

$$\hat{H}(z) = \left[\text{sum of residues of } \frac{a(1 - e^{-pT})}{p(p + a)(1 - z^{-1}e^{pT})} \text{ at poles } p = 0, a\right]$$

$$= \left[\frac{e^{aT} - 1}{1 - z^{-1}e^{-aT}}\right] \tag{6.3-5}$$

From (6.3-4), we the find value of $h(t)$ at $t = 0$, using the initial value theorem of Laplace transforms.

$$h(0) = \lim_{s \to \infty} sH(s) = 0$$

However, $\hat{h}(kT)$ at $k = 0$ does not agree since

$$\hat{h}(0T) = \lim_{z \to \infty} \hat{H}(z) = e^{aT} - 1$$

so we have to subtract this term from (6.3-5) in order that the initial value may be correct. Therefore,

$$H(z) = (1 - e^{-aT})\left[\frac{z^{-1}}{1 - z^{-1}e^{-aT}}\right]$$

which is the z-transform of the sequence $h(nT)$, where

$$h(t) = \mathcal{L}^{-1}\left[\frac{(1 - e^{-sT})a}{s(s + a)}\right]$$

Thus, we have

$$C(z) = (1 - e^{-aT})\left[\frac{z^{-1}}{1 - z^{-1}e^{-aT}}\right]R(z) \tag{6.3-6}$$

For any given input $r(t)$, the output $c(t)$ at $t = kT$ may be evaluated by taking the inverse of (6.3-6). For example, if

$$r(t) = u(t)$$

then

$$r(kT) = 1 \quad \text{for } k \geq 0$$
$$r(kT) = 0 \quad \text{for } k < 0$$

whose z-transform is

$$R(z) = \frac{1}{1 - z^{-1}}$$

$C(z)$ is given by

$$C(z) = (1 - e^{-aT}) \frac{z^{-1}}{(1 - z^{-1}e^{-aT})(1 - z^{-1})}$$

which, when a partial-fraction expansion is made, gives

$$C(z) = (1 - e^{-aT}) \left[\frac{(1 - e^{-aT})^{-1}}{1 - z^{-1}} + \frac{e^{aT}}{1 - e^{aT}} \frac{1}{1 - z^{-1}e^{-aT}} \right]$$

Taking the inverse z-transform of this expression results in

$$c(kT) = 1 - e^{-akT} \quad \text{for } k = 0, 1, 2, \ldots$$

6.4 The Transfer Function of a Digital Computer

Design of modern systems involves the use of a digital computer as an active component. Under those circumstances the digital computer is said to operate on-line with a system. To make a z-transform analysis possible for such a system, we must consider a suitable model describing the digital computer in terms of the z-transform. Chapter 2 presented a difference equation called a linear recursion equation, which is characteristic of the input-output sequence relationship for a digital computer. We shall show here how this linear recursion equation may be expressed as a digital transfer function.

Let the input sequence be $r(k)$ and the output sequence be $c(k)$, as shown in Figure 6.4-1. Equation 2.2-3 gave the relationship, between $c(k)$ and $r(k)$ as

Figure 6.4-1. Digital computer in on-line operation.

$$c(k) = b_0 r(k) + b_1 r(k-1) + \ldots + b_n r(k-n)$$
$$- a_1 c(k-1) - a_2 c(k-2) - \ldots - a_n c(k-n) \quad (6.4\text{-}1)$$

Taking the z-transform yields

$$C(z) = b_0 R(z) + b_1 z^{-1} R(z) + \ldots + b_n z^{-n} R(z)$$
$$- a_1 z^{-1} C(z) - a_2 z^{-2} C(z) - \ldots - a_n z^{-n} C(z)$$

Solving for $C(z)$, we have

$$C(z) = \frac{b_0 + b_1 z^{-1} + \ldots + b_n z^{-n}}{1 + a_1 z^{-1} + a_2 z^{-2} + \ldots + a_n z^{-n}} R(z) \quad (6.4\text{-}2)$$

This expression is representative of a transfer function relation; i.e.,

$$C(z) = D(z) R(z) \quad (6.4\text{-}3)$$

where

$$D(z) = \frac{b_0 + b_1 z^{-1} + \ldots + b_n z^{-n}}{1 + a_1 z^{-1} + \ldots + a_n z^{-n}} \quad (6.4\text{-}4)$$

is called the *digital transfer function*.

With the digital transfer function defined as by (6.4-4), it is possible to apply a z-transform analysis to a discrete-time system that includes a digital computer.

6.5 Typical Sampled-data Systems

With the concepts developed in Sections 6.1 through 6.4, it is possible to analyze many discrete-time systems by the z-transform. A subsystem that appears frequently in sampled-data systems is shown in Figure 6.5-1(a).

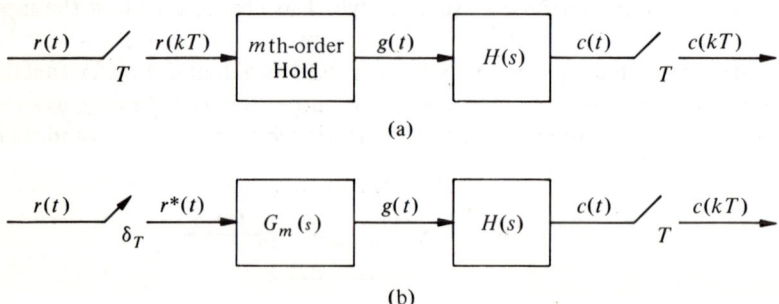

Figure 6.5-1. (a) Typical subsystem; (b) its equivalent.

Under the assumption that the hold circuit is an mth-order polynomial extrapolator as presented in Section 3.2, the fictitious system shown in Figure 6.5-1(b) is mathematically equivalent to the actual system of Figure 6.5-1(a). A comparison of Figure 6.5-1(b) with Figure 6.3-1 shows them to be of the same form. Thus, using the result obtained in (6.3-2), we have

$$C(z) = \mathscr{Z}[G_m(s)H(s)]R(z) \tag{6.5-1}$$

where

$$C(z) = \mathscr{Z}[c(kT)]$$
$$R(z) = \mathscr{Z}[r(kT)]$$
$$\mathscr{Z}[G_m(s)H(s)] = \mathscr{Z}[f(kT)] \quad \text{with } f(t) = \mathscr{L}^{-1}[G_m(s)H(s)]$$

From Theorem (6.3-1), it follows that

$$\mathscr{Z}[G_m(s)H(s)]$$
$$= \left[\text{sum of residues of } \frac{G_m(p)H(p)}{1 - z^{-1}e^{pT}} \text{ at poles of } G_m(p)H(p) \right] + \alpha \tag{6.5-2}$$

Since the zero- and first-order holds are predominantly used in sampled-data systems, we shall now investigate them, using the direct method developed in Section 6.3.

Case 1. Zero-order Hold

$$G_0(s) = \frac{1 - e^{-sT}}{s}$$

Therefore,

$$\mathscr{Z}[G_0(s)H(s)] = \mathscr{Z}\left[\frac{H(s)}{s} - \frac{e^{-sT}H(s)}{s}\right]$$

If

$$h_1(t) = \mathscr{L}^{-1}\left[\frac{H(s)}{s}\right]$$

then, applying the shifting theorem of Laplace transform theory, we have

$$\mathscr{L}^{-1}\left[\frac{e^{-sT}H(s)}{s}\right] = h_1(t - T)$$

so that

$$\mathscr{Z}\left[\frac{e^{-sT}H(s)}{s}\right] = \mathscr{Z}[h_1(kT - T)]$$
$$= z^{-1}\mathscr{Z}[h_1(kT)]$$
$$= z^{-1}\mathscr{Z}\left[\frac{H(s)}{s}\right]$$

Therefore

$$[G_0(s)H(s)] = (1 - z^{-1})\mathscr{Z}\left[\frac{H(s)}{s}\right] \quad (6.5\text{-}3)$$

Case 2. First-order Hold

$$G_1(s) = \frac{Ts + 1}{T}\left(\frac{1 - e^{-sT}}{s}\right)^2$$

Therefore,

$$\mathscr{Z}[G_1(s)H(s)] = \mathscr{Z}\left[\frac{Ts + 1}{Ts^2}(1 - e^{-sT})^2 H(s)\right]$$

Applying the technique used in deriving (6.5-3) results in

$$\mathscr{Z}[G_1(s)H(s)] = (1 - z^{-1})^2 \mathscr{Z}\left[\frac{(Ts + 1)H(s)}{Ts^2}\right] \quad (6.5\text{-}4)$$

Another subsystem that appears frequently in many discrete systems is shown in Figure 6.5-2. The purpose of the digital computer is to operate

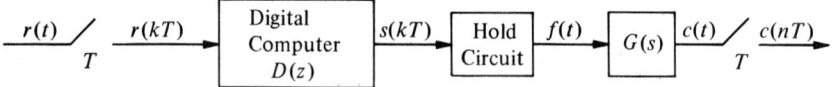

Figure 6.5-2. Computer controlled subsystem.

on the sampled sequence $r(kT)$ in some desirable manner. More will be said in the next chapter about how one synthesizes the transfer function $D(z)$ of a digital computer. It will be assumed that the digital computer operates on the sequence $r(kT)$ in a linear, time-invariant manner to generate $s(kT)$ so that the relationship between $s(kT)$ and $r(kT)$ is of the form

$$s(nT) = \sum_{k=-\infty}^{\infty} d(nT - kT)r(kT) \quad (6.5\text{-}5)$$

If the hold circuit is an mth-order polynomial extrapolator, it may be equivalently replaced by a system with transfer function $G_m(s)$ driven by impulses $s(nT)\delta(t - nT)$. The relationship

$$C(z) = \mathscr{Z}[G_m(s)G(s)]S(z) \quad (6.5\text{-}6)$$

where

$$C(z) = \mathscr{Z}[c(kT)]$$
$$S(z) = \mathscr{Z}[s(kT)]$$

has been established in Section 6.3. Similarly, we have

$$S(z) = D(z)R(z) \tag{6.5-7}$$

Inserting (6.5-7) into (6.5-6) gives the relationship between the input $r(kT)$ and the output $c(kT)$ for the computer-controlled subsystem of Figure 6.5-2; that is,

$$C(z) = D(z)\mathscr{Z}[G_m(s)G(s)]R(z) \tag{6.5-8}$$

It is apparent from (6.5-8) that the presence of the digital computer allows the engineer much design freedom. By properly selecting the poles and zeros of the digital computer transfer function $D(z)$, he may generate a wide variety of input-output characteristics. The discrete transfer function of the system shown in Figure 6.5-2 is

$$D(z)\mathscr{Z}[G_m(s)G(s)]$$

For continuous systems, the use of feedback techniques gives many desirable overall system characteristics. Likewise, sampled-data feedback systems have many similar properties. Consider the error-sampled feedback system shown in Figure 6.5-3.

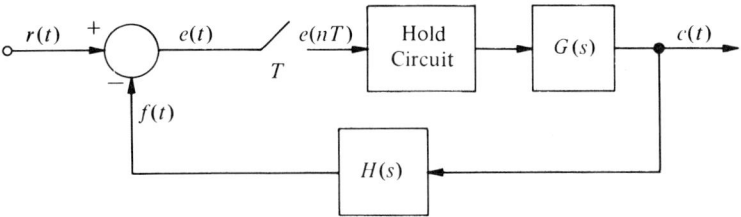

Figure 6.5-3. Error-sampled feedback system.

Using the results previously established in this section, we have

$$C(z) = \mathscr{Z}[G_m(s)G(s)]E(z) \tag{6.5-9}$$
$$F(z) = \mathscr{Z}[G_m(s)G(s)H(s)]E(z) \tag{6.5-10}$$

where

$$F(z) = \mathscr{Z}[f(kT)]$$
$$E(z) = \mathscr{Z}[e(kT)]$$

and $G_m(s)$ is the equivalent transfer function of the hold circuit. Since

$$e(kT) = r(kT) - f(kT)$$

we have

$$E(z) = R(z) - F(z) \qquad (6.5\text{-}11)$$

Combining (6.5-10) and (6.5-11) gives

$$\frac{E(z)}{R(z)} = \frac{1}{1 + \mathscr{L}[G_m(s)G(s)H(s)]} \qquad (6.5\text{-}12)$$

which is the transfer function relating $e(kT)$ and $r(kT)$. Substituting (6.5-12) into (6.5-9) gives the relationship between the input and output signals at the sampling times; that is,

$$\frac{C(z)}{R(z)} = \frac{\mathscr{L}[G_m(s)G(s)]}{1 + \mathscr{L}[G_m(s)G(s)H(s)]} \qquad (6.5\text{-}13)$$

EXAMPLE 6.5-1

For the system shown in Figure 6.5-4, determine the transfer function relating the input and output signals at the sampling times.

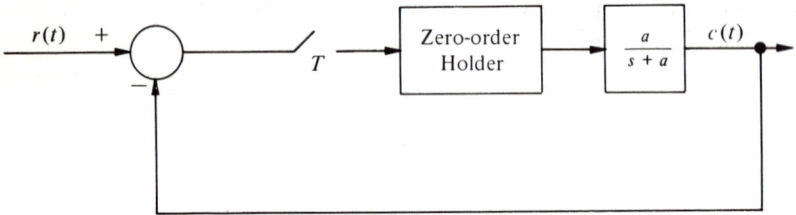

Figure 6.5-4. Sampled-data control system of Example 6.5-1.

By (6.5-13), we have

$$\frac{C(z)}{R(z)} = \frac{\mathscr{L}\left[G_0(s)\dfrac{a}{s+a}\right]}{1 + \mathscr{L}\left[G_0(s)\dfrac{a}{s+a}\right]}$$

and from the results of Section 6.3, we have

$$\mathscr{L}\left[G_0(s)\frac{a}{s+a}\right] = (1 - e^{-aT})\left[\frac{z^{-1}}{1 - z^{-1}e^{-aT}}\right]$$

Therefore,

$$\frac{C(z)}{R(z)} = \frac{(1 - e^{-aT})z^{-1}}{1 + (1 - 2e^{-aT})z^{-1}} \qquad (6.5\text{-}14)$$

EXAMPLE 6.5-2

Calculate the response of the system shown in Figure 6.5-5 to a step input. The sampling rate is $T = 1$ second. Note that this is Example 3.4-1.

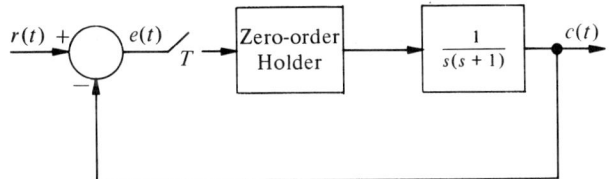

Figure 6.5-5. Sampled system with unity feedback.

By (6.5-13), we have

$$C(z) = \frac{\mathscr{L}\left[G_0(s)\frac{1}{s(s+1)}\right]}{1 + \mathscr{L}\left[G_0(s)\frac{1}{s(s+1)}\right]} R(z) \quad (6.5\text{-}15)$$

By (6.5-3), we have

$$\mathscr{L}\left[G_0(s)\frac{1}{s(s+1)}\right] = (1 - z^{-1})\mathscr{L}\left[\frac{1}{s^2(s+1)}\right]$$

$$= (1 - z^{-1})\mathscr{L}\left\{\frac{1}{s^2} - \frac{1}{s} + \frac{1}{s+1}\right\}$$

From the z-transform table we have

$$\mathscr{L}\left[G_0(s)\frac{1}{s(s+1)}\right] = (1 - z^{-1})\left[\frac{Tz}{(z-1)^2} - \frac{z}{z-1} + \frac{z}{z-e^{-T}}\right]$$

For $T = 1$ this can be simplified to

$$\mathscr{L}\{G_0 G(s)\} = \frac{.368z + .264}{(z-1)(z-.368)} \quad (6.5\text{-}16)$$

Now

$$R(z) = \frac{z}{z-1}$$

Therefore the z-transform of the output is

$$C(z) = \frac{(.368z + .264)z}{(z-1)(z^2 - z + .632)} \quad (6.5\text{-}17)$$

The poles of this transfer function are at

$$z = 1$$
$$z = .5 \pm j.615$$

Because two poles are complex, we expect that the sequence $c(nT)$ will contain an oscillatory component. Furthermore, we can determine the steady-state value of $c(nT)$ by application of the final-value theorem. From (4.5-6) we have

$$c(\infty) = \lim_{z \to 1} (z - 1)C(z)$$
$$= \frac{.368 + .264}{1 - 1 + .632} = 1$$

This indicates that the system will respond to the step input with zero steady-state error. To compute the actual output sequence we expand $C(z)$ into a power series in z^{-1}. This gives

$$C(z) = .368z^{-1} + 1.000z^{-2} + 1.399z^{-3} + 1.399z^{-4}$$
$$+ 1.147z^{-5} + .894z^{-6} + \dots$$

Thus

$$c(nT) = \{.368, 1.000, 1.399, 1.399, 1.147, .894, \dots\}$$

which agrees with Example 3.4-1.

For any input $r(t)$, the output at the sampling instants may be evaluated by first finding $R(z)$ and then determining the inverse z-transform of $C(z)$ as given by (6.5-15).

In order to insert some digital influence into the error-sampled feedback system of Figure 6.5-3, a digital computer is frequently incorporated in the

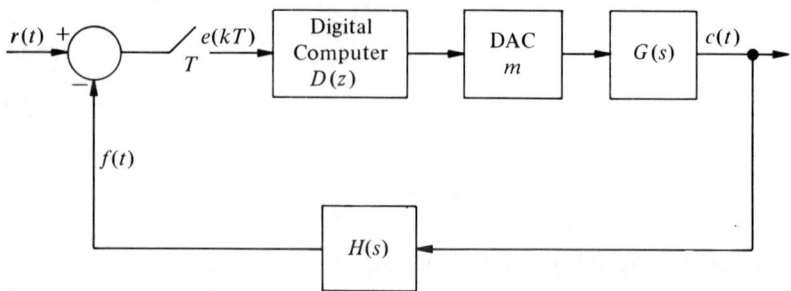

Figure 6.5-6. Digital computer controlled error-sampled feedback system.

feed forward path, as shown in Figure 6.5-6. From (6.5-8), the relationship

$$C(z) = D(z)\mathscr{Z}[G_m(s)G(s)]E(z) \tag{6.5-18}$$

was established. Similarly,

$$F(z) = D(z)\mathscr{Z}[G_m(s)G(s)H(s)]E(z) \tag{6.5-19}$$

and, since $e(kT) = r(kT) - f(kT)$, we find that

$$E(z) = R(z) - F(z) \tag{6.5-20}$$

Combining (6.5-19) with (6.5-20) results in

$$E(z) = \frac{1}{1 + D(z)\mathscr{Z}[G_m(s)G(s)H(s)]} R(z) \tag{6.5-21}$$

which, when inserted into (6.5-18), finally yields

$$\frac{C(z)}{R(z)} = \frac{D(z)\mathscr{Z}[G_m(s)G(s)]}{1 + D(z)\mathscr{Z}[G_m(s)G(s)H(s)]} \tag{6.5-22}$$

Equation (6.5-22) is the transfer function relating the z-transforms of the input and output signals at the sampling instants. Again, the digital computer's transfer function $D(z)$ plays a major role in determining the overall system's dynamics as given by (6.5-22). It is the function of the design engineer properly to select $D(z)$ in order to obtain desirable overall system characteristics. More will be said of this selection process in Chapter 7.

6.6 Stability Analysis

In Section 3.6 we presented a discussion of the stability of discrete systems. It was shown that a discrete system is stable if all the eigenvalues of the state transition matrix lie within the unit circle. This stability condition can now be restated in terms of the location of the poles of the closed-loop transfer function of the system.

Let the input-output relation of a discrete system be given by

$$C(z) = H(z)R(z) \tag{6.6-1}$$

where $H(z)$, the system transfer function, is a rational function of two polynomials in z. Let the poles of $H(z)$ be distinct and different from the poles of $R(z)$ so that the following partial-fraction expansion of (6.6-1) can be made

$$C(z) = \frac{b_1}{z - a_1} + \frac{b_2}{z - a_2} + \cdots + \frac{b_n}{z - a_n} + \text{expansion of poles of } R(z)$$

The corresponding time sequence $c(kT)$ will be

$$c(kT) = b_1 a_1^{k-1} + b_2 a_2^{k-1} + \ldots + b_n a_n^{k-1}$$
$$+ \mathscr{Z}^{-1} \{\text{expansion of poles of } R(z)\} \quad \text{for } k \geq 1 \quad (6.6\text{-}2)$$

The numbers $a_1, a_2, \ldots, a_n$ are the n simple poles of $H(z)$. It is clear from (6.6-2) that if $c(kT)$ is to remain bounded* for $k \geq 1$, the following condition must be satisfied.

$$|a_i| < 1 \quad i = 1, 2, \ldots, n \quad (6.6\text{-}3)$$

i.e., the poles of the system's transfer function must be contained within the unit circle. This condition is equivalent to (3.6-6).

Although we have considered here only the case where all poles are distinct, it can be shown that the stability criterion (6.6-3) is general and applies to systems with multiple poles.

There exist tests for determining the stability of a discrete system without finding the actual numerical values of the poles.† These tests are similar in principle to the Routh criterion for continuous systems in that they will determine whether any poles lie outside the unit circle. However, they are tedious to apply. With the availability of time-sharing computer systems, it is much easier to determine the roots of the characteristic equation directly by means of a root-solving routine. Alternatively, one may employ the digital computer to compute and plot a root-locus diagram that shows the migration of the poles of the system transfer function as one of the parameters is varied.

6.7 Relationship Between z Domain and s Domain

It will be useful to investigate the relationship between the s plane and the z plane. This relationship occurs when the impulse-sampling process is incorporated [e.g., see (6.3-6)] and is given by

$$z = e^{sT} \quad (6.7\text{-}1)$$

which relates the z domain to the s domain. In general, s is the complex variable

$$s = \sigma + j\omega \quad (6.7\text{-}2)$$

where σ is the real part and ω is the imaginary part of s.

---
*We are assuming here that the poles of $R(z)$ do not contribute any unbounded response.
†See, for instance, B. J. Kuo, *Analysis and Synthesis of Sampled-Data Control Systems*, p. 153, Prentice-Hall, Englewood Cliffs, N. J., 1963.

We have then

$$z = e^{(\sigma+j\omega)T} = e^{\sigma T}e^{j\omega T}$$

We see, therefore, that z is a complex variable with magnitude

$$|z| = e^{\sigma T} \quad (6.7\text{-}3)$$

and phase angle

$$\angle z = \omega T \quad (6.7\text{-}4)$$

From (6.7-3) we note that σ bears the following relationship to $|z|$:

$$\begin{array}{ll} \sigma > 0 & |z| > 1 \\ \sigma = 0 & |z| = 1 \\ \sigma < 0 & |z| < 1 \end{array} \quad (6.7\text{-}5)$$

We conclude that the following relationship exists between the s plane and the z plane.

s plane	maps into	z plane
Entire left half plane	$\longleftrightarrow$	Interior of the unit circle
Imaginary axis	$\longleftrightarrow$	Circumference of unit circle
Entire right half plane	$\longleftrightarrow$	Exterior of the unit circle

From (6.7-4) we see that whenever ω changes by some integral multiple of the sampling frequency defined by $\omega_s = 2\pi/T$, the angle of z changes by 2π radians. Thus, for any σ, as ω moves across any one of the strips parallel to the real axis in Figure 6.7-1, the angle of z changes by 2π radians. We con-

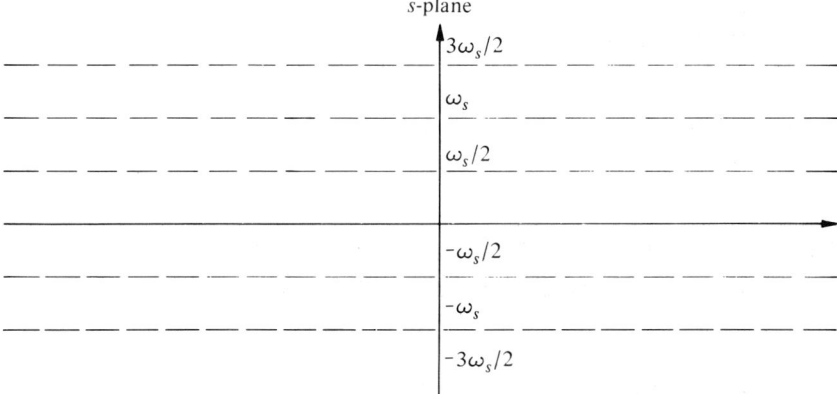

Figure 6.7-1. Periodic strip in the s plane.

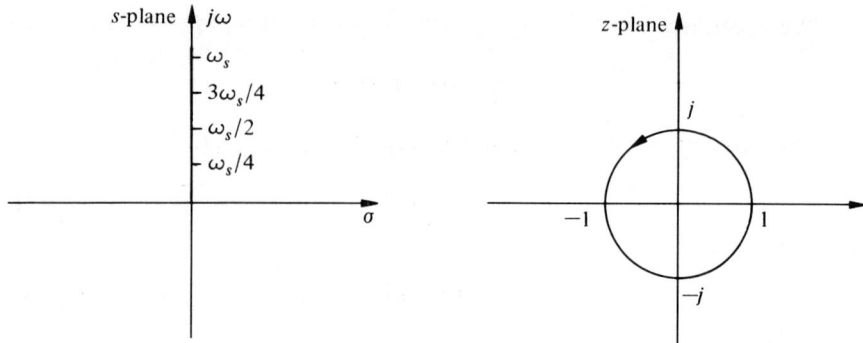

Figure 6.7-2. Mapping the s plane between 0 and $j\omega$ into the unit circle in the z plane.

clude, then, that moving from the real axis to the line ω_s in a vertical direction in the s plane is equivalent to making one counterclockwise revolution in the z plane. Particularly, moving from the origin to ω_s on the imaginary axis of the s plane corresponds to one counterclockwise encirclement on the circumference of the unit circle in the z plane. This is shown in Figure 6.7-2, where, for example, at

$$\omega = 0 \qquad z = 1$$
$$\omega = \frac{\omega_s}{4} \qquad z = j$$
$$\omega = \frac{\omega_s}{2} \qquad z = -1$$
$$\omega = \frac{3\omega_s}{4} \qquad z = -j$$
$$\omega = \omega_s \qquad z = 1$$

It is apparent that as ω increases from ω_s to $2\omega_s$, another counterclockwise encirclement will be made on the unit circle in the z plane. In general, then, as ω varies from $-\infty$ to $+\infty$ along the imaginary axis of the s plane, an infinite number of encirclements is made on the unit circle in the z plane.

We consider now the relation of the location of poles in the z plane to the transient response.

6.8 Effect of Pole-zero Configuration of C(z) in the z Plane upon System Transient Response

From the last section we know that the absolute stability of a discrete system is assured if all the poles of the transfer function of the system are contained in the unit circle. The actual location of the poles inside the unit circle may be effectively used to make important conclusions about the time response of the system. We shall consider several cases of pole locations that are most representative of the typical system transfer functions.

6.8-1 Single Pole on Real Axis

Consider the partial-fraction expansion of a system with a single pole on the real axis. This pole will contribute the term

$$\frac{b}{z-a}$$

This term will give rise to a sequence with elements ba^{k-1}, $k \geq 1$. We can distinguish between six different time sequences that are possible, depending on the location of the pole. (1) If $a > 1$, the sequence diverges. (2) If $a = 1$, the results will be a sequence of b's. (3) If $0 \leq a < 1$, the result will be a sequence of positive numbers, monotonically decreasing. (4) If $0 > a > -1$, we have a sequence of decreasing numbers with alternating signs. (5) If $a = -1$, we obtain a sequence of b's with alternating signs. (5) Finally, if $a < -1$, the sequence is divergent with alternating signs. These six different cases are shown graphically in Figure 6.8-1.

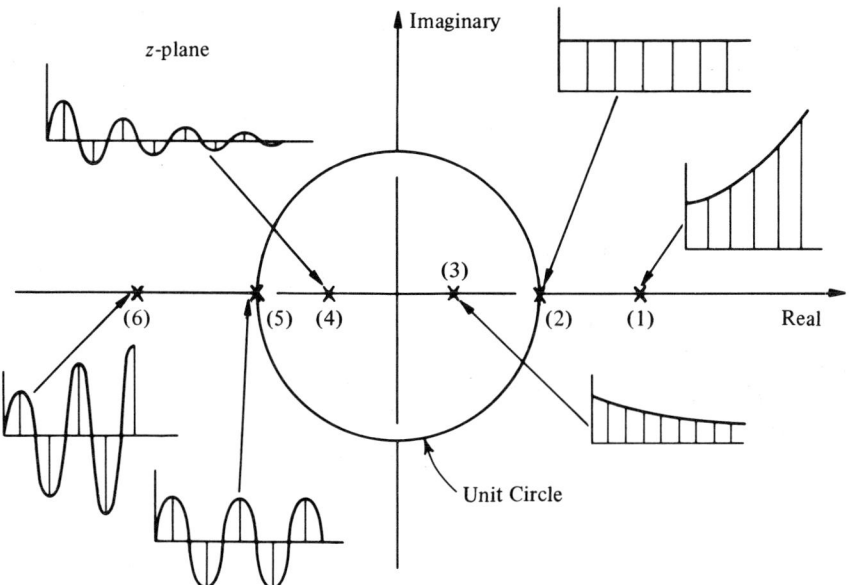

Figure 6.8-1. Transient responses at sampling intervals corresponding to various locations of a real pole.

6.8-2 A Pair of Complex Conjugate Poles

We consider a pair of complex conjugate poles

$$z = a \pm jb$$

It can easily be shown that these poles will give rise to the sequence

$$\alpha r^{k-1} \cos\left[(k-1)\theta + \varphi\right], \quad k \geq 1$$

where α and φ are constants dependent upon the partial-fraction expansion coefficients and

$$r = \sqrt{a^2 + b^2}$$

$$\theta = \tan^{-1} \frac{b}{a}$$

Depending upon the numerical values of a and b that determine the location of the poles, we can distinguish between several cases.

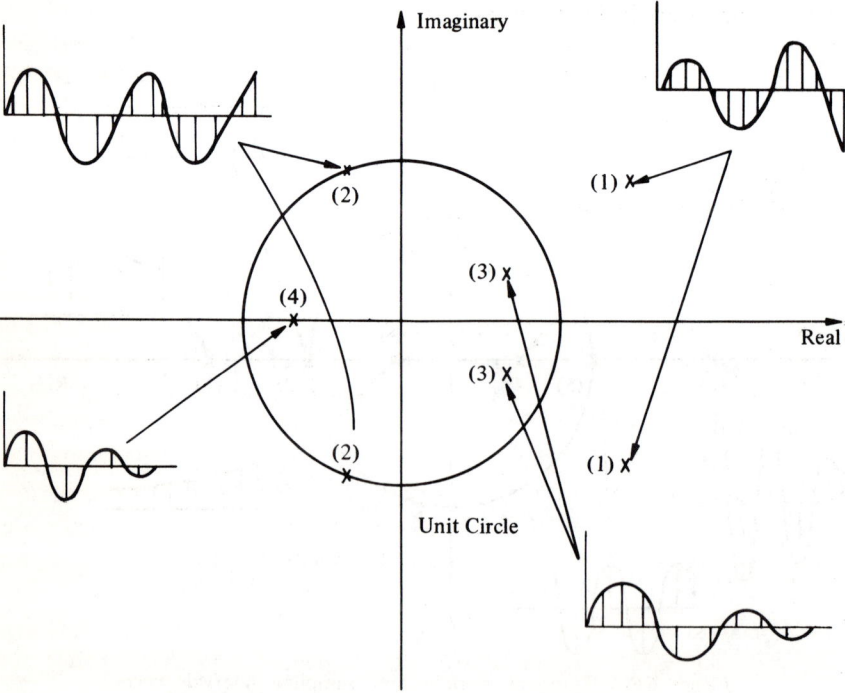

Figure 6.8-2. Transient response of two complex poles.

1. $\sqrt{a^2 + b^2} > 1$, $\theta \neq \pi$. The poles are outside the unit circle. This results in a sequence of numbers that are oscillating and increasing in magnitude.
2. $\sqrt{a^2 + b^2} = 1$, $\theta \neq \pi$. The poles are located on the unit circle. The response is a sequence oscillating at constant amplitude.
3. $\sqrt{a^2 + b^2} < 1$, $\theta \neq \pi$. The poles are inside the unit circle. The response is a sequence defining an oscillation with decreasing amplitude.
4. $\theta = \pi$. The poles are located on the negative real axis, forming a double pole. The response is oscillatory with the frequency of oscillation synchronized with the sampling rate.

These four cases are illustrated in Figure 6.8-2.

6.9 The Root Locus Technique

In the last section we have investigated the relationship between the poles of a system transfer function and the transient response. It was made apparent that an aggregate knowledge of all poles of a system transfer function would permit a fair estimation of the transient response of the system. By means of the root locus technique this task can be made reasonably straightforward. This has been effectively demonstrated for the analysis of continuous-time systems. The root locus technique can also be easily adapted to the study of discrete-time systems when the z-transform is employed. Since the characteristic equation of a single-loop sampled-data system may be represented by the form

$$1 + \mathscr{Z}\{H(s)G(s)\} = 0 \tag{6.9-1}$$

where $\mathscr{Z}\{H(s)G(s)\} = HG(z)$ is a rational function in z, the root locus method may be applied directly to (6.9-1) without modification. Two significant differences between the continuous case and the discrete case exist: (1) The root locus is plotted in the z plane, and (2) the investigation of stability is carried out with respect to the unit circle.

It is clear, then, that the construction of a root locus in the z plane can be accomplished by simply employing the usual rules for construction.

We illustrate the application of the root locus technique by presenting an example that we have previously considered.

EXAMPLE 6.9-1

Construct root locus diagrams for the error-sampled system shown in Figure 6.9-1. This system was previously discussed in Examples 3.4-1 and 6.5-2 with $K = 1$. We wish to compare the root locus diagrams for three values of the sampling interval $T = .1$ $T = 1.0$, and $T = 4.0$.

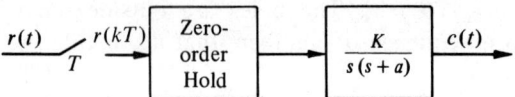

Figure 6.9-1. Sampled-data system for Example 6.9-1.

The z-transform of the open-loop transfer function is calculated first. We have

$$\mathscr{Z}\{G_0(s)G(s)\} = \mathscr{Z}\left\{\frac{1-e^{-sT}}{s}\frac{K}{s(s+1)}\right\}$$

$$= (1-z^{-1})\mathscr{Z}\left\{\frac{K}{s^2(s+1)}\right\}$$

$$= K(1-z^{-1})\mathscr{Z}\left\{\frac{1}{s^2} - \frac{1}{s} + \frac{1}{s+1}\right\}$$

$$= K(1-z^{-1})\left[\frac{Tz}{(z-1)^2} - \frac{z}{z-1} + \frac{z}{z-e^{-T}}\right] \quad (6.9\text{-}2)$$

For the three cases mentioned, the transfer functions are

$T = .1$

$$G_0 G(z) = K\frac{.005(z+.995)}{(z-1)(z-.905)} \quad (6.9\text{-}3)$$

$T = 1.0$

$$G_0 G(z) = K\frac{.368(z+.722)}{(z-1)(z-.368)} \quad (6.9\text{-}4)$$

$T = 4.0$

$$G_0 G(z) = K\frac{2.982(z+.302)}{(z-1)(z-.02)} \quad (6.9\text{-}5)$$

We shall consider the case $T = 1$ in detail.
The open-loop zeros and poles are given by

1 zero at $z = -.722$
2 poles at $z = 1$ and $z = .368$

Asymptotes:
There will be one asymptote ($P - Z = 1$) at -180 degrees.

Branches on the real axis:
There is a branch between the two poles and another one to the left of the zero.

Separation from the real axis:
Using the characteristic equation, we form

$$K = \frac{(z-1)(z-.368)}{z+.722}$$

and

$$\frac{dK}{dz} = 0 = z^2 + 1.44z - 1.35$$

from which the following breakaway points are given.

$$z_{1,2} = -2.09, +.648$$

With this information we may proceed to construct the root locus plot as shown in Figure 6.9-2. This plot shows the part corresponding to a pair of

z-plane

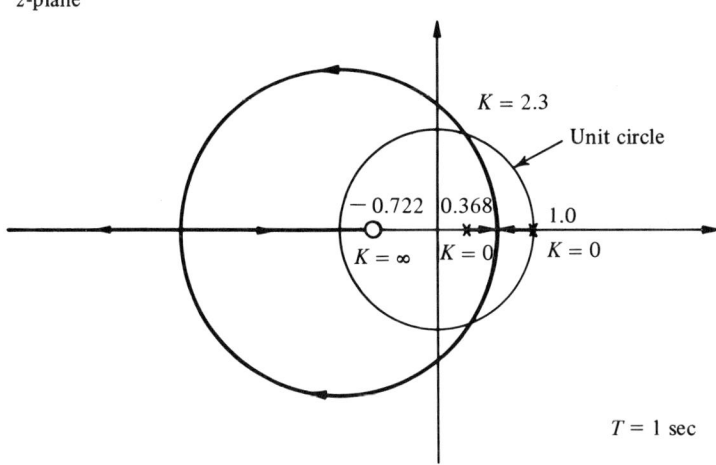

Figure 6.9-2. Root locus plot of error-sampled system.

complex conjugate poles to be a circle with center at the zero $z = -.722$. This is generally true for a root locus plot for a system with two poles and one zero and is proved as follows:

Let

$$z = x + jy$$

Then (6.9-4) becomes

$$G_0 G(z) = \frac{K(x + .722 + jy)}{x^2 - 1.368x + .368 - y^2 + jy(2x - 1.368)}$$

Now from the condition of a root locus plot we have

$$\angle G(z) = (2k+1)\pi$$

Therefore,

$$\tan^{-1}\frac{y}{x+.722} - \tan^{-1}\frac{y(2x-1.368)}{x^2-1.368x+.368-y^2} = (2k+1)\pi$$

Taking the tangent of both sides yields

$$\frac{\dfrac{y}{x+.722} - \dfrac{y(2x-1.368)}{x^2-1.368x+.368-y^2}}{1+\left(\dfrac{y}{x+.722}\right)\left(\dfrac{y(2x-1.368)}{x^2-1.368x+.368-y^2}\right)} = 0$$

Simplifying, we have

$$(x+.722)^2 + y^2 = 1.875 \qquad (6.9\text{-}6)$$

which is the equation of a circle with center at $z = -.722$ and radius $r = 1.37$.

The root locus plots for the cases $T = .1$ and $T = 4.0$ are constructed similarly and are shown in Figures 6.9-3 and 6.9-4, respectively.

We note with interest the magnitude of the gain K at which the system becomes unstable. For the three sampling periods we have

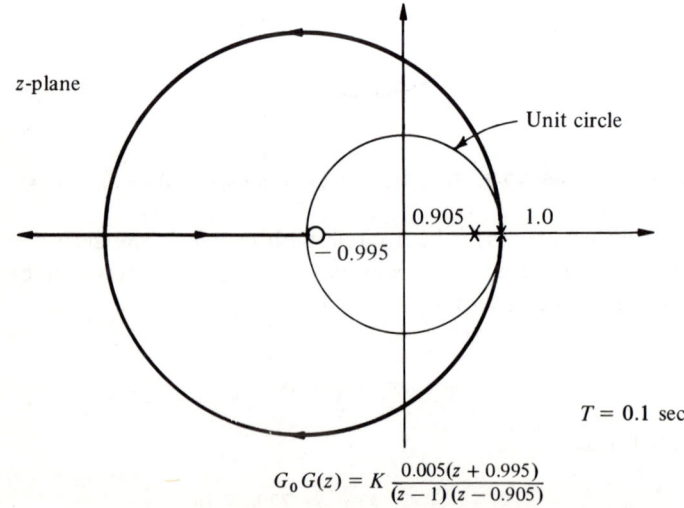

$$G_0 G(z) = K\frac{0.005(z+0.995)}{(z-1)(z-0.905)}$$

$T = 0.1$ sec

Figure 6.9-3. Root locus plot of error-sampled system.

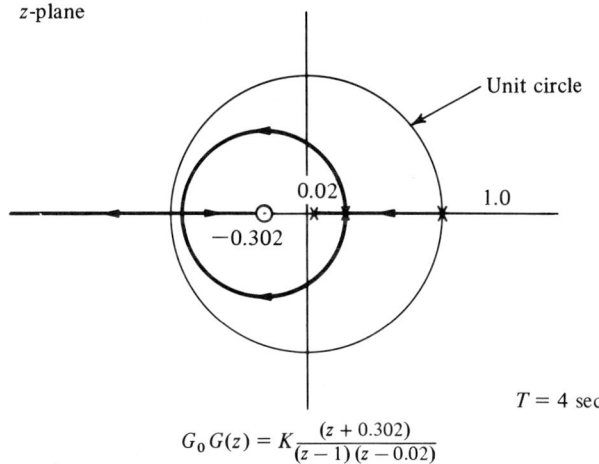

Figure 6.9-4. Root locus plot of error-sampled system.

T	K
.1	19.1
1.0	2.3
4.0	1.08

Thus, it is clear that a smaller sampling period permits a greater margin of stability. This conclusion was originally made available from Example 3.4-1.

The circle that describes the loci of the complex conjugate poles becomes larger as T is reduced; this appears to be in contradiction to the relationship between T and K for marginal stability just described. Indeed, it appears that as $T \to 0$ there will be no margin of stability left as the circle of the complex conjugage poles becomes tangent to the unit circle at $z = 1$. An inspection, however, of the open-loop transfer function clears this question up. This transfer function is given by (6.9-2) and is now simplified to

$$G_0 G(z) = K(T - 1 + e^{-T}) \frac{z - a}{(z - 1)(z - e^{-T})} \quad (6.9\text{-}7)$$

where

$$a = \frac{1 - e^{-T} - T e^{-T}}{T - 1 + e^{-T}}$$

Equation (6.9-7) contains the multiplicative factor $(T - 1 + e^{-T})$, which is seen to go to zero as T approaches zero. Thus (6.9-7) becomes meaningless as a basis for root locus construction.

As T approaches zero, the sampled system approaches a continuous system that is well-known to be stable for all values of K.

The above example illustrates the usefulness of the root locus technique in the analysis of discrete-time systems. It can be seen that considerable effort must go into preparing the open-loop transfer function of a sampled system for the pole-zero format necessary for root locus construction. This fact is particularly evident when the system includes a digital transfer function, which is normally not given in factored form. For these reasons the systems designer turns to the digital computer for construction of a root locus plot. A library of systems programs usually contains a root locus routine for discrete systems. It is capable of handling mixed data systems that have parts both with continuous and discrete data characteristics.

6.10 The Modified z-Transform

In typical sampled-data systems, the output is frequently a continuous signal. It was shown in Section 6.5 how one may determine the relationships existing between the input and output signals at the sampling instants. Because the output signal may be continuous, a method for determining such signals' behavior between sampling instants is clearly desirable. The so-called modified z-transform is such a technique. It parallels in objective the development of Section 3.7, where discrete-time techniques were used to calculate the response of systems between sampling points.

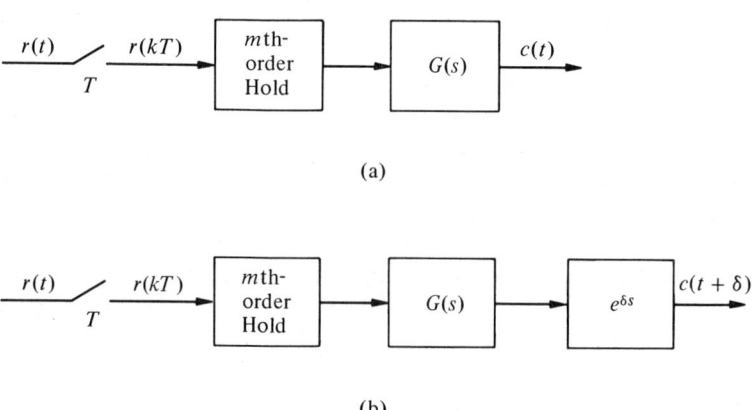

Figure 6.10-1. (a) Original system; (b) system for determining modified z-transform.

Consider the system shown in Figure 6.10-1(a). Since this system is of the same form as that shown in Figure 6.5-1, the relationship between $R(z)$ and $C(z)$ is, by (6.5-1),

$$C(z) = \mathscr{Z}[G_m(s)H(s)]R(z)$$

where

$$C(z) = \mathscr{Z}[c(kT)]$$
$$R(z) = \mathscr{Z}[r(kT)]$$

The response of the systems shown in Figure 6.10-1(a) and (b) are identical except for a shift in time of δ seconds. Therefore, for the system of Figure 6.10-1(b) the output at the sampling instants is $c(nT + \delta)$. Now

$$\hat{C}(z) = \mathscr{Z}[G_m(s)G(s)e^{\delta s}]R(z) \qquad (6.10\text{-}1)$$

where

$$\hat{C}(z) = \mathscr{Z}[c(kT + \delta)]$$

To evaluate the output for all t, we let δ take on values $0 \leq \delta < T$. Another method for evaluating (6.10-1) is obtained by applying Theorem 6.3-1 of Section 6.3. This results in

$$\mathscr{Z}[c(kT + \delta)] = \left(\left[\text{sum of residues of } \frac{G_m(p)G(p)e^{\delta p}}{1 - z^{-1}e^{pT}}\right.\right.$$
$$\left.\left. \text{at poles of } G_m(p)G(p)e^{\delta p}\right] + \alpha \right)R(z) \qquad (6.10\text{-}2)$$

EXAMPLE 6.10-1

Determine the modified z-transform of $c(t)$ when $G_m(s) = G_0(s)$ and $G(s) = a/(s + a)$ by two methods.

(i) From (6.10-1)

$$\mathscr{Z}[c(kT - \delta)] = \mathscr{Z}\left[\frac{1 - e^{-sT}}{s} \frac{a}{s + a} e^{\delta s}\right]R(z)$$

The time function corresponding to $G_0(s)G(s)$ is now evaluated.

$$h(t) = \mathscr{L}^{-1}[G_0(s)G(s)] = \mathscr{L}^{-1}\left[(e^{\delta s} - e^{-(T-\delta)s})\frac{a}{s(s + a)}\right]$$
$$= \mathscr{L}^{-1}\left[(e^{\delta s} - e^{-(T-\delta)s})\left(\frac{1}{s} - \frac{1}{s + a}\right)\right]$$
$$= [1 - e^{-a(t+\delta)}]u(t + \delta) - [1 - e^{-a(t+\delta-T)}]u(t + \delta - T)$$

Therefore,

$$h(kT) = [1 - e^{-a(kT+\delta)}]u(kT + \delta) - [1 - e^{-a(kT+\delta-T)}]u(kT + \delta - T)$$

We shall restrict δ to the values $0 \leq \delta < T$; therefore,

$$h(kT) = e^{-a\delta}e^{-akT}(e^{aT} - 1) \quad k = 1, 2, \ldots \quad (6.10\text{-}3)$$
$$h(0) = 1 - e^{-a\delta}$$

Taking the z-transform of this sequence, we have

$$\mathscr{L}[h(kT)] = \mathscr{L}[G_0(s)G(s)] = 1 - e^{-a\delta} + e^{-a\delta}(e^{aT} - 1)\sum_{k=1}^{\infty} e^{-akT}z^{-k}$$

$$= 1 - e^{-a\delta} + \frac{e^{-a\delta}(1 - e^{-aT})z^{-1}}{1 - e^{-aT}z^{-1}}$$

$$= \frac{1 - e^{-a\delta} + z^{-1}(e^{-a\delta} - e^{-aT})}{1 - e^{-aT}z^{-1}}$$

so that

$$\mathscr{L}[c(kT + \delta)] = \left[\frac{1 - e^{-a\delta} + z^{-1}(e^{-a\delta} - e^{-aT})}{1 - e^{-aT}z^{-1}}\right]R(z) \quad (6.10\text{-}4)$$

(ii) From (6.10-2), we have

$$\mathscr{L}[c(kT + \delta)] = \left(\left[\text{sum of residues of } \frac{a(1 - e^{-pT})e^{\delta p}}{(1 - z^{-1}e^{pT})p(p + a)}\right.\right.$$
$$\left.\left. \text{at poles } p = 0, p = -a\right] + \alpha\right)R(z)$$

$$= \left[\frac{(e^{-aT} - 1)e^{-a\delta}}{(1 - z^{-1}e^{-aT})} + \alpha\right]R(z)$$

To evaluate α, the initial value for $t = 0$ of $\mathscr{L}[G_0(s)G(s)e^{\delta s}]$ must be satisfied. By (6.10-3)

$$\lim_{z \to \infty}\left[\frac{(e^{-aT} - 1)e^{-a\delta}}{(1 - z^{-1}e^{-aT})} + \alpha\right] = 1 - e^{-a\delta}$$

or

$$\alpha = 1 - e^{-a\delta}e^{aT}$$

This gives us

$$\mathscr{L}[c(kT + \delta)] = \left[\frac{1 - e^{-a\delta} + z^{-1}(e^{-a\delta} - e^{-aT})}{1 - z^{-1}e^{-aT}}\right]R(z) \quad (6.10\text{-}5)$$

which is in agreement with the result from part (i), (6.10-4).

When $r(t) = u(t)$, we have

$$\mathscr{Z}[c(kT+\delta)] = \frac{1 - e^{-a\delta} + z^{-1}(e^{-a\delta} - e^{-aT})}{(1 - z^{-1}e^{-aT})(1 - z^{-1})}$$

$$= \frac{1}{1 - z^{-1}} - \frac{e^{-a\delta}}{1 - z^{-1}e^{-aT}}$$

By taking the inverse z-transform, we find

$$c(kT) = 1 - e^{-a(kT+\delta)} \quad \text{for } k = 0, 1, 2$$

where the value of δ is restricted to be $0 \leq \delta < T$. Similar techniques could be used for δ outside this region (including negative values).

The modified z-transform for any function of time $h(t)$ that has a Laplace transform $H(s)$ is determined by evaluating

$$\mathscr{Z}[h(kT+\delta)] = \mathscr{Z}[H(s)e^{\delta s}] \tag{6.10-6}$$

Using residue theory, we find that another form of (6.10-6) is

$$\mathscr{Z}[h(nT+\delta)] = \left[\text{sum of residues of } \frac{H(p)e^{\delta p}}{1 - e^{pT}z^{-1}} \text{ at poles of } H(p)\right] + \alpha \tag{6.10-7}$$

where α is selected so that the initial value is correct. A convenient method for determining the z-transform of a delayed function of time $f(t - \delta)$, where δ is positive, may be obtained by writing δ as

$$\delta = (m - \Delta)T$$

Here m is a positive integer and Δ is a positive number less than unity. Therefore, the function $f(t)$ delayed δ units has the Laplace transform

$$\mathscr{L}[f(t-\delta)] = \mathscr{L}[F(s)e^{-mTs}e^{\Delta Ts}]$$
$$= e^{-mTs}F(s)e^{\Delta Ts}$$

The factor e^{-mTs} gives rise to a delay of m sampling periods to the time function

$$\mathscr{L}^{-1}[F(s)e^{\Delta Ts}]$$

Therefore,

$$\mathscr{Z}[f(t-\delta)] = z^{-m}\mathscr{Z}[F(s)e^{\Delta Ts}] \tag{6.10-8}$$

Let

$$F(z, \Delta) = \mathscr{Z}[F(s)e^{\Delta Ts}] \tag{6.10-9}$$

The z-transform of any $f(t - \delta)$ for all positive δ may be directly evaluated, if it is possible to evaluate the advanced z-transform as given by (6.10-9). Example 6.10-1 illustrates a method whereby $\mathscr{Z}[F(s)e^{\Delta Ts}]$ may be evaluated. By applying (6.10-9), we may generate the abbreviated advanced z-transform table, Table 6.10-1.

Table 6.10-1 Modified z-Transform Pairs

$F(s)$	$F(z, \Delta)$
$\dfrac{1}{s}$	$\dfrac{1}{1 - z^{-1}}$
$\dfrac{1}{s^2}$	$\dfrac{\Delta T + T(1 - \Delta)z^{-1}}{(1 - z^{-1})^2}$
$\dfrac{1}{s^3}$	$T^2\left[\dfrac{z^{-2}}{(1 - z^{-1})^2} + \dfrac{(1 + 2\Delta)z^{-1}}{2(1 - z^{-1})^2} + \dfrac{\Delta^2}{2(1 - z^{-1})}\right]$
$\dfrac{1}{s + a}$	$\dfrac{e^{-a\Delta T}}{1 - e^{-aT}z^{-1}}$
$\dfrac{1}{(s + a)^2}$	$T\left[\dfrac{\Delta e^{-a\Delta T}}{1 - e^{-aT}z^{-1}} + \dfrac{e^{-a(1+\Delta)T}z^{-1}}{(1 - e^{-aT}z^{-1})^2}\right]$
$\dfrac{a}{s(s + a)}$	$\dfrac{1}{1 - z^{-1}} - \dfrac{e^{-a\Delta T}}{1 - e^{-aT}z^{-1}}$
$\dfrac{a}{s^2(s + a)}$	$\dfrac{Tz^{-1}}{(1 - z^{-1})^2} + \dfrac{a\Delta T - 1}{a(1 - z^{-1})} + \dfrac{e^{-a\Delta T}}{a(1 - e^{-aT}z^{-1})}$

REFERENCES

1. Aseltine, J. A., *Transform Method in Linear System Analysis*, McGraw-Hill, New York, 1958.

2. Churchill, R. V., *Introduction to Complex Variables and Applications*, McGraw-Hill, New York, 1948.

3. DeRusso, P. M., R. J. Roy, and C. M. Close, *State Variables for Engineers*, Wiley, New York, 1965.

4. Freeman, H., *Discrete-Time Systems*, Wiley, New York, 1965.

5. Jury, E. I., *Theory and Application of the z-Transform Method*, Wiley, New York, 1964.

6. Monroe, A. J., *Digital Processes for Sampled Data Systems*, Wiley, New York, 1962.

7. Schwarz, R. J., and B. Friedland, *Linear Systems*, McGraw-Hill, New York, 1965.
8. Ragazzini, J. R., and G. F. Franklin, *Sampled-Data Control Systems*, McGraw-Hill, New York, 1958.
9. Tou, J. T., *Digital and Sampled-Data Control Systems*, McGraw-Hill, New York, 1959.

PROBLEMS

6.1 The modified first-order digital-to-analog converter is characterized by

$$h(kT + \tau) = g(kT) + \frac{\alpha\tau}{T}[g(kT) - g(kT - T)] \quad \text{for} \quad 0 \leq \tau < T$$

with $0 \leq \alpha \leq 1$. For $\alpha = 0$, it is a zero-order hold, while it is a first-order hold when $\alpha = 1$. Determine the equivalent transfer function used in conjunction with the fictitious impulse sampler.

6.2 For the system shown in Fig. P6.2, find the response to the input $R(s) = 1/s$.

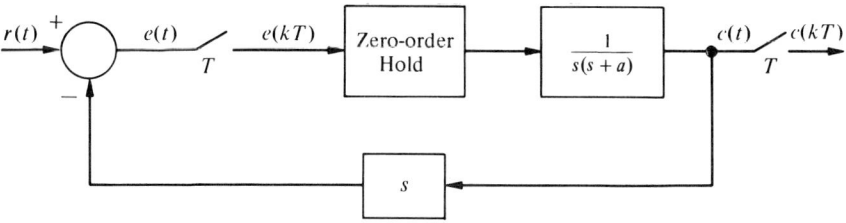

Figure P6.2.

6.3 Find the transfer function for the system shown in Fig. P6.3.

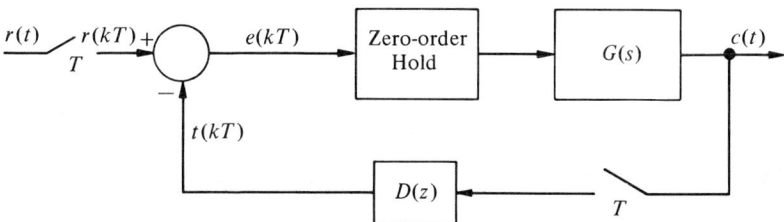

Figure P6.3.

6.4 Determine the z-transform of $H(s)$, where

(a) $H(s) = \dfrac{s}{(s+a)(s+b)}$

(b) $H(s) = \dfrac{1}{s^2(s+a)}$

6.5 Verify the first, second, and fourth entries in the modified z-transform table, Table 6.10-1.

6.6 Design an analog/hybrid simulation circuit for a first-order hold circuit.

6.7 Derive the relations from equations (6.3-1) and (6.3-2).

6.8 Determine the digital transfer function corresponding to the state equations

$$\begin{bmatrix} x_1(k+1) \\ x_2(k+1) \end{bmatrix} = \begin{bmatrix} .5 & -.1 \\ .2 & .3 \end{bmatrix} \begin{bmatrix} x_1(k) \\ x_2(k) \end{bmatrix} + \begin{bmatrix} .25 \\ 0 \end{bmatrix} r(k)$$

$$c(k) = \begin{bmatrix} 1 & -2 \end{bmatrix} \begin{bmatrix} x_1(k) \\ x_2(k) \end{bmatrix}$$

6.9 For the system shown in Figure P6.9, verify that the two representations given below are equivalent.

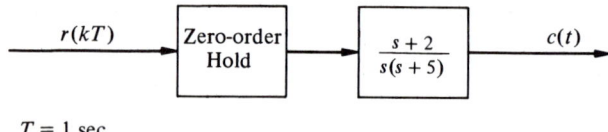

$T = 1$ sec

Figure P6.9

State variable representation:

$$\begin{bmatrix} x_1(k+1) \\ x_2(k+1) \end{bmatrix} = \begin{bmatrix} \mathbf{A}(T) \end{bmatrix} \begin{bmatrix} x_1(k) \\ x_2(k) \end{bmatrix} + \begin{bmatrix} \mathbf{B}(T) \end{bmatrix} r(k) \qquad \text{(P6.9-1)}$$

$$c(k) = [\mathbf{C}] \begin{bmatrix} x_1(k) \\ x_2(k) \end{bmatrix}$$

where the matrices $\mathbf{A}(T)$, $\mathbf{B}(T)$, and $\mathbf{C}$ are to be determined from the transfer function $G(s)$.

z-Transform representation:

$$C(z) = (1 - z^{-1}) \mathscr{Z}\left[\frac{G(s)}{s}\right] R(z) \qquad \text{(P6.9-2)}$$

6.10 Obtain the weighting sequence for the system shown in Figure P6.9 by use of (a) equation (P6.9-1) and (b) equation (P6.9-2).

6.11 Repeat Problems 6.9 and 6.10 if the transfer function is given as

$$G(s) = \frac{s+1}{s+10}$$

6.12 Determine the digital transfer function of a digital PID controller (see Chapter 3).

6.13 Determine the stability of the system shown in Figure P6.13 for $T = 1$ second and $T = .1$ second.

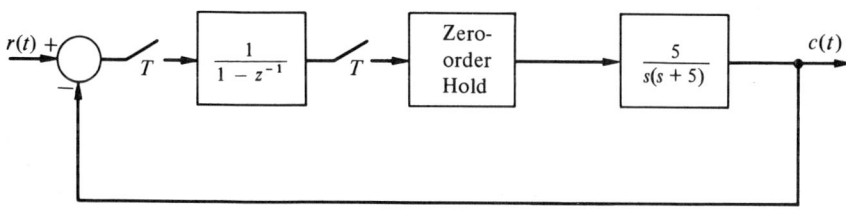

Figure P6.13

6.14 Examine the steady-state error of system P6.13 for a step and ramp input when $T = 1.0$.

6.15 The equations for the so-called α-β tracker were given by equations (1.7-6) through (1.7-8). Derive a digital transfer function for this tracker, relating predicted position to input position. Verify that the α-β tracker is able to follow both a position and a velocity input with zero steady-state error.

6.16 By z-transform methods verify that integral control is able to eliminate offset. Take the problem discussed in Section 3.7 as an example.

6.17 Obtain a root locus plot for the systems shown in Figure P6.17.

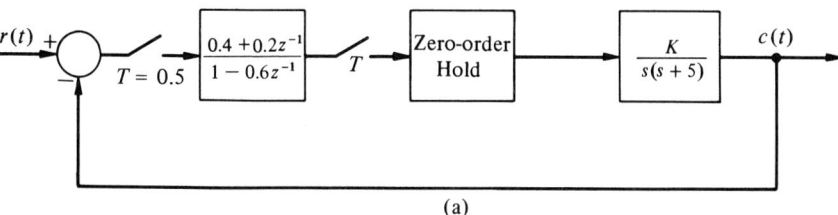

(a)

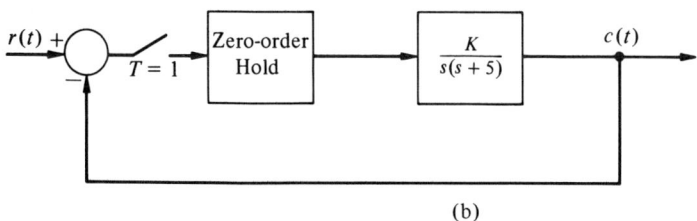

(b)

Figure P6.17

6.18 State the rules of root locus method as it applies to the z domain. Use a text on control theory of continuous-time systems as reference.

6.19 By the method of the modified z-transform, compute the response of the system shown in Figure P6.19 at full and half sampling instants.

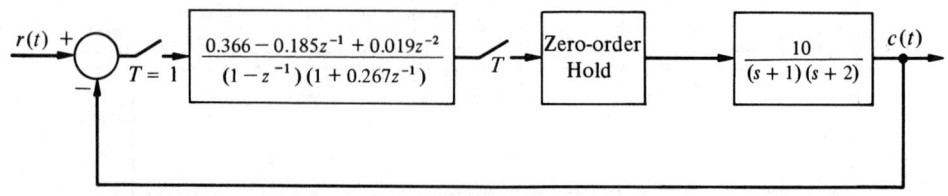

Figure P6.19

6.20 Consider Figure 6.8-2. What is the transient response of a transfer function with two identical poles located on the positive real axis?

6.21 Consider Figure 6.8-2. Calculate the transient response for each of the four cases listed.

Appendix 6A

PROOF OF THEOREM 6.3-1

Proof. Define the infinite impulse sequence $\delta_T(t)$ by

$$\delta_T(t) = \sum_{k=0}^{\infty} \delta(t - kT) \qquad (6\text{A-}1)$$

and take the Laplace transform of the impulse sampled version of $h(t)$, that is,

$$h^*(t) = h(t)\delta_T(t) = \sum_{k=0}^{\infty} h(kT)\gamma(t - kT)$$

to give

$$H^*(s) = \mathscr{L}[h^*(t)] = \sum_{k=0}^{\infty} h(kT)e^{-skT} \qquad (6\text{A-}2)$$

By letting $z = e^{sT}$, (6A-2) simplifies to

$$H^*(s)|_{z=e^{sT}} = \sum_{k=0}^{\infty} h(kT)z^{-k} = H(z) \qquad (6\text{A-}3)$$

which is the z-transform of the sequence $h(0), h(T), h(2T), \ldots$ generated from $h(t)$ at $t = kT$, $k = 0, 1, 2, \ldots$. A method for obtaining $H(z)$ for a given rational $H(s)$ will now be demonstrated. Use of the complex-convolution theorem of Laplace transform theory will be made. This theorem states that the Laplace transform of the product of two time functions $f(t)$ and $g(t)$ is given by

$$\mathscr{L}[f(t)g(t)] = \frac{1}{2\pi j} \int_{c-j\infty}^{c+j\infty} F(p)G(s - p)dp \qquad (6\text{A-}4)$$

where $F(s)$ and $G(s)$ are the Laplace transforms of $f(t)$ and $g(t)$, respectively. The constant c is selected so that all the poles of $F(p)$ lie to the left of the imaginary axis displaced by c units. Applying (6A-4) to (6A-1) with $g(t) = \delta_T(t)$, we have

$$H^*(s) = \mathscr{L}[h^*(t)] = \frac{1}{2\pi j} \int_{c-j\infty}^{c+j\infty} H(p) \frac{1}{1 - e^{-(s-p)T}} dp \qquad (6A\text{-}5)$$

where

$$\mathscr{L}[\delta_T(t)] = \sum_{k=0}^{\infty} e^{-ksT} = \frac{1}{1 - e^{-sT}} \quad \text{for } |e^{-sT}| < 1$$

Letting $z = e^{sT}$ in (6.A-5) as in (6.A-3) and equating the result with (6.A-3), we obtain

$$H(z) = \frac{1}{2\pi j} \int_{c-j\infty}^{c+j\infty} \frac{H(p)}{1 - z^{-1}e^{pT}} dp \qquad (6A\text{-}6)$$

Equation (6A-6) may be evaluated by contour integration. For bounded $h(t)$, the poles of $H(p)$ lie in the left-hand p plane while the factor

$$\frac{1}{1 - e^{-(s-p)T}}$$

has poles located at

$$p_k = s - j\frac{2\pi k}{T} \quad \text{for } k = 0, \pm 1, \pm 2, \ldots$$

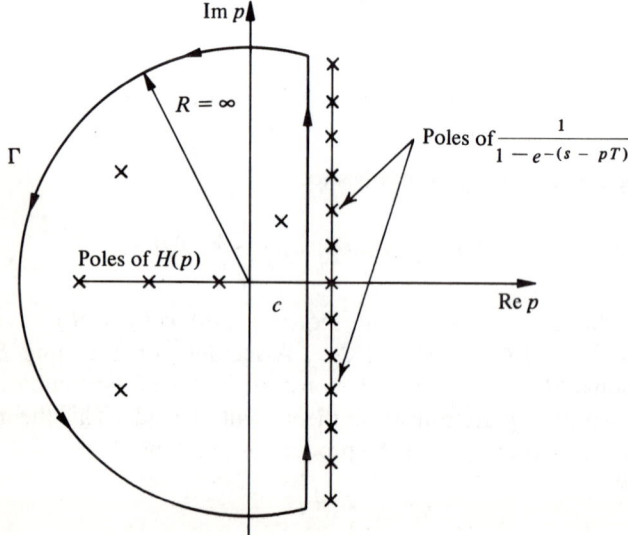

Figure 6A-1

The contour over which the integration is to be made is shown in Figure 6A-1. We choose c so that all the poles of $H(p)$ lie to the left of the path $\{c - j\infty \text{ to } c + j\infty\}$ and all the poles of $1/(1 - e^{-(s-p)T})$ lie to the right of this path. From the residue theorem of complex variables, we have

$$H^*(s) = \frac{1}{2\pi j} \int_\Gamma \frac{H(p)}{1 - e^{-(s-p)T}} dp$$

$$= \left[\text{sum of residue of } \frac{H(p)}{1 - e^{-(s-p)T}} \text{ at poles of } H(p) \right]$$

On Γ, $p = Re^{j\theta}$ and for rational $H(p)$, $H(p) \to bp^m$ as $R \to \infty$. Therefore,

$$\frac{1}{2\pi j} \int \frac{H(p)}{1 - e^{-(s-p)T}} dp = \lim_{R \to \infty} \frac{1}{2\pi j} \int_{\pi/2}^{3\pi/2} \frac{jbR^{m+1} e^{j\theta(m+1)}}{1 - e^{-sT} e^{TRe^{j\theta}}} d\theta$$

$$= 0 \qquad \text{for } m < -1$$

$$= \frac{b}{2} \qquad \text{for } m = -1$$

$$= \text{undefined} \quad \text{for } m \geq -1$$

The constant b is actually the value of $h(t)$ at $t = 0$, as can be seen from the initial-value theorem of Laplace transform theory. Summarizing the results, we have

$$H(z) = \left[\text{sum of residues of } \frac{H(p)}{1 - z^{-1} e^{pT}} \text{ at poles of } H(p) \right] - \frac{h(0)}{2} \quad (6\text{A-}7)$$

Since

$$H(z) = \sum_{k=0}^{\infty} h(kT) z^{-k}$$

the factor $h(0)/2$ in (6.A-7) affects only the $k = 0$ term in the summation. Thus, given a Laplace transform $H(s)$, the z-transform of the corresponding sequence $h(nT)$ {where $h(t) = \mathscr{L}^{-1}[H(s)]$} is obtained by evaluating

$$\hat{H}(z) = \left[\text{sum of residues of } \frac{H(p)}{1 - z^{-1} e^{pT}} \text{ at poles of } H(p) \right] \quad (6\text{A-}8)$$

and checking whether

$$\lim_{z \to \infty} \hat{H}(z) = \lim_{s \to \infty} sH(s) = h(0)$$

If not, the appropriate constant is added to (6A-8) to insure the desired initial value.

7

The Analytical Design of Discrete Systems

7.1 Introduction

The analytical design of discrete systems presents a most interesting challenge. A digital processor, which may be a general-purpose computer or a special-purpose switching circuit, may be effectively employed to generate a control input to a system. The use of a digital computer as a controller of a system is particularly attractive, since the implementation of a control law requires only the preparation of a computer program. A program permits almost unlimited flexibility.

In this chapter we shall investigate a variety of design techniques, all of which are aimed at defining a control algorithm implementable as a computer program. As with all design techniques, it will be necessary to employ a design objective or performance criterion. Usually a design objective implies the optimization or minimization of a certain factor influencing system performance. In so doing, we are applying the concepts of optimal control theory, which has been developed in the last decade. Most of the optimization problems postulated in this chapter have solutions that are obtained by solving a set of linear equations. Such is not the case in the continuous counterpart of optimization theory.

7.2 Time-domain Synthesis with Minimum Settling Time

In this section we will consider procedures that utilize state transition matrices to generate computer algorithms. The design objective will be to achieve minimum settling time in system response. In order to make the presentation of the required calculations possible, it is necessary to restrict the discussion to simple examples. However, in each example given, an attempt will be made to outline a more general case.

A number of different design approaches may be taken to insure that the system's output $c(t)$ will equal its input $r(t)$ in the minimum number of sample times (N). For example, it is possible to synthesize a controller such that $c(t) = r(t)$ at the sampling times but not necessarily in between. Alternatively, it may be desired to have $c(t) = r(t)$ for all $t \geq NT$ for the smallest value of N. A design based on the first approach will be less complex to implement than the latter because of its simpler control task. The complexity of each will depend on the nature of the input signal $r(t)$ and the order of the system being controlled.

Consider the system shown in Figure 7.2-1. It will be our objective to determine the digital transfer function $D(z)$ so that the system may respond with minimum settling time on a closed-loop basis. First we will consider the solution of this problem in the time domain. A later section will present a parallel solution in the z domain.

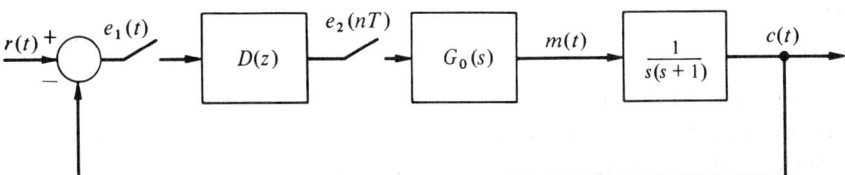

Figure 7.2-1. Digital control system.

7.2-1 Response to a Step Input

For the continuous plant of the system shown in Figure 7.2-1,

$$\frac{d}{dt}\begin{bmatrix} x_1 \\ x_2 \end{bmatrix} = \begin{bmatrix} -1 & 0 \\ 1 & 0 \end{bmatrix}\begin{bmatrix} x_1 \\ x_2 \end{bmatrix} + \begin{bmatrix} 1 \\ 0 \end{bmatrix} m(t) \tag{7.2-1}$$

$$c(t) = \begin{bmatrix} 0 & 1 \end{bmatrix}\begin{bmatrix} x_1 \\ x_2 \end{bmatrix}$$

For $T = 1.0$ second, the discrete state equations are

$$\begin{bmatrix} x_1(k+1) \\ x_2(k+1) \end{bmatrix} = \begin{bmatrix} .368 & 0 \\ .632 & 1 \end{bmatrix} \begin{bmatrix} x_1(k) \\ x_2(k) \end{bmatrix} + \begin{bmatrix} .632 \\ .368 \end{bmatrix} e_2(k) \quad (7.2\text{-}2)$$

$$k = 0, 1, 2, \ldots$$

and

$$c(k) = x_2(k)$$

The sequence $e_2(k)$ represents the input to the zero-order hold.

The objective of minimum settling time requires us to determine the control inputs $e_2(0), e_2(1), \ldots$, which drive the plant from an arbitrary initial state

$$\begin{bmatrix} x_1(0) \\ x_2(0) \end{bmatrix}$$

to a state such that the output $c(t)$ is equal to the input $r(t)$ for $t = t_1 > 0$. Furthermore, the output is to remain equal to the input from that time on. For the case at hand, the input is a step function; i.e.,

$$r(t) = R, \quad t \geq 0 \quad (7.2\text{-}3)$$

Therefore, we require that

and

$$\left. \begin{array}{l} c(t) = R \\ \dot{c}(t) = 0 \end{array} \right\} \quad t \geq t_1 > 0 \quad (7.2\text{-}4)$$

The derivative of the output must be set equal to zero to guarantee that $c(t)$ will not change after it has reached the magnitude of the input. Conditions (7.2-4) may be related to the discrete state variable by use of the state equations (7.2-1); that is,

$$\left. \begin{array}{l} x_2(t) = R \\ x_1(t) = 0 \end{array} \right\} \quad t \geq t_1 > 0 \quad (7.2\text{-}5)$$

The time t_1 at which conditions (7.2-5) are satisfied is unknown at this time. Since the input to the zero-order hold circuit occurs only at the discrete times $0, T, 2T, \ldots$, it is necessary to modify equations (7.2-5) to

$$\left. \begin{array}{l} x_2(NT) = R \\ x_1(NT) = 0 \end{array} \right\} \quad \text{for some integer } N > 0 \quad (7.2\text{-}6)$$

Then the desired system state is synchronized timewise with the input. The integer N for which (7.2-6) holds is unknown. We must, therefore, investigate for what value of N it is possible to satisfy (7.2-6). We will investigate three cases; that is, $N = 1, 2,$ and 3. For $N = 1$, it will be possible for the output to be equal to the input at the sampling instants; however, the system's response shows an undesirable ripple. This type of response is called the *minimal prototype response*. It is generally possible for any order system to exhibit a minimal prototype response ($N = 1$) to a step input. For $N = 2$, we will see that the system responds in a *deadbeat* manner; that is, the output equals the input at all times, not just at the sampling instants. In general, an nth order system requires $N = n$ sampling periods for deadbeat response. For $N = 3$, the system will be seen to respond in a deadbeat manner; in addition, it will be possible to impose some constraints on the magnitude of the control variable. These three cases will be investigated in detail for the second-order system which we are presently considering.

Case a. $N = 1$ (*Minimal Prototype Design*)

The transition equations for the first period are obtained from (7.2-2) by setting $k = 0$. This yields

$$\begin{bmatrix} x_1(1) \\ x_2(1) \end{bmatrix} = \begin{bmatrix} .368 & 0 \\ .632 & 1 \end{bmatrix} \begin{bmatrix} x_1(0) \\ x_2(0) \end{bmatrix} + \begin{bmatrix} .632 \\ .368 \end{bmatrix} e_2(0) \qquad (7.2\text{-}7)$$

If initial conditions are assumed to be zero,* equations (7.2-7) yield

$$x_1(1) = 0 = .632 e_2(0) \qquad (7.2\text{-}8\text{a})$$

and

$$x_2(1) = R = .368 e_2(0) \qquad (7.2\text{-}8\text{b})$$

which have to be satisfied simultaneously for a single $e_2(0)$. This is not possible. We conclude that a single period is not sufficient to accomplish the desired objective. Despite this negative result, it is interesting to pursue this case further to determine what can be accomplished in one period.

Solving (7.2-8b) for $e_2(0)$ yields

$$e_2(0) = \frac{R}{.368} = 2.72 R$$

Setting $e_2(0)$ to this value will assure that $x_2(1) = R$, or $c(1) = R$, but will not produce the desired condition $\dot{c}(1) = 0$. Thus, the system output will

---

*No loss of generality arises by assuming the initial state to be zero.

be equal to the system input at the sampling instant, but it will not stay there. Indeed, for $e_2(0) = 2.72R$

$$x_1(1) = (.632)2.72R = 1.74R$$

Proceeding to the second interval, we have, from the transition equations,

$$\begin{bmatrix} x_1(2) \\ x_2(2) \end{bmatrix} = \begin{bmatrix} .368 & 0 \\ .632 & 1 \end{bmatrix} \begin{bmatrix} 1.74R \\ R \end{bmatrix} + \begin{bmatrix} .632 \\ .368 \end{bmatrix} e_2(1)$$

Again we seek to satisfy the position equation $x_2(2) = R$ and disregard the velocity condition $x_1(2)$. This yields

$$e_2(1) = -2.95R$$

and

$$x_1(2) = -1.225R$$

For the next period a similar calculation yields

$$e_2(2) = 2.12R$$

The remaining terms of the sequence $e_2(k)$ may be calculated accordingly. Since the z-transform of $e_2(k)$ is given by

$$E_2(z) = \sum_{k=0}^{\infty} e_2(k) z^{-k}$$

we can write

$$E_2(z) = 2.72R - 2.95Rz^{-1} + 2.12Rz^{-2} + \cdots \qquad (7.2\text{-}9)$$

From the diagram of Figure 7.2-1 it is clear that

$$E_2(z) = D(z) \cdot E_1(z) \qquad (7.2\text{-}10)$$

Therefore, if $E_1(z)$ is known, $D(z)$ may be specified. Now

$$e_1(t) = r(t) - c(t) \qquad t \geq 0 \qquad (7.2\text{-}11)$$

For $t = kT$

$$e_1(k) = r(k) - c(k) \qquad k = 0, 1, \ldots$$

Thus, for $k = 0$

$$e_1(0) = R - 0 = R$$

and for $k > 0$

$$e_1(k) = 0$$

Thus

$$E_1(z) = R \tag{7.2-12}$$

Now, dividing (7.2-9) by (7.2-12) yields

$$D(z) = \frac{2.72 - 2.95z^{-1} + 2.12z^{-2} + \cdots}{1} \tag{7.2-13}$$

From developments to be introduced later, this transfer function can be shown to be expressible as

$$\begin{aligned} D(z) &= (2.72 - z^{-1})[1 - .717z^{-1} + (.717)^2 z^{-2} - (.717)^3 z^{-3} + \cdots] \\ &= \frac{2.72 - z^{-1}}{1 + .717z^{-1}} \end{aligned} \tag{7.2-14}$$

Equation (7.2-14) prescribes the program for the digital processor as the familiar ratio of two polynomials in z^{-1}. It will drive the system from the zero state to that for which the output of the system matches up with the input at the sampling instants but *not* in between, since the derivative of the output is not zero. The response of the system to a unit step input is shown in Figure 7.2-2. A system that has a step response as shown by this plot is called a minimal prototype system. It shows a pronounced ripple, which is sustained over a large number of periods. Because of this, its practical application is severely limited, and it is primarily of academic interest.

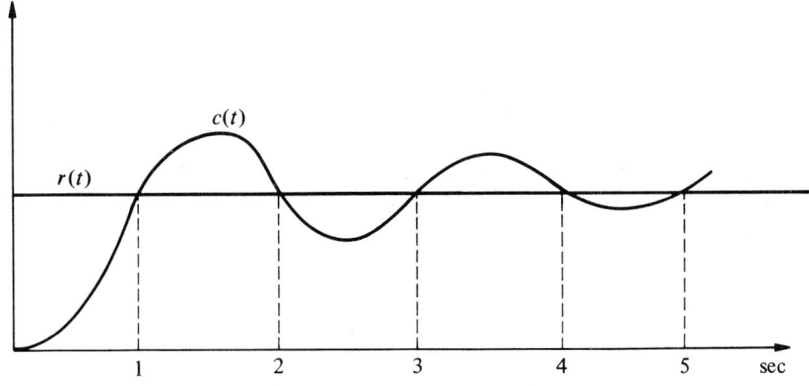

Figure 7.2-2. Response of system.

Case b. $N = 2$ (Ripple-free System)

If two sampling periods are allotted to reach the desired state as given by (7.2-6) it will be possible to insure that the system output will be ripple-free. To show this, we consider the state transition equation over two periods:

$$\begin{bmatrix} x_1(2) \\ x_2(2) \end{bmatrix} = \begin{bmatrix} .368 & 0 \\ .632 & 1 \end{bmatrix}^2 \begin{bmatrix} x_1(0) \\ x_2(0) \end{bmatrix} + \begin{bmatrix} .368 & 0 \\ .632 & 1 \end{bmatrix} \begin{bmatrix} .632 \\ .368 \end{bmatrix} e_2(0) + \begin{bmatrix} .632 \\ .368 \end{bmatrix} e_1(1) \quad (7.2\text{-}15)$$

Since

$$\begin{bmatrix} x_1(2) \\ x_2(2) \end{bmatrix} = \begin{bmatrix} 0 \\ R \end{bmatrix} \quad \text{and} \quad \begin{bmatrix} x_1(0) \\ x_2(0) \end{bmatrix} = \begin{bmatrix} 0 \\ 0 \end{bmatrix}$$

we may simplify (7.2-15) to

$$\begin{bmatrix} 0 \\ R \end{bmatrix} = \begin{bmatrix} .233 & .632 \\ .768 & .368 \end{bmatrix} \begin{bmatrix} e_2(0) \\ e_2(1) \end{bmatrix}$$

which may easily be solved to yield

$$\begin{bmatrix} e_2(0) \\ e_2(1) \end{bmatrix} = \begin{bmatrix} 1.58 \\ -.58 \end{bmatrix} R$$

The numerical values of $e_2(0)$ and $e_2(1)$ uniquely define the first two members of the control sequence $e_2(k)$ required to drive the system from the state

$$\begin{bmatrix} x_1(0) \\ x_2(0) \end{bmatrix} = \begin{bmatrix} 0 \\ 0 \end{bmatrix}$$

to the state

$$\begin{bmatrix} x_1(2) \\ x_2(2) \end{bmatrix} = \begin{bmatrix} 0 \\ R \end{bmatrix}$$

Since the desired state has been reached at the end of the second period, the remaining members of the control sequence will be zero. To verify this fact, apply (7.2-2) with

$$\begin{bmatrix} x_1(2) \\ x_2(2) \end{bmatrix} = \begin{bmatrix} 0 \\ R \end{bmatrix}$$

and $e_2(k) = 0$ for $k \geq 2$. This yields

Sec. 7.2 · Time-domain Synthesis with Minimum Settling Time

$$\begin{bmatrix} x_1(k+1) \\ x_2(k+1) \end{bmatrix} = \begin{bmatrix} .368 & 0 \\ .632 & 1 \end{bmatrix} \begin{bmatrix} 0 \\ R \end{bmatrix} = \begin{bmatrix} 0 \\ R \end{bmatrix} \quad \text{for } k \geq 2$$

The z-transform of the entire sequence is then

$$E_2(z) = R(1.58 - .58z^{-1}) \tag{7.2-16}$$

Equation (7.2-16) establishes the numerator for the digital computer transfer function.

To determine $E_1(z)$, we again make use of (7.2-11). Thus

$$e_1(0) = R$$

and

$$e_1(1) = R - c(1)$$
$$= R - x_2(1)$$

But from (7.2-7) we have

$$x_2(1) = (.368)e_2(0)$$
$$= (.368)1.58R$$
$$= .582R$$

Thus

$$e_1(1) = R(1 - .582) = .418R$$

Furthermore,

$$e_1(k) = 0, \quad k \geq 2$$

since the error is zero.

Collecting terms, we have

$$E_1(z) = (1 + .418z^{-1})R \tag{7.2-17}$$

The digital transfer function is, therefore,

$$D(z) = \frac{1.58 - .58z^{-1}}{1 + .418z^{-1}} \tag{7.2-18}$$

This transfer function will permit a ripple-free response of the system to a step input of any magnitude within two sampling periods of one second each, provided the system is initially at rest. A typical response to a sequence

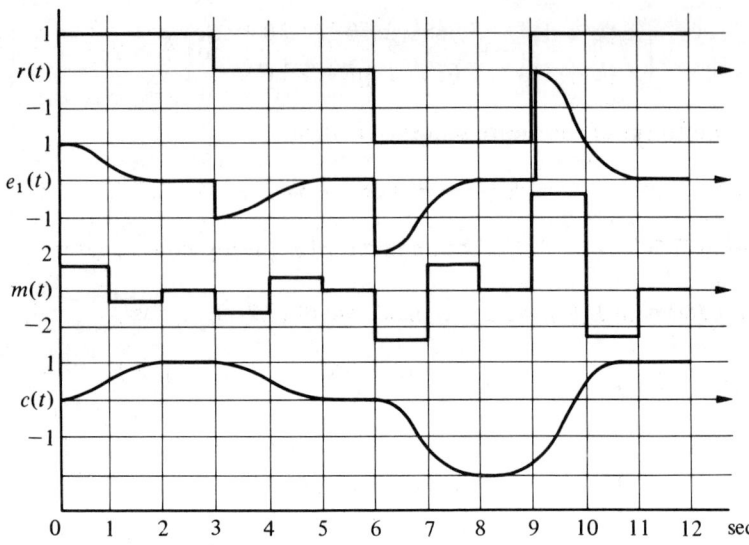

Figure 7.2-3. System response to a sequence of step inputs.

of unit step functions is shown in the response traces of Figure 7.2-3, consisting of input, error, plant input, and output.

In the previous two cases it was shown that for the example presently being discussed each sampling period that is allotted for the transition provides for one degree of freedom. Thus for $N = 1$, one degree of freedom existed; this was used to specify the conditions on one state variable, i.e., $x_2(k)$. For $N = 2$, two degrees of freedom existed; they were used to implement the constraints on two state variables $x_1(k)$ and $x_2(k)$. This resulted in ripple-free response. If more than two periods are available to complete the response to a step input, additional degrees of freedoms are provided which may be used to advantage in incorporating additional constraints. Of interest in this respect are amplitude constraints on the plant input, or on the maximum velocity or acceleration that may be tolerated during a typical transition. In what follows we shall give an illustration of this idea in the form of an input amplitude-constrained system.

Case c. $N \geq 3$ *(Input Amplitude-constrained System)*

The specific case to be considered here is

i. The plant is to respond ripple-free to a unit step input in the smallest number of sampling periods.
ii. The plant input $m(t)$ must satisfy the condition

$$|m(t)| \leq M = 1$$

In general, this problem is extremely difficult to solve. An approximation

Sec. 7.2 Time-domain Synthesis with Minimum Settling Time

to the desired control will now be obtained. The digital computer output $e_2(k)$ is calculated for two consecutive periods starting with $k = 0$. If the desired state can be reached without violating the plant input constraint, the problem is completed. If the constraints are not satisfied, we set $e_2(k)$ equal to ± 1, depending on its polarity, and add one more period to the total response time.

For the first two periods we consider (7.2-15) for $R = 1$.

$$\begin{bmatrix} x_1(2) \\ x_2(2) \end{bmatrix} = \begin{bmatrix} 0 \\ 1 \end{bmatrix} = \begin{bmatrix} .368 & 0 \\ .632 & 1 \end{bmatrix} \begin{bmatrix} .632 \\ .368 \end{bmatrix} e_2(0) + \begin{bmatrix} .632 \\ .368 \end{bmatrix} e_2(1) \quad (7.2\text{-}19)$$

Solving for $e_2(0)$ and $e_2(1)$, we obtain

$$\begin{bmatrix} e_2(0) \\ e_2(1) \end{bmatrix} = \begin{bmatrix} 1.58 \\ -.58 \end{bmatrix}$$

It is seen that $e_2(0) > 1$. This exceeds the allowable limit. Consequently, we set $e_2(0)$ equal to the closest admissible value $+1$, and calculate the state of the plant at the end of the first period in response to this input.

$$\begin{bmatrix} x_1(1) \\ x_2(1) \end{bmatrix} = \begin{bmatrix} .632 \\ .368 \end{bmatrix}$$

Proceeding now to the next two sampling periods, we have

$$\begin{bmatrix} x_1(3) \\ x_2(3) \end{bmatrix} = \begin{bmatrix} 0 \\ 1 \end{bmatrix} = \begin{bmatrix} .368 & 0 \\ .632 & 1 \end{bmatrix}^2 \begin{bmatrix} .632 \\ .368 \end{bmatrix} + \begin{bmatrix} .368 & 0 \\ .632 & 1 \end{bmatrix} \begin{bmatrix} .632 \\ .368 \end{bmatrix} e_2(1) + \begin{bmatrix} .632 \\ .368 \end{bmatrix} e_2(2)$$

Solving for $e_2(1)$ and $e_2(2)$ yields

$$\begin{bmatrix} e_2(1) \\ e_2(2) \end{bmatrix} = \begin{bmatrix} .215 \\ -.215 \end{bmatrix}$$

These values satisfy the constraints. Thus the complete control sequence is

$$e_2(k) = [1, .215, -.215, 0, 0, \ldots]$$

or

$$E_2(z) = 1 + .215z^{-1} - .215z^{-2} \quad (7.2\text{-}20)$$

Having determined the plant input sequence, we can now calculate the error sequence $e_1(k)$. Using (7.2-11), we have, for $k = 0, 1, 2,$

$k = 0$
$$e_1(0) = r(0) - c(0)$$
$$= 1 - 0 = 1$$

$k = 1$
$$e_1(1) = r(1) - c(1)$$
$$= r(1) - x_2(1)$$
$$= 1 - .368$$
$$= .632$$

$k = 2$
$$e_1(2) = r(2) - x_2(2)$$

but $x_2(2)$ is calculated from the transition equations for the second interval

$$\begin{bmatrix} x_1(2) \\ x_2(2) \end{bmatrix} = \begin{bmatrix} .368 & 0 \\ .632 & 1 \end{bmatrix} \begin{bmatrix} x_1(1) \\ x_2(1) \end{bmatrix} + \begin{bmatrix} .632 \\ .368 \end{bmatrix} e_2(1)$$

or

$$\begin{bmatrix} x_1(2) \\ x_2(2) \end{bmatrix} = \begin{bmatrix} .368 & 0 \\ .632 & 1 \end{bmatrix} \begin{bmatrix} .632 \\ .368 \end{bmatrix} + \begin{bmatrix} .632 \\ .368 \end{bmatrix} .215$$
$$= \begin{bmatrix} .368 \\ .847 \end{bmatrix}$$

so that

$$e_1(2) = 1 - .847 = .153$$

$k \geq 3$
$$e_1(k) = 0$$

Consequently,

$$E_1(z) = \mathscr{Z}\{1, .632, .153, 0, \ldots\} = 1 + .632z^{-1} + .153z^{-2} \quad (7.2\text{-}21)$$

With $E_1(z)$ and $E_2(z)$ determined, we can now specify $D(z)$.

$$D(z) = \frac{E_2(z)}{E_1(z)} = \frac{1 + .215z^{-1} - .215z^{-2}}{1 + .632z^{-1} + .153z^{-2}} \quad (7.2\text{-}22)$$

Equation (7.2-22) represents the digital transfer function that will guarantee a ripple-free response of the system shown in Figure 7.2-4 when subjected to a unit step input. The design of this system is limited to a unit step input and a unit input amplitude constraint. Should either of these two conditions be changed, a new digital transfer function will have to be determined. It seems plausible, for instance, that if the magnitude of the step input is increased it would take more than three sampling periods to complete the desired transition, since the plant input would be saturated for more than one period. It will be left as an exercise to explore this point.

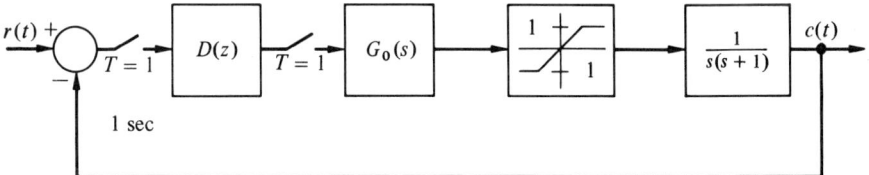

Figure 7.2-4. Block diagram of ripple-free sampled system with input amplitude constraints.

7.2-2 Response to a Ramp Input

An approach similar to the one presented in the previous section may be followed when the input is a ramp function. Let us consider the design of a digital transfer function for the system shown in Figure 7.2-5. The input is assumed to be

$$r(t) = Vt \quad \text{for } t \geq 0$$

and the initial state is taken to be zero.

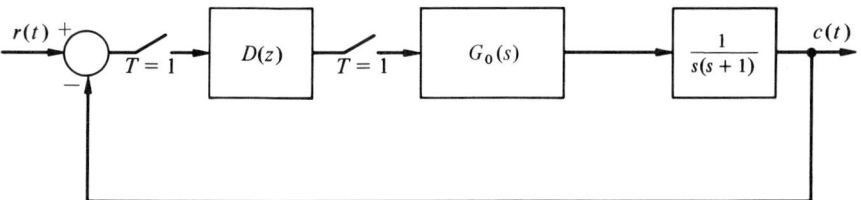

Figure 7.2-5. Sampled system with digital compensation.

The objective, as before, is to determine a digital transfer function $D(z)$ that will provide a ripple-free system response in a minimum number of sampling periods. To realize this response the following terminal conditions must be specified:

$$\begin{bmatrix} x_1(N) \\ x_2(N) \end{bmatrix} = \begin{bmatrix} V \\ VN \end{bmatrix} \quad \text{for some integer } N > 0 \qquad (7.2\text{-}23)$$

These conditions derive from the requirement that

$$\dot{c}(t) = V$$
$$c(t) = Vt$$

and from the relationships between state variables and output variables as given by (7.2-4) through (7.2-6).

The number of sampling periods required to reach steady state, which is indicated by the integer N, is to be minimized. We consider the following cases.

Case a. $N = 1$

This case is meaningless, since no input can be generated to the digital processor during the first period. This input is generated from equation (7.2-11).

$$e_1(t) = r(t) - c(t) \quad \text{for } t = 0, e_1(0) = 0$$

Thus, the first period cannot be used for generating an output from the digital computer. Hence, this problem does not become meaningful until $N \geq 2$.

Case b. $N = 2$ (*Minimal Prototype*)

It is possible to design a digital processor that is functional within two periods. However, it can be shown to suffer from the same shortcoming as demonstrated for Case a of the step input design: only one of the conditions of (7.2-23) can be satisfied. If the condition selected is

$$x_2(N) = VN, \quad N \geq 2$$

then the design will contain a ripple between sampling periods. Such a design represents the minimal prototype for a ramp input. It is left as an exercise to the student to verify this.

Case c. $N = 3$ (*Ripple-free Response*)

When N is selected as 3, it is possible to design a system that will satisfy both conditions of (7.2-23). From the state transition equations (7.2-2) we have

$$\begin{bmatrix} x_1(3) \\ x_2(3) \end{bmatrix} = \begin{bmatrix} V \\ 3V \end{bmatrix} = \begin{bmatrix} .368 & 0 \\ .632 & 1 \end{bmatrix} \begin{bmatrix} .632 \\ .368 \end{bmatrix} e_2(1) + \begin{bmatrix} .632 \\ .368 \end{bmatrix} e_2(2) \quad (7.2\text{-}24)$$

Use has been made of the fact that the initial conditions are zero and that $e_2(0)$ is taken as zero as per discussion of Case a. Solving for $e_2(1)$ and $e_2(2)$, we obtain

$$\begin{bmatrix} e_2(1) \\ e_2(2) \end{bmatrix} = \begin{bmatrix} 3.810 \\ .173 \end{bmatrix} V$$

In determining the remainder of the computer output sequence $e_2(k)$, it is readily seen that

$$e_2(k) = V, \quad k \geq 3$$

For instance, for $k = 3$ we have from the state transition equations,

$$\begin{bmatrix} x_1(4) \\ x_2(4) \end{bmatrix} = \begin{bmatrix} .368 & 0 \\ .632 & 1 \end{bmatrix}\begin{bmatrix} V \\ 3V \end{bmatrix} + \begin{bmatrix} .632 \\ .368 \end{bmatrix}V = \begin{bmatrix} V \\ 4V \end{bmatrix}$$

In summary, then,

$$E_2(z) = (0 + 3.81z^{-1} + .173z^{-2} + z^{-3} + z^{-4} + \cdots)V \quad (7.2\text{-}25)$$

The iterative application of the error equation permits the computation of the input sequence to the digital computer. This yields

$$E_1(z) = (0 + z^{-1} + .6z^{-2})V \quad (7.2\text{-}26)$$

To determine the digital transfer function, $E_2(z)$ is divided by $E_1(z)$.

$$D(z) = \frac{3.81 + .173z^{-1} + z^{-2} + z^{-4} + \cdots}{1 + .6z^{-1}}$$

The numerator may be expressed in closed form by use of the geometric series identity.

$$\begin{aligned} D(z) &= \frac{3.81 + .173z^{-1}}{1 + .6z^{-1}} + \frac{z^{-2}}{(1 - z^{-1})(1 + .6z^{-1})} \\ &= \frac{3.81 + 2.637z^{-1} + .827z^{-2}}{1 - .4z^{-1} - .6z^{-2}} \end{aligned} \quad (7.2\text{-}27)$$

7.2-3 The General Case (for Step Inputs)

A digital control system with minimum settling time as the performance objective may be designed in a very general sense. The examples presented so far were selected subject to two important restrictions. First, the plant was a single input–single output plant; second, the plant contained at least one free integrator. The latter condition provides a simple way of guaranteeing that the output equal the input with all derivatives of the output equal to zero, such as is, for instance, specified by equations (7.2-4). The presence of a free integrator in the plant lets the output of the digital computer in the steady state reach zero. It will now be shown that identical control characteristics may be obtained for plants without a free integrator.

Consider the system shown in the vector block diagram of Figure 7.2-6.

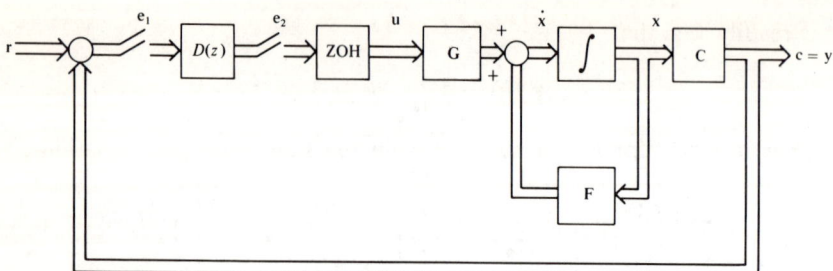

Figure 7.2-6. Multiple input–multiple output computer control system.

It is a multiple input–multiple output system. The plant is described by the linear vector differential equations

$$\frac{d}{dt}\mathbf{x} = \mathbf{Fx} + \mathbf{Gu} \tag{7.2-28}$$

$$\mathbf{c} = \mathbf{y} = \mathbf{Cx} \tag{7.2-29}$$

where $\mathbf{x}$ is an $n \times 1$ state vector
$\mathbf{u}$ is the $m \times 1$ control vector
$\mathbf{y}$ is the $p \times 1$ output vector.

Furthermore, let N be the number of sampling intervals until deadbeat response is achieved.

The objective of this problem is to design $\mathbf{D}(z)$ to cause $\mathbf{c}(t)$ to respond in a deadbeat manner to a step input $\mathbf{r}(t)$.

The computation of $\mathbf{D}(z)$ proceeds in two parts. First, the sequence of vectors for $\mathbf{e}_2(k)$ is computed to obtain the deadbeat response of $\mathbf{y}(t)$ in a minimum number of sampling periods. From the $\mathbf{y}(t)$ response, the sequence of vector $\mathbf{e}_1(k)$ may be computed. $\mathbf{D}(z)$ is to be computed such that $\mathbf{e}_2(k)$ is generated from $\mathbf{e}_1(k)$.

The output at the kth sampling period is related to the $\mathbf{e}_2$ sequence and the disturbance in the following way.

Because of the zero-order hold, $\mathbf{x}(k)$ may be computed as follows:

$$\mathbf{x}(k) = [e^{\mathbf{F}T}]^k \mathbf{x}(0) + \sum_{l=0}^{k-1} [e^{\mathbf{F}T}]^{k-l-1} \int_0^T e^{\mathbf{F}T} dt \, \mathbf{G} \mathbf{e}_2(l) \tag{7.2-30}$$

The initial condition $\mathbf{x}(0)$ is taken as zero for this problem.
Let

$$\mathbf{A} = e^{\mathbf{F}T}$$

and

$$\mathbf{B} = \int_0^T e^{\mathbf{F}T} dt \, \mathbf{G}$$

Since $\mathbf{x}(0) = \mathbf{0}$, $\mathbf{x}(k)$ may be written as

$$\mathbf{x}(k) = \sum_{l=0}^{k-1} \mathbf{A}^{k-l-1}\mathbf{B}\mathbf{e}_2(l)$$

Thus, $\mathbf{y}(k)$ becomes

$$\mathbf{y}(k) = \sum_{l=0}^{k-1} \mathbf{C}\mathbf{A}^{k-l-1}\mathbf{B}\mathbf{e}_2(l) \tag{7.2-31}$$

Following a vector step input of arbitrary size, denoted $\mathbf{r}_0$, we want $\mathbf{e}_1$ to go to zero in the minimum number of sampling periods. Thus,

$$\mathbf{r}_0 = \sum_{l=0}^{N-1} \mathbf{C}\mathbf{A}^{N-l-1}\mathbf{B}\mathbf{e}_2(l) \tag{7.2-32}$$

or, in matrix form,

$$[\mathbf{C}\mathbf{A}^{N-1}\mathbf{B} \quad \mathbf{C}\mathbf{A}^{N-2}\mathbf{B} \quad \ldots \quad \mathbf{C}\mathbf{B}] \begin{bmatrix} \mathbf{e}_2(0) \\ \mathbf{e}_2(1) \\ \vdots \\ \mathbf{e}_2(N-1) \end{bmatrix} = \mathbf{r}_0 \tag{7.2-33}$$

This expression does not guarantee that the response will be deadbeat as required; it only forces the output to $\mathbf{r}_0$. To guarantee a deadbeat response, $\dot{\mathbf{x}}(NT)$ must be zero. Since $\mathbf{u}(t)$ is constant in the interval $NT \leq t < (N+1)T$ and $\dot{\mathbf{x}}(NT) = \mathbf{0}$, then $\mathbf{x}(t)$ cannot change from NT to $(N+1)T$. Thus, $\mathbf{u}(NT)$ will be the control required to effect the step change in the output response.

$$\mathbf{u}(t) = \mathbf{u}(NT) = \mathbf{e}_2(NT) \quad \text{for } t \geq NT \tag{7.2-34}$$

The expression for $\dot{\mathbf{x}}(NT)$ may be computed as follows. From the state equation

$$\dot{\mathbf{x}}(NT) = \mathbf{F}\mathbf{x}(NT) + \mathbf{G}\mathbf{e}_2(NT)$$

Equating $\dot{\mathbf{x}}(NT) = \mathbf{0}$ and substituting the expression for $\mathbf{x}(NT)$, we can write that

$$\mathbf{0} = \mathbf{F}\left[\sum_{l=0}^{N-1} \mathbf{A}^{N-l-1}\mathbf{B}\mathbf{e}_2(l)\right] + \mathbf{G}\mathbf{e}_2(N)$$

or

$$\mathbf{0} = \sum_{l=0}^{N-1} \mathbf{F}\mathbf{A}^{N-l-1}\mathbf{B}\mathbf{e}_2(l) + \mathbf{G}\mathbf{e}_2(N)$$

In matrix form, this equation may be written

$$[FA^{N-1}B \quad FA^{N-2}B \quad \ldots \quad FB \quad G] \begin{bmatrix} e_2(0) \\ e_2(1) \\ \vdots \\ e_2(N-1) \\ e_2(N) \end{bmatrix} = [0] \qquad (7.2\text{-}35)$$

Equations (7.2-33) and (7.2-35) may be combined to form the system of equations that must be solved to determine the e_2 sequence.

$$\begin{bmatrix} CA^{N-1}B & CA^{N-2}B & \ldots & CB & 0 \\ FA^{N-1}B & FA^{N-2}B & \ldots & FB & G \end{bmatrix} \begin{bmatrix} e_2(0) \\ e_2(1) \\ \vdots \\ e_2(N-1) \\ e_2(N) \end{bmatrix} = \begin{bmatrix} r_0 \\ 0 \end{bmatrix} \qquad (7.2\text{-}36)$$

A discussion of the solution to equation (7.2-36) will be given later, but first let us consider computation of the e_1 sequence and of $D(z)$. From the error equation we have that

$$e_1(k) = r_0 - y(k)$$

But $y(k)$ may be computed from equation (7.2-31).

$$e_1(k) = r_0 - \sum_{l=0}^{k-1} C\{A^{k-l-1}Be_2(l)\} \qquad (7.2\text{-}37)$$

On the other hand, $e_2(l)$ is obtained from the solution of equation (7.2-36). Let us denote that solution in the following way:

$$e_2(l) = P(l)r_0, \qquad l = 0, 1, \ldots, N$$

Thus, equation (7.2-37) may be written

$$e_1(k) = \left[I - \sum_{l=0}^{k-1} \{CA^{k-l-1}BP(l)\}\right] r_0 \qquad (7.2\text{-}38)$$

Finally, $D(z)$ may be computed as follows. Taking the z-transform of the $e_1(k)$ and $e_2(k)$ sequences, we get

Sec. 7.2 Time-domain Synthesis with Minimum Settling Time

$$\mathbf{E}_1(z) = \sum_{k=0}^{\infty} \mathbf{e}_1(k) z^{-k}$$

$$= \sum_{k=0}^{N-1} z^{-k} \left[\mathbf{I} - \sum_{l=0}^{k-1} \{\mathbf{CA}^{k-l-1}\mathbf{BP}(l)\} \right] \mathbf{r}_0 \quad (7.2\text{-}39)$$

The infinite series due to the z-transform of $\mathbf{e}_1(k)$ is terminated at $N-1$, since the coefficients $\mathbf{e}_1(k)$ for $k > N-1$ are identically zero.

$$\mathbf{E}_2(z) = \sum_{k=0}^{\infty} \mathbf{P}(k) \mathbf{r}_0 z^{-k}$$

Since the input to the plant is constant after $N-1$ sampling periods, we have $\mathbf{e}_2(k) = \mathbf{P}(N)$ for $k \geq N$, so that

$$\mathbf{E}_2(z) = \left[\sum_{k=0}^{N-1} z^{-k}\mathbf{P}(k) + \mathbf{P}(N) \sum_{k=N}^{\infty} z^{-k} \right] \mathbf{r}_0$$

or

$$\mathbf{E}_2(z) = \left[\sum_{k=0}^{N-1} z^{-k}\mathbf{P}(k) + \mathbf{P}(N) \frac{z^{-N}}{1-z^{-1}} \right] \mathbf{r}_0 \quad (7.2\text{-}40)$$

Since

$$\mathbf{E}_2(z) = \mathbf{D}(z) \mathbf{E}_1(z)$$

we may combine equations (7.2-39) and (7.2-40) to obtain

$$\left[\sum_{k=0}^{N-1} z^{-k}\mathbf{P}(k) + \mathbf{P}(N) \frac{z^{-N}}{1-z^{-1}} \right] \mathbf{r}_0$$

$$= \mathbf{D}(z) \left[\sum_{k=0}^{N-1} z^{-k} \left(\mathbf{I} - \sum_{l=0}^{k-1} \{\mathbf{CA}^{k-l-1}\mathbf{BP}(l)\} \right) \right] \mathbf{r}_0 \quad (7.2\text{-}41)$$

Since equation (7.2-41) must hold for arbitrary $\mathbf{r}_0$, the coefficient matrices premultiplying $\mathbf{r}_0$ must be equal. Thus,

$$\left[\sum_{k=0}^{N-1} z^{-k}\mathbf{P}(k) + \mathbf{P}(N) \frac{z^{-N}}{1-z^{-1}} \right]$$

$$= \mathbf{D}(z) \left[\sum_{k=0}^{N-1} z^{-k} \left(\mathbf{I} - \sum_{l=0}^{k-1} \{\mathbf{CA}^{k-l-1}\mathbf{BP}(l)\} \right) \right] \quad (7.2\text{-}42)$$

Solving equation (7.2-42) for $\mathbf{D}(z)$, we obtain the result

$$\mathbf{D}(z) =$$
$$\left[\sum_{k=0}^{N-1} z^{-k}\mathbf{P}(k) + \mathbf{P}(N) \frac{z^{-N}}{1-z^{-1}} \right] \left[\sum_{k=0}^{N-1} z^{-k} \left(\mathbf{I} - \sum_{l=0}^{N-1} \{\mathbf{CA}^{k-l-1}\mathbf{BP}(l)\} \right) \right]^{-1}$$

$$(7.2\text{-}43)$$

The computation of **P**(k) may be demonstrated in the course of the development of four cases.

7.2-4 Scalar Case: $n = m = p = 1$

State equation

$$\dot{x} = ax + bu$$

Output equation

$$y = cx$$

$$\mathbf{A} = e^{aT}, \qquad \mathbf{B} = \int_0^T e^{at} dt\, b = a^{-1}[e^{aT} - 1]b$$

In this case equation (7.2-36) has two rows (i.e., two equations), so at the end of the first sampling period the output will be zero.

$$\begin{bmatrix} ca^{-1}[e^{aT} - 1]b & 0 \\ [e^{aT} - 1]b & b \end{bmatrix} \begin{bmatrix} e_2(0) \\ e_2(1) \end{bmatrix} = \begin{bmatrix} r_0 \\ 0 \end{bmatrix}$$

Solving for $e_2(0)$ and $e_2(1)$, we determine the inverse of the coefficient matrix, obtaining

$$\begin{bmatrix} e_2(0) \\ e_2(1) \end{bmatrix} = \begin{bmatrix} \frac{a/bc}{e^{aT} - 1} & 0 \\ -\frac{a}{bc} & \frac{1}{b} \end{bmatrix} \begin{bmatrix} r_0 \\ 0 \end{bmatrix}$$

Thus

$$e_2(0) = \frac{a}{b(e^{aT} - 1)c} r_0 \quad \text{and} \quad e_2(1) = -\frac{a}{bc} r_0$$

and

$$\mathbf{P}(0) = \frac{a}{b(e^{aT} - 1)c} \quad \text{and} \quad \mathbf{P}(1) = -\frac{a}{bc}$$

It is seen that $N = 1$ in this case. The digital transfer function is, therefore,

$$D(z) = [z^{-0}(1 - 0)]^{-1} \left[z^{-0} \mathbf{P}(0) + \mathbf{P}(1) \frac{z^{-1}}{1 - z^{-1}} \right]$$

Upon substituting for **P**(0) and **P**(1), we obtain the following:

$$D(z) = \frac{a(1 - e^{aT}z^{-1})}{bc(1 - e^{aT})(1 - z^{-1})}$$

7.2-5 Single Input–Second-order System: $n = 2$, $m = p = 1$

Let the system be represented by the state equations

$$\frac{d}{dt}\begin{bmatrix} x_1 \\ x_2 \end{bmatrix} = \begin{bmatrix} -1 & 0 \\ 1 & 0 \end{bmatrix}\begin{bmatrix} x_1 \\ x_2 \end{bmatrix} + \begin{bmatrix} 1 \\ 0 \end{bmatrix} u$$

with the output given by

$$y = \begin{bmatrix} 0 & 1 \end{bmatrix}\begin{bmatrix} x_1 \\ x_2 \end{bmatrix}$$

An inspection of equation (7.2-36) reveals that $N = 2$. The state transition matrix for the case $T = 1$ is

$$\mathbf{A} = \begin{bmatrix} e^{-1} & 0 \\ 1 - e^{-1} & 1 \end{bmatrix}$$

The input transition matrix is

$$\mathbf{B} = \begin{bmatrix} 1 - e^{-1} \\ e^{-1} \end{bmatrix}$$

Other expressions needed are

$$\mathbf{CAB} = \begin{bmatrix} 0 & 1 \end{bmatrix}\begin{bmatrix} e^{-1} & 0 \\ 1 - e^{-1} & 1 \end{bmatrix}\begin{bmatrix} 1 - e^{-1} \\ e^{-1} \end{bmatrix} = (1 - e^{-1})^2 + e^{-1}$$

$$\mathbf{CB} = e^{-1}$$

$$\mathbf{FAB} = \begin{bmatrix} -1 & 0 \\ 1 & 0 \end{bmatrix}\begin{bmatrix} e^{-1} & 0 \\ 1 - e^{-1} & 1 \end{bmatrix}\begin{bmatrix} 1 - e^{-1} \\ e^{-1} \end{bmatrix} = \begin{bmatrix} -e^{-1}(1 - e^{-1}) \\ e^{-1}(1 - e^{-1}) \end{bmatrix}$$

$$\mathbf{FB} = \begin{bmatrix} -1 & 0 \\ 1 & 0 \end{bmatrix}\begin{bmatrix} 1 - e^{-1} \\ e^{-1} \end{bmatrix} = \begin{bmatrix} e^{-1} - 1 \\ 1 - e^{-1} \end{bmatrix}$$

Thus, equation (7.2-36) becomes

$$\begin{bmatrix} (1 - e^{-1})^2 + e^{-1} & e^{-1} & 0 \\ -e^{-1}(1 - e^{-1}) & e^{-1} - 1 & 1 \\ e^{-1}(1 - e^{-1}) & 1 - e^{-1} & 0 \end{bmatrix}\begin{bmatrix} e_2(0) \\ e_2(1) \\ e_2(2) \end{bmatrix} = \begin{bmatrix} r_0 \\ 0 \\ 0 \end{bmatrix}$$

With the exponentials evaluated, this becomes

$$\begin{bmatrix} .768 & .368 & 0 \\ -.232 & -.632 & 1 \\ .232 & .632 & 0 \end{bmatrix} \begin{bmatrix} e_2(0) \\ e_2(1) \\ e_2(2) \end{bmatrix} = \begin{bmatrix} r_0 \\ 0 \\ 0 \end{bmatrix}$$

When the inverse of the coefficient matrix is computed, the vector e_2 can be calculated.

$$\begin{bmatrix} e_2(0) \\ e_2(1) \\ e_2(2) \end{bmatrix} = \begin{bmatrix} 1.58 \\ -.58 \\ 0 \end{bmatrix} r_0$$

Thus

$$\mathbf{P}(0) = 1.58, \quad \mathbf{P}(1) = 0.58, \quad \mathbf{P}(2) = 0.0$$

$\mathbf{P}(2)$ in this case is zero, because the system contains a free integrator. The digital transfer function becomes, by use of (7.2-43),

$$D(z) = \left[\mathbf{P}(0) + \mathbf{P}(1)z^{-1} + \mathbf{P}(2) \frac{z^{-2}}{1 - z^{-1}} \right] \{z^0(1) + z^{-1}[1 - \mathbf{CBP}(0)]\}^{-1}$$

With the numbers substituted, this is

$$D(z) = \frac{1.58 - .58z^{-1}}{1 + .418z^{-1}}$$

It is seen that this is identical to the expression derived earlier by equation (7.2-18).

7.2-6 Multiple Input–Multiple Output
Fourth-order System: $n = 4$, $m = p = 2$

The state equations of this system are given as

$$\begin{bmatrix} \dot{x}_1 \\ \dot{x}_2 \\ \dot{x}_3 \\ \dot{x}_4 \end{bmatrix} = \begin{bmatrix} 1 & 1 & -5 & -1 \\ 0 & -2 & 0 & 0 \\ 2 & 1 & -6 & -1 \\ -2 & -1 & 2 & -3 \end{bmatrix} \begin{bmatrix} x_1 \\ x_2 \\ x_3 \\ x_4 \end{bmatrix} + \begin{bmatrix} 1 & 1 \\ 0 & 2 \\ 0 & 2 \\ 0 & -1 \end{bmatrix} \begin{bmatrix} u_1 \\ u_2 \end{bmatrix}$$

The output equations of this system are

Sec. 7.2 Time-domain Synthesis with Minimum Settling Time

$$\begin{bmatrix} y_1 \\ y_2 \end{bmatrix} = \begin{bmatrix} 3 & 2 & -3 & 2 \\ 1 & 2 & 1 & 3 \end{bmatrix} \begin{bmatrix} x_1 \\ x_2 \\ x_3 \\ x_4 \end{bmatrix}$$

For $T = .1$ the state transition matrix is computed as

$$\mathbf{A} = \begin{bmatrix} 1.0 & .0779 & -.398 & -.0705 \\ .0 & .819 & .0 & .0 \\ .164 & .0779 & .506 & -.0705 \\ -.164 & -.0779 & .164 & .741 \end{bmatrix}$$

Similarly, the input transition matrix is

$$\mathbf{B} = \begin{bmatrix} .104 & .0734 \\ .0 & .1813 \\ .0088 & .169 \\ -.0088 & -.0861 \end{bmatrix}$$

Because the system has two inputs and two outputs, it follows that $N = 2$. Hence, equations (7.2-36) take on the form

$$\begin{bmatrix} .214 & -.086 & .268 & -.095 & .0 & .0 \\ .064 & .259 & .086 & .346 & .0 & .0 \\ .019 & -.346 & .068 & -.501 & 1.0 & 1.0 \\ .0 & -.297 & .0 & -.362 & .0 & 2.0 \\ .106 & -.432 & .164 & -.597 & .0 & 2.0 \\ -.106 & .211 & -.164 & -.267 & .0 & -1.0 \end{bmatrix} \begin{bmatrix} e_2^1(0) \\ e_2^1(1) \\ e_2^1(2) \\ e_2^2(0) \\ e_2^2(1) \\ e_2^2(2) \end{bmatrix} = \begin{bmatrix} r_0^1 \\ r_0^2 \\ 0 \\ 0 \\ 0 \\ 0 \end{bmatrix}$$

where the corresponding entries are computed by using the matrices $\mathbf{F}$, $\mathbf{A}$, $\mathbf{B}$, and $\mathbf{C}$.

The matrices $\mathbf{P}(l)$ may be obtained upon inverting the coefficient matrix of the last equation. Thus,

$$\begin{bmatrix} \mathbf{P}(0) \\ \mathbf{P}(1) \\ \mathbf{P}(2) \end{bmatrix} = \begin{bmatrix} 24.01 & 1.575 \\ -1.969 & 9.843 \\ -15.750 & .195 \\ .962 & -4.814 \\ .529 & .353 \\ -.118 & .588 \end{bmatrix}$$

Now,

$$\mathbf{P}(0) = \begin{bmatrix} 24.01 & 1.575 \\ -1.969 & 9.843 \end{bmatrix}, \quad \mathbf{P}(1) = \begin{bmatrix} -15.750 & .195 \\ .962 & -4.814 \end{bmatrix},$$

$$\mathbf{P}(2) = \begin{bmatrix} .529 & .353 \\ -.118 & .588 \end{bmatrix}$$

Finally, the digital transfer function may be computed by use of equation (7.2-43) for $k = 2$.

$$\mathbf{D}(z) = \left[\mathbf{P}(0) + z^{-1}\mathbf{P}(1) + \mathbf{P}(2)\frac{z^{-2}}{1 - z^{-1}} \right]\{\mathbf{I} + z^{-1}[\mathbf{I} - \mathbf{CBP}(0)]\}^{-1}$$

Upon substitution of the numerical values of the respective matrices, this becomes a matrix of four transfer functions; that is

$$\mathbf{D}(z) = \begin{bmatrix} \dfrac{24.01 - 39.76z^{-1} + 16.339z^{-2}}{1 - 6.621z^{-1} - 1.08z^{-2}} & \dfrac{1.575 - 1.38z^{-1} + .258z^{-2}}{.515z^{-1} - .515z^{-2}} \\ \dfrac{1.969 + 2.931z^{-1} + 1.08z^{-2}}{1.393z^{-1} - 1.393z^{-2}} & \dfrac{9.843 - 3.542z^{-1} + 5.402z^{-2}}{1 - 3.542z^{-1} + 2.542z^{-2}} \end{bmatrix}$$

7.3 Minimal Prototype Design Using z-Transform Method

We now consider an alternative to the time-domain approach of designing a system with minimum settling time by using methods available to us from the study of sampled-data systems using the z-transform. Namely, design a controlled system such that for a given continuous plant:

1. The overall response and the response of all elements of the system must be nonanticipative.
2. The steady-state error for all polynomial inputs of degree equal to or less than q is zero.
3. The transient response should be as fast as possible, and the settling time should be equal to a finite number of sampling intervals.

Consider the system shown in Figure 7.3-1 for a typical example. From (6.5-13), the relationship

$$\frac{C(z)}{R(z)} = K(z) = \frac{D(z)\mathscr{Z}[G_m(s)G(s)]}{1 + D(z)\mathscr{Z}[G_m(s)G(s)]} \qquad (7.3\text{-}1)$$

Sec. 7.3 Minimal Prototype Design Using z-Transform Method

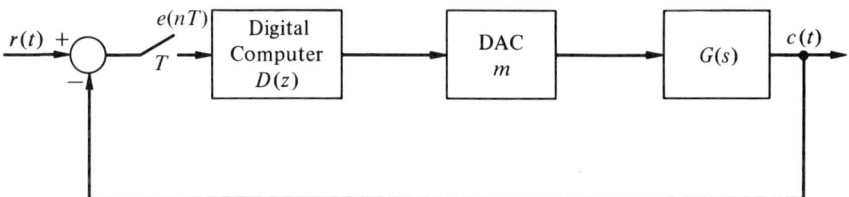

Figure 7.3-1. Sampled-data feedback system.

was established. For rational $G(s)$, we have

$$\mathscr{L}[G_m(s)G(s)] = \frac{p_r z^{-l} + \cdots + p_s z^{-t}}{q_0 + q z^{-1} + \cdots + q_b z^{-b}}$$

$$= c_1 z^{-l} + c_2 z^{-l-1} + \cdots \quad (7.3\text{-}2)$$

In order that $\mathscr{L}[G_m(s)G(s)]$ be nonanticipative, the integer l must be equal to or greater than zero. If this is not true, then the output precedes the input, which corresponds to an anticipative system. This is predicated on the presence of q_0 in (7.3-2), which is a necessary condition for nonanticipativeness.

The transfer function $D(z)$ is of the form

$$D(z) = \frac{a_0 + a_1 z^{-1} + \cdots + a_u z^{-u}}{1 + b_1 z^{-1} + b_2 z^{-2} + \cdots + b_j z^{-j}} \quad (7.3\text{-}3)$$

Inserting (7.3-2) and (7.3-3) into (7.3-1) and simplifying, we obtain

$$K(z) = \frac{k_1 z^{-l} + \cdots + k_p z^{-p}}{l_0 + l_1 z^{-1} + \cdots + l_q z^{-q}} \quad (7.3\text{-}4)$$

In order that $K(z)$ be nonanticipative, its numerator must contain z^{-1} to a power equal to or greater than the lowest power of l appearing in $\mathscr{L}[G_m(s)G(s)]$ in (7.3-2). Again, l_0 must appear for reasons similar to the appearance of q_0 in (7.3-2). The requirement of nonanticipativeness is met if the above properties are satisfied.

It is further desired that $D(z)$ be selected so that the steady-state error at the sampling times is zero. If the desired form of $K(z)$ can be found, then $D(z)$ may be determined by using the identity (7.3-1); that is,

$$D(z) = \frac{K(z)}{\mathscr{L}[G_m(s)G(s)][1 - K(z)]} \quad (7.3\text{-}5)$$

From Figure 7.3-1, the relationship

$$E(z) = R(z) - C(z)$$

follows. Since $C(z) = K(z)R(z)$, we have

$$E(z) = [1 - K(z)]R(z) \qquad (7.3\text{-}6)$$

Since the steady-state error at the sampling time is required to be zero, this implies $e(\infty) = 0$. Using the final value theorem, we have

$$e(\infty) = \lim_{z \to 1} \{(1 - z^{-1})[1 - K(z)]R(z)\}$$

For polynomial inputs of order less than or equal to q, $R(z)$ is of the form

$$R(z) = \frac{A(z)}{(1 - z^{-1})^{q+1}}$$

with $A(z)$ being a polynomial in z^{-1}. Therefore,

$$e(\infty) = \lim_{z \to 1} (1 - z^{-1})[1 - K(z)]\frac{A(z)}{(1 - z^{-1})^{q+1}} \qquad (7.3\text{-}7)$$

To guarantee that the steady-state error is zero for all such polynomials, the term $1 - K(z)$ must be selected so that

$$1 - K(z) = (1 - z^{-1})^{q+1} F(z) \qquad (7.3\text{-}8)$$

where $F(z)$ is a ratio of polynomials in z^{-1} that is analytic at $z = 1$. Substituting (7.3-8) into (7.3-6) gives

$$E(z) = (1 - z^{-1})^{q+1} F(z) R(z)$$
$$= F(z)A(z) = \frac{N(z)}{D(z)}A(z) \qquad (7.3\text{-}9)$$

By dividing $D(z)$ into $N(z)A(z)$, the time history of the error will evolve; that is,

$$E(z) = e_0 + e_1 z + \cdots + e_N z^N + \cdots$$

Requirement 3 is satisfied if this expression is terminated with a finite number of terms and the highest power of z^{-1} is the minimum possible. An investigation of (7.3-9) reveals that these properties are satisfied if $F(z)$ is set equal to a constant; for convenience we choose $F(z) = 1$. From (7.3-8), we have

$$1 - K(z) = (1 - z^{-1})^{q+1} \qquad (7.3\text{-}10)$$

so that the expression for the desired transfer of the digital computer compensator is, by (7.3-5),

$$D(z) = \frac{1-(1-z^{-1})^{q+1}}{(1-z^{-1})^{q+1}\mathscr{Z}[G_m(s)G(s)]} \qquad (7.3\text{-}11)$$

EXAMPLE 7.3-1

Determine $K(z)$ and the settling times for a step, ramp, and acceleration input.

(i) *Step Input*

For a step input $q = 0$; therefore, (7.3-10) gives us

$$K(z) = 1 - (1-z^{-1})^1 = z^{-1}$$

(ii) *Ramp Input*

For ramp inputs $q = 1$; therefore,

$$K(z) = 1 - (1-z^{-1})^2 = 2z^{-1} - z^{-2}$$

(iii) For acceleration inputs $q = 2$; therefore,

$$K(z) = 1 - (1-z^{-1})^3 = 3z^{-1} - 3z^{-2} + z^{-3}$$

The settling times for the various inputs can be determined by noting that for $F(z) = 1$, $E(z)$ is given by

$$E(z) = [1 - K(z)]R(z)$$

For different values of q, we have

Step input of magnitude R:

$$E(z) = [1-z^{-1}]\frac{R}{1-z^{-1}} = R$$

Therefore,

$$e(nT) = 0 \quad \text{for } n = 1, 2, \ldots$$
$$e(0) = R$$

Ramp input of slope V:

$$E(z) = [1-z^{-1}]^2 \frac{VTz^{-1}}{(1-z^{-1})^2} = VTz^{-1}$$

Therefore,

$$e(T) = VT$$
$$e(nT) = 0 \quad \text{for } n = 2, 3, 4, \ldots$$

Acceleration input of magnitude A:

$$E(z) = [(1 - z^{-1})]^2 \frac{ATz^{-1}(1 + z^{-1})}{(1 - z^{-1})^3}$$
$$= ATz^{-1} + ATz^{-2}$$

Therefore,

$$e(T) = AT$$
$$e(2T) = AT$$
$$e(nT) = 0 \quad \text{for } n = 3, 4, 5, \ldots$$

Table 7.3-1 summarizes these results.

Table 7.3-1 Minimal Response Characteristics

Input	$R(t)$	$R(z)$	$K(z) = \dfrac{C(z)}{R(z)}$	Settling Time in Sampling Periods
Step	$\alpha u(t)$	$\dfrac{\alpha}{1 - z^{-1}}$	z^{-1}	T
Ramp	$\alpha t u(t)$	$\dfrac{\alpha T z^{-1}}{(1 - z^{-1})^2}$	$2z^{-1} - z^{-2}$	$2T$
Acceleration	$\alpha t^2 u(t)$	$\dfrac{\alpha T^2 z^{-1}(1 + z^{-1})}{(1 - z^{-1})^3}$	$3z^{-1} - 3z^{-2} + z^{-3}$	$3T$

In the minimal response design, the value of the error signal is driven to zero at the sampling instants. However, between sampling periods, the error signal need not be zero. If this is the case, the actual output signal will tend to oscillate about the desired output, which for unity feedback is the applied input.

EXAMPLE 7.3-2

Design a minimal response digital computer compensator for the system shown in Figure 7.3-1 when a unit step is applied. For this problem

$$G(s) = \frac{1}{s(s + 1)}$$

This is the same problem treated in Section 7.2.

From Table 7.3-1, we have $K(z) = z^{-1}$ for the step input. Now

$$\mathscr{L}\left[G_o(s)\frac{1}{s(s+1)}\right] = (1 - z^{-1})\mathscr{L}\left[\frac{1}{s^2(s+1)}\right] = \frac{z^{-1}[1 + (e - 2)z^{-1}]}{(1 - z^{-1})(e - z^{-1})}$$

Inserting this result along with $K(z) = z^{-1}$ into (7.3-5) gives us the desired digital computer compensator transfer function. Namely,

$$D(z) = \frac{e - z^{-1}}{1 + (e - 2)z^{-1}}$$

which is in agreement with the results (7.2-13) found in Section 7.2.

EXAMPLE 7.3-3

Design a minimal-response digital computer compensator for the system shown in Figure 7.3-2 when ramp inputs are applied.

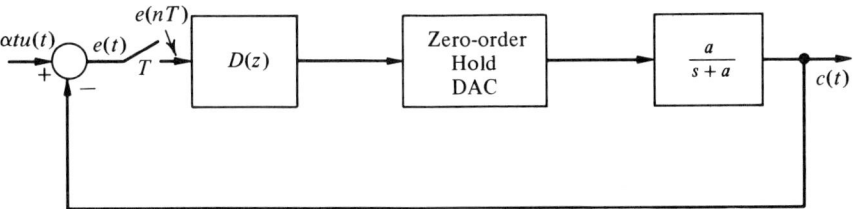

Figure 7.3-2. Minimal response system for step inputs.

From Table 7.3-1, we can find that the desired transfer function $K(z)$ is

$$K(z) = \frac{C(z)}{R(z)} = 2z^{-1} - z^{-2} \qquad (7.3\text{-}12)$$

It can be shown that

$$\mathscr{L}\left[G_0(s)\frac{a}{s+a}\right] = (1 - e^{-aT})\left[\frac{z^{-1}}{1 - z^{-1}e^{-aT}}\right] \qquad (7.3\text{-}13)$$

Inserting (7.3-12) and (7.3-13) into (7.3-5), we find the desired transfer function of the digital computer compensator to be

$$D(z) = \frac{(2 - z^{-1})(1 - z^{-1}e^{-aT})}{(1 - e^{-aT})(1 - z^{-1})^2}$$

A typical response of this system is shown in Figure 7.3-3.

It is possible to design systems that have both finite settling times and zero error for all time after the system has settled. The design of such systems was treated in Section 7.2. We have treated unity feedback systems exclusively for minimal-response design. However, similar methods may be used if the output signal $C(s)$ is fed back through a system with transfer function $H(s)$.

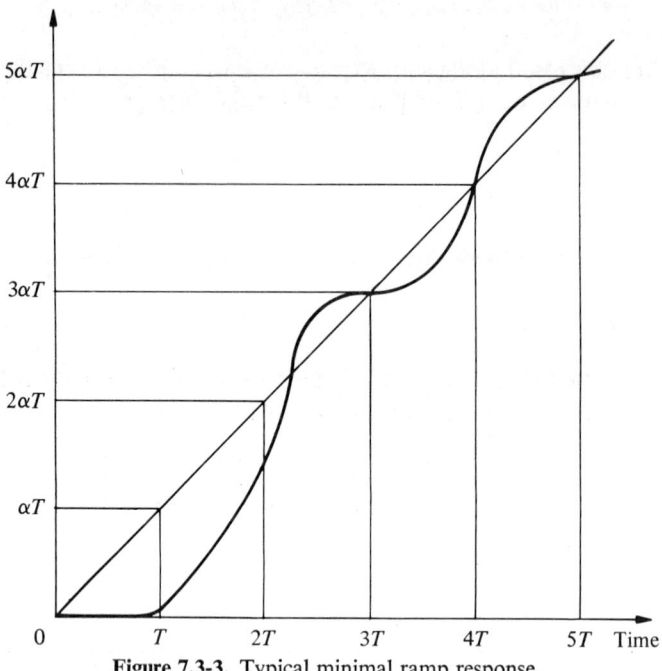

Figure 7.3-3. Typical minimal ramp response.

7.4 Controllability and Observability

Before an in-depth discussion of the design of controllers for linear discrete systems may be made, it is first necessary to introduce the concept of controllability. Consider a linear discrete system characterized by the vector difference equation

$$\mathbf{x}(k+1) = \mathbf{A}\mathbf{x}(k) + \mathbf{B}u(k) \qquad (7.4\text{-}1)$$

where $\mathbf{x}(k)$ is the $n \times 1$ state vector at the kth iteration
$\mathbf{A}$ is the $n \times n$ transition matrix that has an inverse
$\mathbf{B}$ is the $n \times 1$ control vector
$u(k)$ is the scalar input at the kth iteration
k is the discrete iteration time.

7.4-1 Controllability

System (7.4-1) is said to be completely controllable if it is possible to force the system from any arbitrary initial state $\mathbf{x}(0)$ to any arbitrary desired state $\mathbf{x}_D$ in a finite number of iteration times.

Essentially, controllability indicates whether a system may be properly controlled. In order to determine what properties a linear discrete system must satisfy so as to be controllable, we repeatedly apply (7.4-1), obtaining

$$x(1) = Ax(0) + Bu(0)$$
$$x(2) = Ax(1) + Bu(1)$$
$$= A^2x(0) + ABu(0) + Bu(1)$$
$$\cdots \cdots \cdots \cdots \cdots$$
$$x(N) = A^N x(0) + A^{N-1} Bu(0) + \cdots + ABu(N-2) + Bu(N-1) \quad (7.4\text{-}2)$$

According to the definition of controllability, system (7.4-1) is controllable if it is possible to select a value of N and $u(0), u(1), \ldots, u(N-1)$ such that $x(N) = x_D$ with both $x(0)$ and x_D being fixed but arbitrary. Assume that such a selection has been made; that is,

$$x_D - A^N x(0) = h_{N-1} u(0) + h_{N-2} u(1) + \cdots + h_0 u(N-1) \quad (7.4\text{-}3)$$

where

$$h_k = A^k B \quad \text{for } k = 0, 1, 2, \ldots$$

is an $n \times 1$ vector.

For (7.4-3) to be satisfied for arbitrary $x(0)$ and x_D, it is necessary that there be n linearly independent vectors in the set $\{h_0, h_1, \ldots, h_{N-1}\}$. This immediately implies that $N \geq n$. In fact, it may be shown that for the nth-order linear discrete system of (7.4-1) to be controllable, it is necessary and sufficient that the n vectors $h_0, h_1, \ldots, h_{n-1}$ form a set of linearly independent vectors. To help comprehend this fact, rewrite (7.4-3) for $N = n$ as

$$x_D - A^n x(0) = [h_0, h_1, \ldots, h_{n-1}] \begin{bmatrix} u(0) \\ u(1) \\ \vdots \\ u(n-1) \end{bmatrix} = H_n u_n \quad (7.4\text{-}4)$$

where H_n is an $n \times n$ matrix whose jth column is h_{j-1}
u_n is an $n \times 1$ vector whose jth element is $u(j-1)$.

Equation (7.4-4) will have a solution u_n for arbitrary vectors $x(0)$ and x_D if and only if H_n is nonsingular. This immediately implies that the columns h_{j-1} of H_n are linearly independent.

Conventionally, it is said that system (7.4-1) is controllable if the vectors $(f_1, f_2, \ldots, f_n)$ are linearly independent, where

$$f_k = -A^{-k} B \quad (7.4\text{-}5)$$

is an $n \times 1$ vector.

It is an easy exercise to show that if $\mathbf{h}_0, \mathbf{h}_1, \ldots, \mathbf{h}_{n-1}$ are linearly independent then so are $\mathbf{f}_1, \mathbf{f}_2, \ldots, \mathbf{f}_n$ and vice versa.

Fortunately, most practical linear discrete systems are controllable.

EXAMPLE 7.4-1

To illustrate how one determines the controllability characteristics of a specific system, consider a system governed by the second-order differential equation

$$\frac{d^2c}{dt^2} + \frac{dc}{dt} = u \qquad (7.4\text{-}6)$$

as shown in Figure 7.4-1

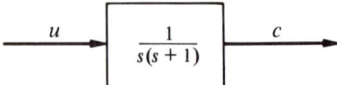

Figure 7.4-1. Second-order system.

The system (7.4-6) may be characterized by

$$\frac{d}{dt}\begin{bmatrix} x_1 \\ x_2 \end{bmatrix} = \begin{bmatrix} -1 & 0 \\ 1 & 0 \end{bmatrix}\begin{bmatrix} x_1 \\ x_2 \end{bmatrix} + \begin{bmatrix} 1 \\ 0 \end{bmatrix}u(t)$$

and

$$c(t) = [0 \quad 1]\begin{bmatrix} x_1 \\ x_2 \end{bmatrix}$$

where $x_1(t) = c$ and $x_2(t) = dc/dt$.

Taking into account the fact that the input $u(t)$ is constrained to be constant over one-second intervals, we find

$$\begin{bmatrix} x_1(k+1) \\ x_2(k+1) \end{bmatrix} = \begin{bmatrix} 1 & .632 \\ 0 & .368 \end{bmatrix}\begin{bmatrix} x_1(k) \\ x_2(k) \end{bmatrix} + \begin{bmatrix} .368 \\ .632 \end{bmatrix}u(k)$$

from which we have

$$\mathbf{A}^{-1} = \begin{bmatrix} 1 & -1.718 \\ 0 & 2.718 \end{bmatrix}, \qquad \mathbf{B} = \begin{bmatrix} .368 \\ .632 \end{bmatrix}$$

The vectors $\mathbf{f}_1$ and $\mathbf{f}_2$ as given by (7.4-5) are

$$\mathbf{f}_1 = \begin{bmatrix} .718 \\ -1.718 \end{bmatrix}, \qquad \mathbf{f}_2 = \begin{bmatrix} 3.671 \\ -4.671 \end{bmatrix}$$

which we may verify to be linearly independent. This system is therefore controllable.

Some simplification, in the thought process at least, is obtained if the desired state x_D is set equal to the zero state (origin in state space). No loss of generality results from this assumption, since it is possible to define a new state vector x_{new} related to the former state vector x_{old} by

$$x_{new} = x_{old} - x_D$$

for which when $x_{old} = x_D$ it follows that $x_{new} = 0$ (the origin in the new state space). This is merely a shifting of the origin of the original state space by the amount x_D.

Unless otherwise noted, the desired state vector x_D will be taken to be equal to the zero vector in all discussions following.

Let us now determine those initial states $x(0)$ that may be forced to the zero state ($x_D = 0$) as a function of the number of iteration times N. Premultiplying both sides of (7.4-2) by A^{-N} and setting $x(N) = 0$, we obtain

$$x(0) = f_1 u(0) + f_2 u(1) + \cdots + f_N u(N-1) \qquad (7.4\text{-}7)$$

where $f_k = -A^{-k}B$. If (7.4-7) is satisfied, then it is guaranteed that $x(N) = 0$.

Specifically, let $N = 1$. Therefore, any initial state that may be forced to the zero state must be representable as

$$x(0) = f_1 u(0) \qquad (7.4\text{-}8)$$

where f_1 is an $n \times 1$ vector and $u(0)$ is a scalar. Any initial state that lies on the line passing through the tip of vector f_1 and the origin in state space may be forced to the zero state (the origin) in one iteration time. All other initial states not positioned on this line cannot be forced to the zero state in one iteration time.

For example, consider the system given in Example 7.4-1. For this system the vector f_1 was given by

$$f_1 = \begin{bmatrix} .718 \\ -1.718 \end{bmatrix}$$

Figure (7.4-2) shows all initial states that may be forced to the origin in one iteration time.

For $N = 2$, equation (7.4-7) becomes

$$x(0) = f_1 u(0) + f_2 u(1) \qquad (7.4\text{-}9)$$

Expression (7.4-9) is simply a linear combination of the two $n \times 1$ vectors f_1 and f_2. The set of initial states that may be forced to the zero state has been expanded (if it is assumed that the order of the system is greater than one)

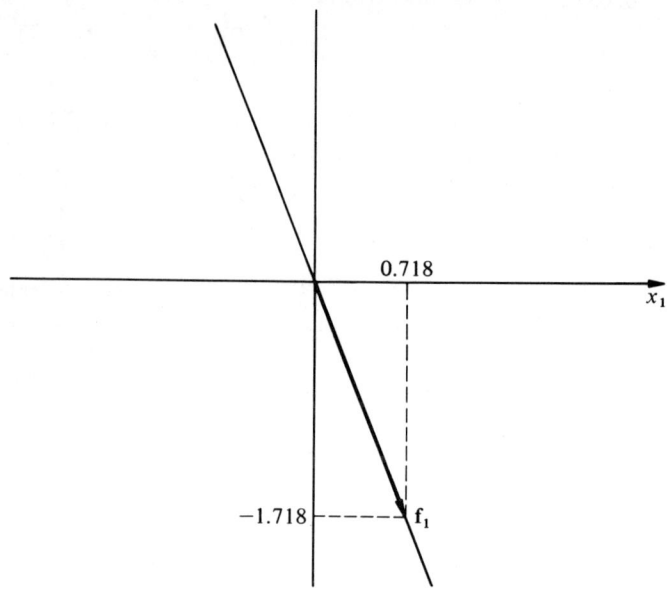

Figure 7.4-2. Initial states lying on line through tip of $\mathbf{f}_1$ and origin may be forced to the zero state in one iteration for the system of Example 7.4-1.

over the case $n = 1$ under the assumption that the vectors $\mathbf{f}_1$ and $\mathbf{f}_2$ are linearly independent. However, if $\mathbf{f}_1$ and $\mathbf{f}_2$ are linearly dependent, that is, if

$$\mathbf{f}_2 = \alpha \mathbf{f}_1 \quad \text{for some scalar}$$

then (7.4-9) becomes

$$\mathbf{x}(0) = \mathbf{f}_1\{u(0) + \alpha u(1)\} = \mathbf{f}_1 \hat{u}(0)$$

and the same set of initial states may be forced to the origin as was the case for $N = 1$. Therefore, it is desirable, from a control viewpoint, to have $\mathbf{f}_1$ and $\mathbf{f}_2$ linearly independent.

The vectors $\mathbf{f}_1$ and $\mathbf{f}_2$ for Example 7.4-1 are shown in Figure 7.4-3 for the system given in Example 7.4-1. We see readily from Figure 7.4-3 that the vectors $\mathbf{f}_1$ and $\mathbf{f}_2$ are linearly independent. Since, in this example, state space is two-dimensional, it follows that *any* initial state $\mathbf{x}(0)$ may be written as a linear combination of the vectors $\mathbf{f}_1$ and $\mathbf{f}_2$. To demonstrate this fact algebraically, rewrite (7.4-9) as

$$\begin{bmatrix} x_1(0) \\ x_2(0) \end{bmatrix} = \begin{bmatrix} .718 \\ -1.718 \end{bmatrix} u(0) + \begin{bmatrix} 3.671 \\ -4.671 \end{bmatrix} u(1) = \begin{bmatrix} .718 & 3.671 \\ -1.718 & -4.671 \end{bmatrix} \begin{bmatrix} u(0) \\ u(1) \end{bmatrix} \quad (7.4\text{-}10)$$

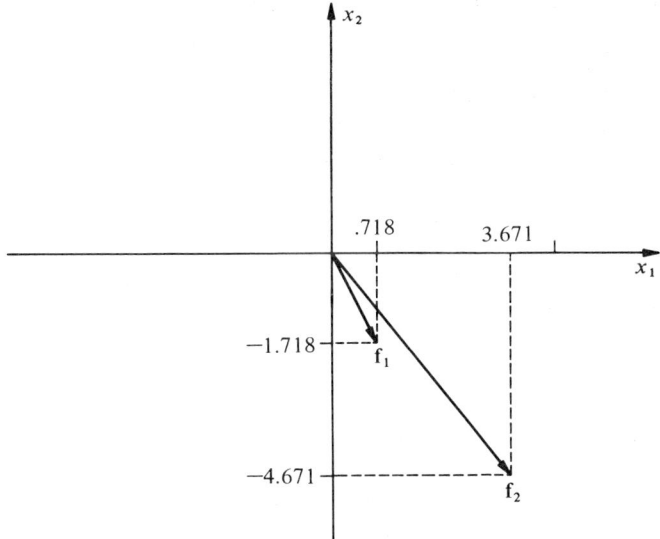

Figure 7.4-3. Vectors $\mathbf{f}_1$ and $\mathbf{f}_2$ for the system of Example 7.4-1.

Now the matrix multiplying the vector

$$\begin{bmatrix} u(0) \\ u(1) \end{bmatrix}$$

is invertible, so (7.4-10) has a unique set of scalars $\{u(0), u(1)\}$, which will satisfy this equation for arbitrary $x_1(0)$, $x_2(0)$.

For systems that are of order exceeding 1 or 2, it follows that not all initial states may be forced to the zero state in one or two iteration times. For example, if (7.4-9) were viewed as a set of n equations in two unknowns [i.e., $u(0)$, $u(1)$], only under very specific choices of $x(0)$ will (7.4-9) have a solution.

We might think of the set of initial states as written in (7.4-8) as forming a one-dimensional subspace of the n-dimensional state space. Similarly, if $\mathbf{f}_1$ and $\mathbf{f}_2$ are linearly independent, the set of vectors written in the form given by (7.4-9) can be thought of as forming a two-dimensional subspace of the n-dimensional state space.

This notion may be extended to $N = 3, 4, 5, \ldots$. The set of initial states that may be forced to the origin in N iteration times is given by

$$\mathbf{R}_N = \left\{ \mathbf{x}(0): \mathbf{x}(0) = \sum_{i=0}^{N-1} \mathbf{f}_{i+1} u(i) \right\} \qquad (7.4\text{-}11)$$

For $N \leq n$, $\mathbf{R}_N$ is a subspace of state space. If the vectors $\mathbf{f}_1, \mathbf{f}_2, \ldots, \mathbf{f}_N$ are linearly independent, subspace $\mathbf{R}_N$ has dimension N.

7.4-2 Control Amplitude Limitations

All practical control systems have some limitation on the amplitude of the control that may be applied. In such cases, the set of initial states that may be forced to the zero state in N iteration times changes drastically. For example, let the control inputs be constrained by

$$-1 \leq u(i) \leq 1 \quad \text{for } i = 0, 1, 2, \ldots \quad (7.4\text{-}12)$$

The set of initial states that may be forced to the origin in N iteration times and this constraint is given by

$$\mathbf{R}_N^s = \left\{ \mathbf{x}(0) \colon \mathbf{x}(0) = \sum_{i=0}^{N-1} \mathbf{f}_{i+1} u(i) \quad \text{with } |u(i)| \leq 1 \right\} \quad (7.4\text{-}13)$$

For the system considered in Example 7.4-1, $\mathbf{R}_1^s$ and $\mathbf{R}_2^s$ are shown in Figures 7.4-4 and 7.4-5, respectively.

An investigation of Figures 7.4-1 through 7.4-4 reveals the control loss one suffers when control amplitude constraints are imposed.

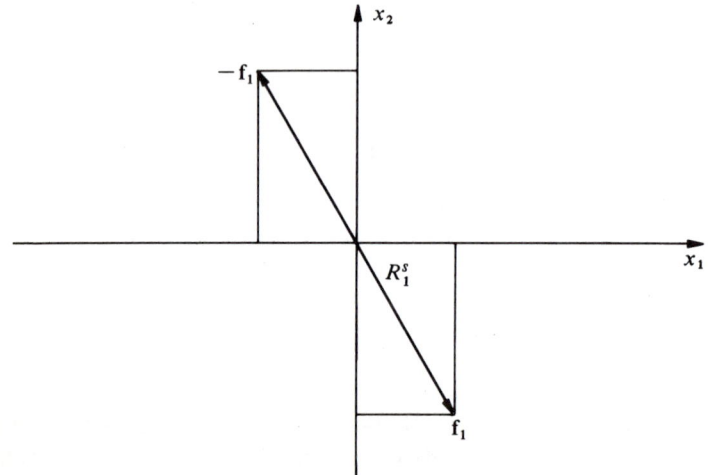

Figure 7.4-4. Region $\mathbf{R}_1^s$.

7.4-3 Observability

To develop some of the basic properties of system observability, we shall study the discrete system

$$\mathbf{x}(k+1) = \mathbf{A}\mathbf{x}(k) + \mathbf{B}\mathbf{u}(k) \quad (7.4\text{-}14)$$
$$y(k) = \mathbf{C}\mathbf{x}(k) \quad (7.4\text{-}15)$$

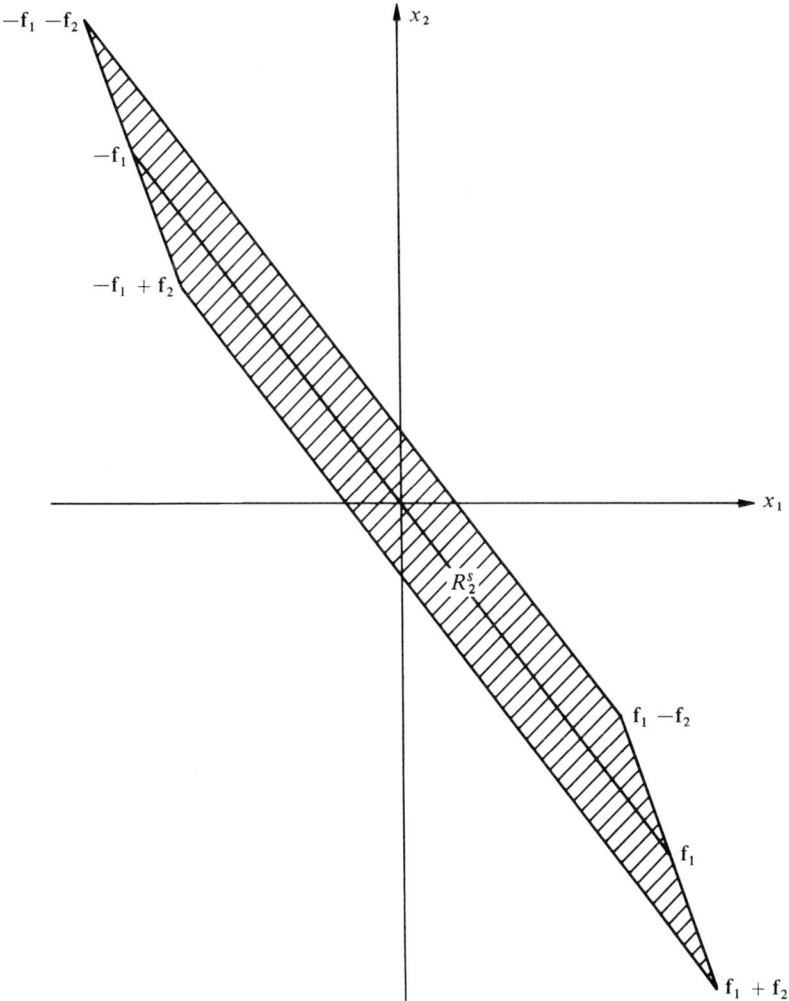

Figure 7.4-5. Region R_2^s.

where $\mathbf{x}(k)$ is a $n \times 1$ state vector, $\mathbf{u}(k)$ is the $m \times 1$ input vector, $y(k)$ is the measurable output signal, while $\mathbf{A}$, $\mathbf{B}$, and $\mathbf{C}$ are $n \times n$, $n \times m$, and $1 \times n$ matrices, respectively. The lack of a $\mathbf{D}\mathbf{u}(k)$ term in (7.4-15) is not very restrictive, since any system whose transfer function has more poles than zeros may always be put into this form.

It is frequently desirable to be able to determine the state of the system $\mathbf{x}(k)$ from a knowledge of the measurable output signal $y(k)$. Since the state vector $\mathbf{x}(k)$ has n components while the output signal is only a scalar, it is apparent that we need many values of $y(k)$, both present and past, in order to determine $\mathbf{x}(k)$. In essence, this is basically the meaning of observability. A system such as (7.4-14) and (7.4-15) is said to be "completely observable" if it is possible to determine the state $\mathbf{x}(k)$ from present and past values of the measurable output signal.

Since the control input vector $\mathbf{u}(k)$ and input matrix $\mathbf{B}$ are known quantities, we may investigate the zero input case [i.e., $\mathbf{u}(k) = 0$] to develop the necessary and sufficient condition for system observability. Namely, we shall be concerned with the system

$$\mathbf{x}(k+1) = \mathbf{A}\mathbf{x}(k) \qquad (7.4\text{-}16)$$
$$\mathbf{y}(k) = \mathbf{C}\mathbf{x}(k) \qquad (7.4\text{-}17)$$

Iteratively applying (7.4-16), we have

$$\mathbf{x}(k-m) = \mathbf{A}^{-m}\mathbf{x}(k) \quad \text{for } m = 0, 1, 2, \ldots$$

so that

$$\mathbf{y}(k-m) = \mathbf{C}\mathbf{A}^{-m}\mathbf{x}(k) = \mathbf{g}_m\mathbf{x}(k)$$

where $\mathbf{g}_m$ is a $1 \times n$ vector. Therefore,

$$\mathbf{y}(k) = \mathbf{g}_0\mathbf{x}(k)$$
$$\mathbf{y}(k-1) = \mathbf{g}_1\mathbf{x}(k)$$
$$\mathbf{y}(k-n+1) = \mathbf{g}_{n-1}\mathbf{x}(k)$$

or, equivalently,

$$\begin{bmatrix} \mathbf{y}(k) \\ \mathbf{y}(k-1) \\ \cdot \\ \cdot \\ \cdot \\ \mathbf{y}(k-n+1) \end{bmatrix} = \mathbf{G}\mathbf{x}(k) \qquad (7.4\text{-}18)$$

where $\mathbf{G}$ is an $n \times n$ matrix whose first row is $\mathbf{g}_0$, whose second row is $\mathbf{g}_1$, etc. A sufficient condition that (7.4-18) have a unique solution is that the matrix $\mathbf{G}$ be invertible. If this be the case, then

$$\mathbf{x}(k) = \mathbf{G}^{-1} \begin{bmatrix} \mathbf{y}(k) \\ \mathbf{y}(k-1) \\ \cdot \\ \cdot \\ \cdot \\ \mathbf{y}(k-n+1) \end{bmatrix} \qquad (7.4\text{-}19)$$

Expression (7.4-19) indicates that if the value of the output signal at times $k, k-1, \ldots, k-n+1$ is retained, then the state $\mathbf{x}(k)$ may be directly determined. Therefore, the invertibility of $\mathbf{G}$ is a sufficient condition for "complete observability," and it turns out to be a necessary condition also. $\mathbf{G}$ is invertible if and only if it has independent rows, which implies that an alternate necessary and sufficient condition for "complete observability" is that the set of $1 \times n$ vectors

$$\{\mathbf{A}, \mathbf{CA}, \mathbf{CA}^2, \ldots, \mathbf{CA}^{n-1}\}$$

must be linearly independent. This gives us a very straightforward method for determining the "complete observability" of a system.

With the concepts of controllability and observability established, we shall now treat some of the more important optimal control problems.

7.5 Regulator Problem

The most basic control problem is one of regulation. Suppose that we wish to control a system governed by the vector difference equation

$$\mathbf{x}(k+1) = \mathbf{A}\mathbf{x}(k) + \mathbf{B}u(k) \tag{7.5-1}$$

in such a manner as to transfer the system from some arbitrary initial state $\mathbf{x}(0)$ to a desired state $\mathbf{x}_D$. What is the form of the required control sequence and the control law that generates this sequence?

It will be assumed that this nth-order discrete system with one control input is completely controllable; that is, the set of vectors $\{\mathbf{f}_1, \mathbf{f}_2, \ldots, \mathbf{f}_n\}$, where

$$\mathbf{f}_k = -\mathbf{A}^k\mathbf{B} \quad (\text{an } n \times 1 \text{ vector for } k \geq 1)$$

forms a set of linearly independent vectors.

Repeatedly applying equation (7.5-1) gives

$$\mathbf{x}(N) = \mathbf{A}^N\mathbf{x}(0) + \mathbf{A}^{N-1}\mathbf{B}u(0) + \mathbf{A}^{N-2}\mathbf{B}u(1)$$
$$+ \cdots + \mathbf{A}\mathbf{B}u(N-2) + \mathbf{B}u(N-1)$$

If it is possible to select a control sequence $u(0), u(1), \ldots, u(N-1)$ so that the state $\mathbf{x}(N) = \mathbf{x}_D$, this implies that the control sequence must satisfy the relationship

$$\mathbf{x}(0) - \mathbf{A}^{-N}\mathbf{x}_D = \mathbf{f}_1 u(0) + \mathbf{f}_2 u(1) + \cdots + \mathbf{f}_N u(N-1)$$

or in matrix form

$$\mathbf{x}(0) - \mathbf{A}^{-N}\mathbf{x}_D = \mathbf{F}_N\mathbf{u}_N \qquad (7.5\text{-}2)$$

where $\mathbf{F}$ is an $n \times N$ matrix with columns $\mathbf{f}_1, \mathbf{f}_2, \ldots, \mathbf{f}_N$ while $\mathbf{u}_N$ is an $N \times 1$ vector whose first element is $u(0)$, whose second element is $u(1)$, etc. Equation (7.5-2) is, in fact, a system of n equations in N unknowns [the $u(i)$'s]. If N is less than n, that is, if the number of control sequences is less than the order of the system, it may not be possible to find a control vector $\mathbf{u}_N$ to satisfy (7.5-2). With this in mind, let $N = n$; now the $n \times n$ matrix $\mathbf{F}_n$ has an inverse, since it has n linearly independent columns (the $\mathbf{f}_k$'s) because of the assumption of system controllability. Premultiplying both sides of (7.5-2) by $\mathbf{F}_n^{-1}$ gives the required control vector to effect the desired state transformation; that is,

$$\mathbf{u}_n = \mathbf{F}_n^{-1}[\mathbf{x}(0) - \mathbf{A}^{-N}\mathbf{x}_D] \qquad (7.5\text{-}3)$$

The control law as given by (7.5-3) is an open-loop control law, since the required control vector $\mathbf{u}_n$ depends only on the initial state $\mathbf{x}(0)$. Therefore, once the initial state $\mathbf{x}(0)$ is monitored and the desired state $\mathbf{x}_D$ given, the control sequence $u(0), u(1), \ldots, u(n-1)$ is immediately calculated from (7.5-3). This sequence is unique because for $N = n$, equation (7.5-2) is a set of n equations in n unknowns and, since $\mathbf{F}_n$ is a nonsingular matrix, there exists one solution, which is given by (7.5-3).

For controllable systems, expression (7.5-3) gives the control vector that will drive the system from any initial state to any desired state in n iteration times. If $N < n$, it is not always possible to select the control sequence $\{u(k)\}$ so that $\mathbf{x}(N) = \mathbf{x}_D$. When $N > n$, the control engineer has some design freedom. In this latter case (i.e., $N > n$), there exists an infinite number of different control sequences to effect the desired control, and the designer may select from this multitude of choices one that may satisfy secondary constraints such as minimum energy regulation, minimum amplitude regulation, etc.

EXAMPLE 7.5-1

Consider the system studied in Example 7.4-1. Suppose that it is desired to drive this system from any arbitrary initial state to the zero state in two iteration (sample times) times. From (7.5-2) with $\mathbf{x}_D = \mathbf{0}$ and $N = 2$, we have

$$\begin{bmatrix} x_1(0) \\ x_2(0) \end{bmatrix} = \begin{bmatrix} .7183 & 3.6708 \\ -1.7183 & -4.6708 \end{bmatrix} \begin{bmatrix} u(0) \\ u(1) \end{bmatrix} \qquad (7.5\text{-}4)$$

Premultiplying both sides of (7.5-4) by $\mathbf{F}_2^{-1}$ yields

$$\begin{bmatrix} u(0) \\ u(1) \end{bmatrix} = \begin{bmatrix} -1.5820 & -1.2433 \\ .5820 & .2433 \end{bmatrix} \begin{bmatrix} x_1(0) \\ x_2(0) \end{bmatrix}$$

which, when we set $x_1(0) = c(0)$ and $x_2(0) = dc(0)/dt$, results in

$$u(0) = -1.5820c(0) - 1.2433\frac{dc(0)}{dt}$$
$$u(1) = .5820c(0) + .2433\frac{dc(0)}{dt} \quad (7.5\text{-}5)$$

The control law as given by (7.5-5) reveals the open-loop nature of this type of control.

In the regulation problem, it is not necessary to equate the number of control iterations (N) to the order of the linear discrete system under control (n). Let us now investigate the case when $N \geq n$ for controllable systems. Equation (7.5-2) is a set of n equations in N [the $u(i)$'s] unknowns, and when $N > n$, we have more unknowns than equations. When these equations are consistent (have at least one solution), there exists an infinite number of different solutions. Because the system under control is assumed controllable, this set of n equations in N unknowns is *always* consistent for $N \geq n$. This fact is best illustrated by means of an example.

EXAMPLE 7.5-2

For the system investigated in Example 7.4-1, determine the properties of the control sequence needed to force $\mathbf{x}(3) = \mathbf{0}$.

In this case, for $N = 3$, we find that

$$\mathbf{f}_1 = \begin{bmatrix} .7183 \\ -1.7183 \end{bmatrix}, \quad \mathbf{f}_2 = \begin{bmatrix} 3.6708 \\ -4.6708 \end{bmatrix}, \quad \mathbf{f}_3 = \begin{bmatrix} 11.6965 \\ -12.6965 \end{bmatrix}$$

so that (7.5-3) becomes

$$\begin{bmatrix} x_1(0) \\ x_2(0) \end{bmatrix} = \begin{bmatrix} .7183 & 3.6708 & 11.6965 \\ -1.7183 & -4.6708 & -12.6965 \end{bmatrix} \begin{bmatrix} u(0) \\ u(1) \\ u(2) \end{bmatrix}$$

or, in equation form,

$$x_1(0) = .7183u(0) + 3.6708u(1) + 11.6965u(2)$$
$$x_2(0) = -1.7183u(0) - 4.6708u(1) - 12.6965u(2)$$

that is, $2(n = 2)$ equations in $3(N = 3)$ unknowns. We may easily verify that this set of equations has at least one solution [e.g., let $u(2) = 0$ and see Example 7.5-1].

7.6 Minimum Energy Control

In many applications, we desire to accomplish a given control task, using the minimum amount of control energy necessary. Such a problem will now be formulated. (See also Ref. 5.)

The system under study is governed by the vector difference equation

$$\mathbf{x}(k+1) = \mathbf{A}\mathbf{x}(k) + \mathbf{B}u(k) \qquad (7.6\text{-}1)$$

A linear, time-invariant vector difference equation of the form given by (7.6-1) will occur whenever a continuous system that is characterized by a linear, time-invariant differential equation is driven by a single input that is constant over fixed intervals of time (sampled-data systems). This was demonstrated in Chapters 2 and 3.

We desire to drive such a system from any arbitrary initial state $\mathbf{x}(0)$ to a desired state $\mathbf{x}_D$ in N iteration times, using a minimum of control energy. Control energy will be measured by the quantity

$$E_N = \sum_{k=0}^{N-1} u^2(k) \qquad (7.6\text{-}2)$$

If the alotted number of control iterations N is smaller than the order of the system n, it will not always be possible to accomplish the desired control action. In order to take this factor into account, a slightly different control problem is postulated.

Given a discrete system governed by (7.6-1), design a controller that generates the control input $u(0), u(1), \ldots, u(N-1)$ that

1. Takes the system from any initial state $\mathbf{x}(0)$ to a desired state $\mathbf{x}_D$ in N iterations while minimizing control energy [as measured by (7.6-2)] and, if this is not possible,

2. Minimizes the Euclidean distance of the state of the discrete system from the desired state at the end of N iteration times; that is,

$$(\mathbf{x}_D - \mathbf{x}(N))^T(\mathbf{x}_D - \mathbf{x}(N)) \qquad (7.6\text{-}3)$$

Expression (7.6-3) gives a measure of the distance that the error vector $\mathbf{x}_D - \mathbf{x}(N)$ is from the zero vector. For the remainder of this section, the desired state is taken to be the zero state. No loss in generality is incurred under this assumption, since if the zero state can be reached in N iteration times, then so can any other state.

Minimum Energy Control

It is desired to force $x(N) = 0$ by a proper selection of control inputs $u(0), u(1), \ldots, u(N-1)$. If this is to be possible for any initial state $x(0)$, then it must be possible to select the parameters (control inputs) $u(0), u(1), \ldots, u(N-1)$ so that

$$x(N) = 0 = A^N x(0) + A^{N-1} B u(0) + \cdots + A B u(N-2) + B u(N-1)$$

or, if we premultiply both sides by A^{-N} and use the identity $f_k = -A^{-k}B$,

$$x(0) = u(0)f_1 + u(1)f_2 + \cdots + u(N-1)f_N \quad (7.6\text{-}4)$$

Equation (7.6-4) indicates that if $x(N) = 0$, then the initial state $x(0)$ must be expressible as a linear combination of the vectors $f_1, f_2, \ldots, f_N$. For controllable systems, the vectors $f_1, f_2, \ldots, f_n$ form a set of linearly independent vectors, so that any $n \times 1$ vector may be expressed as a linear combination of such a set. This suggests that for controllable systems it is always possible to force $x(N) = 0$ for $N \geq n$ for arbitrary initial states $x(0)$. When $N < n$, it is not possible to force all initial states to the zero state. Two cases will then be considered: (i) $N \geq n$, and (ii) $N < n$.

Case (i) $N \geq n$

For $N \geq n$, the state vector after N iteration times may always be forced to the zero state.

Thus, expressing 7.6-4 in matrix form,

$$x(0) = Fu \quad (7.6\text{-}5)$$

where $F = $ the $n \times N$ matrix with columns f_i
$u = $ the $N \times 1$ control sequence vector with elements u_i

The optimal control sequence must satisfy the matrix equation (7.6-5) and also must be a minimum energy solution. It should be noted that if the control sequence satisfies equation (7.6-5), then $x(N) = 0$ is guaranteed.

Case (ii) $N < n$

The case when the plant is controllable and $N < n$ is treated analogously to the case when $N \geq n$, but with one basic difference. The assumption that any initial state can be forced to the zero state after N sampling periods is not valid. If $x(N)$ can be forced to the zero state, the control sequence required for $N < n$ is unique, and if it is not possible to force $x(N) = 0$ then the Euclidean distance

$$\|x(N)\|^{1/2} = \left(\sum_{i=1}^{n} x_i^2(N) \right)^{1/2} \quad (7.6\text{-}6)$$

must be minimized. It will be observed that no mention of minimum energy is made for $N < n$. Thus the minimum energy problem in this case is equivalent to selecting a control sequence that minimizes $\|\mathbf{x}(N)\|$. The solution to the minimum energy problem for controllable systems is illustrated in Table 7.6-1.

Table 7.6-1 Solution of Minimum Energy Problem for Controllable Plants

Case	Solution
$N \geq n$	Minimum energy solution of $\mathbf{x}(0) = \mathbf{Fu}$
$N < n$	Minimize $\|\mathbf{x}(N)\|$

Before we consider the techniques of obtaining the solution to this problem, a few developments in inverse matrix theory will be presented.

7.6-1 Right and Left Inverse Matrix Theory

To demonstrate the use of inverse matrix theory in the solving of a system of n linear equations in m unknowns, the following set of equations will be considered.

$$a_{11}x_1 + a_{12}x_2 + \cdots + a_{1m}x_m = b_1$$
$$a_{21}x_1 + a_{22}x_2 + \cdots + a_{2m}x_m = b_2$$
$$\cdots \cdots \cdots \cdots \cdots$$
$$a_{n1}x_1 + a_{n2}x_2 + \cdots + a_{nm}x_m = b_n$$

or in its equivalent matrix form

$$\mathbf{Ax} = \mathbf{b} \qquad (7.6\text{-}7)$$

where

$$\mathbf{A} = \begin{bmatrix} a_{11} & a_{12} & \cdots & a_{1m} \\ a_{21} & a_{22} & \cdots & a_{2m} \\ \cdot & \cdot & & \cdot \\ a_{n1} & a_{n2} & \cdots & a_{nm} \end{bmatrix}$$

$$\mathbf{x} = \begin{bmatrix} x_1 \\ x_2 \\ \cdot \\ \cdot \\ x_m \end{bmatrix}, \quad \mathbf{b} = \begin{bmatrix} b_1 \\ b_2 \\ \cdot \\ \cdot \\ b_n \end{bmatrix}$$

Consider the special case when $m = n$; it can be easily shown that if the rank of matrix $\mathbf{A}$ is n, then equation (7.6-7) has a unique solution given by

$$\mathbf{x} = \mathbf{A}^{-1}\mathbf{b}$$

where the $n \times n$ matrix $\mathbf{A}^{-1}$ is that unique matrix that satisfies the two properties

$$\mathbf{A}\mathbf{A}^{-1} = \mathbf{I}_n$$
$$\mathbf{A}^{-1}\mathbf{A} = \mathbf{I}_n$$

$\mathbf{I}_n$ is the $n \times n$ identity matrix, and $\mathbf{A}^{-1}$ is the inverse matrix of $\mathbf{A}$.

An alternate necessary and sufficient condition for a square matrix $\mathbf{A}$ to have an inverse matrix $\mathbf{A}^{-1}$ is for det $(\mathbf{A}) \neq 0$. If the determinant of $\mathbf{A}$ is equal to zero, then equation (7.6-7) may have infinitely many solutions or in fact may have no solutions at all.

When $m \neq n$, matrix theory can still play an important role in the seeking of solutions to a system of linear equations.

Definition. An $n \times m$ matrix $\mathbf{A}$ is said to have a right inverse $\mathbf{A}^R$ if $\mathbf{A}\mathbf{A}^R = \mathbf{I}_n$. Similarly, the $n \times m$ matrix $\mathbf{A}$ is said to have a left inverse $\mathbf{A}^L$ if $\mathbf{A}^L\mathbf{A} = \mathbf{I}_m$.

Theorem 7.6-1. If the $n \times m$ matrix $\mathbf{A}$ has a right inverse matrix $\mathbf{A}^R$, then

$$\mathbf{x} = \mathbf{A}^R\mathbf{b}$$

is a solution to the consistent matrix equation $\mathbf{A}\mathbf{x} = \mathbf{b}$. The term *consistent* as used above indicates that the matrix equation has at least one solution.

Proof. Substitute the assumed solution into the matrix equation; if it is indeed a solution it will satisfy the matrix equation

$$\mathbf{A}\mathbf{x} = \mathbf{b}$$

Let

$$\mathbf{x} = \mathbf{A}^R\mathbf{b}$$
$$\mathbf{A}\mathbf{x} = \mathbf{A}\mathbf{A}^R\mathbf{b} = \mathbf{b}$$

Theorem 7.6-2. If the $n \times m$ matrix $\mathbf{A}$ has a right inverse matrix $\mathbf{A}^R$, then

$$\mathbf{x} = (\mathbf{I}_m - \mathbf{A}^R\mathbf{A})\mathbf{y}$$

is a solution to the homogeneous matrix equation $\mathbf{A}\mathbf{x} = \mathbf{0}$, where $\mathbf{y}$ is "any" $m \times 1$ vector.

Proof. $\mathbf{Ax} = \mathbf{0}$

Let
$$\mathbf{x} = (\mathbf{I}_m - \mathbf{A}^R\mathbf{A})\mathbf{y}$$
$$\mathbf{Ax} = \mathbf{A}(\mathbf{I}_m - \mathbf{A}^R\mathbf{A})\mathbf{y} = (\mathbf{A} - \mathbf{A}\mathbf{A}^R\mathbf{A})\mathbf{y} = \mathbf{0}$$

Corollary. If the $n \times m$ matrix $\mathbf{A}$ has a right inverse matrix $\mathbf{A}^R$, then

$$\mathbf{x} = \mathbf{A}^R\mathbf{b} + (\mathbf{I}_m - \mathbf{A}^R\mathbf{A})\mathbf{y} \tag{7.6-8}$$

is a general solution to the consistent matrix equation $\mathbf{Ax} = \mathbf{b}$, where $\mathbf{y}$ is any $m \times 1$ vector.

It will be noted that the solution space of the matrix equation $\mathbf{Ax} = \mathbf{b}$ is of dimension equal to the rank of $\mathbf{I}_m - \mathbf{A}^R\mathbf{A}$. So that, in general, if the matrix of $\mathbf{A}$ has a right inverse, there is an infinite number of solutions all contained in the subset specified by

$$\mathbf{x} = \mathbf{A}^R\mathbf{b} + (\mathbf{I}_m - \mathbf{A}^R\mathbf{A})\mathbf{y}$$

where the vector $\mathbf{A}^R\mathbf{b}$ is fixed (for a given $\mathbf{A}^R$) and the vector $\mathbf{y}$ is allowed to span the m-dimensional space.

It should be pointed out that a right inverse matrix $\mathbf{A}^R$, if it exists, is in general not unique, as will be demonstrated in the next example.

EXAMPLE 7.6-1

Consider the set of equations

$$\begin{aligned} x_1 + 2x_2 &= 3 \\ x_2 + 2x_3 &= 3 \end{aligned} \tag{7.6-9}$$

Therefore,

$$\mathbf{A} = \begin{bmatrix} 1 & 2 & 0 \\ 0 & 1 & 2 \end{bmatrix}, \quad \mathbf{b} = \begin{bmatrix} 3 \\ 3 \end{bmatrix}$$

and, as is easily verified, one right inverse of $\mathbf{A}$ is given by

$$\mathbf{A}^R = \begin{bmatrix} 1 & 2 \\ 0 & -1 \\ 0 & 1 \end{bmatrix} \tag{7.6-10}$$

The general solution using equation (7.6-8) becomes

$$\begin{bmatrix} x_1 \\ x_2 \\ x_3 \end{bmatrix} = \begin{bmatrix} 9 \\ -3 \\ 3 \end{bmatrix} + \begin{bmatrix} 0 & -4 & -4 \\ 0 & 2 & 2 \\ 0 & -1 & -1 \end{bmatrix} \begin{bmatrix} y_1 \\ y_2 \\ y_3 \end{bmatrix}$$

The rank of $\mathbf{I}_m - \mathbf{A}^R\mathbf{A}$ in this example is one so that in general any solution to equations (7.6-9) is located on a line in three-dimensional real space which passes through the point $(9, -3, 3)$. The equation of this line is specified by $(\mathbf{I}_m - \mathbf{A}^R\mathbf{A})\mathbf{y}$ and is obtained in the following manner:

$$\begin{bmatrix} 0 & -4 & -4 \\ 0 & 2 & 2 \\ 0 & -1 & -1 \end{bmatrix} \begin{bmatrix} y_1 \\ y_2 \\ y_3 \end{bmatrix} = \begin{bmatrix} -4(y_2 + y_3) \\ 2(y_2 + y_3) \\ -(y_2 + y_3) \end{bmatrix} = \begin{bmatrix} -4 \\ 2 \\ -1 \end{bmatrix}(y_2 + y_3)$$

The range of $y_2 + y_3$ is $(-\infty, \infty)$. The solutions to equations (7.6-9) become

$$\begin{bmatrix} x_1 \\ x_2 \\ x_3 \end{bmatrix} = \begin{bmatrix} 9 - 4a \\ -3 + 2a \\ 3 - a \end{bmatrix}, \quad -\infty < a < \infty$$

To demonstrate that the right inverse of matrix $\mathbf{A}$ as given by equation (7.6-10) is not unique, we may easily verify that the following matrix is also a right inverse of $\mathbf{A}$.

$$\mathbf{A}^R = \begin{bmatrix} 3 & 2 \\ -1 & -1 \\ \frac{1}{2} & 1 \end{bmatrix}$$

It can be shown that an $n \times m$ matrix $\mathbf{A}$ has a right inverse if and only if $\mathbf{A}$ is of rank n and has a left inverse if and only if the rank of $\mathbf{A}$ is m.

7.6-2 Minimal Right and Left Inverse Matrices

If the $n \times m$ matrix $\mathbf{A}$ has a right inverse, then the solutions to the matrix equation $\mathbf{A}\mathbf{x} = \mathbf{b}$ are given by

$$\mathbf{x} = \mathbf{A}^R\mathbf{b} + (\mathbf{I}_m - \mathbf{A}^R\mathbf{A})\mathbf{y}$$

Of all the solutions existent, we are particularly interested in the solution $\mathbf{x}^0$, which is smallest in the Euclidean norm sense; i.e.,

$$\sum_{i=1}^n (x_i^0)^2 = \|\mathbf{x}^0\| = \min_{\mathbf{y}} [\|\mathbf{A}^R\mathbf{b} + (\mathbf{I}_m - \mathbf{A}^R\mathbf{A})\mathbf{y}\|] \qquad (7.6\text{-}11)$$

The problem to be considered here is the determination of that right inverse of $\mathbf{A}$ (which will be denoted by $\mathbf{A}^{RM}$) that will yield the solution $\mathbf{x}^0$ given by equation (7.6-11); i.e.,

$$\mathbf{x}^0 = \mathbf{A}^{RM}\mathbf{b}$$

Definition. $\mathbf{x}^0$ is the "minimal Euclidean" solution to the consistent matrix equation $\mathbf{Ax} = \mathbf{b}$ if

(i) $\mathbf{Ax}^0 = \mathbf{b}$
(ii) $\|\mathbf{x}^0\| \leq \|\mathbf{x}\|$ for all $\mathbf{x}$ that satisfy $\mathbf{Ax} = \mathbf{b}$

Theorem 7.6-3. If the $n \times m$ matrix $\mathbf{A}$ is of rank n, then the "minimal Euclidean" solution to the consistent matrix equation $\mathbf{Ax} = \mathbf{b}$ is given by

$$\mathbf{x}^0 = \mathbf{A}^{RM}\mathbf{b} \qquad (7.6\text{-}12)$$

where $\mathbf{A}^{RM} = \mathbf{A}^T(\mathbf{AA}^T)^{-1}$ and $\mathbf{A}^T$ is the transpose of $\mathbf{A}$. $\mathbf{A}^{RM}$ will be called the minimal right inverse of $\mathbf{A}$.

EXAMPLE 7.6-2

The "minimal Euclidean" solution to Example 7.6-1 will now be determined.

We recall that

$$\mathbf{A} = \begin{bmatrix} 1 & 2 & 0 \\ 0 & 1 & 2 \end{bmatrix}, \qquad \mathbf{b} = \begin{bmatrix} 3 \\ 3 \end{bmatrix}$$

Therefore,

$$\mathbf{AA}^T = \begin{bmatrix} 5 & 2 \\ 2 & 5 \end{bmatrix}$$

$$\mathbf{A}^{RM} = \mathbf{A}^T(\mathbf{AA}^T)^{-1} = \frac{1}{21}\begin{bmatrix} 5 & -2 \\ 8 & 1 \\ -4 & 10 \end{bmatrix}$$

and

$$\mathbf{x}^0 = \mathbf{A}^{RM}\mathbf{b} = \begin{bmatrix} \frac{3}{7} \\ \frac{9}{7} \\ \frac{6}{7} \end{bmatrix}$$

Definition. $\mathbf{x}^0$ is the "minimal Euclidean" approximation solution to the matrix equation $\mathbf{Ax} = \mathbf{b}$ if

$$\|\mathbf{Ax} - \mathbf{b}\| > \|\mathbf{Ax}^0 - \mathbf{b}\| \quad \text{for all } \mathbf{x} \neq \mathbf{x}^0$$

Theorem 7.6-4. If the $n \times m$ matrix $\mathbf{A}$ is of rank m, then the "minimal Euclidean" approximation solution to the matrix equation $\mathbf{Ax} = \mathbf{b}$ is given by

$$\mathbf{x}^0 = \mathbf{A}^{LM}\mathbf{b} \qquad (7.6\text{-}13)$$

where $\mathbf{A}^{LM} = (\mathbf{A}^T\mathbf{A})^{-1}\mathbf{A}^T$ and will be called the minimal left inverse of $\mathbf{A}$.

7.6-3 Application of the Minimal Right Inverse Matrix to the Minimum Energy Problem for Controllable Plants

In the case where the plant is controllable and $N \geq n$, it was shown that the minimum energy problem was obtained by satisfying the matrix equation

$$\mathbf{x}(0) = \mathbf{Fu} \qquad (7.6\text{-}14)$$

and the control sequence vector $\mathbf{u}$ which satisfies this equation must be a minimal energy solution. As $N \geq n$, this control sequence will guarantee that the state of the plant after N sampling periods will be the zero state vector.

Since the plant is assumed controllable, the $n \times N$ matrix has rank n; thus it has a minimal right inverse. Using Theorem 7.6-3, we find that the matrix $\mathbf{F}$ takes the place of matrix $\mathbf{A}$ and the "minimal Euclidean" solution to equation (7.6-14) becomes

$$\mathbf{u}^0 = \mathbf{F}^T(\mathbf{FF}^T)^{-1}\mathbf{x}(0)$$

That $\mathbf{u}^0$ is the minimal energy solution follows from the fact that $\mathbf{u}^0$ is the "minimal Euclidean" solution to equation (7.6-14); i.e.,

$$\|\mathbf{u}^0\| = \left(\sum_{i=0}^{N-1}(\mathbf{u}_i^0)^2\right)$$

which is a measure of the energy. Since $\mathbf{u}^0$ satisfies equation (7.6-14), it follows, by the remarks leading to Theorem 7.6-3, that $\mathbf{u}^0$ is the minimal energy control sequence that forces $\mathbf{x}(N) = \mathbf{0}$.

The matrix $\mathbf{F}^T(\mathbf{FF}^T)^{-1}$ is an $N \times n$ matrix and right-multiplying it by $\mathbf{x}(0)$ requires Nn multiplications to generate the minimal energy control sequence vector $\mathbf{u}^0$. When the optimal controller first senses the initial disturbance $\mathbf{x}(0)$, it performs the n multiplications needed to generate the control function for the first sampling period; i.e.,

$$u(0) = \sum_{j=1}^{n} \alpha_{ij} x_j \qquad (7.6\text{-}15)$$

where α_{ij} = the (i, j) element of $\mathbf{F}^T(\mathbf{FF}^T)^{-1}$
x_j = the jth component of $\mathbf{x}(0)$.

The optimal controller then applies $u(0)$ during the first sampling period and simultaneously determines the remaining components $u(1), u(2), \ldots, u(N-1)$. The time between when the initial disturbance is detected and when $u(0)$ is applied to the plant is essentially the time required to carry out the n multiplications as given by equation (7.6-15). If this time is small in comparison to the sampling period and the plant time constants, then a real-time minimum energy control strategy is feasible.

It is felt that the case when the plant is controllable and $N \geq n$ is the most practical situation a control engineer will meet. This is because the great majority of plants encountered in practice are controllable and by selecting $N \geq n$ it is guaranteed that any initial disturbance will always be completely extinguished in N sampling periods.

The optimal controller for this case is illustrated in block diagram form in Figure 7.6-1.

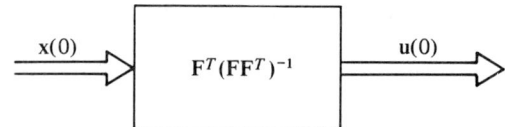

Figure 7.6-1. Optimal controller for a controllable plant, $N \geq n$.

7.6-4 Numerical Example

The plant under study is characterized by the transfer function

$$G(s) = \frac{1}{s(s+1)}$$

For this plant we have

$$\mathbf{x} = \begin{bmatrix} x \\ \dot{x} \end{bmatrix}$$

$$\mathbf{A} = \begin{bmatrix} 0 & 1 \\ 0 & -1 \end{bmatrix}$$

$$\mathbf{B} = \begin{bmatrix} 0 \\ 1 \end{bmatrix}$$

The fundamental matrix $\mathbf{A}$ is evaluated by standard techniques, and for a sampling period of one second it is given by

$$\mathbf{A} = \begin{bmatrix} 1 & 1 - e^{-1} \\ 0 & e^{-1} \end{bmatrix} \qquad (7.6\text{-}16)$$

B is given by

$$\mathbf{B} = \begin{bmatrix} e^{-1} \\ 1 - e^{-1} \end{bmatrix} \qquad (7.6\text{-}17)$$

The vectors $\mathbf{f}_k$ which are the columns of **F**, are given by

$$\mathbf{f}_k = -\mathbf{A}^{-k}\mathbf{B} \qquad (7.6\text{-}18)$$

It can be shown that

$$\mathbf{A}^{-1} = \begin{bmatrix} 1 & 1-e \\ 0 & e \end{bmatrix}$$

Evaluating $\mathbf{f}_1, \mathbf{f}_2, \mathbf{f}_3,$ and $\mathbf{f}_4$ as given by equation (7.6-18) we obtain

$$\mathbf{f}_1 = \begin{bmatrix} .7183 \\ -1.7183 \end{bmatrix}, \quad \mathbf{f}_2 = \begin{bmatrix} 3.6708 \\ -4.6708 \end{bmatrix}, \quad \mathbf{f}_3 = \begin{bmatrix} 11.6965 \\ -12.6965 \end{bmatrix},$$

$$\mathbf{f}_4 = \begin{bmatrix} 33.5126 \\ -34.5126 \end{bmatrix}$$

Since $\mathbf{f}_1$ and $\mathbf{f}_2$ are linearly independent, the plant is controllable.

Problem. Find the control sequence vector for the minimal energy problem for $N = 2, 3, 4$.

Since the order of the system under study is $n = 2$, the minimal energy solution for $N = 2, 3, 4$ will force any initial disturbance to zero. The results of Section 7.6 with $N \geq n$ are used.

Solution. The problem will be solved in detail for $N = 4$. The results for $N = 2$ and $N = 3$ will be given without detail.

$$\mathbf{Fu} = \mathbf{x}(0)$$

for $N = 4$, **F** becomes

$$\mathbf{F} = \begin{bmatrix} .7183 & 3.6708 & 11.6965 & 33.5126 \\ -1.7183 & -4.6708 & -12.6965 & -34.5126 \end{bmatrix}$$

Thus

$$\mathbf{FF}^T = \begin{bmatrix} 1273.8924 & -1323.4916 \\ -1323.4916 & 1377.0897 \end{bmatrix}$$

Therefore,

$$(\mathbf{FF}^T)^{-1} = \begin{bmatrix} .5225 & .5022 \\ .5022 & .4833 \end{bmatrix}$$

and the minimal right inverse becomes

$$\mathbf{F}^{RM} = \begin{bmatrix} -.4876 & -.4698 \\ -.4275 & -.4143 \\ -.2643 & -.2632 \\ .1794 & .1473 \end{bmatrix}$$

The minimal energy solution for $N = 4$ becomes

$$\begin{bmatrix} u(0) \\ u(1) \\ u(2) \\ u(3) \end{bmatrix} = \begin{bmatrix} -.4876 & -.4698 \\ -.4275 & -.4143 \\ -.2643 & -.2632 \\ .1794 & .1473 \end{bmatrix} \begin{bmatrix} x(0) \\ \dot{x}(0) \end{bmatrix} \qquad (7.6\text{-}19)$$

Similarly, for $N = 3$

$$\begin{bmatrix} u(0) \\ u(1) \\ u(2) \end{bmatrix} = \begin{bmatrix} -.7910 & -.7191 \\ -.5000 & -.4738 \\ .2910 & .1929 \end{bmatrix} \begin{bmatrix} x(0) \\ \dot{x}(0) \end{bmatrix} \qquad (7.6\text{-}20)$$

and finally for $N = 2$

$$\begin{bmatrix} u(0) \\ u(1) \end{bmatrix} = \begin{bmatrix} -1.5820 & -1.2433 \\ .5820 & .2433 \end{bmatrix} \begin{bmatrix} x(0) \\ \dot{x}(0) \end{bmatrix} \qquad (7.6\text{-}21)$$

A specific initial condition will now be considered.

$$\mathbf{x}(0) = \begin{bmatrix} x(0) \\ \dot{x}(0) \end{bmatrix} = \begin{bmatrix} -40.9067 \\ 43.5067 \end{bmatrix}$$

Substituting this initial condition into the matrix equations (7.6-19), (7.6-20), and (7.6-21), we obtain

$N = 4$

$u(0) = -.4963, \quad u(1) = -.5352,$
$u(2) = -.6408, \quad u(3) = -.9277$

$N = 3$

$$u(0) = 1.0732, \quad u(1) = -.1602, \quad u(2) = -3.5130$$

$N = 2$

$$u(0) = 10.6225, \quad u(1) = -13.2225$$

Although each of the above control sequences will force the plant with the specific initial condition to the zero state vector, it will be noted that by making N progressively larger the energy requirements are drastically reduced; i.e.,

$N = 2$

$$\sum_{i=0}^{1} u_i^2 = 287.6720$$

$N = 3$

$$\sum_{i=0}^{2} u_i^2 = 13.5186$$

$N = 4$

$$\sum_{i=0}^{3} u_i^2 = 1.8040$$

7.6-5 Minimum Energy Control with Amplitude Constraint

There are many practical control systems in which there is an amplitude limitation on the control that may be applied. A problem is now proposed for a sampled-data control system with a control amplitude limitation that may be solved by utilizing the techniques developed in this section.

Problem. Force any controllable plant from some initial state to the zero state in the minimum number of sampling periods subject to the condition that for that number of sampling periods it is the minimum energy solution and in addition none of the components of u_i exceeds a certain positive number a in absolute value; e.g.,

$$|u_i| \leq a$$

Solution. Since we are to force the plant to the zero state, use of the minimal right inverse matrix will be made. Precalculate $\mathbf{F}_N^{RM}$ for $N = n$, $n + 1, \ldots$ and store these matrices in the optimal controller. First compute

$$\mathbf{u}_n^0 = \mathbf{F}_n^{RM} \mathbf{x}(0)$$

If none of the components of **u** exceeds a in absolute value, then the optimal control sequence is obtained. If some do, then compute

$$\mathbf{u}^0_{n+1} = \mathbf{F}^{RM}_{n+1}\mathbf{x}(0)$$

and continue this process until a $\mathbf{F}^{RM}_k\mathbf{x}(0)$ is obtained which satisfies the absolute magnitude constraint. The $\mathbf{u}^0_k$ obtained is the solution to the problem.

As a practical example, consider the system studied in section 7.6-4 with constraints

$$|u_i| \leq 1, \quad i = 1, 2, \ldots$$

and the initial condition

$$\mathbf{x}(0) = \begin{bmatrix} -40.9067 \\ 43.5067 \end{bmatrix}$$

It was shown that the optimal solution for the problem proposed in this section is given by

$$u_1 = -.4693, \quad u_2 = -.5352, \quad u_3 = -.6408, \quad u_4 = -.9277$$

7.7 Tracking Test Inputs

A frequent requirement of discrete systems is the ability to track certain deterministic test input signals. The standard test input for such purposes is the discrete step of amplitude Q; that is

$$\begin{aligned} r(k) &= Q \quad \text{for } k \geq 0 \\ r(k) &= 0 \quad \text{for } k < 0 \end{aligned} \quad (7.7\text{-}1)$$

By *tracking*, we mean the ability of the system to respond to test inputs, such as the discrete step, so that the system's output is equal to the system's input with possibly some delay involved. Expressing this mathematically, we have

$$c(k) = r(k - m) \qquad (7.7\text{-}2)$$

where $c(k)$ denotes the system's output and the integer m is the delay interval given in discrete-time iterations. If $m = 0$, this implies that the system's output exactly equals its input when the test signal is applied.

As most practical discrete systems have nonzero time constants, equalities such as that given in (7.7-2) are possible only after the transient terms have decayed to zero. Therefore, we shall initially investigate the case when (7.7-2) holds for very large values of k (i.e., as $k \to \infty$). It is assumed that the system to which the input is applied is linear, so the relationship between $C(z)$ and $R(z)$ is of the form

$$C(z) = H(z)R(z) \qquad (7.7\text{-}3)$$

The error signal is defined by

$$e(k) = c(k) - r(k - m) \qquad (7.7\text{-}4)$$

and measures the amount by which the desired relationship (7.7-2) is incorrect. Taking the z-transform of (7.7-4) gives

$$E(z) = C(z) - z^{-m}R(z)$$

and using (7.7-3) results in

$$E(z) = [H(z) - z^{-m}]R(z) \qquad (7.7\text{-}5)$$

Since we wish $e(k)$ to be zero for large k, we apply the final value theorem to (7.7-5) and set the result to zero; that is,

$$e(\infty) = \lim_{z \to 1} (1 - z^{-1})[H(z) - z^{-m}]R(z) = 0 \qquad (7.7\text{-}6)$$

Relationship (7.7-6) is the necessary condition that must be satisfied in order for the discrete system with transfer function $H(z)$ to track the input $r(k)$.

An investigation of the properties required by $H(z)$ in order to track specific test input will now be made.

7.7-1 Step Input

This test input is characterized by equation (7.7-1) so that

$$R(z) = \frac{R}{1 - z^{-1}}$$

which when inserted into (7.7-6) yields

$$\lim_{z \to 1} [H(z) - z^{-m}] = 0 \qquad (7.7\text{-}7)$$

Therefore, $H(1) - 1 = 0$, which implies that a linear discrete system will properly track a step input only if its transfer function evaluated at $z = 1$ equals unity. Since $H(z)$ is related to its weighting sequence $h(n)$ by

$$H(z) = \sum_{k=0}^{\infty} h(k) z^{-k}$$

requirement (7.7-7) is seen to be equivalent to

$$\sum_{k=0}^{\infty} h(k) = 1 \tag{7.7-8}$$

that is, the sum of the terms in the weighting sequence equals unity. For example, the discrete system with weighting sequence

$$h(0) = 1$$
$$h(1) = -2$$
$$h(2) = 2$$
$$h(k) = 0 \quad \text{for } k \neq 0, 1, 2$$

will properly track a step input.

To demonstrate the potential control applications available, consider the digitally controlled system shown in Figure 7.7-1. Suppose it is desired that this control system track a step input of amplitude R; that is,

$$c(kT) \longrightarrow R \quad \text{for large } k$$

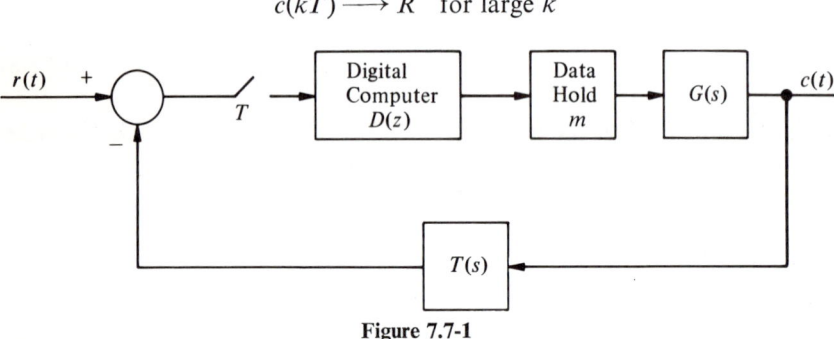

Figure 7.7-1

What constraint does this imply on the digital computer? From (6.5-13) it was previously shown that

$$H(z) = \frac{C(z)}{R(z)} = \frac{D(z)\mathscr{Z}[G_m(s)G(s)]}{1 + D(z)\mathscr{Z}[G_m(s)G(s)T(s)]} \tag{7.7-9}$$

For proper tracking of a step input, we must have $H(1) = 1$, which from (7.7-9) gives

$$D(z)\bigg|_{z=1} = \sum_{k=0}^{\infty} d(k) = \frac{1}{\mathscr{Z}[G_m(s)G(s)] - \mathscr{Z}[G_m(s)G(s)T(s)]}\bigg|_{z=1} \tag{7.7-10}$$

There is obviously an uncountable number of discrete systems that satisfy

(7.7-10), all of which will guarantee $c(nT) \rightarrow R$ for sufficiently large n. It is up to the design engineer to select that $D(z)$ which satisfies (7.7-10) and meets other criteria. For example, since it is desired to force $c(t) = R$ identically for sufficiently large time t (and not just at the sampling times nT) we select a $D(z)$ that both satisfies (7.7-10) and meets this requirement. Techniques for making such a selection are treated in references 2, 3, and 4.

An interesting special case occurs when $T(s) = 1$ (i.e., unity feedback). In this case

$$D(z)|_{z=1} = \infty$$

which implies that the transfer function of the digital computer has a pole at $z = 1$.

We return to expression (7.7-7); a linear system with transfer function $H(z)$ will track a step input if $H(z) - z^{-m}$ has a zero of at least order one at $z = 1$. Putting this into an explicit form gives

$$H(z) = z^{-m} + (1 - z^{-1})S(z)$$

where $S(z)$ is a ratio of polynomials in z which has a finite value at $z = 1$ [i.e., $S(z)$ has no pole at $z = 1$].

7.7-2 Ramp Input

A test ramp input signal is characterized by

$$\begin{aligned} r(k) &= Vk \quad \text{for } k \geq 0 \\ r(k) &= 0 \quad \text{for } k < 0 \end{aligned} \tag{7.7-11}$$

which has the z-transform

$$R(z) = \frac{Vz^{-1}}{(1 - z^{-1})^2} \tag{7.7-12}$$

Inserting (7.7-12) into (7.7-6) gives

$$\lim_{z \to 1} \left[\frac{H(z) - z^{-m}}{1 - z^{-1}} \right] = 0 \tag{7.7-13}$$

Expression (7.7-13) is the necessary condition that the discrete system with transfer function $H(z)$ must satisfy in order to track a ramp input. An inspection of (7.7-13) reveals that $H(z) - z^{-m}$ must have a zero of at least order two at $z = 1$; that is,

$$H(z) - z^{-m} = (1 - z^{-1})^2 S(z) \tag{7.7-14}$$

where $S(z)$ is a ratio of polynomials which is analytic (no pole) at $z=1$. Evaluating (7.7-14) at $z=1$ reveals that $H(1)=1$, which implies that a discrete system that tracks a ramp input will also track a step input.

If it is desired that the discrete system perfectly follow a ramp input, i.e.,

$$c(k) = r(k) \quad \text{as } k \longrightarrow \infty$$

we set $m=0$ in (7.7-14), obtaining

$$H(z) = 1 - (1 - z^{-1})^2 S(z)$$

Examples of two systems that will track a ramp input are
(a) $H(z) = 1$; set $S(z) = 0$.
(b) $H(z) = 2z^{-1} - z^{-2}$; set $S(z) = 1$.

7.7-3 Acceleration Input

An acceleration input of amplitude A is characterized by the time sequence

$$r(k) = \frac{A}{2}k^2 \quad \text{for } k \geq 0$$
$$r(k) = 0 \quad \text{for } k < 0$$

Therefore,

$$R(z) = \frac{A}{2} \frac{z^{-1}(1 + z^{-1})}{(1 - z^{-1})^3}$$

The necessary condition for proper tracking becomes

$$\lim_{z \to 1} \left[\frac{H(z) - z^{-m}}{(1 - z^{-1})^2} \right] = 0$$

which indicates that $H(z) - z^{-m}$ must have a zero of order at least three at $z=1$. Therefore, the form of $H(z)$ is given by

$$H(z) = z^{-m} + (1 - z^{-1})^3 S(z)$$

with $S(z)$ being a ratio of polynomials in z analytic at $z=1$.

7.8 Controller with a Quadratic Performance Index

A useful criterion for measuring the performance of a control system is the quadratic performance index. This index, typically, will have the form

Sec. 7.8　Controller with a Quadratic Performance Index

$$J = \sum_{k=0}^{N-1} [\tfrac{1}{2}\mathbf{x}^T(k)\mathbf{Q}\mathbf{x}(k) + \tfrac{1}{2}\mathbf{u}^T(k)\mathbf{R}\mathbf{u}(k)] \tag{7.8-1}$$

where $\mathbf{x}(k)$ is the $n \times 1$ state vector and $\mathbf{u}(k)$ is the $p \times 1$ control vector at the kth iteration time. Q is an $n \times n$ positive semidefinite symmetric matrix, and $\mathbf{R}$ is a $p \times p$ positive definite symmetric matrix. These matrices may be selected to weight the magnitudes of the state vector and control vector.

The quadratic control problem is the following: Given a system characterized by the vector difference equation

$$\mathbf{x}(k+1) = \mathbf{A}\mathbf{x}(k) + \mathbf{B}\mathbf{u}(k) \tag{7.8-2}$$

which is at some arbitrary initial state $\mathbf{x}(0)$, determine the control sequence $\mathbf{u}(0), \mathbf{u}(1), \ldots, \mathbf{u}(N-1)$ that minimizes the quadratic performance (7.8-1).

This problem may be treated as a minimization problem involving a function of several variables. The objective here is to minimize J of (7.8-1) subject to the constraint equations specified by (7.8-2). By use of a set of Lagrange multipliers $\boldsymbol{\lambda}(0), \boldsymbol{\lambda}(1), \ldots, \boldsymbol{\lambda}(N)$, we may recast this problem as one in which the augmented performance index

$$H = \sum_{k=0}^{N-1} \{\tfrac{1}{2}\mathbf{x}^T(k)\mathbf{Q}\mathbf{x}(k) + \tfrac{1}{2}\mathbf{u}^T(k)\mathbf{R}\mathbf{u}(k) \\ + \boldsymbol{\lambda}^T(k+1)[\mathbf{A}\mathbf{x}(k) + \mathbf{B}\mathbf{u}(k) - \mathbf{x}(k+1)]\} \tag{7.8-3}$$

is to be minimized. It is known that the recast problem is equivalent to the original problem.

The minimization of (7.8-3) is carried out as an ordinary problem of finding the maximum or minimum of a function of several variables. We merely obtain the partial derivatives of J with respect to $\mathbf{x}(k)$, $\mathbf{u}(k)$, and $\boldsymbol{\lambda}(k)$, for all values of k, and equate these relations to zero.

$$\frac{\partial H}{\partial \mathbf{x}_k} = 0 \quad k = 0, 1, \ldots, N-1 \tag{7.8-4}$$

$$\frac{\partial H}{\partial \mathbf{u}_k} = 0 \quad k = 0, 1, \ldots, N-1 \tag{7.8-5}$$

$$\frac{\partial H}{\partial \boldsymbol{\lambda}_k} = 0 \quad k = 0, 1, \ldots, N-1 \tag{7.8-6}$$

Equations (7.8-4) through (7.8-6) constitute the necessary conditions for H to have a minimum.

The individual relations involve the differentiation of such quadratic expressions as $\mathbf{z}^T\mathbf{W}_1\mathbf{v}$, $\mathbf{v}^T\mathbf{W}_2$, and $\mathbf{z}^T\mathbf{W}_3\mathbf{z}$. Each one of these expressions is a scalar, but is differentiated with respect to the vector variables $\mathbf{z}$ and $\mathbf{v}$.

For the purpose of this problem, we are interested in the following results:

$$\frac{\partial \mathbf{z}^T \mathbf{W}_1 \mathbf{v}}{\partial \mathbf{v}} = \mathbf{W}_1^T \mathbf{z} \qquad (7.8\text{-}7)$$

$$\frac{\partial \mathbf{v}^T \mathbf{W}_2 \mathbf{z}}{\partial \mathbf{v}} = \mathbf{W}_2 \mathbf{z} \qquad (7.8\text{-}8)$$

$$\frac{\partial \mathbf{z}^T \mathbf{W}_3 \mathbf{z}}{\partial \mathbf{z}} = \mathbf{W}_3^T \mathbf{z} + \mathbf{W}_3 \mathbf{z} \qquad (7.8\text{-}9)$$

It is left as an exercise for the reader to verify these results.

The differentiation as indicated by the necessary conditions may now be carried out.

For $k = 0$ we have

$$\frac{\partial H}{\partial \mathbf{x}(0)} = \mathbf{Q}\mathbf{x}(0) + \mathbf{A}^T \boldsymbol{\lambda}(1) - \boldsymbol{\lambda}(0) = \mathbf{0} \qquad (a)$$

$$\frac{\partial H}{\partial \mathbf{u}(0)} = \mathbf{R}\mathbf{u}(0) + \mathbf{B}^T \boldsymbol{\lambda}(1) = \mathbf{0} \qquad (b) \quad (7.8\text{-}10)$$

$$\frac{\partial H}{\partial \boldsymbol{\lambda}(0)} = \mathbf{A}\mathbf{x}(0) + \mathbf{B}\mathbf{u}(0) - \mathbf{x}(1) = \mathbf{0} \qquad (c)$$

For $k = 1, 2, \ldots, N-1$

$$\frac{\partial H}{\partial \mathbf{x}(k)} = \mathbf{Q}\mathbf{x}(k) + \mathbf{A}^T \boldsymbol{\lambda}(k+1) - \boldsymbol{\lambda}(k) = \mathbf{0} \qquad (a)$$

$$\frac{\partial H}{\partial \mathbf{u}(k)} = \mathbf{R}\mathbf{u}(k) + \mathbf{B}^T \boldsymbol{\lambda}(k+1) = \mathbf{0} \qquad (b) \quad (7.8\text{-}11)$$

$$\frac{\partial H}{\partial \boldsymbol{\lambda}(k)} = \mathbf{A}\mathbf{x}(k) + \mathbf{B}\mathbf{u}(k) - \mathbf{x}(k+1) = \mathbf{0} \qquad (c)$$

Note that equations (c) in (7.8-10) and (7.8-11) are the system equations.

When $k = N - 1$, a term with the index N is present in H. We must also include this in our conditions.

$$\frac{\partial H}{\partial_x N} = \boldsymbol{\lambda}(N) = \mathbf{0} \qquad (7.8\text{-}12)$$

This condition specifies a fixed value for the last number of the set of Lagrange multipliers which will help us in the solution of equations (7.8-10) and (7.8-11) for the control vector $\mathbf{u}(k)$.

From (7.8-10a) and (7.8-11a), we obtain

$$\boldsymbol{\lambda}(k) = \mathbf{Q}\mathbf{x}(k) + \mathbf{A}^T \mathbf{x}(k+1) \qquad (7.8\text{-}13)$$

Next we solve for $\mathbf{u}(k)$ in equations (7.8-10b) and (7.8-11b)

$$\mathbf{u}(k) = -\mathbf{R}^{-1}\mathbf{B}^T\boldsymbol{\lambda}(k+1) \tag{7.8-14}$$

and substitute this result into equations (7.8-2).

$$\mathbf{x}(k+1) = \mathbf{A}\mathbf{x}(k) - \mathbf{B}\mathbf{R}^{-1}\mathbf{B}^T\boldsymbol{\lambda}(k+1) \tag{7.8-15}$$

We next stipulate an important relationship between the state vector and the Lagrange multiplier.

$$\boldsymbol{\lambda}(k) = \mathbf{P}(k)\mathbf{x}(k) \tag{7.8-16}$$

This linear transformation is called a Riccati transformation and is of fundamental importance in the solution of this problem. An investigation of the validity of this transformation is beyond the scope of this book. Related discussions may be found in any advanced text on optimal control theory. Utilizing the Riccati transformation in equations (7.8-13) and (7.8-15) enables us to eliminate $\boldsymbol{\lambda}(k)$ with the result

$$\mathbf{P}(k)\mathbf{x}(k) = \mathbf{Q}\mathbf{x}(k) + \mathbf{A}^T\mathbf{P}(k+1)\mathbf{x}(k+1) \tag{7.8-17}$$

$$\mathbf{x}(k+1) = \mathbf{A}\mathbf{x}(k) - \mathbf{B}\mathbf{R}^{-1}\mathbf{B}^T\mathbf{P}(k+1)\mathbf{x}(k+1) \tag{7.8-18}$$

Solving for $\mathbf{x}(k+1)$ in (7.8-18) yields

$$\mathbf{x}(k+1) = [\mathbf{I} + \mathbf{B}\mathbf{R}^{-1}\mathbf{B}^T\mathbf{P}(k+1)]^{-1}\mathbf{A}\mathbf{x}(k) \tag{7.8-19}$$

Substituting equation (7.8-19) into (7.8-17) yields

$$\mathbf{P}(k)\mathbf{x}(k) = \mathbf{Q}\mathbf{x}(k) + \mathbf{A}^T\mathbf{P}(k+1)[\mathbf{I} + \mathbf{B}\mathbf{R}^{-1}\mathbf{B}^T\mathbf{P}(k+1)]^{-1}\mathbf{A}\mathbf{x}(k) \tag{7.8-20}$$

Since equation (7.8-20) must hold for all $\mathbf{x}(k)$, it simplifies to

$$\mathbf{P}(k) = \mathbf{Q} + \mathbf{A}^T\mathbf{P}(k+1)[\mathbf{I} + \mathbf{B}\mathbf{R}^{-1}\mathbf{B}^T\mathbf{P}(k+1)]^{-1}\mathbf{A} \tag{7.8-21}$$

This is a recursive relationship for the matrix $\mathbf{P}(k)$ used in the Riccati transformation. It must be solved backwards, starting with $\mathbf{P}(N)$. Since $\boldsymbol{\lambda}(N) = \mathbf{P}(N)\mathbf{x}(N)$ and $\boldsymbol{\lambda}(N) = \mathbf{0}$, we have $\mathbf{P}(N) = \mathbf{0}$. With $\mathbf{P}(k)$ determined the problem is essentially solved.

To compute the control vector, we eliminate $\boldsymbol{\lambda}(k+1)$ from equations (7.8-14) and (7.8-13). This results in

$$\begin{aligned}\mathbf{u}(k) &= -\mathbf{R}^{-1}\mathbf{B}^T(\mathbf{A}^T)^{-1}[\boldsymbol{\lambda}(k) - \mathbf{Q}\mathbf{x}(k)] \\ &= -\mathbf{R}^{-1}\mathbf{B}^T(\mathbf{A}^T)^{-1}[\mathbf{P}(k) - \mathbf{Q}]\mathbf{x}(k)\end{aligned} \tag{7.8-22}$$

This is the desired expression for the optimal control law. We note that it is of the form

$$\mathbf{u}(k) = \mathbf{H}(k)\mathbf{x}(k) \tag{7.8-23}$$

which indicates the components of the control vector are proportional to the state vector. Indeed, this expression prescribes a feedback control law with time-varying feedback gains $\mathbf{H}(k)$. This is a very useful and convenient result.

EXAMPLE 7.8-1

Compute the feedback gain matrix $\mathbf{H}(k)$ for the following problem: The system to be controlled is

$$\begin{bmatrix} x_1(k+1) \\ x_2(k+1) \end{bmatrix} = \begin{bmatrix} 0.8 & 1.0 \\ 0 & 0.5 \end{bmatrix} \begin{bmatrix} x_1(k) \\ x_2(k) \end{bmatrix} + \begin{bmatrix} 1.0 \\ 0.5 \end{bmatrix} u(k)$$

The performance index to be minimized is

$$J = \sum_{k=0}^{4} [\tfrac{1}{2}(x_1^2(k) + x_2^2(k)) + \tfrac{1}{2}u^2(k)]$$

The initial conditions are given as

$$\begin{bmatrix} x_1(0) \\ x_2(0) \end{bmatrix} = \begin{bmatrix} 100.0 \\ 0.0 \end{bmatrix}$$

Thus it is seen that

$$\mathbf{Q} = \begin{bmatrix} 1 & 0 \\ 0 & 1 \end{bmatrix}, \quad R = 1 \quad \text{and} \quad N = 5$$

The required calculations to determine the feedback gain matrix may best be carried out by a computer program such as that presented in Appendix 7B of this chapter. This program solves equations (7.8-2), (7.8-21), and (7.8-22). Let the solution for the control law be represented as

$$u(k) = h_1(k)x_1(k) + h_2(k)x_2(k)$$

Table 7.8-1 shows the values of the control variable, the feedback gains, and the state variables.

Table 7.8-1

k	$h_1(k)$	$h_2(k)$	$u(k)$	$x_1(k)$	$x_2(k)$
0	−.395	−.687	−39.5	100.0	.0
1	−.395	−.687	−2.4	40.46	19.7
2	−.395	−.677	3.48	10.19	−11.9
3	−.355	−.555	1.42	.54	−3.8
4	.0	.0	.0	−1.45	−9.42

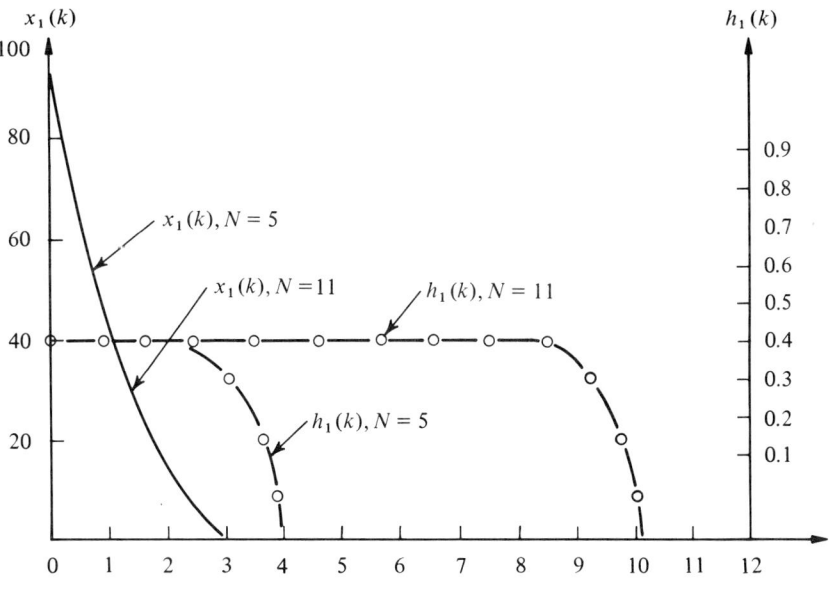

Figure 7.8-1. Response of linear regulator.

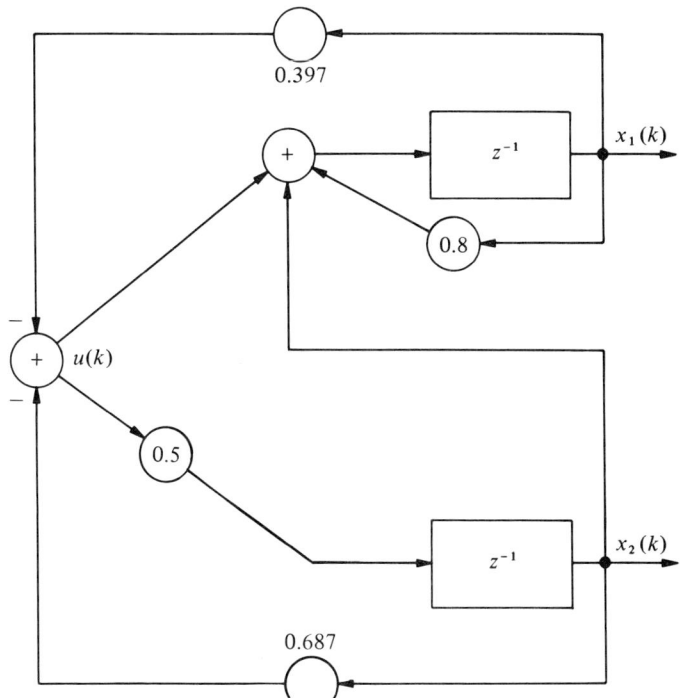

Figure 7.8-2. Block diagram for linear regulator, where $N \to$ large.

An interesting property of the linear regulator is revealed by plotting these variables as a function of k. This is shown in Figure 7.8-1. The feedback gains [only $h_1(k)$ is shown] are constant for low values of k and decrease rapidly to zero as k approaches $N-1$. Note that they are a function of $\mathbf{P}(k)$, which is solved backwards in time. The solutions for $N=5$ and $N=11$ are shown. In both cases identically shaped curves result for $h_1(k)$, except for a lateral displacement, while the responses of $x_1(k)$ are identical. This result carries the important implication that if N is made sufficiently large, the feedback gains in the linear regulator become constants, permitting a closed-loop control system design with constant feedback gains, as displayed by Figure 7.8-2.

7.9 DC Gain of a Discrete System

A meaningful characteristic in the design of linear continuous systems is the DC gain. It applies to systems that generate a steady-state constant output in response to a step input. The DC gain is defined as the ratio of the system's steady-state output to the amplitude of the input step.

The DC gain concept may be extended to discrete systems. For example, consider a system whose transfer is $H(z)$ and which has applied to it a discrete step input of amplitude R. What is the system's resultant response $c(n)$? Writing the familiar transfer function relationship gives

$$C(z) = H(z)R(z) = H(z)\left(\frac{R}{1-z^{-1}}\right) \tag{7.9-1}$$

Expanding (7.9-1) by partial-fraction expansion gives

$$C(z) = \frac{RH(1)}{1-z^{-1}} + \hat{C}(z) \tag{7.9-2}$$

where $H(1) = \lim_{z \to 1} H(z)$

$$\hat{C}(z) = \frac{R}{1-z^{-1}}[H(z) - H(1)]$$

The term $\hat{C}(z)$ contains the transient response terms that depend on the poles of $H(z)$. If the system is stable, these terms will decay to zero for sufficiently large values of discrete time n. It is, therefore, possible to express the system's steady-state response to a discrete step of amplitude R as

$$c_{ss} = \lim_{n \to \infty} c(n) = RH(1) \tag{7.9-3}$$

The DC gain of this system is defined as

$$K = \frac{\text{steady-state response}}{\text{amplitude of step input}} = \frac{RH(1)}{R} = H(1)$$

Some comments on the partial-fraction expansion (7.9-2) should now be made. In this expansion, it has been assumed that the transfer function $H(z)$ has no poles at $z = 1$. If this is not the case, then the expansion as given is incorrect. To demonstrate this, assume that $H(z)$ has a simple pole at $z = 1$. The proper partial-fraction expansion of $C(z)$ would be of the form

$$C(z) = \frac{RH(1)}{(1-z^{-1})^2} + \frac{A}{1-z^{-1}} + \hat{C}(z) \qquad (7.9\text{-}4)$$

Expression (7.9-4) indicates that the system's response, in part, will be a ramp of slope $RH(1)$. This ramp was generated because the $H(z)$ has a simple pole at $z = 1$.

If, as is standard, a stable discrete system is defined as one whose transfer function has all its poles inside the unit circle, then the partial-fraction expansion as given by (7.9-2) will be proper for all stable systems. With this in mind, the DC gain of a stable system with transfer function $H(z)$ is given by

$$K = H(1) = \lim_{z \to 1} H(z) \qquad (7.9\text{-}5)$$

Rewriting (7.9-5) in the standard expansion of $H(z)$ in terms of its weighting sequence $h(n)$, we have

$$K = \lim_{z \to 1} H(z) = \lim_{z \to 1} \left\{ \sum_{n=0}^{\infty} h(n) z^{-n} \right\} = \sum_{n=0}^{\infty} h(n) \qquad (7.9\text{-}6)$$

It has been shown in the section on the tracking of test signals that the value of the DC gain, as given by (7.9-6), plays an important role in the system's tracking ability.

EXAMPLE 7.9-1

Determine the DC gain for the system characterized by

$$c(n+2) + \tfrac{3}{4} c(n+1) + \tfrac{1}{8} c(n) = r(n+1) + 2r(n)$$

This system has the transfer function

$$H(z) = \frac{z+2}{z^2 + \tfrac{3}{4} z + \tfrac{1}{8}}$$

which has poles at $z = -\frac{1}{4}$, $z = -\frac{1}{2}$, so it is stable. Utilizing (7.9-5), we find

$$K = H(1) = \tfrac{8}{5}$$

Conclusions

Several selected optimization problems have been investigated in this chapter. Many of the optimal control laws that resulted involved the solution of a system of linear equations. Under very minor assumptions, it is guaranteed that this system of equations has a unique solution.

The reader is reminded that only a very limited number of design techniques has been treated here. For example, the concepts of dynamic programming and the discrete maximum principle have not been discussed. For a more extensive treatment of optimal design techniques for discrete systems numerous texts are available.

REFERENCES

1. Sage, A., *Optimum Systems Design*, Prentice-Hall, Englewood Cliffs, N. J., 1968.
2. Ragazzini J. R. and G. F. Franklin, *Sampled-data Control Systems*, McGraw-Hill, New York, 1960.
3. Tou, J. T., *Digital and Sampled-data Control Systems*, McGraw-Hill, New York, 1960.
4. Dorf, R. C., *Time Domain Analysis and Design of Control Systems*, Addison-Wesley, New York, 1964.
5. Cadzow, J. A., "A Study of Minimum Norm Control for Sampled-data Systems," 1965, Joint Automatic Control Conference, pp. 545–550.

PROBLEMS

7.1 For the plant with transfer function

$$G(s) = \frac{1}{s^2}$$

and step input make a time domain:
(a) Minimal prototype design
(b) Ripple-free design

Let the sampling period T be an arbitrary parameter in the design process. The initial state is zero.

7.2 For the plant with transfer function

$$G(s) = \frac{1}{(s+a)(s+1)}$$

and step input make a time domain:
(a) Minimal prototype design
(b) Ripple-free design
with $T = 1$ second and initial state zero. Investigate the comparison of results with those of Section 6.2 as $a \to 0$.

7.3 For the system in Problem 7.1 and ramp input make a time domain:
(a) Minimal prototype design
(b) Ripple-free design

Check to verify that it has desirable characteristics for a unit step input.

7.4 For the system in Problem 7.1, carry out a z-domain synthesis for a minimal prototype design. Check with the results of Problem 7.1.

7.5 For the system in Problem 7.3, carry out a z-domain synthesis for a minimal prototype design. Check with the results of Problem 7.3.

7.6 For the system of Problem 7.1 and a step input make a time domain:
(a) Minimal prototype design
(b) Ripple-free design
under the assumption that the initial state is not zero.

7.7 Determine the controllability characteristics of the discrete systems with transfer functions

(a) $\dfrac{C(z)}{U(z)} = \dfrac{z+2}{(z+1)(z+3)}$

(b) $\dfrac{C(z)}{U(z)} = \dfrac{z}{(z+1)^2}$

7.8 Design a digital regulator for the system characterized by

$$\frac{C(s)}{U(s)} = \frac{1}{s^3}$$

if the input $u(t)$ is constrained to be constant over one-second time intervals.

7.9 Repeat Problem 7.8 for the system with transfer function

$$\frac{C(s)}{U(s)} = \frac{1}{s(s+a)}$$

Check the results with (7.5-5) by letting $a = 1$.

7.10 Design a minimum energy regulator (Section 7.6) for the system with transfer function

$$\frac{C(s)}{U(s)} = \frac{1}{s^2}$$

when the inputs are constrained to be constant over one-second intervals. Carry out the design for $N = 1, 2, 3, 4$.

7.11 For the discrete plant

$$\begin{bmatrix} x_1(k+1) \\ x_2(k+1) \end{bmatrix} = \begin{bmatrix} \frac{1}{2} & 0 \\ 0 & \frac{1}{4} \end{bmatrix} \begin{bmatrix} x_1(k) \\ x_2(k) \end{bmatrix} + \begin{bmatrix} 1 \\ 2 \end{bmatrix} u(k)$$

$$y(k) = \begin{bmatrix} 1 & 1 \end{bmatrix} \begin{bmatrix} x_1(k) \\ x_2(k) \end{bmatrix}$$

determine the optimal control law for the performance index

$$J = \sum_{k=0}^{N} [y^2(k) + ru^2(k)]$$

7.12 For the simple system characterized by

$$c_{n+1} = \tfrac{1}{2} c_n + u_n$$

it is desired to minimize the functional

$$J = \sum_{n=0}^{N} [(c_{n+1} - b)^2 + pu_{n^2}]$$

subject to the constraint

$$\sum_{n=0}^{N} u_{n^2} = L$$

Determine the optimal control sequence for any initial condition $c(0)$.

7.13 Using the techniques of Section 7.11, find the optimal control law for Problem 7.11.

7.14 Derive equations (7.8-7), (7.8-8), and (7.8-9).

7.15 Apply a suitable computer algorithm to design a minimum settling time controller (one-second sampling interval) for the system described by the equations

$$\frac{d}{dt} \begin{bmatrix} x_1 \\ x_2 \\ x_3 \\ x_4 \end{bmatrix} = \begin{bmatrix} -1 & 1 & 0 & 1 \\ 2 & .5 & 3 & 0 \\ 0 & .0 & -1 & 1 \\ 0 & 1 & 2 & -4 \end{bmatrix} \begin{bmatrix} x_1 \\ x_2 \\ x_3 \\ x_4 \end{bmatrix} + \begin{bmatrix} 1 & 0 \\ 0 & 1 \\ 1 & 1 \\ 1 & -1 \end{bmatrix} \begin{bmatrix} u_1 \\ u_2 \end{bmatrix}$$

$$\begin{bmatrix} c_1 \\ c_2 \end{bmatrix} = \begin{bmatrix} 1 & 0 & 1 & 0 \\ 0 & 1 & 0 & 1 \end{bmatrix} \begin{bmatrix} x_1 \\ x_2 \\ x_3 \\ x_4 \end{bmatrix}$$

7.16 Apply a suitable computer algorithm to design a quadratic performance index controller for the discrete process ($N = 5$).

$$\begin{bmatrix} x_1(k+1) \\ x_2(k+1) \end{bmatrix} = \begin{bmatrix} .2 & -.1 \\ .1 & -.2 \end{bmatrix} \begin{bmatrix} x_1(k) \\ x_2(k) \end{bmatrix} + \begin{bmatrix} 0 \\ 1 \end{bmatrix} u(k)$$

7.17 Repeat Problem 7.16 for $N = 10, 20, 100, \infty$.

7.18 Discuss the stability of a closed-loop system with a controller designed according to the quadratic performance criterion.

7.19 Derive the transfer function for a digital computer such that a system defined by Figure 7.7-1 will properly track (1) a step input and (2) a ramp input.

$$G(s) = \frac{1}{s(s+2)}, \qquad T(s) = 1, \, m = 0$$

Compare the transients of the system designed under (1) and (2) when the input to both designs is a step input.

Appendix 7A

COMPUTER PROGRAM FOR DESIGN OF DEADBEAT CONTROLLER

The synthesis of a general deadbeat controller as given by Equation (7.2-43) is implemented by a computer program. The program is limited to the specific case in which the total number of sampling periods to reach steady state is two. The program may be readily modified to accommodate other cases.

It is written as a subroutine and assumes that the **F**, **G**, and **C** matrices of the linear plant to be controlled are read into the core storage prior to calling the subroutine. The call name is

SUBROUTINE DDBT2 (F, G, C, N, M, T),

where N is the order of **F**, M the number of inputs and outputs, and T is the sampling interval.

The program requires the use of three other subroutines:

1. SUBROUTINE MATEXP (F, G, PHI, DELT, N, M, T)

    This program is discussed in Appendix 2A.

2. SUBROUTINE MULTIQ (X, Y, Z, N1, N2, N3)

    A listing is shown in Appendix 2A.

3. SUBROUTINE MINV (A, N, D, L, M)

    This program obtains the inverse of the matrix **A** of order N and stores the result in **A** with determinant D. The arrays L and M are working vectors of order N; they are included in the calling vector because the subroutine used here is of variable dimensioning. Any matrix inversion routine that is available through a library may be used. An illustrative problem is solved here.

The system is characterized by

$$\mathbf{F} = \begin{bmatrix} 1. & 1. & -5. & -5. \\ 0. & -2. & 0. & 0. \\ 2. & 1. & -6. & -1. \\ -2. & -1. & 2. & -3. \end{bmatrix}$$

$$\mathbf{G} = \begin{bmatrix} 1. & 1. \\ 0. & 2. \\ 0. & 2. \\ 0. & -1. \end{bmatrix}, \quad \mathbf{H} = \begin{bmatrix} 3.0 & 2.0 & -3.0 & 2.0 \\ 1.0 & 2.0 & 1.0 & 3.0 \end{bmatrix}$$

The sampling period is $T = 1$. Since it is a fourth-order system with two inputs and two outputs, two sampling periods are required for deadbeat control; i.e., $k = 2$.

```
      SUBROUTINE DDBT2(F,G,C,N,M,T)
      DIMENSION F(5,5), G(5,5), C(5,5), DELT(5,5), PHI(5,5),
     * CDELT(5,5), ADELT(5,5) PHIDE(5,5), CPHIDE(5,5), APHIDE(5,5),
     * INTPHI(5,5) STORGE(5,5), FF(100),F1(5,5), F2(5,5),
     * F10(5,5), F21(5,5), CDF0N(5,5), CDF01(5,5),
     * P(10,10),CDF0(5,5), F0(5,5), L(10),MM(10)
C
C*****************************************************************
C
C     THIS IS FOR K = 2
C
C*****************************************************************
C
      CALL MATEXP(F,G,PHI,DELT,N,M,T)
      CALL MULTIQ (F, DELT, ADELT, N, N, M)
      CALL MULTIQ(C, DELT, CDELT, M, N, M)
      CALL MULTIQ (PHI, DELT, PHIDE, N, N, M)
      CALL MULTIQ (F, PHIDE, APHIDE, N, N, M)
      CALL MULTIQ(C, PHIDE, CPHIDE, M, N, M)
      DO 19 I1 = 1,M
      DO 19 J1 = 1,M
   19 P(I1, J1+N) = 0.0
      DO 20 I2 = 1,N
      DO 20 J2 = 1,M
   20 P(M+I2, N+J2) = G(I2,J2)
      DO 21 I3 = 1,M
      DO 21 J3 = 1,M
   21 P(I3, J3+M) = CDELT(I3,J3)
```

```
         DO  22   I4  =  1,N
         DO  22   J4  =  1,M
   22    P(I4+M, J4+M)  =  ADELT(I4, J4)
         DO  23   I5  =  1,M
         DO  23   J5  =  1,M
   23    P(I5, J5)  =  CPHIDE(I5, J5)
         DO  24   I6  =  1,N
         DO  24   J6  =  1,M
   24    P(I6+M, J6)  =  APHIDE(I6, J6)
         WRITE  (6, 100)
  100    FORMAT (1H1, 10X, 24HTHE  P  MATRIX  IS  EQUAL  TO   //)
         MN  =  M+N
         DO  101  II  =  1,MN
         WRITE  (6, 102)  (P(II,JJ),  JJ  =  1,  MN)
  102    FORMAT(5X,  6(E10.3,  6X)//)
  101    CONTINUE
         NN  =  1
         DO  5  I  =  1,MN
         DO  5  J  =  1,MN
         FF(NN)  =  P(I,J)
    5    NN  =  NN+1
         CALL  MINV(FF,MN,D,L,MM)
         NN  =  1
         DO  6  I  =  1,MN
         DO  6  J  =  1,MN
         P(I,J)  =  FF(MN)
    6    NN  =  NN+1
         WRITE(6, 105)
  105    FORMAT(////,10X, 25HTHE  P  INVERSE  IS  EQUAL  TO   //)
         DO  103  I  =  1,MN
         WRITE(6, 104)(P(I,J),J  =  1,MN)
  104    FORMAT(5X,  6(F10.3,6X)//)
  103    CONTINUE
         DO  25  I7  =  1,M
         DO  25  J7  =  1,M
   25    F0(I7, J7)  =  P(I7, J7)
         CALL  MULTIQ (CDELT, F0, CDF0, M, M, M)
         DO  57  I  =  1,M
         DO  57  J  =  1,M
   57    F1(I,J)  =  P(I+M,J)
         DO  58  I  =  1,M
         DO  58  J  =  1,M
         M2  =  2*M
   58    F2(I,J)  =  P(I+M2,J)
         DO  59  I  =  1,M
         DO  59  J  =  1,M
   59    F10(I,J)  =  F1(I,J)+F0(I,J)
         DO  60  I  =  1,M
         DO  60  J  =  1,M
   60    F21(I,J)  =  F2(I,J)+F1(I,J)
         DO  61  I  =  1,M
         DO  61  J  =  1,M
```

```
   61 CDF0N(I,J) = -1*CDF0(I,J)
      DO 62 I = 1,M
      DO 62 J = 1,M
   62 CDF01(I,J) = CDF0(I,J)
      DO 50 I = 1,M
   50 CDF01(I,I) = CDF01(I,I)-1.
      WRITE (6,200)
  200 FORMAT(1H1,///,5X,*THE DIGITAL TRANSFER FUNCTION IS E2(Z)/E1(Z)*/)
      WRITE(6,201)
  201 FORMAT(//5X, *WHERE* /)
      WRITE(6,202)
  202 FORMAT(5X,45HE2(Z)  =  (F0+Z**(-1)*F1+Z**(-2)*F2+ . . . ) /)
      WRITE(6,203)
  203 FORMAT(5X,44HE1(Z)  =  (1+Z**(-1)*G1+Z**(-2)*G2+ . . .)   //)
      WRITE(6,204)
  204 FORMAT(5X,*TRANSFER  FUNCTION  ORDER  IS  2*//)
      WRITE(6,205)
  205 FORMAT(5X,*THE  COEFFICIENTS  ARE* /)
      WRITE(6,206)
  206 FORMAT(5X,*F0*)
      DO 207 I = 1,M
      WRITE(6, 208) (F0(I,J), J = 1,M)
  207 CONTINUE
      WRITE(6,209)
  208 FORMAT(10X,5(E12.5,5X))
  209 FORMAT(/5X, *F1*)
      DO 210 I = 1,M
      WRITE(6,208) (F10(I,J), J = 1,M)
  210 CONTINUE
      WRITE(6,211)
  211 FORMAT(/5X,*F2*)
      DO 212 I = 1,M
      WRITE (6,208) (F21(I,J),J = 1,M)
  212 CONTINUE
      WRITE(6,213)
  213 FORMAT(/5X, *G1*)
      DO 214 I = 1,M
      WRITE(6,208) (CDF0N(I,J), J = 1,M)
  214 CONTINUE
      WRITE(6,215)
  215 FORMAT(/5X, *G2*)
      DO 216 I = 1,M
      WRITE(6,208) (CDF01(I,J), J = 1,M)
  216 CONTINUE
      RETURN
      END
```

EXAMPLE OUTPUT
THE DIGITAL TRANSFER FUNCTION IS E2(Z)/E1(Z)

WHERE
E2(Z) = (F0+Z**(−1)*F1+Z**(−2)*F2+ . . .)
E1(Z) = (1+Z**(−1)*G1+Z**(−2)*G2+ . . .)
TRANSFER FUNCTION ORDER IS 2
THE COEFFICIENTS ARE
F0
    2.40115E+01   1.57514E+00
   −1.96863E+00   9.84316E+00
F1
   −3.97571E+01   −1.38028E+00
    2.93139E+00   −1.46570E+01
F2
    1.62750E+00   1.58080E−01
   −1.08041E+00   5.40204E+00
G1
   −6.62124E+00   5.14642E−01
   −1.39309E+00   −3.54289E+00
G2
    5.62124E+00   −5.14642E−01
    1.39309E+00   2.54289E+00

Appendix 7B

COMPUTER PROGRAM FOR DISCRETE LINEAR REGULATOR

This program represents a computer implementation of equations (7.8-2), (7.8-21), and (7.8-22), describing the solution to the linear regulator problem. The program requires the matrix multiplication and matrix inversion routines introduced in Appendix 7A.

For an illustration of the use of this program see Section 7.8.

```
      COMMON  A(5,5),B(5,5),R(5,5),RI(5,5),BT(5,5),AT(5,5),S(5,5),
     1SSUM(100),SUM(5,5),ATT(5,5),P(5,5,100),HZ(5,5),HN(5,5),HH(5,5)
     2H2(5,5),H3(5,5),H4(5,5),U1(5,5),U2(5,5),U3(5,5),U(5,5),UI(5,5),
     3UJ(5,100),B1(5,5),B2(5,5),Y(5,100),N,L,KF,Q(5,5),X(5),XX(5,5),
     4U4(5,5)
      COMMON  H5(5,5)
C     READ IN DIMENSIONS OF MATRICES AND FINAL TIME KF
999   READ(5,10)  N,L,KF
  10  FORMAT(3I3)
C     A MATRIX N*N  B MATRIX N*L
C     READ IN ALL MATRICES BY ROWS
      DO 1 I = 1,N
      READ(5,11)(A(I,J),J = 1,N)
    1 CONTINUE
   11 FORMAT(5E14.7)
      DO 2 I = 1,N
      READ(5,12)  (B(I,J),J = 1,L)
    2 CONTINUE
   12 FORMAT(F14.7)
      DO 3 I = 1,L
      READ(5,12)(R(I,J),J = 1,L)
```

```
      3 CONTINUE
        DO 4 I = 1,N
        READ(5,11)(Q(I,J),J = 1,N)
      4 CONTINUE
        DO 5 I = 1,N
        READ(5,11)(S(I,J),J = 1,N)
      5 CONTINUE
C       COMPUTE R INVERSE
        DO 6 I = 1,L
        DO 6 J = 1,L
      6 SUM(I,J) = R(I,J)
        CALL PINV
        DO 7 I = 1,L
        DO 7 J = 1,L
      7 RI(I,J) = SUM(I,J)
C       COMPUTE A TRANSPOSE AND A INVERSE TRANSPOSE
        DO 8 I = 1,N
        DO 8 J = 1,N
        AT(I,J) = A(J,I)
      8 SUM(I,J) = AT(I,J)
        CALL PINV
        DO 9 I = 1,N
        DO 9 J = 1,N
      9 ATT(I,J) = SUM(I,J)
C       COMPUTE B TRANSPOSE
        DO 20 I = 1,N
        DO 20 J = 1,L
     20 BT(J,I) = B(I,J)
C
C       READ IN INITIAL CONDITIONS ON STATE VARIABLES
C
        READ(5,11)(X(I),I = 1,N)
        DO 800 I = 1,N
C       AT THIS POINT WE HAVE AVAILABLE A,AT,AIT,B,BT,R,RT,S,X(0)
C
C       COMPUTE B*BI*BT = HN
    800 Y(I,1) = X(I)
        CALL MULTIQ(RI,BT,HZ,N,L,N)
        CALL MULTIQ(B,HZ,HN,N,L,N)
C
C       NOW COMPUTE ELEMENTS IN RICCATI EQUATION-P(I,J,K)
C
        K = KF+1
        DO 21 I = 1,N
        DO 21 J = 1,N
     21 P(I,J,K) = S(I,J)
    102 DO 100 I = 1,N
        DO 100 J = 1,N
    100 HH(I,J) = (I,J,K)
        CALL MULTIQ(HN,HH,H2,N,N,N)
        DO 22 I = 1,N
        H2(I,I) = H2(I,I)+1.0
```

```
              DO 22 J = 1,N
         22   SUM(I,J) = H2(I,J)
              CALL  PINV
              CALL  MULTIQ(SUM,A,H3,N,N,N)
              CALL  MULTIQ(HH,H3,H4,N,N,N)
              CALL  MULTIQ(AT,H4,H5,N,N,N)
              K = K-1
              DO 101 I = 1,N
              DO 101 J = 1,N
        101   P(I,J,K) = Q(I,J)+H5(I,J)
              IF(K.NE.1) GO TO 102
C
C             PRINT OUT P MATRIX
C
              WRITE(6,900)
        900   FORMAT(//10X,8HP  MATRIX/)
              K = KF+1
              DO 901 I = 1,K
              DO 901 J = 1,N
              WRITE(6,903) I
        903   FORMAT(I3)
              WRITE(6,902)(P(J,JJ,I),JJ = 1,N)
        901   CONTINUE
        902   FORMAT(5E14.7)
C
C             COMPUTE TRAJECTORIES AND CONTROL
C
              CALL  MULTIQ(BT,ATT,U1,L,N,N)
              CALL  MULTIQ(RI,U1,U2,L,L,N)
        850   FORMAT(2X,3HU1 = ,E14.7,3X,3HU2 = ,E14.7)
              K = 1
              WRITE (6,307)
        307   FORMAT (3X,1HK,10X,5HGAIN1,13X,5HGAIN2)
        300   DO 201 I = 1,N
              XX(I,I) = X(I)
              DO 201 J = 1,N
        201   U3(I,J) = P(I,J,K)-Q(I,J)
              CALL  MULTIQ(U2,U3,U4,L,N,N)
              WRITE  (6,310) K,(U4(1,I),I = 1,N)
        310   FORMAT (3X,I3,2(E14,7))
              CALL  MULTIQ(U4,XX,U,L,N,1)
              DO 202 I = 1,L
              UI(I,1) = -U(I,1)
        202   UJ(I,K) = -U(I,1)
        965   FORMAT(2X,2E14.7)
              CALL  MULTIQ(B,UI,B1,N,L,1)
              CALL MULTIQ(A,XX,B2,N,N,1)
        966   FORMAT(2X,5(E11.4,3X))
              K = K+1
              DO 203 I = 1,N
              X(I) = B2(I,1)+B1(I,1)
        203   Y(I,K) = X(I)
```

```
          IF(K .NE. KF+2) GO TO 300
          WRITE(6,500)
    500   FORMAT(3X,1HK,10X,2HX1,20X,2HX2,20X,1HU)
          K = KF+1
          DO 700 I = 1,K
    700   WRITE(6,600) I,Y(1,I),Y(2,I),UJ(1,I)
    600   FORMAT(1X,I3,3(4X,E14.7))
          GO TO 999
    998   CALL EXIT
          END

          SUBROUTINE PINV
          COMMON A(5,5),B(5,5),R(5,5),RI(5,5),BT(5,5),AT(5,5),S(5,5),
         1SSUM(100),SUM(5,5),ATT(5,5),P(5,5,100),HZ(5,5),HN(5,5),HH(5,5),
         2H2(5,5),H3(5,5),H4(5,5),U1(5,5),U2(5,5),U3(5,5),U(5,5),UI(5,5),
         3UJ(5,100),B1(5,5),B2(5,5),Y(5,100),N,L,KF,Q(5,5),X(5),XX(5,5),
         4U4(5.5)
          COMMON H5(5,5)
          DIMENSION LL(50),MM(50)
          M = N
          D = 0.0
          NN = 1
          DO 5 I = 1,M
          DO 5 J = 1,M
          SSUM(NN) = SUM(I,J)
      5   NN = NN+1
          CALL MINV(SSUM,M,D,LL,MM)
          NN = 1
          DO 6 I = 1,M
          DO 6 J = 1,M
          SUM(I,J) = SSUM(NN)
      6   NN = NN+1
          RETURN
          END
```

8

Engineering Characteristics of Computer Control Systems

The study of discrete systems requires coverage of both theoretical and practical aspects since the design of a system is based upon both analytical and engineering considerations. One of the prime responsibilities of the practicing engineer, indeed, consists of generating a design that is sound analytically and poses reasonable demands on the present state of technology. In the design of a computer control system, which is probably the most prominent example of a discrete system, the engineer's ability is put to a particularly hard test. The analytical aspects involve some of the more advanced concepts and require the support of mathematical tools only recently developed; at the same time the computer technology is an extremely rapidly changing field. We therefore appreciate the need for a balanced understanding of discrete system theory and technology. Inasmuch as the preceding chapters have dealt exclusively with theory, we shall now devote this chapter to a survey-like treatment of the physical characteristics of digital components that are typically employed in discrete systems. To properly prepare the reader for the understanding of the functional characteristics of these components, we first present some background in number systems and digital logic.

8.1 Number Systems

Basic to the use of computers and processing of information in and out of computers is an understanding of the interrelationship of several number systems. The "natural" number system of the user is the decimal system, whereas the computer operates in the binary system. For general-purpose computer utilization the user sees the computer as a decimal machine, since all input and output information is automatically processed by special input-output routines that perform the necessary binary-decimal conversion.

These routines are usually part of the computing system's software, and the user need not concern himself with their internal structure. However, the system designer is required to view the computer as an electronic component that functions in response to electrical pulse signals. Although programmable subroutines and specially designed hardware interface systems are available, an understanding of the basic operations is essential. For instance, the system designer must be versed in storing information directly in the computer's memory as well as through the use of software channels. The latter can be accomplished only if the binary equivalent of the decimal number is known or can be generated externally to the computer.

For this reason we now introduce some of the characteristics of the pertinent number systems.

8.1-1 Decimal Numbers

The decimal number system uses ten symbols representing the quantities 0 through 9. All numbers of the decimal system are constructed by assigning different weights to the position of the symbol relative to the decimal point. Each position in a decimal number has a value that is ten times the value of the position to its right. Thus positions are associated with integer powers of ten. The ones position is 10^0, the tens position is 10^1, the hundreds position is 10^2, etc. This progression of increasing exponents may be carried as far as desired to the left of the decimal point. To the right of the decimal point a progression of negative exponents applies. For example, the first position to the right of the decimal point is the tenths position, with weight 10^{-1}. The general skeleton of a decimal number is made up as shown in Figure 8.1-1.

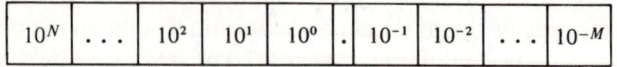

Figure 8.1-1. Structure of a decimal number.

8.1-2 The Radix of a Number System

A number system is classified according to the number of different symbols used. This number is called the *radix*. The decimal number system has a radix of ten. A number system with a different radix has the weights of the

Table 8.1-1 Counting in Different Number Systems

Decimal		Hexadecimal		Octal		Binary				
tens	ones	sixteens	ones	eights	ones	sixteens	eights	fours	twos	ones
	0		0		0					0
	1		1		1					1
	2		2		2				1	0
	3		3		3				1	1
	4		4		4			1	0	0
	5		5		5			1	0	1
	6		6		6			1	1	0
	7		7		7			1	1	1
	8		8	1	0		1	0	0	0
	9		9	1	1		1	0	0	1
1	0		A	1	2		1	0	1	0
1	1		B	1	3		1	0	1	1
1	2		C	1	4		1	1	0	0
1	3		D	1	5		1	1	0	1
1	4		E	1	6		1	1	1	0
1	5		F	1	7		1	1	1	1
1	6	1	0	2	0	1	0	0	0	0
1	7	1	1	2	1	1	0	0	0	1
1	8	1	2	2	2	1	0	0	1	0
1	9	1	3	2	3	1	0	0	1	1
2	0	1	4	2	4	1	0	1	0	0
2	1	1	5	2	5	1	0	1	0	1
2	2	1	6	2	6	1	0	1	1	0
2	3	1	7	2	7	1	0	1	1	1
2	4	1	8	3	0	1	1	0	0	0
2	5	1	9	3	1	1	1	0	0	1
2	6	1	A	3	2	1	1	0	1	0
2	7	1	B	3	3	1	1	0	1	1
2	8	1	C	3	4	1	1	1	0	0
2	9	1	D	3	5	1	1	1	0	1
3	0	1	E	3	6	1	1	1	1	0
3	1	1	F	3	7	1	1	1	1	1
3	2	2	0	4	0	1 0	0	0	0	0

different position to the right or left of the point given by integer powers of the radix. The structure of an arbitrary number with radix R is thus shown in Figure 8.1-2. Number systems with four different values of R are presently

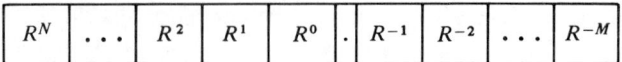

Figure 8.1-2. Structure of a number with radix R.

used in connection with digital computers: $R = 2$, $R = 8$, $R = 10$, $R = 16$. These are called the binary, octal, decimal, and hexadecimal number systems, respectively. Examples of counting in these number systems are shown in Table 8.1-1.

Notice that the number of symbols used for these systems is equal to the radix. The hexadecimal system uses six alphabetic characters in addition to 10 numerals, for lack of additional symbols denoting the numbers "10," "11," ..., "15."

8.1-3 The Binary Number System

The binary number system has a radix of two and uses two symbols. Its positional weights are powers of two. Arithmetic operations in binary numbers obey the same rules as do numbers in the decimal system, as is illustrated in Table 8.1-2.

Table 8.1-2 Binary Arithmetic

Binary Addition	Binary Subtraction
$0 + 0 = 0$	$0 - 0 = 0$
$0 + 1 = 1$	$1 - 0 = 1$
$1 + 0 = 1$	$0 - 1 = 1$ and 1 to borrow
$1 + 1 = 0$ and 1 to carry	$1 - 1 = 0$

Binary Multiplication	Binary Division
$0 \times 0 = 0$	$0 \div 0 = $ ⎱ undefined
$0 \times 1 = 0$	$1 \div 0 = $ ⎰
$1 \times 0 = 0$	$1 \div 1 = 1$
$1 \times 1 = 1$	$0 \div 1 = 0$

EXAMPLE 8.1-1

Perform the following binary arithmetic operations:
Addition:

$$\begin{array}{r} 101101 \\ +1010 \\ \hline 110111 \end{array} \qquad \begin{array}{r} 101101 \\ +1100 \\ \hline 111001 \end{array}$$

Subtraction:

$$\begin{array}{r} 101101 \\ -1010 \\ \hline 100011 \end{array} \qquad \begin{array}{r} 101101 \\ -1111 \\ \hline 11110 \end{array}$$

Multiplication: *Division:*

```
       101101              1001
    ×    101         101 )101101
       101101              101
       000000              ----
       101101              0001
       ------              0000
      11100001             ----
                             10
                             00
                             ---
                            101
                            101
                            ---
                              0
```

All digital computers presently manufactured are binary machines; that is, arithmetic operations are carried out in the binary number system. Furthermore, all storage devices, such as tape, disk, card core, drum, etc., function on a binary basis. Consequently, the binary system is of prime importance in the study of computer control systems. Thus, the computer's natural number system is binary, whereas the user's natural system is usually decimal. Since computer inputs and outputs must often be in decimal notation (for convenience of the user) a variety of codes is available by which decimal numbers may be presented by binary symbols. Machines that use codes are usually called *decimal machines*. Binary machines use the pure binary number system. A discussion of possible coding techniques is presented in a later section dealing with encoders.

8.1-4 The Octal Number System

The octal number system, as the name implies, has a radix of 8. It uses the symbols 0 through 7. The positional weights in the octal number system are powers of 8.

The octal number system is involved in digital computer work because it offers an efficient way of dealing with binary numbers. One may look upon it as a binary shorthand. The binary-octal conversion is a very simple process, due to the fact that 8 is the third power of 2. This relation permits a direct correspondence between successive three-bit groups of a binary number and an octal symbol. A table for octal-to-binary conversion is shown in

Table 8.1-3. Using this table, we find that it is straightforward to represent a binary number as an octal number.

Table 8.1-3 Octal-to-Binary Conversion

Octal	Binary
0	000
1	001
2	010
3	011
4	100
5	101
6	110
7	111

EXAMPLE 8.1-2

The following are illustrations of octal-to-binary and binary-to-octal conversions.

Binary	Octal
101 001 101 111	5157
111 111 111 111	7777
1 110 110	166

Because of the ease with which octal-to-binary conversions are carried out, the octal system is widely employed by digital system users.

8.1-5 Number System Conversion

Quite frequently it becomes necessary to change the radix of a number. Such conversion is relatively straightforward if we recall the structure of a number. In the decimal system, for instance, a number such as 279 has the meaning "9 ones +7 tens +2 hundreds." To convert a number to the radix 10 from any other radix we proceed similarly.

A conversion process could be viewed as evaluating the polynomial

$$N = a_n R^n + a_{n-1} R^{n-1} + \cdots + a_1 R^1 + a_0 R^0 \qquad (8.1\text{-}1)$$

where R is the radix or base and a_i are the weights.

EXAMPLE 8.1-3

(a) $100010111_2 = 279_{10}$

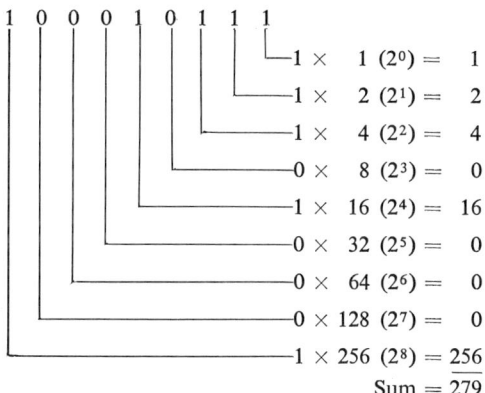

(b) $427_8 = 279_{10}$

$$-7 \times 8^0 = 7$$
$$-2 \times 8^1 = 16$$
$$-4 \times 8^2 = 256$$
$$\text{Sum} = 279$$

A convenient algorithm used to convert an integer from the radix 10 to any other radix is based upon the following procedure. To convert 279 into a binary number, for instance, we divide 279 by 2, obtaining 139 and a remainder of 1. The remainder is the least significant bit of the binary number. The next least significant bit is obtained by dividing 139 by 2, yielding 69 and a remainder of 1. Succeeding bits are obtained by continuing this procedure. This and other examples are illustrated below.

EXAMPLE 8.1-4

(a) $279_{10} = 100010111_2$

$$279 = 2 \cdot (139) + 1$$
$$139 = 2 \cdot (69) + 1$$
$$69 = 2 \cdot (34) + 1$$
$$34 = 2 \cdot (17) + 0$$
$$17 = 2 \cdot (8) + 1$$
$$8 = 2 \cdot (4) + 0$$
$$4 = 2 \cdot (2) + 0$$
$$2 = 2 \cdot (1) + 0$$
$$1 = 2 \cdot (0) + 1$$

Binary Number

(b) $279_{10} = 427_8$

$$279 = 8 \cdot (34) + 7$$
$$34 = 8 \cdot (4) + 2$$
$$4 = 8 \cdot (0) + 4$$

Octal Number

The principle of this algorithm may be understood by successively substituting the remainder terms into the nearest equation above and comparing the result with the polynomial (8.1-1).

8.1-6 Two's Complement Binary Representation

A special way of representing a decimal number in binary format is the so-called *two's complement binary*. To illustrate this format, we consider as an example a binary number with nine bits. With so many bits it is possible to distinguish between $2^9 = 512_{10}$ different integers. If the range of integers is evenly divided between positive and negative integers, a two's complement representation is patterned according to Table 8.1-4.

Table 8.1-4 Two's Complement Nine-bit Numbers

Pairing	Binary	Signed Integer
	011 111 111	+255
	011 111 110	+254
	⋮	⋮
	000 000 010	+ 2
	000 000 001	+ 1
	000 000 000	0
	111 111 111	− 1
	111 111 110	− 2
	⋮	⋮
	100 000 010	−254
	100 000 001	−255

The binary numbers in this table have the property that the sum of a positive and a negative binary number of the same absolute value yields 1 000 000 000. Therefore, they form complements of one another. Notice that the leftmost bit is 0 for all positive numbers and 1 for all negative numbers. It represents the sign bit of numbers. Thus positive and negative numbers are easily recognized.

The relationship between positive and negative numbers provides a straightforward way to negate either a positive or negative number by changing all its bits from 1 to 0 or vice versa (a process called *complementing*) and adding 1 to the result.

EXAMPLE 8.1-5

Represent 1251 and -1251 as 12-bit two's complement binary numbers. First 1251 is converted into 12-bit natural binary.

2	1251	
2	625	1
2	312	1
2	156	0
2	78	0
2	39	0
2	19	1
2	9	1
2	4	1
2	2	0
2	1	0
2	0	1

Hence, $1251_{10} = 010\ 011\ 100\ 011_2$.
To obtain -1251 we complement and add 1, obtaining

$$101\ 100\ 011\ 101$$

Two's complement representation is widely used in present-day computers, because it facilitates adding.

8.2 Digital Encoding

The digital computer is an assembly of two-state devices that manipulate the 1 and 0 of the binary number system. The user of the digital computer is accustomed to the decimal number system. To facilitate rapid interchange between the binary and decimal number systems special input/output computing devices are necessary. These may consist of mechanical input/output devices such as card readers and punches, digital shaft encoders, or "thumbwheel" switches. These and similar devices have a decimal input and a binary output.

To the person who uses a digital computer for general-purpose computing (through FORTRAN, for instance), it matters little what relations exist between the binary numbers inside the computer and the decimal numbers outside the computer. He sees only the input and output, both of which are in the decimal system. However, the systems engineer who employs computers for on-line applications requires a more thorough understanding of how a decimal number is represented inside the computer. In this section we shall study methods for encoding decimal numbers for binary machine representation.

The purpose of a code is the representation of decimal digits by binary digits, or bits. The most compact codes require four bits. Some codes involve more than four bits. The usefulness of a code is determined primarily by the ease with which the decimal equivalent is read and arithmetic operations are carried out.

8.2-1 Four-bit Codes

Shown in Table 8.2-1 are several four-bit codes. The first one, which is called the 8-4-2-1 code, uses the same weights as the binary number system. It is readily interpreted, and arithmetic operations, such as addition and subtraction, are easily performed by using the same basic method as in the binary system, since the number sequence is the same. Machine arithmetic that uses codes is called *binary-coded decimal arithmetic*. Arithmetic operations resemble decimal operations, since "carry" exists on the binary nine.

In the excess 3 code, a decimal number D is represented by the binary equivalent of the number $D + 3$. The excess 3 code is not a weighted code, but since it follows the same number sequence as binary, it is useful in arithmetic operations.

Other examples of four-bit weighted codes include the 2421, the 5421, and the 5311 code. Of these the 2421 is commonly employed in counting systems.

Table 8.2-1 Four-bit Decimal Codes

Decimal	8421	Excess 3	Gray Code	2421	5421	5311
0	0000	0011	0000	0000	0000	0000
1	0001	0100	0001	0001	0001	0001
2	0010	0101	0011	0010	0010	0011
3	0011	0110	0010	0011	0011	0100
4	0100	0111	0110	0100	0100	0101
5	0101	1000	0111	1011	1000	1000
6	0110	1001	0101	1100	1001	1001
7	0111	1010	0100	1101	1010	1011
8	1000	1011	1100	1110	1011	1100
9	1001	1100	1101	1111	1100	1101

Observe that the exceess 3 and 2421 codes are designed to be self-com-

Sec. 8.2 Digital Encoding

plementing; 0 is the one's complement of 9, 1 is the one's complenent of 8, etc.

The Gray code is used in shaft encoders to eliminate ambiguity in reading brushes. The Gray code is obtained from the natural binary number with the following change: proceeding from left to right, copy the natural binary number but complement any bit immediately preceded by a 1. For example, 1110 binary equals 1001 Gray code. The use of the Gray code in shaft encoding will be presented in the next section. Note that in the Gray code only one digit changes between adjacent numbers.

8.2-2 Codes Involving More than Four Bits

Codes greater than four bits enjoy greater simplicity in decoding and are often used for error detection. Three commonly used codes are shown in Table 8.2-2. The biquinary is a seven-bit weighted code in which two ones and five zeros appear in the representation of any digit; thus, by counting the number of zeros and ones it is usually possible to detect errors in the representation of a number.

Table 8.2-2a Codes Greater than Four Bits

Decimal	Biquinary	Ring Counter Code
0	0100001	0000000001
1	0100010	0000000010
2	0100100	0000000100
3	0101000	0000001000
4	0110000	0000010000
5	1000001	0000100000
6	1000010	0001000000
7	1000100	0010000000
8	1001000	0100000000
9	1010000	1000000000

A ten-bit code, called a *ring counter code*, allows any of the ten symbols of the decimal system to be represented with a single 1 and nine 0's. This code is used on IBM punch cards. It is obviously the least complicated, but it requires ten binary symbols to be implemented. This code is also often used in counting operations; the counter is a ten-stage shift register with the final stage connected to the initial stage, giving it the name ring counter.

As a final example of a code utilizing more than four bits we consider the so-called *switch-tail ring counter*. It is more efficient than a simple ring counter code, since only five stages are required to implement it. It may be decoded in a straightforward manner. Any state may be recognized by a two-input gate, which is conditioned by two neighboring flip-flops (see Section 8.4).

Table 8-2.2b Switch-tail Ring Counter Code

Decimal	Switch-tail Ring Counter Code
0	00000
1	00001
2	00011
3	00111
4	01111
5	11111
6	11110
7	11100
8	11000
9	10000

8.3 Shaft Encoders

A shaft encoder is an electromechanical transducer that is attached to a rotating shaft to produce a series of pulses to indicate angular shaft position. When the output is differentiated numerically, it may also be used as a tachometer. The shaft encoder contains a disc with a printed pattern of alternating conducting and nonconducting surfaces called *segments*.

Encoders have an accuracy that is inherently better than analog devices. A typical analog device is a rotational potentiometer with infinite resolution. However, since it is read by a bridge or meter circuit its resolution is finite, and the overall accuracy is limited to $\frac{1}{2}$ of one percent. On the other hand, encoders may contain a disc two inches in diameter with segments .005 inch wide, permitting a resolution of $(\pi \times 2)/.005 = 1256$ segments on a single circumference, an accuracy of one part in 1256, or roughly .1 percent. Accuracy may be further improved by either increasing the radius of the disc or reducing the width of the segments.

Shaft encoders can be obtained in basically three different kinds: brush, optical, and magnetic. We shall present a brief discussion of their construction and elaborate on the digital coding techniques used to relate shaft position to binary number.

8.3-1 Types of Encoders

Brush Encoders

Brush encoders have two main components: the encoding disc and the brush assembly. The encoder disc contains a set of concentric tracks of segments, as shown in Figure 8.3-1, which may number from two to 30. One track is not segmented. The brush assembly contains a brush for each track plus one for the solid track, which functions as an electrical return

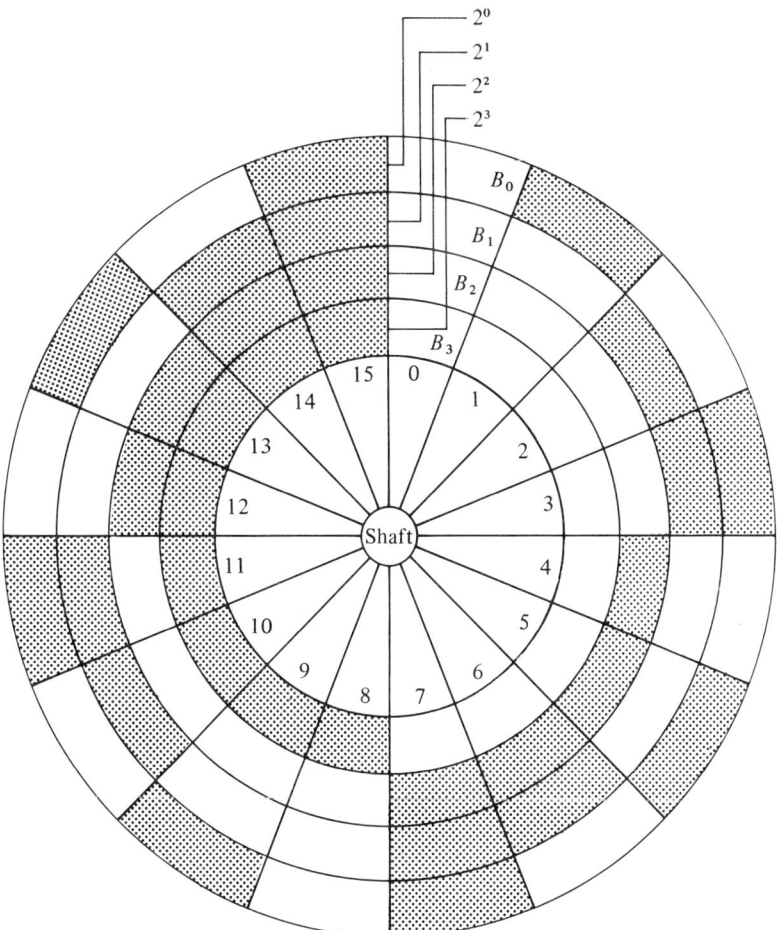

Figure 8.3-1. Encoder disc: four tracks represent binary codes consisting of 2^0, 2^1, 2^2, 2^3. Associated brushes are B_0, B_1, B_2, B_3. Numerals 0–15 represent least significant digits. This unit does not contain a solid common track.

path. As the disc rotates, a square wave pattern is generated at each brush. Brush encoders are limited in shaft speed and size of segments, the latter being about .010 inch.

Optical Encoders

Optical encoders are composed of three principal elements: the segmented disc, a light source, and a light detector. The tracks consist of alternating transparent and opaque segments (see Figure 8.3-1). Optical encoders eliminate the speed and resolution limitations imposed by brushes. Tracks can be produced successfully with segments as small as .0005 inch.

Magnetic Encoders

Magnetic encoders represent a recent entry into the encoder field. Operation of this type of encoder is based upon magnetic saturation of ferritic material cemented onto the disc. The magnetic state of these spots may be interrogated by a square loop ferritic core, called a *magnetic head*, positioned close to the surface of the disc. A second winding on these magnetic heads is used to change the state of the disc spot.

8.3-2 Coding

An encoder may be either incremental or direct reading.

Incremental

The pattern on the rotating disc is a series of uniformly spaced marks. Each pulse of the theoretically square wave obtained from reading these marks represents a small fraction of a revolution of the shaft. The pulses are counted, and the total count corresponds to the total angle through which the shaft is rotated relative to an arbitrary starting point.

Direct Reading

The pattern on the rotating disc is like that shown in Figure 8.3-1. The encoder reads absolute angle. The direct reading encoders are the more frequently used because of this obvious advantage over incremental encoders.

The direct reading encoder requires the use of a code by which the angular position is read out from the pattern of binary signals originating from the disc. Three four-bit codes are widely in use. They are natural binary, Gray code, and excess 3. Whatever code is used, it must be interpreted or translated into natural binary if the encoder data is fed into a computer or into decimal if the encoder data is used for direct readout or other analog applications.

Readout ambiguity represents a formidable problem in direct reading encoders. To put this problem into focus we consider the four-bit binary encoded pattern shown in Figure 8.3-2. Assume the brushes stationary and the code pattern moving right to left. At brush position A the readout is 0010. At B the brushes are close to the halfway point between 1000 and 1001. Either number may be indicated, depending upon whether the 2^0 brush, which reads the least significant bit, is conducting or not. Either reading is equally correct and within the achievable accuracy of ± 1 bit. At C, the brushes can for a brief moment read any one of 16 possible binary numbers due to minute code pattern boundary irregularities and finite brush misalignments. The temporary ambiguity thus caused is a realistic problem,

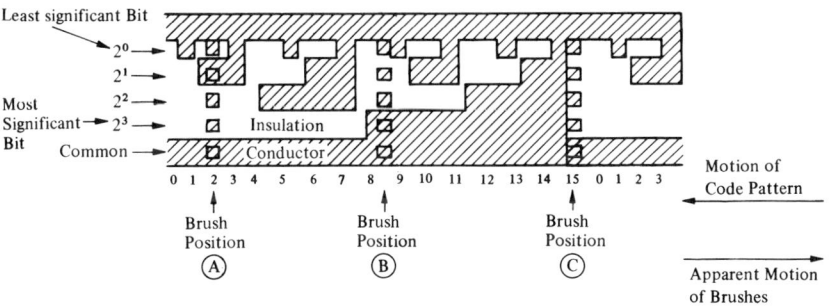

Figure 8.3-2. Ambiguities in reading of binary code pattern.

contributing considerably to the noise characteristics of the encoder. This problem is usually overcome by using a Gray code pattern.

A rectilinear section of a Gray code pattern is shown in Figure 8.3-3. Any minor brush misalignment or code pattern boundary irregularity is incapable of causing a temporary position indication error in excess of one bit. The Gray code is arranged so that only one bit changes state between adjacent counts. The encoder is thus accurate to within ± 1 bit at any position. It requires a Gray code converter when used as a direct readin to a computer. This task is easily handled by use of suitable subroutines.

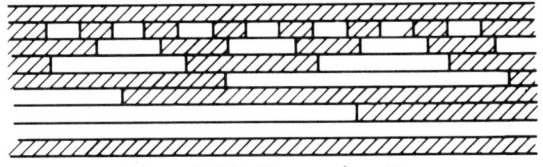

Figure 8.3-3. Gray code pattern.

8.3-3 Application of Encoders

Most encoders sold today are utilized in the digital automation field. Typical applications include the use of encoders as a feedback element in a servo loop, such as that shown in Figure 8.3-4. The comparator used in this loop has two digital inputs and an analog output.

Other applications involve an encoder as a digital readout device for angular position.

Incremental shaft encoders may be used as a pulse source for the generation of a signal proportional to angular velocity.* A typical circuit for such an application is shown in Figure 8.3-5. The circuit will provide a continuous indication of the pulse rate averaged over a selected time interval. The circuit utilizes a shift register containing n bits and a bidirectional counter (see

---

*J. T. Beckett, "Analysis of Incremental Digital Positioning System with Digital Rate Feedback," ASME paper 64-WA/AUT-3.

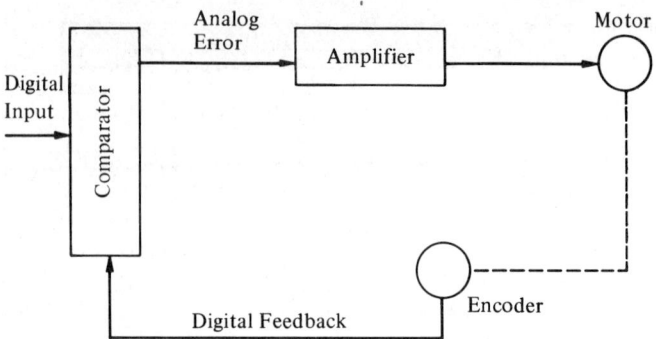

Figure 8.3-4. Feedback system employing an encoder.

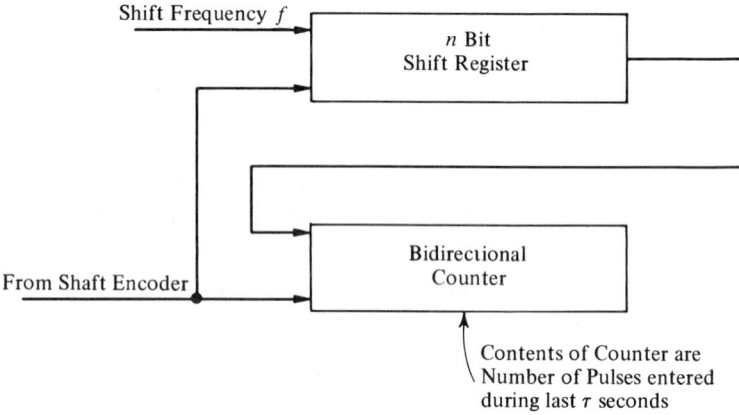

Figure 8.3-5. Pulse rate measurement technique.

Section 8.4). The contents of the shift register are shifted at a constant frequency f producing a delay between the information input and the output of the shift register given by

$$\tau = \frac{n}{f} \tag{8.3-1}$$

The input pulse train coming from the incremental shaft encoder at a frequency f_i also enters the shift register. If $f > f_i$, the number of pulses in the shift register will be the number of pulses that entered during the preceding time interval τ. The average input pulse rate is, therefore,

$$R_{av} = \frac{P}{\tau} \tag{8.3-2}$$

where P is the number of pulses counted in the interval τ.

Since each input pulse enters the shift register immediately, and since its contents are shifted at a constant frequency, the average pulse rate will be continuously monitored, and any change in the input pulse rate will have an immediate effect on the value of R_{av}.

The bidirectional counter is employed to indicate the number of pulses that are in the shift register. In order to accomplish this task, the input pulses that enter the shift register are also used to increase the value of the counter, and the pulses that exit from the shift register are used to decrease the value of the counter. At a given time $t = T$ the value of the counter is

$$R_T = \sum_{i=0}^{T} P_i - \sum_{i=0}^{T-\tau} P_i = \sum_{i=T-\tau}^{T} P_i = P \qquad (8.3\text{-}3)$$

where P_i is the instantaneous pulse rate, or, if (8.3-2) is used,

$$R_T = \tau \cdot R_{av} \qquad (8.3\text{-}4)$$

Thus we see that the value of the counter gives a reading that is proportional to the average input pulse rate.

8.4 Digital Circuit Modules

An important class of components in the field of digital computer engineering is represented by digital circuit modules. These are packaged transistorized circuits that are designed to carry out a specific logic function. Although modules find their widest application as building blocks of digital computers, they may be effectively utilized singly or in combinations of less sophisticated complexity to implement simple logic processors. From the point of view of the systems designer three areas of utilization of logic circuits are important: (1) design of timing circuits for the automatic sequencing of the operation of systems, (2) generation of decision signals needed for the automatic operation of systems, and (3) design of interface equipment between analog and digital systems. One noteworthy system to illustrate these three uses is a hybrid computer, which is discussed in more detail in Chapter 9.

We begin the discussion of logic circuit modules with a brief definition of switching functions pertinent to the discussion of logic circuits.

8.4-1 Switching Operations

Switching operations are involved in dealing with functions which take on only two possible values. The two values are 0 and 1. We shall define

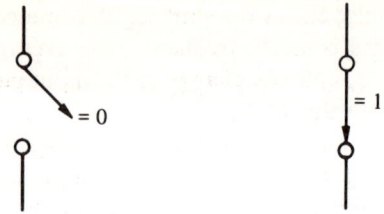

Figure 8.4-1. Switch analogy of binary states.

here two important switching operations, the OR operation and the AND operation. It is helpful to use a switch analogy in discussing these operations. Figure 8.4-1 shows a switch in two positions, open and closed, corresponding to the binary values 0 and 1, respectively.

OR operation

A circuit implementing an OR operation is formed by connecting two switches in parallel, as shown in Figure 8.4-2. The OR gate has output 1 if A or B or both are in state 1.

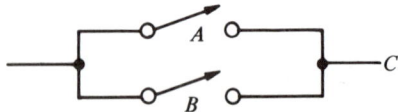

Figure 8.4-2. Switch analogy of OR gate.

The OR operation defines a *logical sum* $C = A + B$ according to the following table:

A	B	C
0	0	0
1	0	1
0	1	1
1	1	1

AND operation

An AND operation circuit is illustrated in Figure 8.4-3 by its switch analogy. Both A and B must be in state 1 for the output to be 1. The AND operation defines a logical product $C = A \cdot B$ according to the following table:

Figure 8.4-3. Switch analogy of AND gate.

A	B	C
0	0	0
1	0	0
0	1	0
1	1	1

OR and AND operation circuits consist of diode and transistor circuits whose inputs and outputs operate at preselected, transistor-circuit level compatible voltages. Only two voltage levels ($\pm$tolerance) are recognized by these circuits. These vary from manufacturer to manufacturer.

For convenience, the two symbols for OR gates and AND gates shown in Figure 8.4-4 have been adopted widely.

Figure 8.4-4.

Gates may also be designed with two outputs, the normal output C and an inverted output $\bar{C}$, which is the complement of C. The inverted output is frequently required for interconnections with other logic modules.

A special version of the AND gate, called a *diode-capacitor-diode gate* (DCD gate) has been designed by the Digital Equipment Corporation for use with flip-flops and other logic elements. The DCD gate has two inputs, one of which is the *pulse input* or ordinary signal input. The other input is a delayed input with a delay longer than an ordinary logic signal pulse. This input is called the *level input*. For example, the logic pulse is of 100 nsec duration, and the delay is about 400 nanoseconds. The DCD gate is shown schematically in Figure 8.4-5. The operation of a DCD gate is displayed by

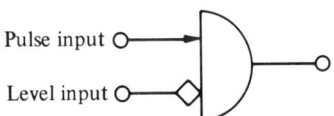

Figure 8.4-5. DCD gate.

the logic signal traces of Figure 8.4-6. The DCD gate provides logical isolation between pulse and level inputs and produces a logical delay that is essential for sampling flip-flops at the same time they are being changed. For the DCD gate to have a logical output of 1, the level input must precede the pulse input by at least the amount of its built-in delay.

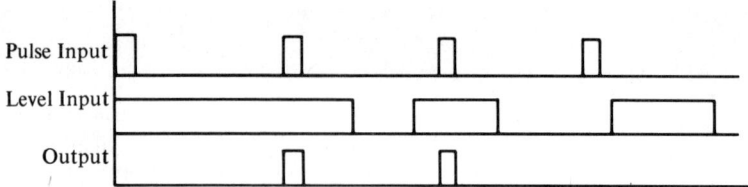

Figure 8.4-6. Logic signals in a DCD gate.

8.4-2 Flip-flops

The flip-flop provides a means of storing logical conditions within a digital system. It is, therefore, a one-bit memory element. It has two stable states representing 0 and 1, and remains in one of these states until an appropriate command to change states is received. The input to a flip-flop may be one of three signals: CLEAR, also called RESET, which puts the flip-flop in the 0 state; SET, which puts the flip-flop in the 1 state; and COMPLEMENT, which changes its state regardless of its previous state. A flip-flop usually has two outputs, which consist of the "true" output and the "complementary"; these are also called the 1 output and the 0 output, respectively. A typical flip-flop circuit is made up of resistors, diodes, and transistors.

A flip-flop used for direct CLEAR and SET operations only is shown schematically in Figure 8.4-7. In order to reduce erratic state changes in a flip-flop the inputs are gated to the outputs, as shown in Figure 8.4-8. Thus,

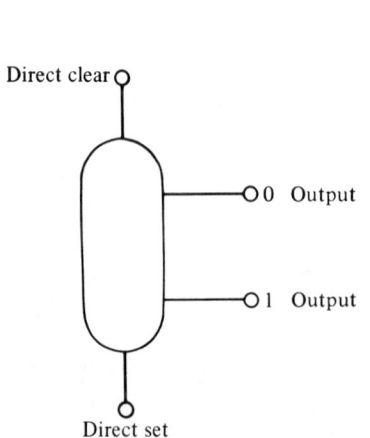

Figure 8.4-7. Direct CLEAR-SET flip-flop.

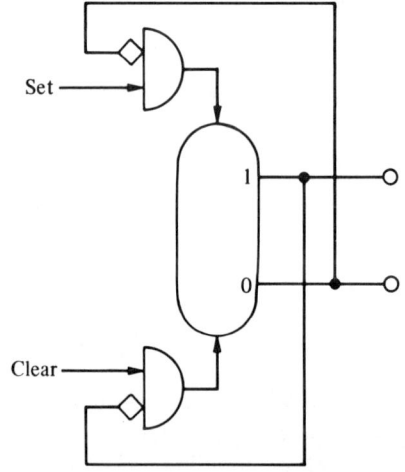

Figure 8.4-8. Flip-flop with input conditioned by output.

Sec. 8.4   Digital Circuit Modules   343

a set signal will reach the flip-flop only if the flip-flop was previously in the 0 state. Similarly, a clear command reaches a flip-flop only if it was previously in the 1 state.

For the COMPLEMENT command, the gated set and clear inputs are connected together as shown in Figure 8.4-9(a). The terminals marked E,

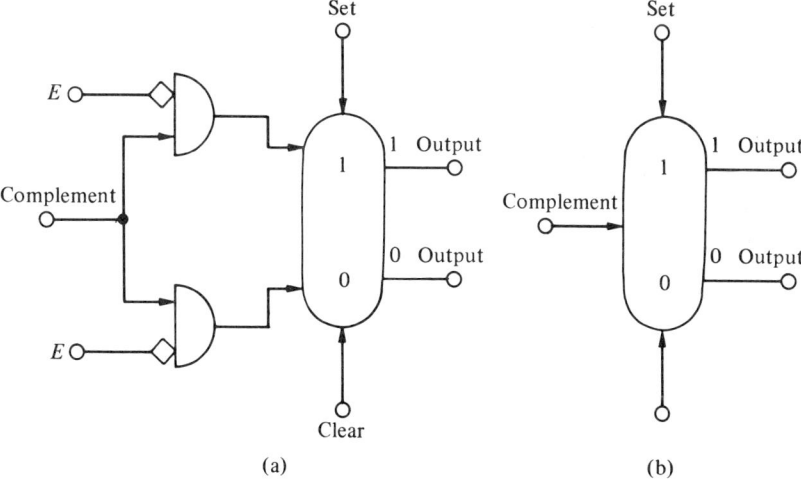

Figure 8.4-9. (a) Circuit for complete flip-flop; (b) simplified.

which are normally connected to the output, are available for external enables in this mode. Normally, the enable input is 1; if not, the flip-flop does not change state. The technique of employing these two gates allows a flip-flop to be used in such varied applications as up counters, down counters, up-down counters, shift registers, ring counters, BCD counting, etc.

For the sake of operational efficiency a gated flip-flop is used in a simplified manner, as shown in Figure 8.4-9(b). Only three inputs are shown: SET and CLEAR, which are level-type inputs, and COMPLEMENT, which is a pulse-type input.

EXAMPLE 8.4-1

A Three-bit Binary Up-counter. An illustration of using DCD gated flip-flops in counting is shown by the three-bit binary up-counter in Figure 8.4-10. Three flip-flops in COMPLEMENT mode are connected in a counting chain. When a flip-flop in this chain changes from the 1 to the 0 state, it complements the succeeding flip-flop. Flip-flop C, the first in the chain, complements on each input pulse. Flip-flop B complements when C changes from 1 to 0, and so on through the counter. In this arrangement a flip-flop complements only if all preceding flip-flops are in the 1 state. Counting is according

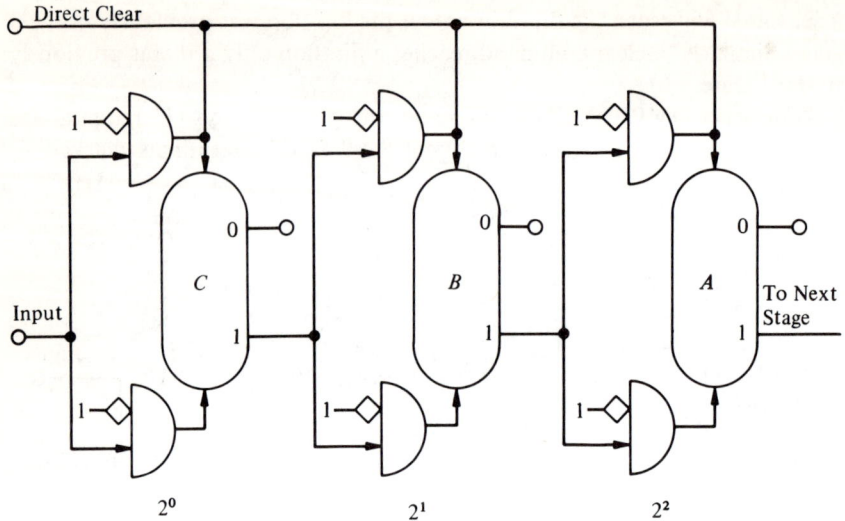

Figure 8.4-10. Three-bit binary up-counter.

to Table 8.4-1. The direct clear may be used at any time to reset all flip-flops to 0.

Table 8.4-1 Counting Sequence

Decimal	Binary		
	A	B	C
0	0	0	0
1	0	0	1
2	0	1	0
3	0	1	1
4	1	0	0
5	1	0	1
6	1	1	0
7	1	1	1
0	0	0	0

If the complementing input connection between stages comes from the 0 output of each flip-flop, a binary down counter results. It is left as an exercise for the reader to investigate this configuration.

EXAMPLE 8.4-2

A Three-bit Bidirectional Counter. The bidirectional counter is a combination of the up counter and down counter. This configuration is shown in Figure 8.4-11. The input consists of a pulse signal applied in complement

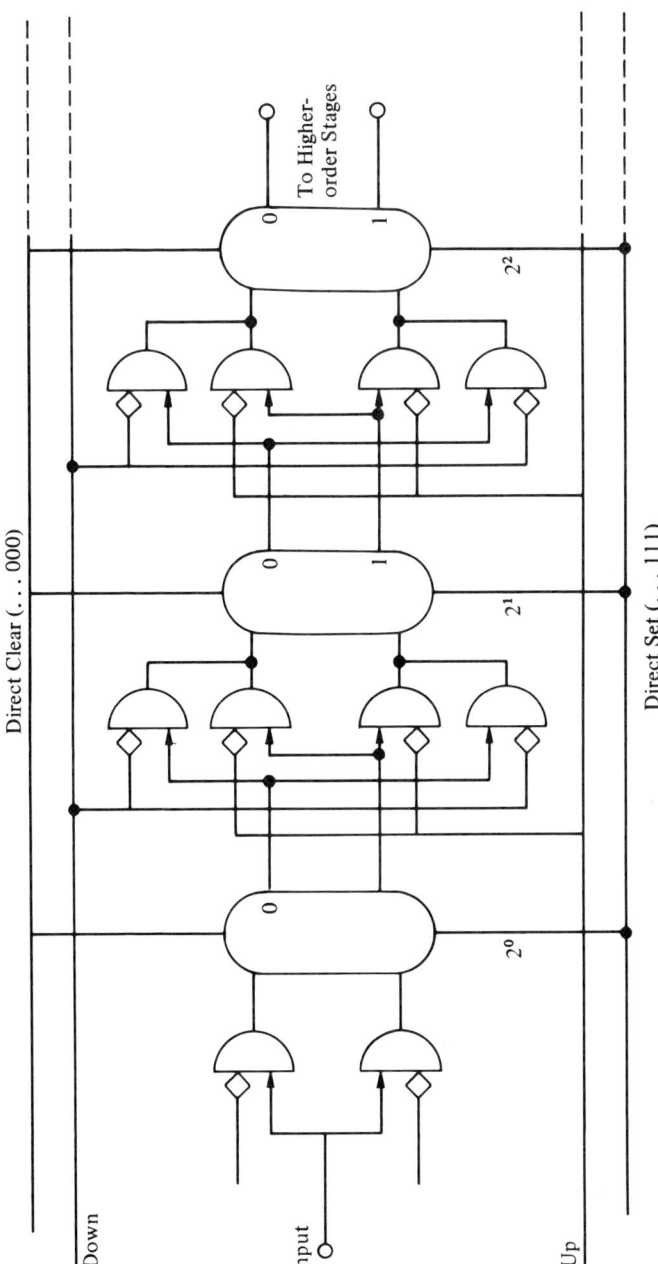

Figure 8.4-11. Schematic of a three-bit bidirectional counter.

mode to the 2^0 flip-flop, and a level signal to enable the gates of the successive counter stages. For up-counting UP = 1 and DOWN = 0; for down-counting UP = 0 and DOWN = 1. The up and down enables must not both be 1. The respective counting process is identical to that of the up-counter discussed in Example 8.4-1. All bits in the counter may be set to 0 by a pulse into the direct CLEAR input or to 1 by a pulse into the direct SET input.

EXAMPLE 8.4-3

Binary Coded Decimal Counter. A count of 10 can be produced by the circuit illustrated in Figure 8.4-12. It is generated by counting twice to 5. The counting sequence of this counter is indicated by Table 8.4-2.

Table 8.4-2. Counting Sequence of Decimal Counter

Decimal	A	B	C	D
0	0	0	0	0
1	0	0	0	1
2	0	0	1	0
3	0	0	1	1
4	0	1	0	0
5	0	1	0	1
6	0	1	1	0
7	0	1	1	1
8	1	0	0	0
9	1	0	0	1
0	0	0	0	0

EXAMPLE 8.4-4

A Three-bit Buffer Register. Flip-flops are widely used in the design of binary registers for the temporary storage of data. Such a circuit is called a *buffer register*. As an illustration, consider the schematic in Figure 8.4-13 showing a simple buffer register. In normal operation the register is first cleared; that is, all flip-flops receive a pulse through the direct CLEAR input. The three-bit word to be stored is read in by enabling the gates of the respective stages and then pulsing the reading input. Then each flip-flop is set in accordance with the state of the level input to its gate.

The preceding examples demonstrate the use of flip-flops as memory elements. They are but a small sample of the many applications that the flip-flop finds in the digital systems field.

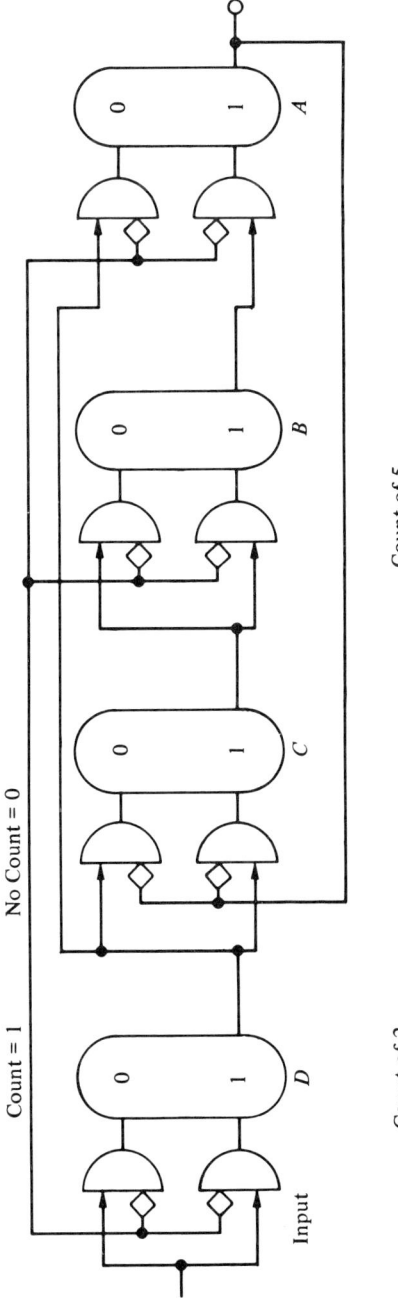

Figure 8.4-12. Decimal counter using the 8-4-2-1 code.

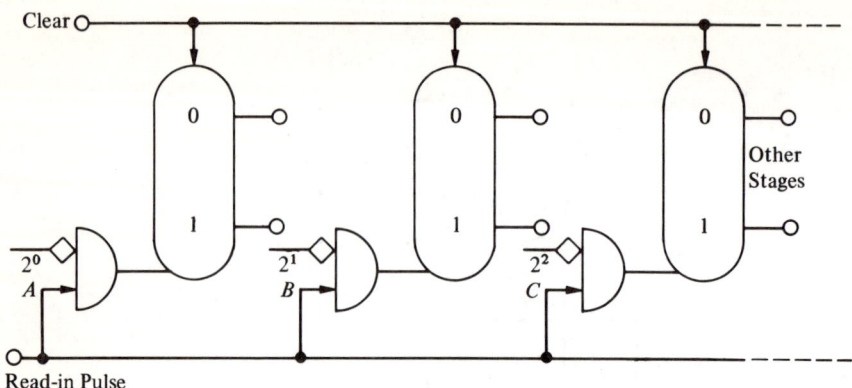

Figure 8.4-13. A three-bit buffer.

8.4-3 Delays (Monostable Multivibrator)

The delay is a basic timing element. Its circuit is a monostable multivibrator. When the input to the delay changes from 0 to 1, the output of the delay also changes from 0 to 1 for a preset, but adjustable period of time. At the end of this period the output returns to the state of 0.

A monostable may be designed by the use of a flip-flop and an *RC* time delay feedback as shown in Figure 8.4-14. The output, obtained from the

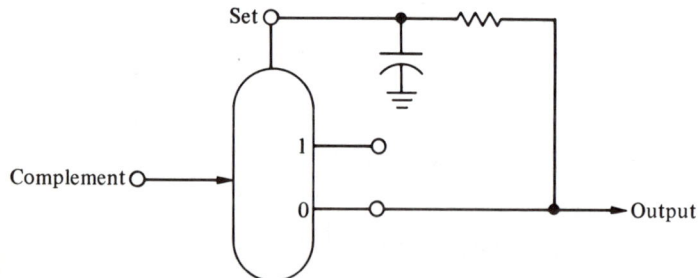

Figure 8.4-14.

complementary terminal, is normally in the 0 state. Upon receiving a complement pulse the output goes to the 1 state until the voltage in the capacitor is sufficiently high to trigger the SET input, whereupon the output returns to the 0 state. A delay is schematically represented by Figure 8.4-15. For many applications it is convenient to check the input to a delay with a gate.

Delay units are most frequently used in generating delayed pulses or signals of arbitrary width. A typical delay circuit may be adjusted to give delay intervals ranging from microseconds to seconds. Longer delays may

Sec. 8.4   Digital Circuit Modules   **349**

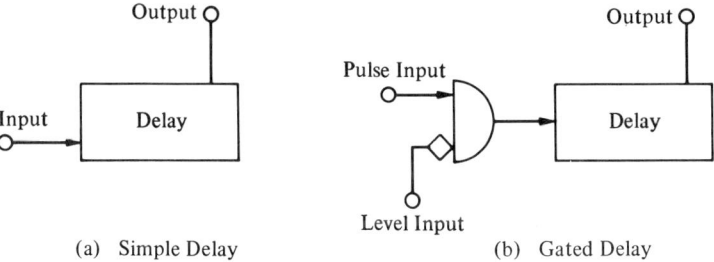

(a) Simple Delay   (b) Gated Delay

Figure 8.4-15. Schematic diagrams of delays.

be obtained by simply cascading two or more delays. An application of a delay is presented next.

EXAMPLE 8.4-5

Pulse Rate Comparator. This example discusses a circuit designed to measure the period of a square wave which monotonically decreases. The square wave is the output of an incremental shaft encoder, or pulse disc, which is coupled to a servo motor. The servo motor is to be accelerated from standstill to a preselected velocity. When this velocity is reached it is to be maintained. The proposed pulse rate comparator circuit is to indicate when the running speed is reached.

A logic circuit schematic of a proposed pulse rate comparator is shown in Figure 8.4-16. It utilizes a flip-flop, an AND gate, a delay module, and a Schmidt trigger.*

The Schmidt trigger converts the square wave into a pulse train of frequency equal to the input frequency and timing characteristics compatible with the logic modules used. It assumes the role of an interface element.

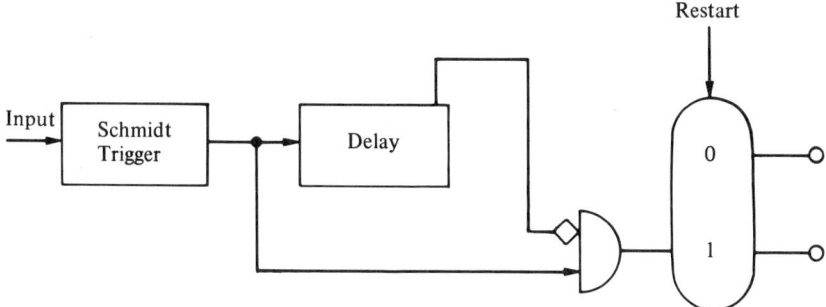

Figure 8.4-16. Schematic of pulse rate comparator.

*A Schmidt trigger produces in response to a positive going input a short square pulse that is of the proper electrical characteristics to trigger other logic modules.

Initially, the pulse train is of very low frequency when the motor starts. Also, the flip-flop is set to 0 by the restart pushbutton. The delay interval is set to be less than the low-speed pulse train period.

The operational principle of the pulse rate comparator is based upon changing the state of a flip-flop. When a pulse is delivered from the Schmidt trigger it sets the delay to 1, and it also goes directly to the DCD gate. The flip-flop stays in the 0 state because the output of the delay is connected to the level input of the gate. Because of the delay in the level input of the DCD gate, the gate does not trigger. This is shown in Figure 8.4-17. As the input

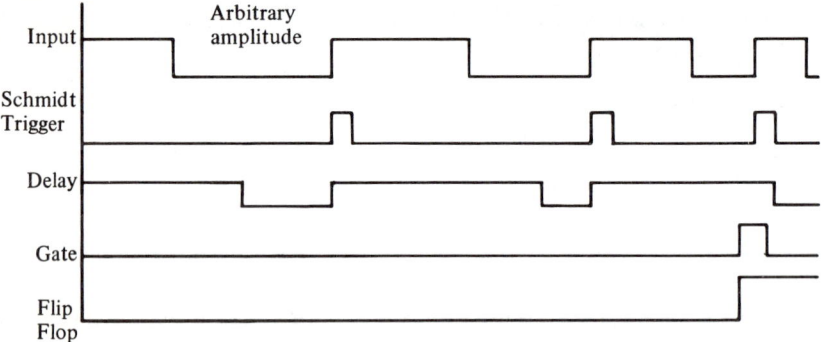

Figure 8.4-17. Operation of pulse rate comparator.

frequency increases, the output of the delay stays at the logic 1 level sufficiently long to hold the level input at the gate at 1 and to permit the pulse from the pulse generator to set the flip-flop. When this state is achieved, the desired high speed of the motor has been detected.

8.5 Data Converters

An important and absolutely essential component in a digital system is a device that links the analog and digital parts of the system. These are called *converters*, specifically, analog-to-digital and digital-to-analog converters. Typically, converters are located in a digital system as shown by the schematics of Figure 8.5-1.

In this section we shall discuss the principles of operation of data converters.

8.5-1 Digital-to-analog Converter

The basis of a digital-to-analog converter is a simple resistive network, as shown in Figure 8.5-2. The digital word to be converted is transferred into a

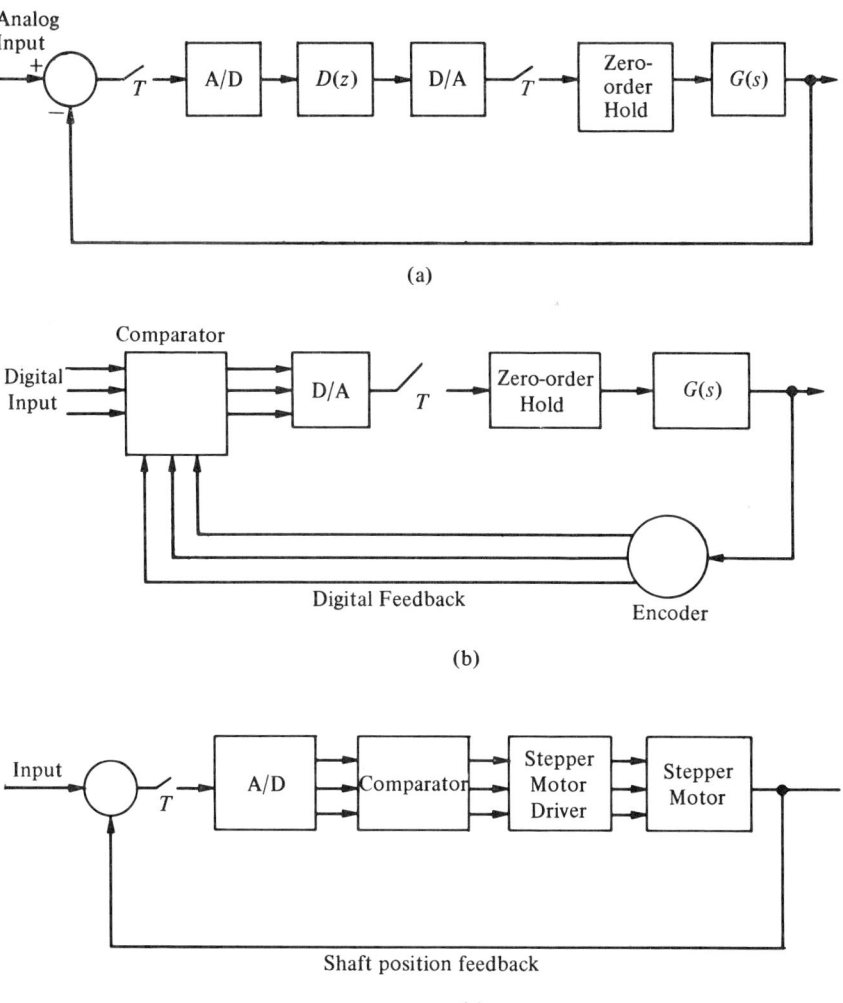

Figure 8.5-1. Typical systems employing data converters.

binary register. The states of the flip-flops in the register determine whether the resistors are connected to ground or to a reference level v_{ref}. The resistors form the input resistors to an operational amplifier. The resistors are weighted in a binary manner. For instance, the first produces twice as much output as the second. In general, the output of the operational amplifier is

$$v_0 = -\left[1(\text{FF1}) + \frac{1}{2}(\text{FF2}) + \frac{1}{4}(\text{FF3}) + \cdots + \frac{1}{n}(\text{FFN})\right]v_{\text{ref}} \quad (8.5\text{-}1)$$

where FF1, FF2, ... are the respective states of the flip-flops.

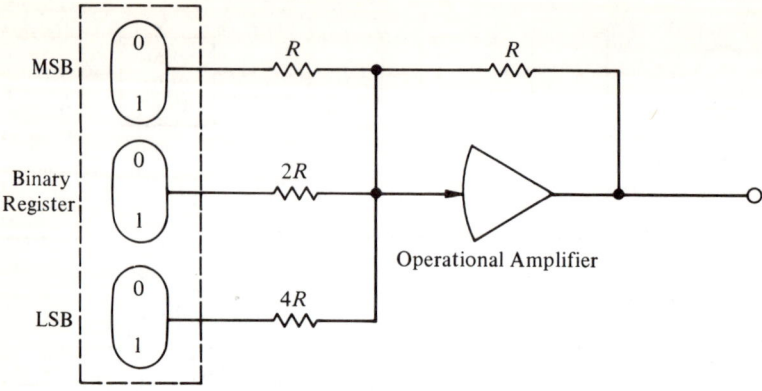

Figure 8.5-2. Digital-to-analog converter.

In the configuration shown in Figure 8.5-2 the reference voltage v_{ref} corresponds to the voltage level of logic 1 of the digital modules. However, because digital logic voltage levels are not usually as precise as required in an analog system, level amplifiers are placed between the flip-flops and the resistive divider network. The amplifiers switch the divider network between ground and reference voltage supplied by a precision voltage supply. The digital-to-analog converter is, therefore, limited to the range between ground and twice the reference level. This range can be, for instance, 0 to -10 volts. A complete digital-to-analog converter is composed as illustrated by the schematic of Figure 8.5-3.*

Consider a three-bit converter with a -4-volt reference supply and resistive network arranged so that the most significant bit represents -4 volts. Then the relationship between representative binary words and analog

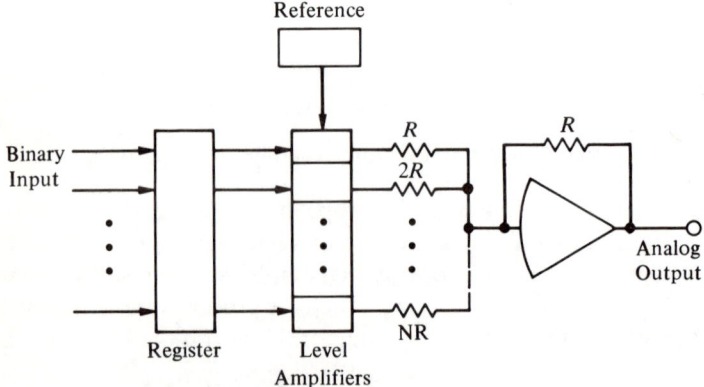

Figure 8.5-3. Complete digital-to-analog converter.

*MSB means most significant bit; LSB means least significant bit.

outputs is given by Table 8.5-1. It is seen that for all bits equal to zero the output is zero, but that for all bits equal to 1, the output is 7, one quantization level short of 8 volts.

Table 8.5-1 Digital-to-analog Conversion of Binary Numbers

Binary Number	DAC Output Voltage
0 0 0	0
0 0 1	1
0 1 0	2
0 1 1	3
1 0 0	4
1 0 1	5
1 1 0	6
1 1 1	7

To convert sign-sensitive binary numbers, such as two's complement binary numbers, a slight modification may be made in the internal connection of the DAC. Recall that in two's complement binary representation the most significant bit is used as the sign bit. Consider the conversion of three-bit two's complement numbers with a converter that complements the most significant bit by taking the output from the 0 terminal instead of from the 1 terminal of the MSB flip-flop, as is shown in Figure 8.5-4. Table 8.5-2 then shows in the column marked Modified Two's Complement Binary what the binary number is that is actually being converted. The remainder of the table shows the analog output. It is seen that the output is just the reverse of

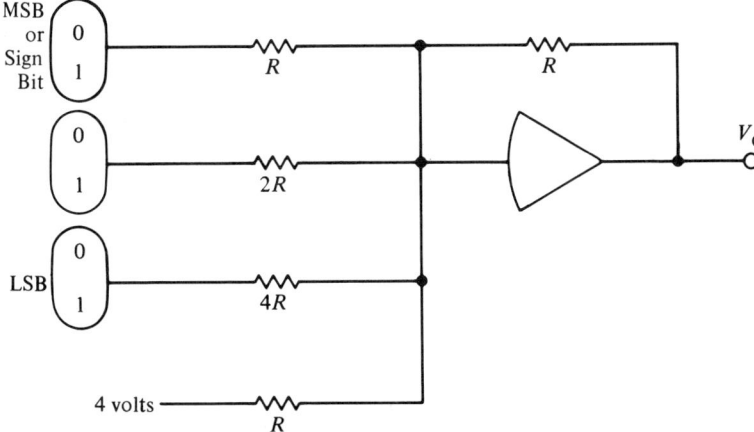

Figure 8.5-4. Three-bit DAC with sign bit inverted and output level shifted.

that shown in Table 8.5-1. If the output is shifted by −4 volts by adding a 4-volt source into the DAC summing network, an adjusted DAC output is generated which agrees with the decimal equivalent of the two's complement number that is being converted.

Table 8.5-2 Digital-to-analog Conversion of Two's Complement Numbers

Decimal	Two's Complement Binary	Modified Two's Complement Binary	DAC Output	Adjusted DAC Output
+3	011	111	7	7 − 4 = +3
+2	010	110	6	6 − 4 = +2
+1	001	101	5	5 − 4 = +1
0	000	100	4	4 − 4 = 0
−1	111	011	3	3 − 4 = −1
−2	110	010	2	2 − 4 = −2
−3	101	001	1	1 − 4 = −3

Conversion Time

The conversion time of a DAC is given by the total time required to load the register, switch the level amplifiers, and add the respective binary weighted voltages. This time rarely exceeds a few microseconds.

Multiplying DAC

A digital-to-analog converter may be used as a high-accuracy electronic multiplier by using the reference voltage to manipulate the converter output, as shown in Figure 8.5-5. If the reference voltage of the converter is con-

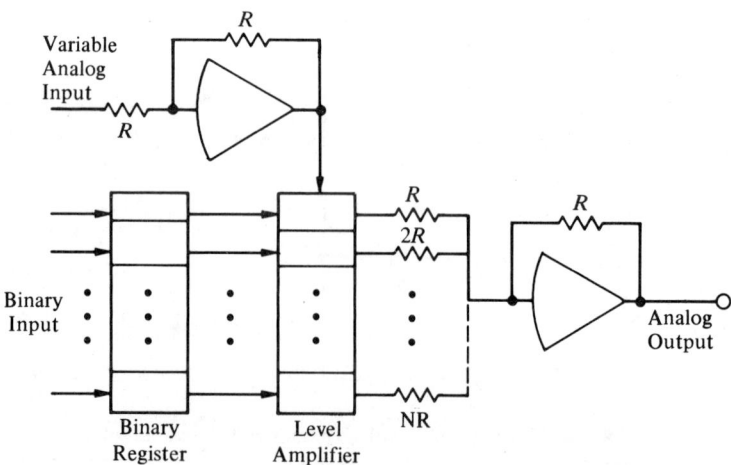

Figure 8.5-5. Schematic of multiplying DAC.

Sec. 8.5   Data Converters   355

trolled by an operational amplifier within the range -5 to $+5$ volts, the DAC output will be between -10 and $+10$ volts. Multiplying DAC's are conveniently used in hybrid computers because of the availability of a large number of digital-to-analog converters.

Zero-order Hold

The presence of a full bit register in a digital-to-analog converter automatically provides for a zero-order hold. In a typical operation the analog output voltage remains constant as long as the digital word in the binary register is left unchanged.

8.5-2 Analog-to-digital Converter

The process of converting analog data into digital form may be carried out in a variety of ways. There are basically two factors entering into the design of an analog-to-digital converter: speed and cost. Methods that are extremely fast tend to become unwieldy in the amount of equipment involved and are costly. On the other hand, simplicity in equipment implies slow conversion speeds. A method that presents a fair compromise between speed and complexity is the successive approximation conversion. This will be presented in what follows.

The successive approximation converter is basically a feedback system, as the schematic of Figure 8.5-6 shows. The main components are a comparator, a gating and control circuit, a control register, and a package of

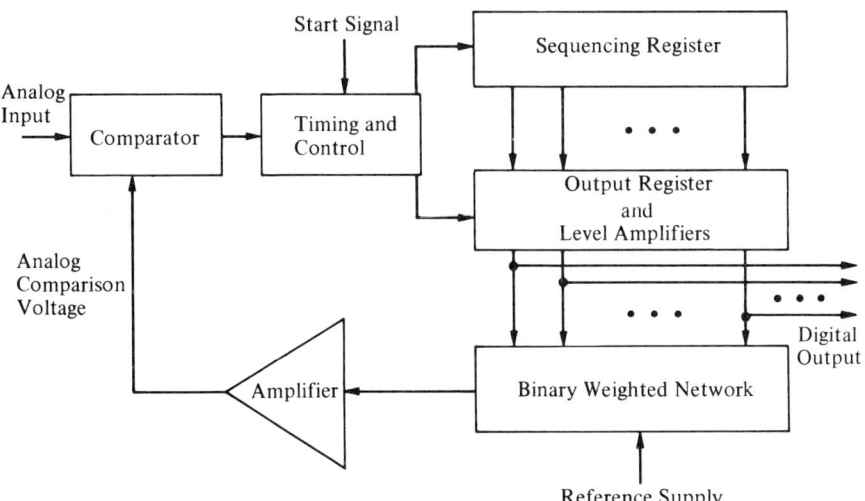

Figure 8.5-6. Structure of successive approximation converter.

components that are the same as those making up a digital-to-analog converter.

The conversion is initiated by a starting pulse received by the timing and control network. The output register is then set at one-half the maximum input voltage, which is binary 1000 ... (see Table 8.5-1). This binary word, which represents the most significant bit, is then converted to an analog voltage through the binary weighted ladder network. This voltage is compared with the analog input voltage. The output of the comparator (1 or 0) indicates whether the input voltage is in the upper or lower half of the range. If it is in the upper half, the most significant bit is left at 1. If it is in the lower half, the MSB is reset to 0.

The sequencing register (which is actually a ring counter with a single 1) is then advanced to the next bit to permit adjustment of the next significant bit, and the output register is set at binary 1100 ... (upper half) or binary 0100 ... (lower half). Again, the contents of the output register are converted and compared with the input voltage. The comparator then indicates which quarter of the range contains the analog input voltage. This process is repeated for as many bits as are available in the output register. The number of bits desired may be adjustable in a typical converter. Thus, when the preselected bit accuracy is obtained, the conversion process is terminated, and the converted digital word is contained in the output register. Figure 8.5-7

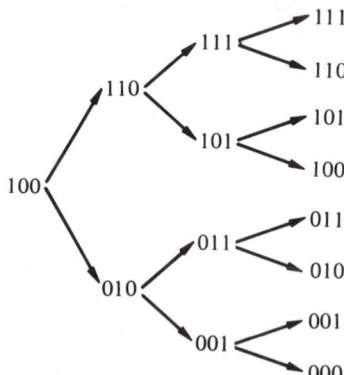

Figure 8.5-7. Three-bit conversion sequence.

illustrates the possible conversion sequences that could be followed for a three-bit converter.

Conversion Time

Conversion time depends on the number of bits desired. Typical figures are 9 microseconds for 6 bits to 35 microseconds for 12 bits. Proportionately more time is required for less significant bits because higher-accuracy ladder networks are required, which have a longer settling time. Because conversion time is variable, the converter will set a flag when conversion is done. This can be tested by a program instruction.

When the analog signal varies rapidly so that significant changes are possible within a given conversion time, a high response sample-and-hold amplifier is required to hold the analog signal constant during conversion.

Conversion Accuracy

It is obvious that no matter what the word length, the final analog-to-digital voltage can never approximate the input voltage with an accuracy greater than one-half the value of the least significant bit. Therefore, the overall converter accuracy is equal to plus or minus one-half the voltage value of the least significant bit. Table 8.5-3 gives an indication of the maximum conversion error for various bit lengths.

Table 8.5-3 Conversion Accuracy

Bit Accuracy	Conversion Error (percent)
1	±50
2	±25
3	±12.5
.	.
.	.
.	.
6	± 1.6
7	± .8
8	± .4
.	.
.	.
12	± .025

8.6 The Stepper Motor

One of the most interesting developments in the field of digital systems is a mechanical output device that may be controlled directly by a digital processor. This is the stepper motor. It is an electromechanical transducer whose input consists of binary coded voltage level commands and whose output consists of quantized angular movements that are in direct correspondence with the input. The stepper motor thus represents a unique development in the technology of servo components, whose inputs heretofore have been variable voltages and whose output has been a continuously changing shaft position.

In this section we shall consider the design and operating characteristics of the stepper motor. Also discussed are suitable drive amplifiers, which act as interface between the low-power logic signals of the digital processor and the high-power pulse signals applied to the stepper motor.

8.6-1 Stepper Motor Types

A variety of high-performance stepper motors has been designed. Some types are unidirectional and some bidirectional. Some units provide separate input terminals for clockwise or counterclockwise rotation. In other types the direction of rotation is dependent upon the sequencing of the pulse inputs.

Stepper motors meeting this description fall into two categories:

1. Permanent magnet or synchronous inductor type, which works on the reaction between an electromagnetic field and a permanent magnet.
2. Variable reluctance type, which works on the reaction between an electromagnetic field and a soft iron rotor.

Generally, stepper motors are recognizable as AC motors whose stator windings are sequentially energized by DC voltages with the result of driving the rotor through discrete angles of equal quantization.

The stepping angle is determined by design and may range from approximately 1 degree to no more than 120 degrees. It is possible to make the stepping angle any value of $2\pi/n$ radians where $n \geq 3$. The smaller the angle, however, the larger the number of windings. Using a large number of coils has certain disadvantages, since only a small fraction of the windings is energized, thus requiring a large motor package to deliver the same torque as a motor with fewer coils.

8.6-2 Variable Reluctance Motors

Variable reluctance stepper motors can achieve high stepping speeds and torques while being relatively small in size. VR motors employ electromagnetic, low-retentivity rotors that cannot simultaneously align all poles with stator poles. As an illustration, consider the schematic of Figure 8.6-1, showing a 12-pole stator and 8-pole rotor. Each stator pole has a winding. These windings are grouped into three sets of four poles each (see pole sets marked A, B, and C). One set is energized at a time. In the example shown, four rotor poles are aligned with stator poles A, which are energized, while the other four rotor poles are placed midway between the remaining stator poles. Since these rotor poles are 15 degrees out of alignment with the stator poles marked B and C, energizing either poles B or poles C will cause a 15-degree clockwise or counterclockwise turn of the rotor as it seeks to close

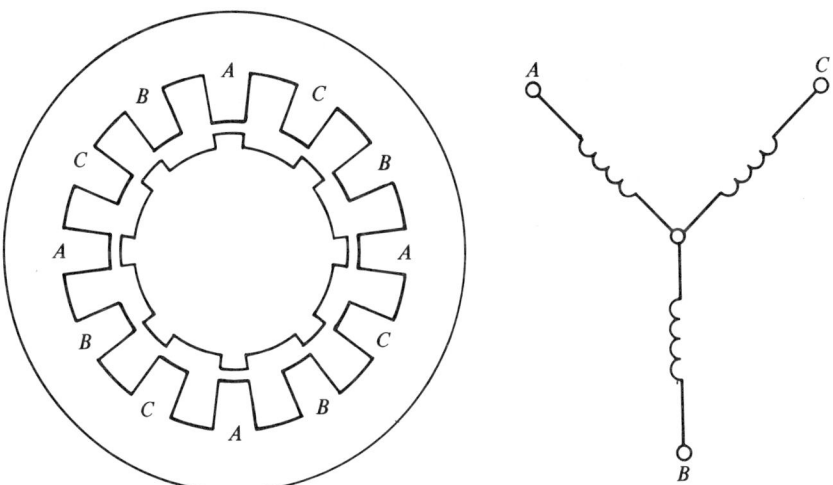

Figure 8.6-1. Variable reluctance stepper motor and simplified wiring diagram.

the shortest flux path. Thus the sequence and step rate in which stator windings are energized determines direction of rotation and speed, respectively.

8.6-3 Permanent Magnet Motors

Permanent magnet motors that are used as stepper motors are synchronous inductor motors. In a synchronous inductor motor, torque is developed as a result of the interaction of a phase-stepped magnetic field on a stationary multiphase stator and a unidirectional flux on the rotor. The construction is usually confined to a two-phase stator winding and a permanent magnet rotor.

The synchronous inductor motor is built as a slow-speed device. The slow speed is attributable to rotor and stator construction characterized by numerous toothed projections. The speed of the rotor is given by

$$s = \frac{60f}{n} \qquad (8.6\text{-}1)$$

where s = rpm of rotor
n = number of rotor teeth
f = stepping rate in cycles per second, or complete sequence of voltage pulses per second.

The speed equation may be explained by use of Figure 8.6-2, which shows a schematic of a synchronous inductor motor. The stator has a two-phase, four-pole winding with a total of eight poles, and the rotor has ten projec-

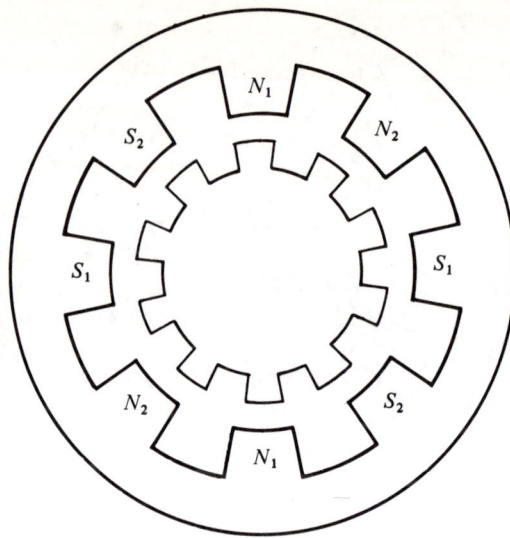

Figure 8.6-2. Synchronous inductor stepper motor.

tions or teeth. The magnetic structure of the rotor is made up of two separate identical discs separated by a cylindrical permanent magnet, which is magnetized axially, making one rotor section a north pole and the other a south pole. The teeth on the two sections are offset by one-half a rotor tooth-pitch to permit the use of a common stator magnetic structure and winding.

To permit the energizing of the windings by a single-polarity DC source, the two phases are connected as bifilar windings and switched as shown by the connection diagram in Figure 8.6-3.

Consider now the operation of this motor if a square wave sequence is applied to the windings. The four-part schematic of Figure 8.6-4 illustrates this. In step 1 windings A_1 and B_1 are energized, producing maximum flux at pole N_1; the rotor is positioned at N_1. In step 2, a quarter cycle, the maxi-

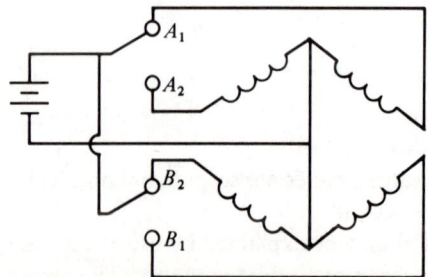

Figure 8.6-3. Connection diagram of motor with bifilar windings.

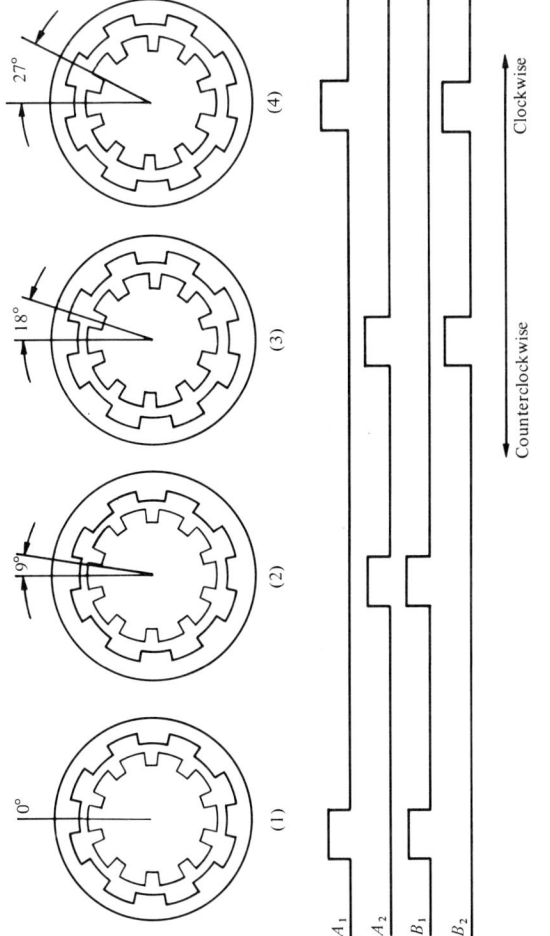

Figure 8.6-4. Sequence of step motor cycling.

mum flux occurs at N_2 and the rotor is rotated by 9 degrees. It is clear that a full sequence cycle moves the rotor by 36 degrees for this 10-tooth motor. In general, the angular step size for a two-phase motor and n teeth is $\pi/2n$ radians; four steps make up one cycle. The motor may be moved clockwise by following a 12341 sequence or counterclockwise by following a 43214 sequence.

The bifilar winding is of particular advantage when the switching is carried out by electronic circuits employing a single-ended power supply. Instead of the current's being reversed in a winding, a current of the same polarity is switched to an identical winding wound in the opposite directon.

The use of the permanent magnet rotor provides for magnetic detenting when all windings are de-energized.

Of the two types of motors discussed, the permanent magnet or synchronous inductor motor has the wider application and potential for growth.

8.6-4 Stepper Motor Dynamics

A stepper motor is unique in that it can provide dual operational capabilities. Stepper motors can operate bidirectionally in synchronism with the input pulse rate up to a certain maximum response rate or, when the maximum response rate is exceeded, as a unidirectional synchronous motor. To study these types of motion we derive a mathematical model of the stepper motor. To this end it will be sufficient to restrict the attention to conditions corresponding to the application of a step voltage and the ensuing discrete shaft motion. We shall further assume that the step size is small so that linear relations dominate.

With the above limitations it is appropriate to begin with the differential equations governing the motion of an electro-dynamometer about an equilibrium point:

$$v = Ri + K_g \dot{\varphi} \qquad (8.6\text{-}2)$$

$$\tau = -K_m i + \left(J \frac{d}{dt} + B \right) \dot{\varphi} \qquad (8.6\text{-}3)$$

where $v =$ voltage of winding
  $i =$ current through winding
  $\tau =$ shaft torque
  $\dot{\varphi} =$ shaft velocity
  $R =$ winding resistance
  $J, B =$ mechanical rotor constants
  $K_g =$ generator constant, back emf
  $K_m =$ torque constant.

The differential equations for the dynamics of the motor must further take into account the magnetic detenting of the motor. This is characterized by a torque that tends to restore the shaft position to the equilibrium point, the commanded position. For small angular motions this restoring torque is proportional to the deviation from the desired position and may be expressed as

$$\tau_r = K_r(\varphi_c - \varphi) \qquad (8.6\text{-}4)$$

where τ_r = restoring torque due to magnetic detenting
K_r = magnetic torque constant
φ_c = commanded position.

The restoring torque is directly attributable to the voltage v applied to the winding. Therefore, we must set $v = 0$ to correctly define the conditions of the motion about the equilibrium point; otherwise the applied voltage would produce an offset.

Combining (8.6-2) through (8.6-4) yields the final equation

$$\left(J\frac{d^2}{dt^2} + \left(\frac{K_m K_g}{R} + B\right)\frac{d}{dt} + K_r\right)\varphi = K_r\varphi_c \qquad (8.6\text{-}5)$$

The commanded position may be taken as zero.

Equation (8.6-5) may be used to study the motion of the rotor from one position to the next. It is a typical regulator equation of second order. Note that the damping characteristics consist of viscous mechanical damping and electrical back emf. Of particular interest is the role the winding resistance R plays in the damping factor.

On the basis of (8.6-5) we may infer that at low pulsing rates and no shaft load, the rotor indexes from one position to the next with an oscillatory motion as shown in Figure 8.6-5. The oscillation is well damped and subsides rapidly for a typical stepper motor.

When the frequency of the pulse rate is increased to a value for which the oscillation cannot damp out, the motor has reached the maximum pulse rate.

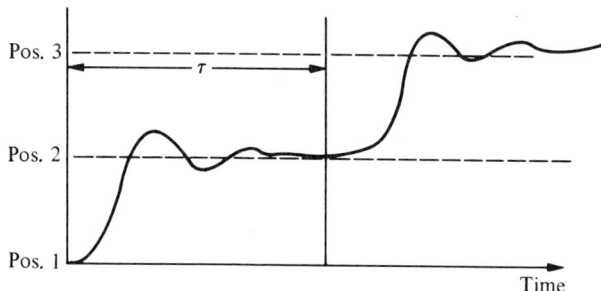

Figure 8.6-5. Response to a slow stepping rate.

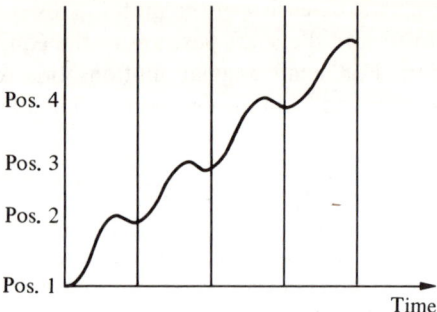

Figure 8.6-6. Slewing response.

This is indicated by Figure 8.6-6. This type of response is called *slewing*. At pulse frequencies causing slewing the stepper motor runs as a synchronous motor. When the motor is in a slewing mode, it may not be stopped within one period. Also, it requires several periods to reestablish its stepping mode of operation if the input pulse rate is reduced below the slewing rate. Thus, when in slew, a motor does not lose synchronism with the input pulse rate, but cannot stop, or reverse on a given command. If the motor is required to stop precisely, the pulse rate must be reduced to the stepper pulsing range first.

8.6-5 Driver Circuits for Stepper Motors

It was stated earlier that a stepper motor requires logic signals of relatively high power. Typically, the voltage levels of these signals fall into the range from 3 to 30 volts DC with current requirements in the .1 to 10 ampere range. Since command signals for stepper motors are generated in low-power digital circuits as pulse trains, a driver circuit is required to provide the necessary power and translate the pulse train into a binary code consistent with the sequence design of the stepper motor.

The driver circuit requirements are easily met by use of digital modules and solid-state switches. As an illustration, consider the schematic of Figure 8-6.7, showing a driver circuit for a two-phase permanent magnet stepper motor that requires a command signal sequence, as shown in Figure 8.6-4. The input consists of two pulse trains, for clockwise and counterclockwise rotation. Two flip-flops, which are arranged in switch-tail ring counter configuration, generate the sequence required. The outputs of the flip-flops drive the four solid-state switches.

To explain the operation of the switch-tail ring counter, let the states of the flip-flops be such that $D = 1$ and $C = 1$. Then the next clockwise pulse will switch flip-flop D so that $D = 0$. The next clockwise pulse will cause

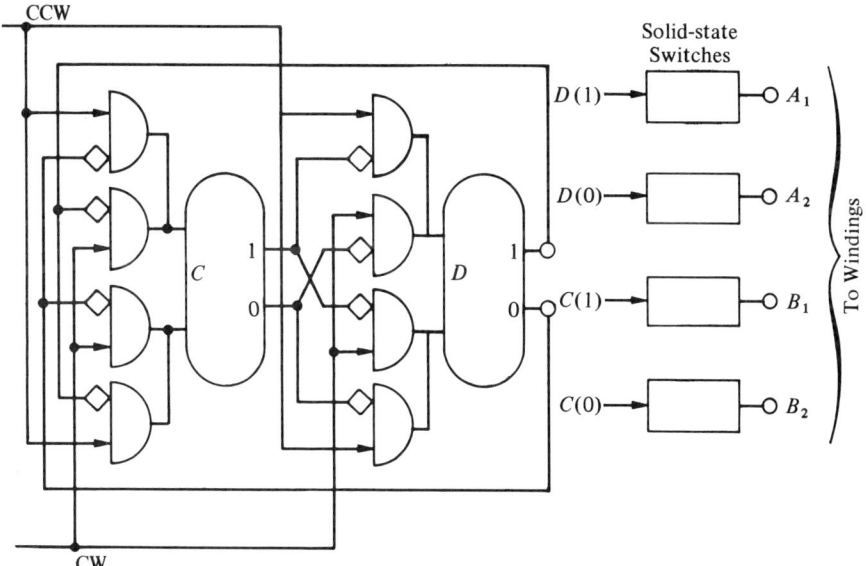

Figure 8.6-7. Pulse to sequence translator and switch connections.

flip-flop C to change state. In general, a sequence as tabulated in Table 8.6-1 is generated, corresponding to the desired sequence. The sequence is reversed in response to a counterclockwise pulse.

Table 8.6-1 Driver Circuit Sequence

Count	CW-Sequence		CCW-Sequence	
	C	D	C	D
1	1	1	1	1
2	0	1	1	0
3	0	0	0	0
4	1	0	0	1
5	1	1	1	1

8.7 The Stepper Motor in Control Applications

Because of its digital behavior, the stepping motor has been widely used in open-loop control applications. If the stepping rate is slow enough and the inertia load on the shaft is small enough, the stepper motor is able to translate digital pulses into discrete angular positions. Because the stepper motor can achieve a one-to-one correspondence between digital signal and shaft position with no accumulative error, it has been effectively employed in a variety of instrumentation applications such as digital plotters and pen recorders. In these and similar applications, load conditions are constant and speed requirements are predictable. However, only a portion of the potential performance of the stepper motor is realized in such applications, since severe restrictions exist as to large variations in the stepping rates and load inertias. For instance, an examination of the second-order response characteristics of the motor in the stepping mode reveals a significant oscillatory behavior as the motor moves from one step to the next. This section will discuss a number of possible ways in which the performance of stepper motors can be improved by means of feedback.

8.7-1 A Conventional Feedback System

As a first illustration of a feedback system employing a stepper motor, we consider the schematic of Figure 8.7-1. A stepper motor with a matching

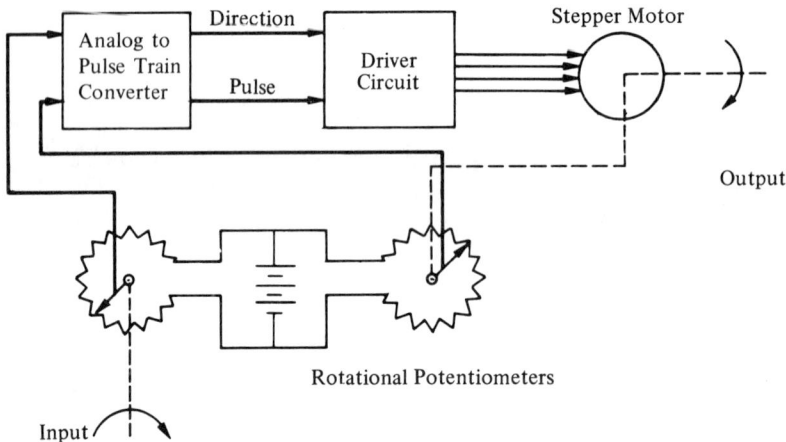

Figure 8.7-1. Simple stepper motor feedback system.

driving circuit is connected in closed-loop operation. The command signal is generated from an analog-to-pulse converter, which converts the position error signal into a pulse train whose frequency is proportional to the magnitude of the analog signal. In addition, the converter senses the sign of the error and provides a directional logic signal for the driver circuit. It is clear, then, that this feedback arrangement controls the speed (really, stepping rate) of the motor, which is proportional to the error signal—a conventional feedback system. During a typical transient, the motor will be stepped at a rate directly proportional to the magnitude of the error signal. Since a drive signal will be generated as long as an error signal exists, the system will act like a type 1 position control. Of course, a final position error will remain, which may be as large as the stepping angle of the motor.

In order to avoid hunting about the commanded position the pulse train from the converter to the driver circuit may be gated to a monostable. The monostable is triggered by the pulse train, and its delay time is adjusted to be slightly shorter than the period of the hunting pulses.

The performance of the system is similar to an equivalent analog system with one important exception— it is considerably more damped. During large transients the motor falls into the slewing range because of the large error signal. But as the error is reduced in magnitude, the corresponding pulse rate will enable the motor to be stepped in direct correspondence with the drive signal. Thus no overshoot occurs unless the inertia load on the motor is so large that the motor cannot be adequately stepped. Figure 8.7-2 shows

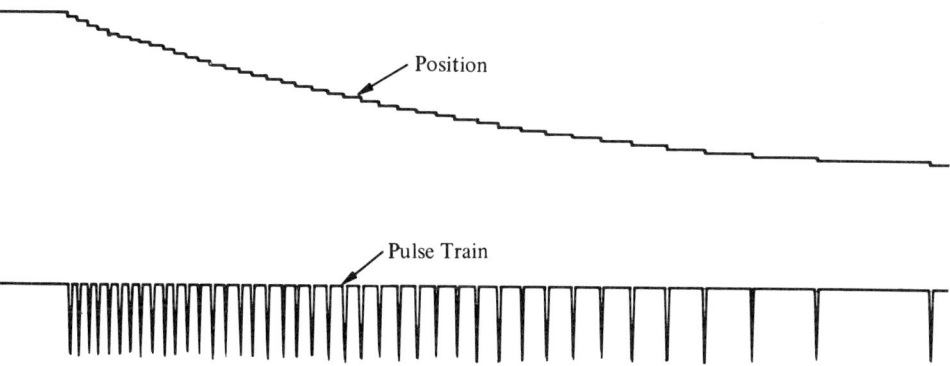

Figure 8.7-2. Response of feedback system, showing shaft position and pulse train.

the response of a system of this configuration with a motor having a 1.8 degree angular step size. The overdamped nature of the response is very clearly evident.

8.7-2 Digital Feedback for Stepper Motors

The closed-loop digital control of stepper motors presents some intriguing possibilities. We shall investigate here a novel feedback method that greatly enhances the motor's utility as a rotating actuator.* This development deals with the generation of driving signals from the motor's own output subject only to external control modulation.

For the detailed discussion we consider a motor with 200 steps per revolution and bifilar windings, wired in a two-phase arrangement such as that shown in Figure 8.6-3. As was shown earlier, for this motor four valid energized winding combinations exist. These are listed in Table 8.7-1 and defined as inputs 1, 2, 3, and 4. Each energized position represents one of the 200 possible positions.

*T.R. Fredriksen, "Closed-Loop Stepping-Motor Application," 1965 Joint Automatic Control Conference, pp. 531–538.

Table 8.7-1 Windings Energized for Various Positions

Input	A_1	A_2	B_1	B_2	Position
1	on		on		$0° + 7.2°n$
2	on			on	$1.8° + 7.2°n$
3		on		on	$3.6° + 7.2°n$
4		on	on		$5.4° + 7.2°n$
1	on		on		$7.2° + 7.2°n$

$$n = 1, 2, \ldots, 50$$

The steady-state response to any one of the inputs is one out of 200 shaft positions, each 1.8 degrees apart. When an input is applied, the motor will position itself at the nearest corresponding position. Although there are 200 possible shaft locations, only four input-output relationships are possible. These are named and defined as follows:

Stop: The motor is energized for a step it presently occupies.
CW: The motor is energized for a step located 1.8 degrees in a clockwise direction.
CCW: The motor is energized for a step located 1.8 degrees in a counterclockwise direction.
Double step: The motor is energized for a step located 3.6 degrees in either direction.

The last relationship, double step, cannot be restricted in direction, since

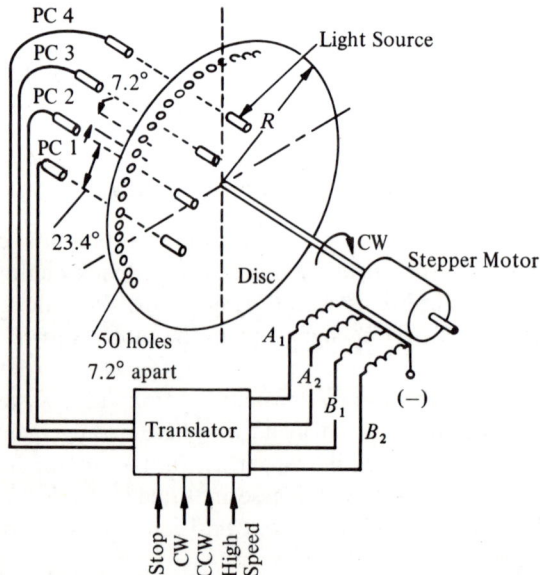

Figure 8.7-3. Feedback control of stepper motor.

Sec. 8.7   *The Stepper Motor in Control Applications*

the motor ideally develops zero torque for this command. The direction of rotation in response to this input will depend upon velocity conditions.

We consider now the control of this stepper motor by means of the feedback circuit shown in Figure 8.7-3. An optical shaft encoder is connected to the stepper motor shaft. Its disc contains 50 holes spaced 7.2 degrees apart. Four photodiodes with accompanying light sources are located 23.4 degrees apart. This arrangement will generate a shaft encoder signal output from the four photodiodes according to Figure 8.7-4. The shaft encoder's main function consists of providing information of which step number the shaft is presently holding. The wave pattern of Figure 8.7-4 clearly can discriminate between the four basic positions.

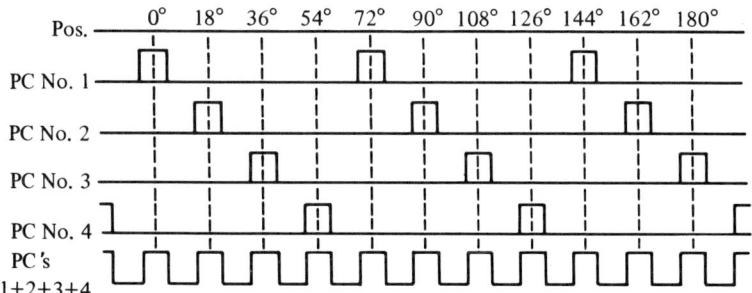

Figure 8.7-4. Shaft encoder output.

The loop is closed by a logic circuit called a *translator*. Its purpose is to combine the present position information with the input commands defined earlier to generate appropriately timed motor winding signals. Table 8.7-2 completely describes the translator functions. For example, if the shaft encoder indicates a present position of 3 and the input command calls for a counterclockwise step, the mode is energized for a move to position 2.

Table 8.7-2 Translator Function

Command Input	Shaft Encoder Output			
	1	2	3	4
Stop	1	2	3	4
CW	2	3	4	1
CCW	4	1	2	3
High Speed	3	4	1	2

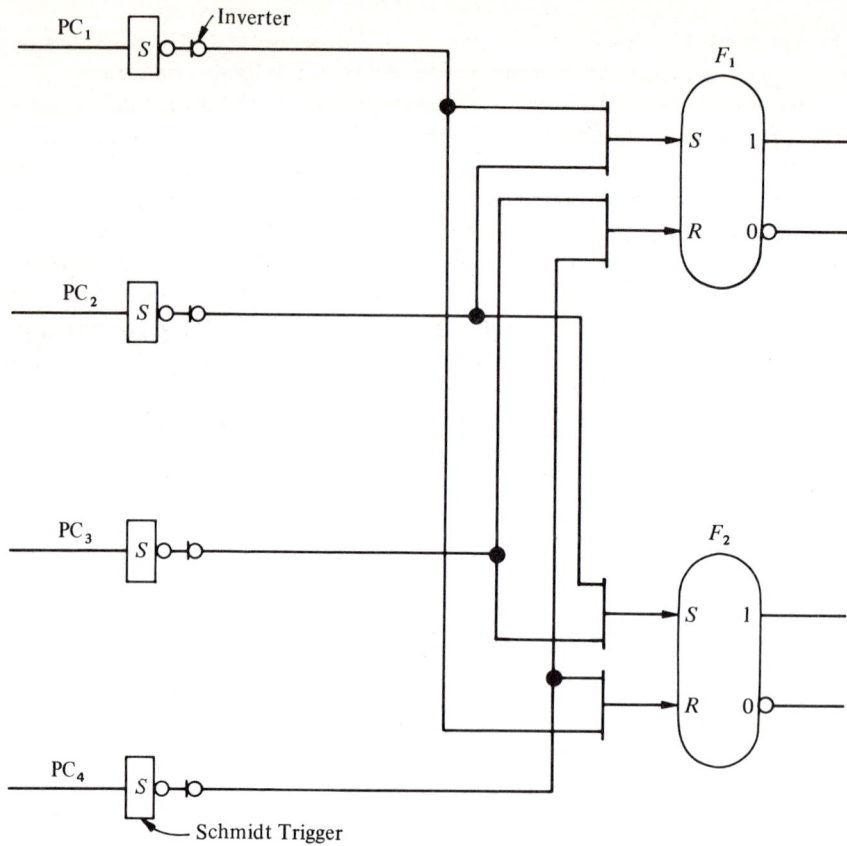

Figure 8.7-5. Logic control circuit.

The feedback loop thus described accomplishes the control of the stepper motor in one of two modes: (1) When the command inputs are pulses, the four inputs are executed one step at time; (2) when the command inputs are level inputs, the response is a continuous execution of the command input. Under mode 1 the motor is operated as a regular stepper motor, whereas under mode 2 it functions like a synchronous inductor motor.

The implementation of the translator's functions may be readily obtained by use of digital logic modules. A possible configuration is shown in Figure 8.7-5. The logic circuit determines the appropriate signals to the driver circuit in accordance with the desired response. It consists of standard logic modules. The operation of this circuit may be best understood by considering a typical situation. It remains as an exercise for the reader to verify the design of this circuit.

Present Position: 1

Command Input: Stop

Sec. 8.7   The Stepper Motor in Control Applications   371

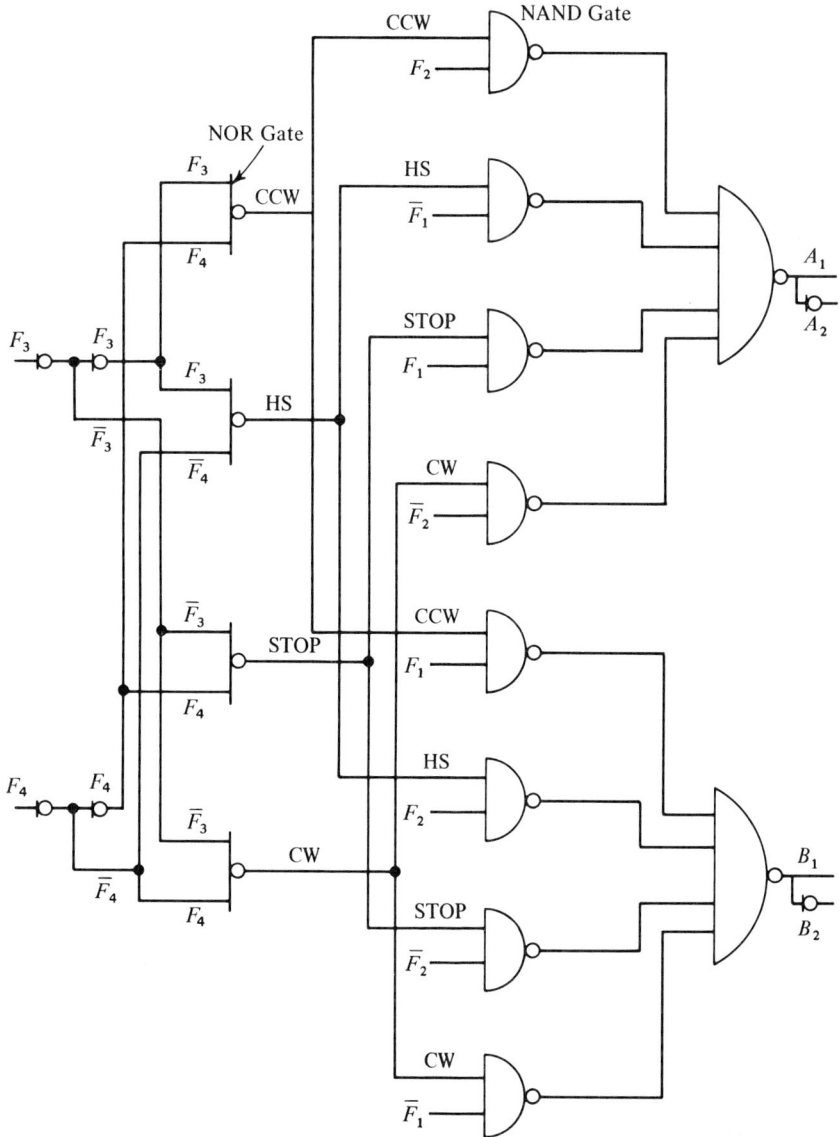

Figure 8.7-5. Logic control circuit (*cont.*).

PC_1 conducts; hence $F_1 = 1$, $F_2 = 0$. With $F_3 = 1$ and $F_4 = 0$, corresponding to a STOP command, the group of four NOR gates makes the STOP outputs equal to 1 and all other outputs equal to 0. The group of eight 2-input NAND gates will have the outputs (1 1 0 1 1 1 0 1), going from top to bottom. With these inputs the two 4-input NAND gates will yield $A_1 =$

1 and $B_1 = 1$, which corresponds to energizing the motor windings for position 1. The motor will remain in the present position.

The digital feedback thus described changes the operating characteristics of the stepper motor considerably. The principal benefit derived from the digital feedback is the ability of the stepper motor to be stepped at a rate that is properly matched to its present speed. For example, with a double step command the motor is now capable of smoothly accelerating up to 2000 steps per second. In contrast, under open-loop operation the maximum stepping rate for this motor is at best 200 steps per second. Furthermore, in closed-loop operation it is feasible to operate the motor under varying load conditions. Since it automatically selects its own stepping rate, it will always recover to run at some speed. This is not possible in open-loop operation; it would stall. Under digital feedback operation the stepper motor becomes an electromechanical transducer that is directed by four exclusive input commands: stop, CW, CCW, and double step.

Experimental tests of this system show that the motor can accelerate from 0 to approximately 280 steps per second in approximately 25 milliseconds when the input is changed from stop to CW or CCW. Similarly, an input command from stop to double step produces a maximum speed of roughly 2300 steps per second in .5 second. It is evident that the motor behaves like a two-speed device. Low-speed operation is achieved by a CW or CCW input, while the double step command produces a high speed.

The motor may be stopped from both of these speed conditions by a stop input. The motor decelerates smoothly and shows the characteristic magnetic detenting action. The responses described above are summarized by the graphs shown in Figure 8.7-6. It is clear that the magnitude of the resultant

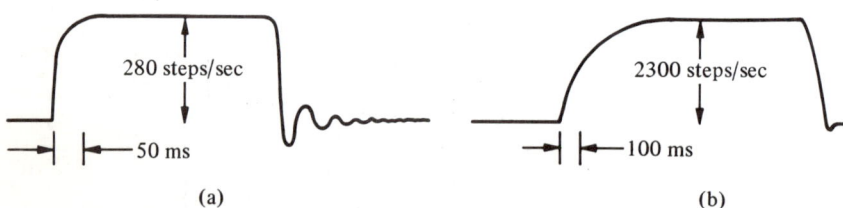

Figure 8.7-6. Closed-loop response of motor: (a) low speed; (b) high speed.

speeds is not unique, but rather an equilibrium speed directly dependent on the motor characteristics and load friction.

Stepping motors provide precise positioning without the use of feedback sensors, but under open-loop control their usefulness is limited by low torque output and low stepping rates. When operated in open-loop fashion the pulse input rate must be matched to the step response of the motor when started from standstill. However, much better performance can be achieved if a feedback system is used to control the input pulse rate according to the motor speed and load conditions. Feedback makes possible the application of the maximum instantaneous pulse rate during acceleration and constant speed control.

8.8 Summary

Digital control techniques are rapidly evolving into an important area in the field of automatic control. Because of the increasing availability of small-scale digital computers and more complete understanding of the operational aspects of computer control systems, the feasibility of assigning a digital computer to direct control is a reality.

In this chapter we have brought into focus a number of digital components and instrumentation that typically see application in computer control systems. We have discussed digital logic modules, which form the basic ingredients of all digitally operating devices, including computers. We have studied the basis of operation of data converters, which form an essential link between a digital computer and surrounding equipment, providing a means of communication. We have also considered the operational characteristics of electromechanical devices that have a digital connection, i.e., the shaft encoder and the stepper motor.

The entire chapter has a common denominator in digital signals, signals that operate at two discrete levels, signals that represent the means of translating and processing information in digital computers. A proper understanding of the material in this chapter cannot be stressed strongly enough, for the systems designer depends on it for the successful realization of digital control.

REFERENCES

1. *Small Computer Handbook*, Digital Equipment Corporation, Maynard, Massachusetts.
2. *Logic Handbook*, Digital Equipment Corporation, Maynard, Massachusetts.
3. *Handbook of Analog Computation*, Systron Donner Corporation, Concord, California, June 1967.
4. Hannauer, G., *Basics of Parallel Hybrid Computers*, Electronic Associates, Inc., Princeton, N. J., 1967.
5. *Systems Designer's Handbook*, Benwill Publishing Corporation, 167 Corey Road, Brookline, Massachusetts (a biannual publication).

6. McCluskey, E. J., *Introduction to the Theory of Switching Circuits*, McGraw-Hill, New York, 1965.
7. Susskind, A. K., *Notes on Analog Digital Conversion Techniques*, Wiley, New York, 1957.
8. Foss, F. A., *The Use of Digital Codes in Digital Control Systems*, IRE Transactions of Electronic Computer, Vol. EC-7, No. 4, December 1954.

PROBLEMS

8.1 Determine the following conversions:
  (a) Decimal to binary, octal, and hexadecimal
      157, 67, 99
      15.35, .25
  (b) Decimal to two's complement binary for a 12-bit binary number
      1999, −1653, −9
  (c) Octal to binary
      7435, 0762
  (d) Binary to decimal
      1011101, 111000110
  (e) Octal to decimal
      40,000, 100, 7777
  (f) Two's complement binary to decimal (nine-bit word)
      111 010 111, 011 111 000, 110 111 111

8.2 Write a subroutine program to convert an eight-bit Gray code into
  (a) Natural binary
  (b) Two's complement binary

8.3 Design a hardware decoder to convert a switch-tail ring counter code into pure binary.

8.4 Determine the accuracy of an eight-bit shaft encoder using a one-inch disc.

8.5 Design a comparator circuit that compares two six-bit digital signals (coded in pure binary) and indicates the comparison with a binary output.

8.6 Repeat Problem 8.5 with the modification that the output is an analog signal that is proportional to the difference of the digital signals.

8.7 Design a logic circuit that can be used as a three-bit down-counter according to the combining sequence

	Decimal	0	1	2	3	4	5	6	7	0
	FF1	1	0	1	1	0	0	0	0	1
Binary	FF2	1	1	0	0	1	1	0	0	1
	FF3	1	1	1	0	1	0	1	0	1

8.8 Design a three-bit down-counter that may be initialized at binary numbers other than 111_2.

8.9 Discuss the characteristics of the network shown in Figure P8.9 as a digital-to-analog converter. The level amplifiers are switched between 0 and V volts, depending on the state of a binary register driving it. Identify the MSB.

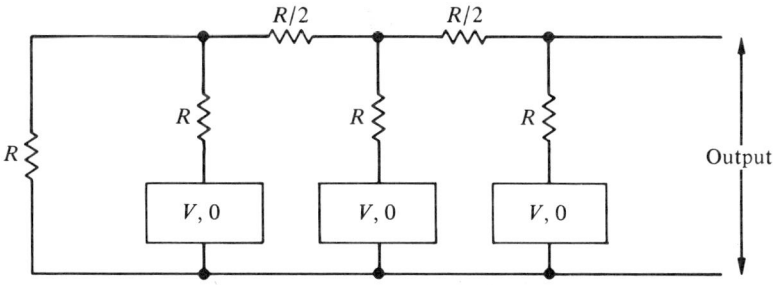

Figure P8.9. Three-bit DAC.

8.10 Construct a table similar to Table 8.5-2 for a four-bit converter and a -5-volt reference supply.

8.11 The logic circuit shown in Figure P8.11 is called a leading edge differentiator. Explain its operation.

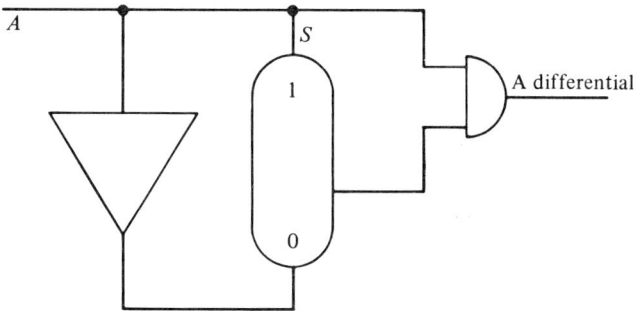

Figure P8.11.

8.12 Construct a logic circuit that will perform falling edge differentiation.

8.13 What is the magnitude of the step of a synchronous inductor stepper motor with a two-phase, three-pole stator and an eight-tooth rotor?

8.14 What are the operational characteristics of a variable reluctance stepper motor with a two-phase, eight-pole stator and a four-pole rotor?

8.15 Design a sequencing logic circuit suitable to drive a three-phase stepper motor with a 12-pole stator and an eight-pole rotor.

8.16 Outline for each successive bit the conversion of the analog voltage 43.5 into a seven-bit binary word. Use the successive approximation A/D converter. What is the decimal equivalent of the binary answer?

9

Computer Methods in Systems Studies

9.1 Introduction

In the preceding eight chapters, we have given an introduction to the study of discrete systems. It has been continually demonstrated that the electronic computer is integrally involved in this study. Not only is the computer required as an essential tool in the analysis, simulation, and design of discrete systems, but we also see the computer assume a significant role in the control and general operation of discrete systems and mixed continuous-discrete systems. Throughout the material presented in the eight chapters, computer techniques are used. Mention has been made of the existence of computer routines to carry out a number of computational tasks. Several computer techniques have been recommended for the analysis of sampled-data systems and for the optimum design of discrete systems.

Because of the rather important role played by the modern computer in the analysis, design, development, and control of a great variety of systems, we consider it highly appropriate to introduce the reader to some of the detailed aspects of computer methods. We will concern ourselves primarily with the programming and application of computers, not with their construction. Furthermore, attention will be given to a discussion of a selected number of numerical techniques that are useful in system simulations.

Historically, it has been customary to divide computers into analog and digital. In recent years, a third important kind of computer has been added to the computer family—the *hybrid computer*. This particular variety of computer has found extensive use in systems studies of all kinds (particularly simulation, optimum design, high-speed iterative solutions, etc.). We will assume a background in basic analog and digital programming (FORTRAN) for the purpose of studying the material presented in this chapter. This will be sufficient to follow the development on hybrid programming.

To a large extent computer programs are prepared by expert programmers to handle a great many different tasks. However, it is often very helpful for the systems designer, the user of these programs, to have an elementary understanding of some of the techniques upon which these programs are based. It is for this purpose that this chapter has been prepared.

9.2 Numerical Methods of Simulating System Dynamics

In almost every digital simulation of dynamic systems the programming of the solution of differential and/or difference equations is required. These equations characterize the dynamics of the system under study. The digital machine performs its basic operations involving arithmetic, memory, and logic operations, in terms of variables which are always represented in discrete form. All continuous mathematical operations must be converted into a corresponding discrete form before they can be processed by the computer. Subject to this requirement are the frequently occurring operations such as integration and differentiation, for example. Also affected, although somewhat differently, are the computations of basically continuous response curves such as root-locus and frequency response.

To handle these and a great many other tasks, suitable digital programs, called *subroutines*, are available for use by the systems designer. In this section we will examine a number of routines which are frequently used in the solution of differential equations. We will direct our attention to a presentation of the discrete approximation technique and errors that are incurred by it. Briefly, two basic types of errors are introduced: the truncation error, which is due to finite approximations of infinite expansions, and the round-off error, which is due to the finite bit capacity of the digital computer. Furthermore, consideration has to be given to the possibility of an unstable discrete representation.

In the solution of differential equations by a digital computer it is assumed that the system to be simulated is characterized in first-order differential equation form, or state variable equations, i.e.,

$$\frac{d}{dt} x_i = f_i(x_1, x_2, \ldots, x_n, t) \qquad i = 1, 2, \ldots, n \qquad (9.2\text{-}1)$$

or, in the case of a linear system,

$$\dot{\mathbf{x}} = \mathbf{F}\mathbf{x} + \mathbf{G}\mathbf{r} \qquad (9.2\text{-}2)$$

Techniques dealing with the derivation of these equations have been presented in numerous texts on modern system theory.

Numerical integration of these equations is based upon a recursive evaluation of their discrete equivalent, which is essentially of the form

$$x_i(k+1) = g_i(x_1(k), x_2(k), \ldots, x_n(k), kT) \quad i = 1, 2, \ldots, n \qquad (9.2\text{-}3)$$

where $x_i(k)$ is the ith state variable of the simulated system at time $t = kT$. We shall now present four methods of generating a digital simulation model and discuss their relative merits.

9.2-1 Euler Method

Probably the simplest approximation to (9.2-1) is obtained from a first-order Taylor series expansion of the state variables $x_i(t)$. Such an expansion yields

$$x_i(t+T) = x_i(t) + T\dot{x}_i(t) \qquad (9.2\text{-}4)$$

If we let

$$x_i(t+T) = x_i(k+1) \quad \text{and} \quad x_i(t) = x_i(k)$$

where k denotes the kth evaluation, corresponding to $t = kT$, and also substitute (9.2-1) into (9.2-4), we obtain

$$x_i(k+1) = x_i(k) + Tf_i(x_1, x_2, \ldots, x_n, k) \qquad (9.2\text{-}5)$$

This approximation is generally referred to as the Euler approximation. It is simple and can be programmed in a straightforward fashion. As an illustration, consider the following differential equation.

EXAMPLE 9-2.1

Determine the Euler approximation of the differential equation

$$\ddot{x} + a\dot{x} + bx = 1 \quad t \geq 0 \qquad (9.2\text{-}6)$$

Sec. 9.2 Numerical Methods of Simulating System Dynamics

First, a set of state equations is given by

$$\dot{x}_1 = -ax_1 - bx_2 + 1$$
$$\dot{x}_2 = x_1 \tag{9.2-7}$$

Consequently, the Euler approximation is

$$x_1(k+1) = x_1(k) + T(-ax_1(k) - bx_2(k) + 1)$$
$$x_2(k+1) = x_2(k) + Tx_1(k) \tag{9.2-8}$$

where T is the incremental time used in the numerical integration of (9.2-6).

The Euler approximation works both for linear and nonlinear system models. Despite these attractive features, however, it suffers serious shortcomings in accuracy and stability and is therefore rarely used. Discussion of these drawbacks will serve well to point out some of the important considerations that go into the selection of a numerical technique in general.

Truncation Error

The series expansion (9.2-4) is truncated after the first derivative term. The error in the solution which is due to this effect is called the *truncation error*.

The significance of the truncation error can be further emphasized by the fact that it tends to accumulate with time. However, differential equations which characterize stable dynamic systems contain built-in feedback which is preserved in modified form in their discrete approximation. This feedback acts in support of the accuracy of the solution. In addition, the numerical approximation solution will seek the same steady-state as the original continuous model, provided, of course, that the stability of the equivalent discrete model is not adversely affected. Let us illustrate this problem by reconsidering Example 9.2-1.

EXAMPLE 9.2-2

Investigate the steady-state and stability characteristics of the Euler approximation of Example 9.2-1.

It is quite clear that the steady-state of (9.2-6)—i.e., $\ddot{x} = \dot{x} = 0$, is $1/b$ with no problems of stability, provided both a and b are positive. From (9.2-8) it can be seen as $k \to \infty$ that $x_1(k) \to 0$ and $x_2(k) \to 1/b$. Thus, the discrete approximation has the same steady-state value.

We next examine the stability of the discrete approximation. For this purpose we determine eigenvalues of the system matrix of (9.2-8). In matrix format we have

$$\begin{bmatrix} x_1(k+1) \\ x_2(k+1) \end{bmatrix} = \begin{bmatrix} 1 - aT & -bT \\ T & 1 \end{bmatrix} \begin{bmatrix} x_1(k) \\ x_2(k) \end{bmatrix} + \begin{bmatrix} T \\ 0 \end{bmatrix} \tag{9.2-9}$$

The eigenvalues are

$$\lambda_{1,2} = 1 - \frac{aT}{2} \pm \sqrt{T\left(\frac{a^2}{4} - b\right)}$$

If, for instance, $a = 2$ and $b = 10$, then the eigenvalues are given by $\lambda_1 = 1 - 4T$ and $\lambda_2 = 1 + 2T$. It is seen that the Euler approximation is unstable for any choice of T. If, however, $a = .2$ and $b = .1$, then $\lambda_{1,2} = 1 - .1T \pm .3jT$. For $T = .1$, $\lambda_{1,2} = .99 \pm j.03$. For this choice of a, b and T we have $|\lambda_{1,2}| < 1$; hence, the Euler approximation is stable. However, if T is increased to 1, one may readily verify that the discrete simulation would be unstable.

The effect of the truncation error may be partially countered by making the time interval T sufficiently small. This is accompanied, however, by an increase in the number of steps in the solution requiring more computer time and eventually making another type of error, round-off error, prominent. Similarly, the stability of the solution may be controlled by the selection of the step size.

Round-off Error

Round-off error is the second basic error encountered in digital computation. It results as a consequence of the finite number of digits with which a digital computer can carry out arithmetic operations. This number may range from four digits for digital machines with 12-bit words to 15 digits for digital machines with 60-bit words. Round-off affects the last digit of a given digital word. For instance, in a word with eight digits the first seven digits are exact, while the last digit is rounded off. By this process it is possible to represent a real number by an eight-digit word that conceivably could have an infinite number of digits for exact definition. The first seven digits match, while the eighth digit is adjusted up if the ninth digit is larger than .5; otherwise it stays the same.

Round-off error affects every digital computation. In numerical integration it assumes particular importance when the integration step has to be selected as very small. Under this condition the significant changes in the values of dynamic systems variables occur more and more in the last few digits of the computer words. An illustration of the influence of round-off error is given by the next example.

EXAMPLE 9.2-3

Consider the numerical solution of the differential equation

$$\ddot{x} + \omega^2 x = 0, \quad \begin{aligned} x(0) &= A \\ \dot{x}(0) &= B \end{aligned} \quad (9.2\text{-}10)$$

This is the equation of a simple harmonic oscillator with resonant frequency ω. When $x(t)$ and $\dot{x}(t)$ are computed and plotted against one another in $(x, \dot{x})$ space, the resulting curve, for instance, is a circle with radius $R = \sqrt{A^2 + B^2}$ when $\omega = 1$. Our objective here is to use an Euler approximation to obtain $x(t)$ and $\dot{x}(t)$ and to determine how close the result agrees with a circle.

Results of this experiment are shown at the end of this section and compared with other methods.

9.2-2 Tustin Method

A technique that offers some appeal in the simulation of linear systems is the Tustin method. It is basically an approximation of differentiation by a difference equation. Consider the following relationships.

The definition of the discrete transform variable z is given by

$$z = e^{sT} \qquad (9.2\text{-}11)$$

Solving for s yields

$$s = \frac{1}{T} \ln z \qquad (9.2\text{-}12)$$

The logarithmic term may be approximated by the series

$$\ln z = 2(u + \tfrac{1}{3}u^3 + \tfrac{1}{5}u^5 + \ldots) \qquad (9.2\text{-}13)$$

where

$$u = \frac{1 - z^{-1}}{1 + z^{-1}}$$

Truncating the series after the first term and substituting into (9.2-12), we obtain the so-called Tustin approximation for the derivative operator s.

$$s \approx \frac{2}{T} \frac{1 - z^{-1}}{1 + z^{-1}} \qquad (9.2\text{-}14)$$

This relationship is used in converting the continuous-time models of linear systems into difference equations. This difference equation may be readily solved recursively on a digital computer, which yields a fairly accurate digital simulation of the linear system.

The substitution (9.2-14) may be applied to an nth-order linear differential equation, to n first-order linear differential equations, or to an nth-order transfer function. The result can usually be arranged into one of two forms, (1) n first-order difference equations or (2) one nth-order difference equation. This may best be illustrated by two examples.

EXAMPLE 9.2-4

Derive the difference equation that represents a Tustin simulation of the transfer function

$$G(s) = \frac{as + b}{s(s + c)}$$

Upon substitution of the approximation given by equation (9.2-14) we obtain

$$G(z) = \frac{a\dfrac{2}{T}\dfrac{1 - z^{-1}}{1 + z^{-1}} + b}{\left(\dfrac{2}{T}\dfrac{1 - z^{-1}}{1 + z^{-1}}\right)^2 + c\dfrac{2}{T}\dfrac{1 - z^{-1}}{1 + z^{-1}}}$$

which is simplified to

$$G(z) = \frac{2aT + bT^2 - 2bT^2 z^{-1} + (bT^2 - 2aT)z^{-2}}{4 + 2cT - 8z^{-1} + (4 - 2cT)z^{-2}}$$

This second-order digital transfer function may be readily programmed for computer execution.

EXAMPLE 9.2-5

Derive a difference equation for the harmonic oscillator of Example 9.2-3, using the Tustin approximation.

The differential equation of the harmonic oscillator is given by (9.2-10), or in state variable representation

$$\dot{x}_1 = -\omega^2 x_2$$
$$\dot{x}_2 = x_1$$

Using (9.2-14), we have

$$\frac{2}{T}\frac{1 - z^{-1}}{1 + z^{-1}} X_1(z) = -\omega^2 X_2(z)$$

and

$$\frac{2}{T}\frac{1 - z^{-1}}{1 + z^{-1}} X_2(z) = X_1(z)$$

Converting into difference equations, we obtain

$$x_1(k) = x_1(k-1) - \frac{\omega^2 T}{2}[x_2(k) + x_2(k-1)] \qquad (9.2\text{-}15)$$

$$x_2(k) = x_2(k-1) + \frac{T}{2}[x_1(k) + x_1(k-1)] \qquad (9.2\text{-}16)$$

In order to evaluate these difference equations recursively, we must eliminate $x_2(k)$ from the first equation. This yields

$$x_1(k) = \left(\frac{4 - \omega^2 T^2}{4 + \omega^2 T^2}\right) x_1(k-1) + 2\left(\frac{2 - \omega^2 T}{4 + \omega^2 T^2}\right) x_2(k-1) \qquad (9.2\text{-}17)$$

The digital simulation of the harmonic oscillator now involves (9.2-17) and (9.2-16), to be recursively evaluated, in that order.

This simulation is carried out for various values of T and is compared with the results of the Euler and Runge-Kutta methods. The results of this comparison are summarized at the end of this section.

The Tustin method obviously requires a considerable amount of manipulation to make the substitution (9.2-14) and to rearrange the result into a suitable form. This job can be easily handled by a digital computer subroutine. For instance, one program may start with the description of the system by a transfer function and generate a digital transfer function of the form

$$D(z) = \frac{b_0 + b_1 z^{-1} + \ldots + b_n z^{-n}}{1 + a_1 z^{-1} + \ldots + a_n z^{-n}}$$

This may then be easily programmed as a digital recursion equation.

9.2-3 Runge-Kutta Method

The Runge-Kutta method and its various modifications are the most widely employed of the single-step methods. These are methods of a class for which $x_i(k+1)$ can be obtained from $x_i(k)$ alone and the differential equations. The methods are self-starting and are not difficult to program. The basic Runge-Kutta method is partially an extension of the Euler method, which, as was mentioned, is based upon the first-order Taylor series (9.2-4). This series expansion suggests inclusion of higher-order terms to reduce the truncation error. Although the first derivative is available from the differential equation, higher-order derivatives must be evaluated separately, possibly by difference methods. This would result in an effective method, except that the evaluation of the higher-order derivatives can be very tedious. The Runge-Kutta method makes use of the higher-order Taylor approximations indirectly so as to avoid this problem.

The Runge-Kutta methods used in practice are based on fourth-order Taylor approximations. Although these are simple in their use, their derivations entail complicated developments.* It will suffice here to present only the computational algorithm.

Let the system of equations to be solved be given in the familiar form

$$\dot{x}_i = f_i(x_1, x_2, \ldots, x_n, t) = f_i(\mathbf{x}, t)$$
$$x_i(t_0) = x_{i0} \quad i = 1, 2, \ldots, n \quad (9.2\text{-}18)$$

Let $x_i(k)$ be the value of x_i at $t = t_k$ and $f_i(\mathbf{x}(k), t_k)$ the derivative of x_i at $t = t_k$. If T is the increment (step-size) of the time variable t, the Runge-Kutta fourth-order method uses the formulas

$$K_{1i} = Tf_i(\mathbf{x}(k), t_k)$$
$$K_{2i} = Tf_i[\mathbf{x}(k) + .5\mathbf{K}_1, t_k + .5T]$$
$$K_{3i} = Tf_i[\mathbf{x}(k) + .5\mathbf{K}_2, t_k + .5T] \quad (9.2\text{-}19)$$
$$K_{4i} = Tf_i[\mathbf{x}(k) + \mathbf{K}_3, t_k + T]$$
$$x_i(k+1) = x_i(k) + \tfrac{1}{6}[K_{1i} + 2K_{2i} + 2K_{3i} + K_{4i}] \quad i = 1, 2, \ldots, n$$

where $\mathbf{K}_1, \mathbf{K}_2, \mathbf{K}_3$, and $\mathbf{K}_4$ are $n \times 1$ vectors determined by (9.2-19).

For each step of integration in (9.2-19) the Runge-Kutta method requires four evaluations of the functions $f_i(x_1, x_2, \ldots, x_n, t)$, thus requiring considerable amount of machine time. The Runge-Kutta method is characterized by a high degree of accuracy, which compares well with analytical methods of solution.

EXAMPLE 9.2-6

Consider a Runge-Kutta solution to the problem of the harmonic oscillator presented in Example 9.2-3. The results are shown for various step sizes at the end of this section.

Selection of Step Size and Control of Truncation Error

Essentially, the only factor left to the user's discretion is the time increment T. For obvious reasons this should be selected as large as possible to keep the machine running time as small as possible. Because the truncation error is kept small, in most cases the only risk incurred by making the step size too large is the possibility of numerical instability. For most applications numerical instability is a more predominant problem than inaccuracy caused by truncation error. The analytical complexity of the method prohibits a direct stability analysis, even for the simplest case. A good rule of thumb is to keep the step size approximately at a tenth of the value of the smallest time constant of the differential equations to be solved.

*See, for example, Hildebrand, F. B., *Introduction to Numerical Analysis*, McGraw-Hill, 1958.

The typical dynamic transient is associated with large values of derivatives during the early stages, which rapidly decrease during the final stages. Considerable computer time may be saved if the step is adjusted to be small when the derivatives are large and vice versa.

The step size may be adjusted automatically during a calculation provided there exists an explicit expression for the upper bound of the truncation error.

With an explicit knowledge of the instantaneous, or local, truncation error one can always select that step size which keeps the truncation error just below an acceptable upper bound. Although it is known that the local truncation error is roughly proportional to the fifth power of the step size for a fourth-order Runge-Kutta integration method, it is not possible to obtain a running estimate of the truncation error during the course of an integration.

A modification of the basic fourth-order Runge-Kutta method which has a slightly smaller truncation error and provides an estimate of the truncation error is the Runge-Kutta-Merson method. It requires five iterations per step size and step size adjustments may be made during the course of an integration. The Runge-Kutta-Merson* algorithm is

$$K_{1i} = \tfrac{1}{3}Tf_i[\mathbf{x}(k), t_k]$$

$$K_{2i} = \tfrac{1}{3}Tf_i\left[\mathbf{x}(k) + \mathbf{K}_1, t_k + \frac{T}{3}\right]$$

$$K_{3i} = \tfrac{1}{3}Tf_i\left[\mathbf{x}(k) + .5\mathbf{K}_1 + .5\mathbf{K}_2, t_k + \frac{T}{3}\right] \quad (9.2\text{-}20)$$

$$K_{4i} = \tfrac{1}{3}Tf_i(\mathbf{x}(k) + \tfrac{3}{8}\mathbf{K}_1 + \tfrac{9}{8}\mathbf{K}_3, t_k + .5T)$$

$$K_{5i} = \tfrac{1}{3}Tf_i(\mathbf{x}(k) + \tfrac{3}{2}\mathbf{K}_1 - \tfrac{9}{2}\mathbf{K}_3 + 6\mathbf{K}_4, t_k + T)$$

$$x_i(k+1) = x_i(k) + .5(K_{1i} + 4K_{4i} + K_{5i}) \quad i = 1, 2, \ldots, n \quad (9.2\text{-}21)$$

The estimate of the error is given by

$$\text{truncation error} = \max(K_{1i} - \tfrac{9}{2}K_{3i} + 4K_{4i} - \tfrac{1}{2}K_{5i}) \quad (9.2\text{-}22)$$

The operation of automatic step size adjustment in the Runge-Kutta-Merson algorithm would be to maximize the step size while the truncation is kept within a specified bound. A computer program called subroutine INTFUN utilizing this algorithm in contained in Appendix 9A.

9.2-4 Adams-Moulton Predictor-corrector Method

A third important numerical technique for the integration of differential equations is the Adams-Moulton method. It employs a predictor-corrector principle and uses the following recursive algorithm:

*Lance, G. N., *Numerical Methods for Highspeed Computers*, Iliffe, London, 1960.

$$x_i^p(k+1) = x_i(k) + \frac{T}{24}[55f_i(k) - 59f_i(k-1) + 37f_i(k-2) - 9f_i(k-3)] \quad (9.2\text{-}23)$$

$$x_i(k+1) = x_i(k) + \frac{T}{24}[9f_i^p(k+1) + 19f_i(k) - 5f_i(k-1) + f_i(k-2)]$$
$$i = 1, 2, \ldots, n \quad (9.2\text{-}24)$$

The method requires two evaluations of the differential equations for each step. One involves $f_i(k)$ in the predictor equation (9.2-23), while the other is $f_i^p(k+1)$ in the corrector equations (9.2-24). The principle of the Adams-Moulton method is based on the determination of a first estimate $x_i^p(k+1)$ by (9.2-23) using values of the derivatives at four successive time instants, with a subsequent correction by (9.2-24) using values of the derivatives at three successive time instants and $f_i^p(k+1)$, the derivative at the first estimate.

The Adams-Moulton method has a time-saving advantage over the Runge-Kutta method. However, it is not self-starting. For the first three time intervals, a one-step method like the Runge-Kutta method is used to obtain the needed starting values. Thus a combination of the Runge-Kutta and Adams-Moulton methods is employed.

Control of Step Size and Truncation Error

The Adams-Moulton is a fourth-order method, and hence the truncation error is of the order of T^5. An explicit expression for an estimate of the truncation error is available, thus making automatic step size adjustments possible. It is given by

$$\text{truncation error} = \max_i \frac{|x_i^p(k+1) - x_i(k+1)|}{14 D_i} \quad (9.2\text{-}25)$$

where

$$D_i = \max_i \{x_i^p(k+1), \alpha\} \quad i = 1, 2, \ldots, n$$

and where α is a positive constant used to prevent unnecessary reductions in T. It is usually set equal to 1.

As is the case with the various Runge-Kutta methods, the Adams-Moulton method may be used in the integration of differential equations with arbitrary expressions for $f_i(\mathbf{x}, t)$. In fact, the execution of the program is totally indifferent to the nature of $f_i(x_1, x_2, \ldots, x_n, t)$.

In choosing between a Runge-Kutta method or the Adams-Moulton method, obviously time is of prime consideration. An important additional criterion affecting a choice is the presence of discontinuities in the right-hand side functions. If the derivatives contain no discontinuities, a predictor-corrector method using only two function evaluations per step is likely to be faster than a one-step method. If, however, they do have discontinuities, as is so frequently the case in engineering problems, a one-step method is more accurate and more efficient.

The remaining three techniques of simulation to be presented in this section are restricted to linear systems, or at least piecewise linear systems with constant coefficients.

9.2-5 State Transition Method

A numerical technique of simulating linear systems that is rapidly gaining widespread acceptance by systems designers is based upon state variable techniques. A linear system may be described by the equations

$$\frac{d}{dt}\mathbf{x} = \mathbf{Fx} + \mathbf{G}u \qquad (9.2\text{-}26)$$

$$y = \mathbf{Cx} + du \qquad (9.2\text{-}27)$$

When the input $u(t)$ can be adequately represented by a piecewise constant equivalent

$$u(t): \quad u(kT + \tau) = u(kT) \quad 0 \leq \tau < T \qquad (9.2\text{-}28)$$

it is possible to determine $y(t)$ at the discrete times $t = 0, T, 2T, \ldots$ by means of the discrete state equations

$$\mathbf{x}[(k+1)T] = e^{\mathbf{F}T}\mathbf{x}(kT) + \int_0^T e^{\mathbf{F}\tau} d\tau \mathbf{G}u(kT) \qquad (9.2\text{-}29)$$

$$y(kT) = \mathbf{Cx}(kT) + du(kT) \qquad (9.2\text{-}30)$$

These relations were derived in Chapter 2 with the following notation:

$$\mathbf{A}(T) = e^{\mathbf{F}T} \quad \text{and} \quad \mathbf{B} = \int_0^T e^{\mathbf{F}\tau} d\tau \mathbf{G} \qquad (9.2\text{-}31)$$

The matrices $\mathbf{A}$ and $\mathbf{B}$ have to be evaluated only once for any given time interval T.

The most remarkable feature of the state transition method is that it offers a discrete simulation of a continuous-time system which is almost completely exact. There are only two sources of error. One is introduced by sampling the input. The other one is caused by the iterative procedure for evaluating the matrices $\mathbf{A}$ and $\mathbf{B}$. But no error is introduced by the actual discrete model. It is exact.

Evaluation of $\mathbf{A}$ *and* $\mathbf{B}$

We shall present two procedures for approximating $\mathbf{A}$ and $\mathbf{B}$ by series techniques. Because of the relatively recent development of these techniques an extensive development on the truncation error will be given.

The first of the techniques is due to Liou.* The transition matrix **A** is expressed by the infinite series

$$\mathbf{A} = \sum_{i=0}^{\infty} \frac{\mathbf{F}^i T^i}{i!}, \qquad \mathbf{F}^0 = \mathbf{I} \qquad (9.2\text{-}32)$$

This series is uniformly convergent in a finite interval. It is, therefore, possible to evaluate **A** within prescribed accuracy. If the series is truncated at $i = L$, then we may write

$$\mathbf{A} = \sum_{i=0}^{L} \frac{\mathbf{F}^i T^i}{i!} + \sum_{i=L+1}^{\infty} \frac{\mathbf{F}^i T^i}{i!} = \mathbf{M} + \mathbf{R} \qquad (9.2\text{-}33)$$

The first term in (9.2-33) represents the series approximation, while the second term corresponds to the remainder term.

If each element in **A** is required to be within an accuracy of at least d significant figures, then

$$|r_{ij}| \leq 10^{-d} |m_{ij}| \qquad (9.2\text{-}34)$$

where r_{ij} and m_{ij} correspond to the elements of **R** and **M**, respectively. Let the norm of **F** be

$$\|\mathbf{F}\| = \max_{i} \left(\sum_{j=1}^{n} |a_{ij}| \right)$$

Then, it can be shown that

$$\|\mathbf{F}^k\| \leq \|\mathbf{F}\|^k \qquad k = 1, 2, \cdots$$

Hence each element of the matrix $\mathbf{F}^k$ is less than or equal to $\|\mathbf{F}\|^k$. It follows that

$$|r_{ij}| \leq \sum_{i=L+1}^{\infty} \frac{\|\mathbf{F}\|^i T^i}{i!} \qquad (9.2\text{-}35)$$

Let the ratio of the second term to the first term of the series (9.2-35) be ϵ, that is

$$\epsilon = \frac{\|\mathbf{F}\| T}{L + 2} \qquad (9.2\text{-}36)$$

from which we conclude that

$$\frac{\|\mathbf{F}\| T}{2} \leq \epsilon$$

Substituting the last relation into (9.2-35) we have

---

*M. L. Liou, "A novel method of evaluating transient responses," *Proceedings of IEEE*, Vol. 54, No. 1, January 1966, pp. 20–23.

$$|r_{ij}| \leq \frac{\|\mathbf{F}\|^{L+1}T^{L+1}}{(L+1)!}(1 + \epsilon + \epsilon^2 + \epsilon^3 + \cdots)$$

$$= \frac{\|\mathbf{F}\|^{L+1}T^{L+1}}{(L+1)!} \frac{1}{1-\epsilon} \qquad (9.2\text{-}37)$$

Thus the matrix **A** can be evaluated approximately according to the following iterative procedure:

1. Choose an initial value of L.
2. Evaluate m_{ij} by (9.2-33).
3. Determine ϵ by (9.2-36).
4. Find the upper bound of $|r_{ij}|$ by (9.2-37).
5. Compare each element of M obtained from (2) with the upper bound of $|r_{ij}|$ obtained from (4); if (9.2-34) is not satisfied, increase L and repeat the iteration; otherwise, the iteration is complete.

The matrix **B** may be evaluated in a similar manner. From the properties of exponential matrices it can be shown that

$$\mathbf{B} = (e^{\mathbf{F}T} - \mathbf{I})\mathbf{F}^{-1}\mathbf{G} \qquad (9.2\text{-}38)$$

Thus, using (9.2-30),

$$\mathbf{B} = T \sum_{j=0}^{\infty} \frac{\mathbf{F}^j T^j}{(j+1)!} \mathbf{G} \qquad (9.2\text{-}39)$$

Since the series expression for **B** converges faster than (9.2-32), it suffices to determine L for $e^{\mathbf{F}T}$ as outlined above and apply the same value for **B**.

A second method of evaluating the matrices **A** and **B** also uses the truncated series approximation (9.2-33);* however, it differs in the method of evaluating the series and in the manner in which the series is terminated. The finite power series **M** is related to the identity

$$\mathbf{M} = \sum_{i=0}^{L} \frac{(\mathbf{F}T)^i}{i!}$$

$$= \left[\mathbf{I} + \mathbf{F}T\left(\mathbf{I} + \frac{\mathbf{F}T}{2}\left\{\mathbf{I} + \frac{\mathbf{F}T}{3}\left[\mathbf{I} + \cdots + \frac{\mathbf{F}T}{L-1}\left(\mathbf{I} + \frac{\mathbf{F}T}{L}\right)\right] \cdots \right\}\right)\right]$$

$$(9.2\text{-}40)$$

Starting with the innermost factor, this nested product expansion lends itself very well to digital programming. Since the evaluation starts with the last term first the value of L, the number of terms of the series approximation, must be determined beforehand. The number of terms to be included is

---

*S. G. Hoppe et al., "A Feasibility Study of Self-learning Adaptive Flight Control for High Performance Aircraft," Report AFFDL-TR-67-18, Cornell Aeronautical Laboratory, February, 1967.

related empirically to the norm of the matrix $\mathbf{F}T$; that is,

$$L = \min\{3\,\|\mathbf{F}T\| + 6,\, 100\} \tag{9.2-41}$$

This relation assures that no more than a 100 terms are included. By experimental verification it can be demonstrated that the series $e^{\mathbf{F}T}$ is accurate to at least six significant figures.

The matrix $\mathbf{B}$ may also be computed by a similar expression by combining (9.2-40) with (9.2-35). This yields

$$\mathbf{B} = T\left(\mathbf{I} + \frac{\mathbf{F}T}{3}\left\{\mathbf{I} + \frac{\mathbf{F}T}{3}\left[\mathbf{I} + \ldots + \frac{\mathbf{F}T}{L-1}\left(\mathbf{I} + \frac{\mathbf{F}T}{L}\right)\ldots\right]\right\}\right)\mathbf{G} \tag{9.2-42}$$

Because of the great similarity between (9.2-40) and (9.2-42) the evaluation of the two series may be easily combined into a single computer routine.

EXAMPLE 9.2-7

Determine the state transition matrix $\mathbf{A}$ for the simple harmonic oscillator. The state equations are given by Example 9.2-5.

$$\dot{x}_1 = -\omega^2 x_2$$
$$\dot{x}_2 = x_1$$

By analytical techniques we determine that

$$e^{\mathbf{F}t} = \begin{bmatrix} \cos\omega t & -\omega \sin\omega t \\ \dfrac{1}{\omega}\sin\omega t & \cos\omega t \end{bmatrix}$$

so that the difference equations are

$$\begin{bmatrix} x_1(k+1) \\ x_2(k+1) \end{bmatrix} = \begin{bmatrix} \cos\omega T & -\omega \sin\omega T \\ \dfrac{1}{\omega}\sin\omega T & \cos\omega T \end{bmatrix}\begin{bmatrix} x_1(k) \\ x_2(k) \end{bmatrix}$$

The results of this example are discussed next.

9.2-6 Comparison of Techniques

The Euler and Runge-Kutta methods are applicable to nonlinear systems, while the Tustin and state transition methods are restricted to linear systems. The Euler and Runge-Kutta methods require the preparation of an identical subprogram for the evaluation of derivatives; this subprogram is used concurrently with the execution of the integration programs. The Tustin and state transition methods require intermediate programs for the preparation of the difference equations: the Tustin method a program to implement the Tustin substitution, and the state transition method one to determine the exponential matrices e^{FT} and $\int e^{FT}$. These programs are run prior to the integration program. To use the Tustin method the description of the dynamic system must be available in transfer function form, whereas the state transition method requires a state model.

It is probably fair to state that with respect to convenience of application no method is particularly disadvantageous, as long as the digital computer is effectively utilized to carry out the computations.

Speed of Computation

The relative speed of computation of the four methods depends entirely upon the total number of instructions that require execution during the course of the program. Table 9.2-1 shows a listing of computer time elapsed during

Table 9.2-1

	Euler	Runge-Kutta	Tustin	State Transition
Compile time	0:57	1:40	0:59	1:46
Execution time	2:55	12:53	4:26	5:34
Load time	0:19	0:21	0:19	0:22
Total time	4:16	15:18	5:49	7:42

compile, load, and execution stages for identical problems* for which the four methods were used. Time is given in minutes and seconds. It is easily seen that the Euler method is the fastest and the Runge-Kutta method is the slowest, while the Tustin and state transition methods rank second and third. Since the Runge-Kutta method requires four times as many calculations (it is a four-step method) as the Euler method, it is easily explained that roughly four times as much time is required. The total time for the Tustin method is approximately 25 percent longer than the Euler, and the state transition method requires about 75 percent more time than the Euler method.

To draw any meaningful conclusion from this time summary one must also take into consideration the accuracy factor, for a method may require little time for implementation but may be marked by poor accuracy. This we shall examine now.

Accuracy

It goes without saying that accuracy is the most important factor in considering the selection of a numerical method for system simulation. In

---

*The problem referred to here is the linear oscillator and the computer used is an IBM 7044.

Table 9.2-2 Numerical Results of Four Methods

Time	Runge-Kutta	Euler	Tustin	State Transition
0	5.417	10.00	6.0	5.403
2	−4.010	0.	−2.8	−4.161
3	−9.695	−20.	−9.36	−9.900
4	−6.542	−40.	−8.432	−6.536
5	2.491	−40.	−7.584	2.837
6	9.161	0.	7.522	9.602
7	7.463	80.	9.785	7.540
8	−9.638	160.	4.220	−1.455
9	−8.417	160.	−4.721	−9.111
10	−8.166	0.	−9.885	−8.40

(a) $T = 1$ second

Time	Runge-Kutta	Euler	Tustin	State Transition
1	5.403	5.708	5.410	5.403
2	−4.161	−4.530	−4.146	−4.161
3	−9.900	−1.148	−9.896	−9.900
4	−6.536	−8.097	−6.562	−6.536
5	2.837	3.434	2.797	2.837
6	9.602	1.286	9.588	9.602
7	7.539	1.089	7.577	7.540
8	−1.455	−1.775	−1.389	−1.455
9	−9.111	−1.406	−9.080	−9.111
10	−8.391	−1.409	−8.436	−8.391

(b) $T = .1$ second

Time	Runge-Kutta	Euler	Tustin	State Transition
1	5.403	5.43	5.403	5.403
2	−4.161	−4.203	−4.161	−4.161
3	−9.900	−10.05	−9.900	−9.900
3.14	−10.000	−10.16	−10.00	−10.00
4	−6.536	−6.669	−6.536	−6.536
5	2.837	2.907	2.836	2.836
6	9.602	9.893	9.602	9.602
6.28	10.000	10.32	10.00	10.00
7	7.539	7.809	7.539	7.539
8	−1.455	−1.512	−1.454	−1.454
9	−9.111	−9.529	−9.111	−9.111
10	−8.391	—	—	—

(c) $T = .01$ second

the earlier paragraphs of this section we considered the solution of the differential equation

$$\ddot{x} + \omega^2 x = 0 \tag{9.2-10}$$

via the four methods under consideration. It is now our intention to compare the results.

Time	Runge-Kutta	Euler	Tustin	State Transition
1	5.403	5.405	5.403	5.403
2	−4.161	−4.165	−4.161	−4.161
3	−9.900	−9.915	−9.900	−9.900
3.14	−10.00	−10.02	−10.00	−10.00
4	−6.536	−6.549	−6.536	−6.536
5	2.837	2.844	2.837	2.837
6	9.601	9.630	9.601	9.601
6.28	10.00	10.03	10.00	10.00
7	7.539	7.565	7.539	7.539
8	−1.455	−1.461	−1.455	−1.455
9	−9.111	−9.152	−9.111	−9.111
10	−8.390	—	−8.390	−8.390

(d) $T = .001$ second

Time	Runge-Kutta	Euler	Tustin	State Transition
1	5.403	5.403	5.403	5.403
2	−4.161	−4.161	−4.160	−4.160
3	−9.898	−9.900	−9.897	−9.896
3.14	−9.998	−10.00	−9.997	−9.996
4	−6.535	−6.536	−6.534	−6.533
5	2.836	2.836	2.835	2.835
6	9.598	9.601	9.596	9.594
6.28	9.996	9.999	9.994	9.991
7	7.536	7.539	7.534	7.532
8	−1.454	−1.455	−1.454	−1.454
9	−9.106	−9.111	−9.104	−9.100
10	—	—	—	—

(e) $T = .0001$ second

Time	Runge-Kutta	Euler	Tustin	State Transition
1	5.400	5.400	5.401	5.393
2	−4.157	−4.157	−4.147	−4.147
3	−9.882	−9.882	−9.864	−9.848
3.14	−9.982	−9.982	−9.963	−9.945
4	−6.521	−6.521	−6.510	−6.490
5	2.828	2.828	2.815	2.812
6	9.568	9.568	9.533	9.501
6.28	9.963	9.963	9.926	9.889
7.00	7.508	7.508	7.481	7.446

(f) $T = .00001$ second

The solution of (9.2-10) represents a circle (for $\omega = 1$) when $\dot{x}$ is plotted versus x. The period of this circle is 2π seconds. Plotting the result therefore can offer a quick visual check on the quality of the solution.

Of interest here are the effects of truncation error and round-off on the accuracy of the solution. For this purpose we show the solution $x(t)$ for six choices of T in Table 9.2-2. The initial conditions are chosen as $x(0) = 10.0$

and $\dot{x}(0) = 0.0$ so that a circle of radius 10.0 results. The solution is run for roughly 10 seconds of time, sufficient to cover one full period. Solution values are shown at full second intervals and for $T \leq .01$, also for $t = 3.14$ and $t = 6.28$, which correspond to half and full periods of the circle.

We recall that error due to truncation of series approximation is present in the Euler, Runge-Kutta, and Tustin methods, but not in the state transition method. Therefore, for the largest value of the increment of integration ($T = 1.0$) we can expect that truncation will be a prominent factor in the Euler, Runge-Kutta, and Tustin methods, but not in the state transition method. Reviewing Table 9.2-2(a), we see that the Euler method produces an unstable solution; the Runge-Kutta and Tustin methods produce stable solutions but with considerable truncation error, while the solution under the state transition method is accurate to within the decimal places shown. We recall that the accuracy of the last method depends in this case only on the accuracy to which the series expansion of e^{FT} is computed; each entry in the matrix e^{FT} is accurate to within 10^{-6}. Thus, the solution generated by the state transition method may serve as a reference.

When the increment of integration is reduced to $T = .1$, the Runge-Kutta solution becomes exact to the places shown and the Tustin method is improved considerably. The solution generated by the Euler method appears stable but still suffers considerable truncation error.* These results are shown in Table 9.2-2(b).

When T is further reduced to $T = .01$ Runge-Kutta and Tustin methods are identical to the state transition method. The Euler method, however, still shows truncation error effects. See Table 9.2-2(c). Also shown are values of the solution at $t = 3.14$ and $t = 6.28$ for which the exact solutions are -10.00 and $+10.00$, respectively.

Table 9.2-2(d) shows the solutions for $T = .001$. The results indicate no error for the Runge-Kutta, Tustin, and state transition methods. The Euler method is now accurate to within two places.

A further reduction in the time increment to $T = .0001$ permits the generation of an exact solution (four significant figures) by the Euler method. On the other hand, the other three methods are beginning to show the effects of round-off. See Table 9.2-2(e).

When the increment is selected as small as $T = .00001$, round-off error becomes a significant influence in determining the quality of the solution. Table 9.2-2(f) demonstrates this. The results further indicate that the Runge-Kutta and Euler methods generate identical solutions. This supports the fact that the truncation error for $T = .00001$ is completely negligible in both these methods, and they provide equally good Taylor series approximations.

*Actually, the Euler method does not yield a stable difference equation for this case.

9.3 Use of the State Transition Method in Simulation Studies

The discrete transition method may be employed to handle a variety of problems. Its use is most suitable in the simulation of discrete-time systems. We refer here to two special cases in connection with simulation studies.

9.3-1 Conversion of Nonhomogeneous to Homogeneous State Transition Equations

Normally, the discrete state equivalent of a linear continuous plant with piecewise constant inputs is of the form

$$\mathbf{x}(k+1) = \mathbf{A}(T)\mathbf{x}(k) + \mathbf{B}(T)r(k) \tag{9.3-1}$$

An alternate approach to the solution of (9.3-1) may be followed when this equation represents the discrete-time equation of a transfer function, and the input to the transfer function is derived from a hold element, as shown in Figure 9.3-1. An example is now used to illustrate the procedure to be followed.

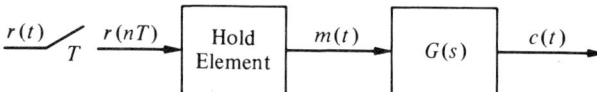

Figure 9.3-1. Linear system driven by hold element.

EXAMPLE 9.3-1

Consider the case when

$$G(s) = \frac{s+1}{(s+2)(s+10)} = \frac{s+1}{s^2 + 12s + 20} \tag{9.3-2}$$

and a zero-order hold circuit is employed.

To derive $\mathbf{A}$ we represent this system using the nested programming method. The state equations are

$$\begin{bmatrix} \dot{x}_1 \\ \dot{x}_2 \end{bmatrix} = \begin{bmatrix} -12 & 1 \\ -20 & 0 \end{bmatrix} \begin{bmatrix} x_1 \\ x_2 \end{bmatrix} + \begin{bmatrix} 1 \\ 1 \end{bmatrix} m(t) \tag{9.3-3}$$

$$c(t) = x_1$$

Now $m(t)$ is given by

$$m(nT + \tau) = r(nT) \quad \text{for } 0 \leq \tau < T \tag{9.3-4}$$

i.e., $m(t)$ is a piecewise constant input. During any sampling period it is therefore possible to view $m(t)$ as the output of an integrator whose input is zero and whose initial condition is set to $r(nT)$ at the beginning of the nth sampling period. This is shown by the state variable diagram in Figure 9.3-2.

Consider now the combination of the state variable diagram for (9.3-3) and Figure 9.3-2, as shown in Figure 9.3-3. The state equations corresponding to this figure are

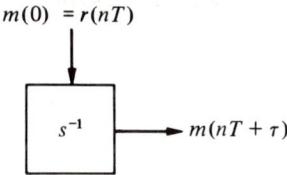

Figure 9.3-2. State variable diagram of zero-order hold.

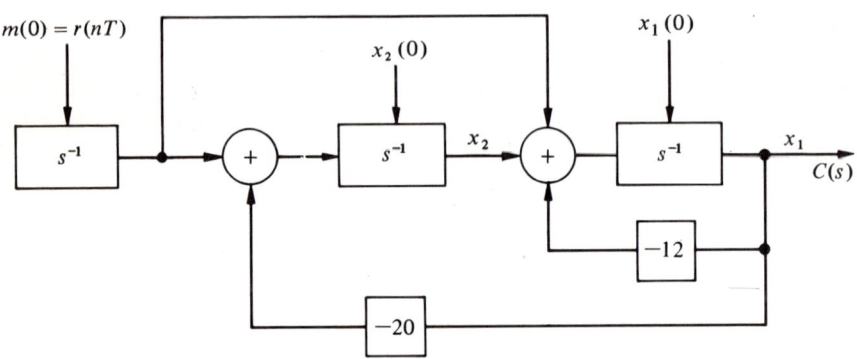

Figure 9.3-3. State variable diagram of $G(s)$ with zero-order hold.

$$\begin{bmatrix} \dot{m} \\ \dot{x}_1 \\ \dot{x}_2 \end{bmatrix} = \begin{bmatrix} 0 & 0 & 0 \\ 1 & -12 & 1 \\ 1 & -20 & 0 \end{bmatrix} \begin{bmatrix} m \\ x_1 \\ x_2 \end{bmatrix} \quad 0 \leq t < T \tag{9.3-5}$$

$$c = x_1$$

The solution at the end of the first sampling period is

$$\begin{bmatrix} m(T) \\ x_1(T) \\ x_2(T) \end{bmatrix} = e^{\mathbf{F}_a T} \begin{bmatrix} m(0) \\ x_1(0) \\ x_2(0) \end{bmatrix} = e^{\mathbf{F}_a T} \begin{bmatrix} r(0) \\ x_1(0) \\ x_2(0) \end{bmatrix} \tag{9.3-6}$$

where $\mathbf{F}_a$ is the augmented system matrix [e.g. (9.3-5)].

Repeated application of (9.3-6) leads to the recursion relation

$$\begin{bmatrix} m(k+1) \\ x_1(k+1) \\ x_2(k+1) \end{bmatrix} = e^{F_aT} \begin{bmatrix} r(k) \\ x_1(k) \\ x_2(k) \end{bmatrix} \qquad (9.3\text{-}7)$$

$$c(k) = x_1(k)$$

In general, by augmenting the matrix $\mathbf{F}$ such that

$$\mathbf{F}_a = \begin{bmatrix} 0 & 0 \\ \mathbf{G} & \mathbf{F} \end{bmatrix}$$

the nonhomogeneous recursion equation (9.3-1) is changed into a homogeneous equation. The obvious advantage is that in the digital computer evaluation of (9.3-7) only one exponential series needs evaluation. It is pointed out, however, that (9.3-6) is completely equivalent to using the nonhomogeneous equation

$$\begin{bmatrix} x_1(k+1) \\ x_2(k+1) \end{bmatrix} = e^{FT} \begin{bmatrix} x_1(k) \\ x_2(k) \end{bmatrix} + \left\{ \int_0^T e^{F\tau} \begin{bmatrix} 1 \\ 1 \end{bmatrix} d\tau \right\} r(k)$$

or

$$\begin{bmatrix} x_1(k+1) \\ x_2(k+1) \end{bmatrix} = \mathbf{A}(T) \begin{bmatrix} x_1(k) \\ x_2(k) \end{bmatrix} + \mathbf{B}(T)r(k) \qquad (9.3\text{-}8)$$

9.3-2 Evaluation of System Response Between Sampling Periods

Suppose that it is necessary to determine the output of a system that receives piecewise constant inputs at intervals of T seconds at times corresponding to subintervals of T_s seconds such that

$$T = \alpha T_s, \quad \alpha \text{ an integer}$$

It is possible to use either (9.3-1) or (9.3-7) to accomplish this. For instance, let $\alpha = 5$; then $T_s = .2T$.

Approach A

Evaluate $\mathbf{A}(T_s) = e^{FT_s}$ and $\mathbf{B}(T_s)$. Then at time $t = nT$ use the recursion equation (9.3-1) five times; e.g.,

$$x(nT + T_s) = A(T_s)x(nT) + B(T_s)r(nT)$$
$$x(nT + 2T_s) = A(T_s)x(nT + T_s) + B(T_s)r(nT)$$
$$x(nT + 3T_s) = A(T_s)x(nT + 2T_s) + B(T_s)r(nT) \qquad (9.3\text{-}9)$$
$$x(nT + 4T_s) = A(T_s)x(nT + 3T_s) + B(T_s)r(nT)$$
$$x[(n+1)T] = x(nT + 5T_s) = A(T_s)x(nT + 4T_s) + B(T_s)r(nT)$$

These five equations are repeated every main sampling interval with a new input supplied at times $t = nT$, $n = 1, 2, \ldots$.

Approach B

Evaluate $A_a(T_s) = e^{F_a T_s}$. Then use the recursion (9.3-7) five times; e.g.,

$$\begin{bmatrix} m(nT + T_s) \\ x(nT + T_s) \end{bmatrix} = e^{F_a T_s} \begin{bmatrix} r(nT) \\ x(nT) \end{bmatrix}$$

$$\begin{bmatrix} m(nT + 2T_s) \\ x(nT + 2T_s) \end{bmatrix} = e^{F_a T_s} \begin{bmatrix} r(nT) \\ x(nT + T_s) \end{bmatrix} \qquad (9.3\text{-}10)$$

$$\vdots$$

$$\begin{bmatrix} m[(n+1)T] \\ x[(n+1)T] \end{bmatrix} = \begin{bmatrix} m(nT + 5T_s) \\ x(nT + 5T_s) \end{bmatrix} = e^{F_a T_s} \begin{bmatrix} r(nT) \\ x(nT + 4T_s) \end{bmatrix}$$

These equations are repeated every main sampling interval with a new input applied at times given by $t = nT$, $n = 1, 2, \ldots$. Of course, the variable $m(nT + T_s)$, $m(nT + 2T_s)$, etc. is not used.

9.4 Digital Computer Simulation of a Digital Control System

As an illustration of the use of numerical integration techniques, we consider the digital simulation of a computer control system. Consider the system shown in Figure 9.4-1. The transfer function of the continuous system is given as

$$G(s) = \frac{k_0(s+1)}{s(s^2 + 4s + 10)} \qquad (9.4\text{-}1)$$

A zero-order hold element is used. The digital recursion equation is given by

$$e_2(kT) = e_1(kT) + .2e_1[(k-1)T] - .2e_1[(k-2)T]$$
$$- .6e_2[(k-1)T] + .15e_2[(k-2)T] \qquad (9.4\text{-}2)$$

Sec. 9.4 Digital Computer Simulation of a Digital Control System

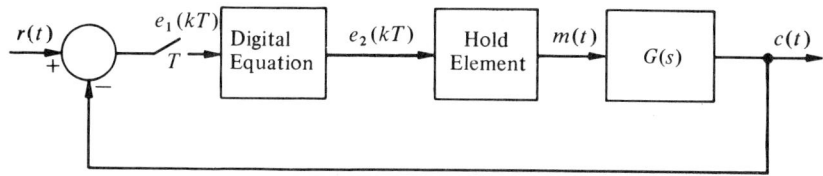

Figure 9.4-1. Computer control system.

It is desired to set up a digital computer simulation to analyze the response of the system for a variety of sampling periods, system gains, and step inputs. It is, therefore, necessary to provide built-in flexibility in the program to accommodate an analysis with respect to these three factors.

Using direct programming, we form the state equations of (9.4-1), which are

$$\begin{bmatrix} \dot{x}_1 \\ \dot{x}_2 \\ \dot{x}_3 \end{bmatrix} = \begin{bmatrix} -4 & -10 & 0 \\ 1 & 0 & 0 \\ 0 & 1 & 0 \end{bmatrix} \begin{bmatrix} x_1 \\ x_2 \\ x_3 \end{bmatrix} + \begin{bmatrix} 1 \\ 0 \\ 0 \end{bmatrix} m(t) \qquad (9.4\text{-}3)$$

$$c(t) = k_0(x_2 + x_3)$$

Thus

$$\mathbf{F} = \begin{bmatrix} -4 & -10 & 0 \\ 1 & 0 & 0 \\ 0 & 1 & 0 \end{bmatrix} \quad \text{and} \quad \mathbf{G} = \begin{bmatrix} 1 \\ 0 \\ 0 \end{bmatrix}$$

To compute the transition matrices $\mathbf{A}(T)$ and $\mathbf{B}(T)$ we may use a computer routine such as

SUBROUTINE MATEXP (F,G,A,B,N,M,T)

described in Appendix 2A.

The linear recursion equation of the digital computer can be simulated in the very form it is given.

A general flow chart for the simulation is shown in Figure 9.4-2. After the flow chart, the program of the simulation is presented.

Definition of symbols:

T = sampling period
GAIN = gain k_0 of system
NRUN = number of runs made
TIME = time of response
XIN = magnitude of step input
NDATA = total number of runs to be made
TIMEFN = total length of each run

```
          DIMENSION  F(3,3),G(3),  A(3,3),  B(3),E1(3),E2(3),X(3)
          READ  (5,1)  F,G,NDATA,TIMEFN
C         PARAMETERS  FOR  THIS  RUN
   21     READ  (5,2)  T,GAIN,XIN
C         COMPUTE  A(T)  AND  B(T)
          DO  10  I  =  1,3
          B(I)  =  G(I)
          DO  10  J  =  1,3
   10     A  (I,J)  =  F(I,J)
          CALL  TRANS  (A  ,  B  ,3,T,15)
C         INITIALIZE  NEW  RUN
          NRUN  =  NRUN  +  1
          C  =  0.0
          TIME  =  0.0
          XM  =  0.0
          DO  11  I  =  1,3
          E1(I)  =  0.0
          E2(I)  =  0.0
   11     X(I)  =  0.0
C         COMPUTE  COMPUTER  INPUT
   20     E1(3)  =  E1(2)
          E1(2)  =  E1(1)
          E1(1)  =  XIN-C
C         COMPUTE  COMPUTER  OUTPUT
          E2(3)  =  E2(2)
          E2(2)  =  E2(1)
          E2(1)  =  E1(1)+.2*E1(2)-.2*E1(3)-.6*E2(2)+.15*E2(3)
C         COMPUTE  NEW  PLANT  INPUT
          XM  =  E2(1)
C         COMPUTE  NEW  PLANT  OUTPUT
          DO  12  I  =  1,3
          X(I)  =  X(I)+B(I)*XM
          DO  12  J  =  1,3
   12     X(I)  =  X(I)+A(I,J)*X(J)
          C  =  GAIN*(X(2)+X(3))
C         PRINT  OUTPUT
          TIME  =  TIME+T
          WRITE(6,3)  TIME,E1(1),XM,C
C         TEST  IF  THIS  RUN  IS  COMPLETED
          IF(TIME.LT.TIMEFN)  GO  TO  20
C         TEST  IF  A  NEW  RUN  IS  TO  BE  INITIATED
          IF(NRUN. LT. NDATA)  GO  TO  21
          STOP
```

The program is written in FORTRAN. Not shown are FORMAT statements.

The digital simulation illustrated above can also be carried out by using

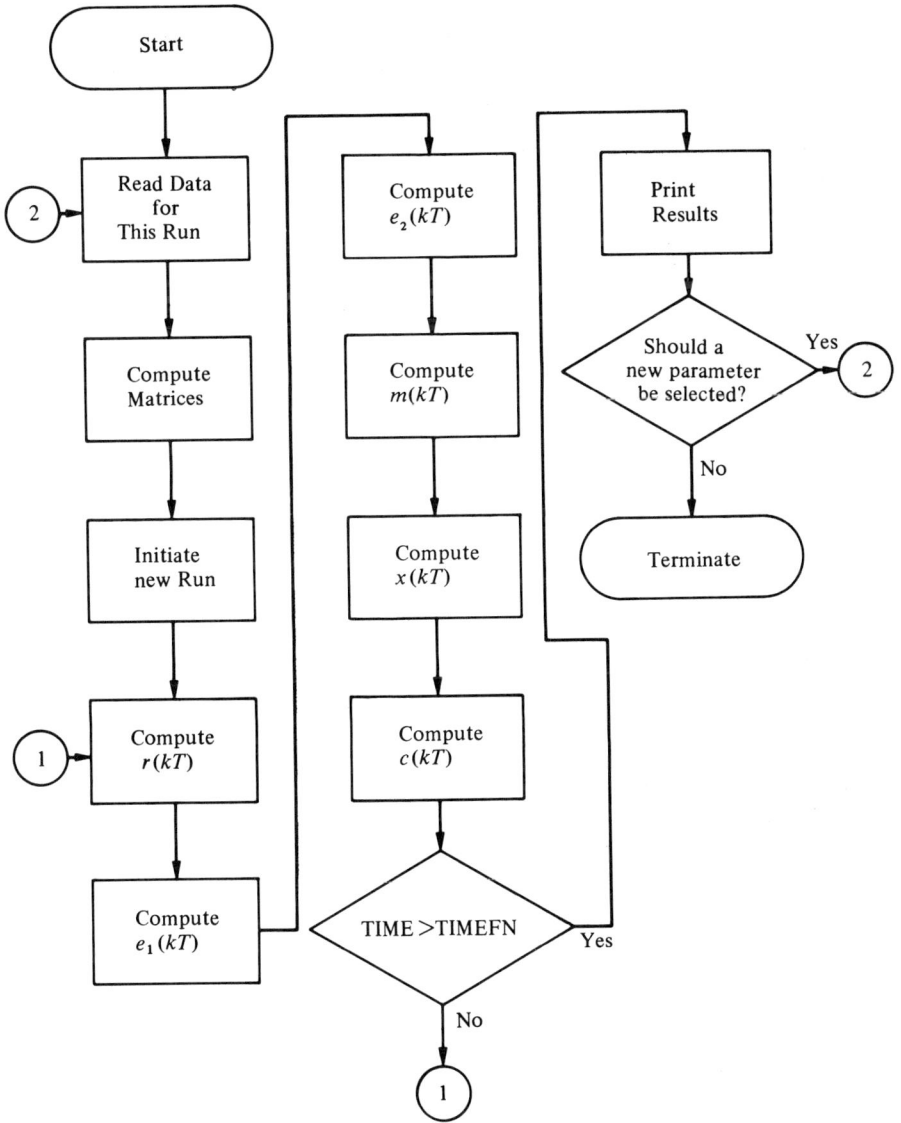

Figure 9.4-2. Flow chart for digital simulation.

the augmented system matrix $\mathbf{F}_a$, since the input to the plant $G(s)$ is a piecewise constant input. The augmented matrix is

$$\mathbf{F}_a = \begin{bmatrix} 0 & 0 & 0 & 0 \\ 1 & -4 & -10 & 0 \\ 0 & 1 & 0 & 0 \\ 0 & 0 & 1 & 0 \end{bmatrix}$$

9.4-1 Other Ways of Implementing a Digital Computer Simulation of a Discrete-time System

The last section demonstrated a simulation procedure of computer control systems by representing the continuous-time parts of the system by discrete-time state transition matrices for the purpose of calculating their response at discrete-time intervals. An alternate approach to the computation of the response of the continuous-time part of a mixed system is the utilization of numerical integration techniques such as the Euler or Runge-Kutta procedures. To follow this approach it is convenient to construct a subroutine that will generate a solution of the differential equations representing the plant over a single sampling interval or some subinterval. SUBROUTINE INTFUN is such a program. (See Appendix 9A.)

The use of the subroutine is illustrated by repeating the simulation of the problem of the previous section. The flow chart is given as shown in Figure 9.4-3. The program follows on page 403.

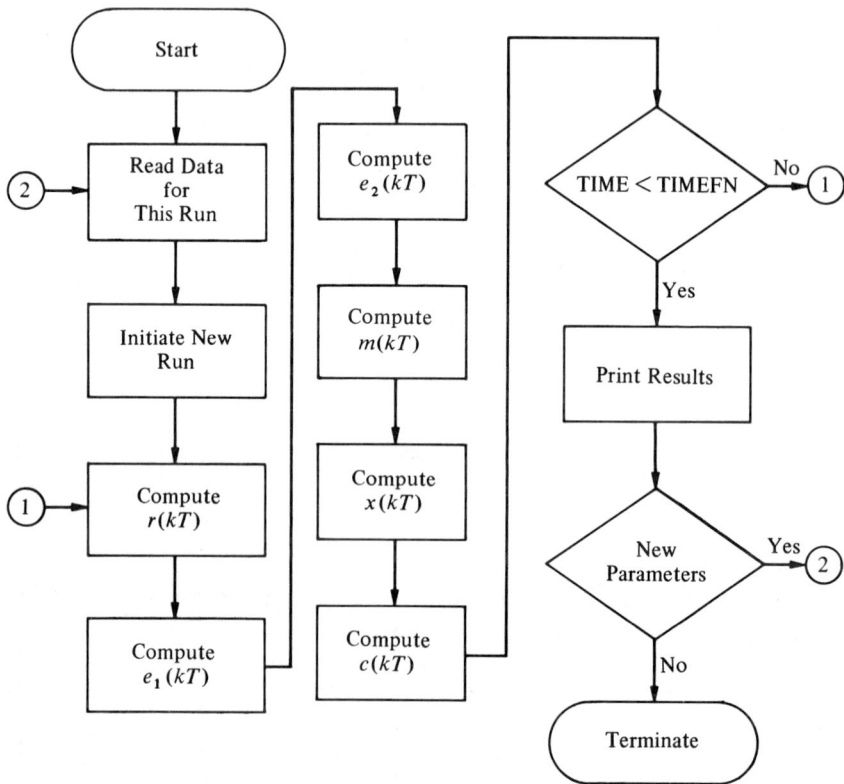

Figure 9.4-3. Flow chart for digital simulation.

```
              DIMENSION  E1(3),E2(3),  X(4)
              READ  (5,1)  NDATA,TIMEFN
C             PARAMETERS  FOR  THIS  RUN
           21 READ  (5,2)  T,GAIN,XIN
C             INITIALIZE  NEW  RUN
              NRUN  =  NRUN+1
              C  =  0.0
              FIVET  =  5.*T
              TIME  =  0.0
              DO  11  I  =  1,3
              E1(I)  =  0.0
              E2(I)  =  0.0
           11 X(I)  =  0.0
C             NEW  COMPUTER  INPUT
           20 E1(3)  =  E1(2)
              E1(2)  =  E1(1)
              E1(1)  =  XIN-C
C             NEW  COMPUTER  OUTPUT
              E2(3)  =  E2(2)
              E2(2)  =  E2(1)
              E2(1)  =  E1(1)+.2*E1(2)-.2*E1(3)-.6*E2(2)+.15*E2(3)
C             NEW  PLANT  INPUT
              X(4)  =  E2(1)
C             INTEGRATE  PLANT  FOR  NEXT  INTERVAL  WITH  5  POINTS
              CALL  INTFUN(X,TIME,FIVET,N)
C             NEW  PLANT  OUTPUT
              C  =  GAIN*(X(2)+X(3))
C             PRINT  OUTPUT
              TTOTAL  =  TTOTAL+T
              WRITE  (6,3)  TTOTAL,E1(1),E2(1),C
C             TEST  IF  THIS  RUN  IS  COMPLETE
              IF  (TTOTAL.LT.TIMEFN)  GO  TO  20
C             TEST  IF  A  NEW  RUN  IS  TO  BE  INITIATED
              IF  (NRUN.IT.NDATA)  GO  TO  21
              END
              STOP
```

In addition to the above program we need the statements for subroutine DERIV; these are

```
              SUBROUTINE  DERIV(X,TIME,DX)
              DIMENSION  X(4),DX(3)
              DX(1)  =  -4.*X(1)-10.*X(2)+X(4)
              DX(2)  =  X(1)
              DX(3)  =  X(2)
              RETURN
              END
```

The use of the Runge-Kutta method to provide a solution for the state equations of the plant is particularly appropriate when these equations are nonlinear. In that case, a state transition matrix cannot be obtained.

9.5 Analog Computer Simulation of Systems

The analog computer has been widely employed in the simulation of complex dynamic systems. Simulation is closely related to the solution of state equations by analog computer. However, simulation implies the solution of dynamic systems in a broader sense: It involves the solution of the state equations representing appropriately chosen subsystems of the system with interconnections that are in a one-to-one correspondence with the topology of the system. Let us consider the simulation of the feedback control system shown in Figure 9.5-1.

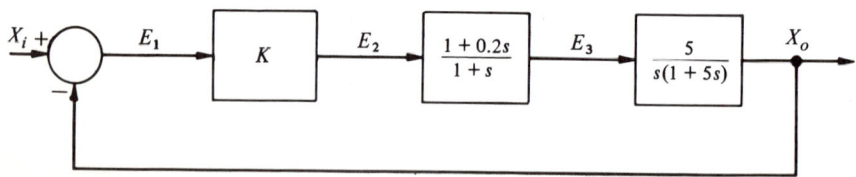

Figure 9.5-1. Block diagram of a control system.

We can dissect the system into four subassemblies.

1.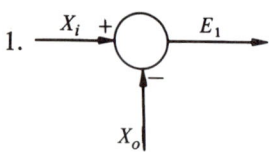

$$E_1 = X_i - X_o$$

In terms of a computer diagram,

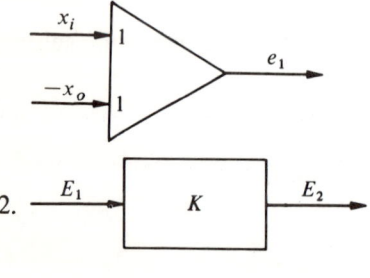

2. $\quad \xrightarrow{E_1} \boxed{K} \xrightarrow{E_2}$

$$E_2 = KE_1$$

Sec. 9.5 Analog Computer Simulation of Systems 405

and

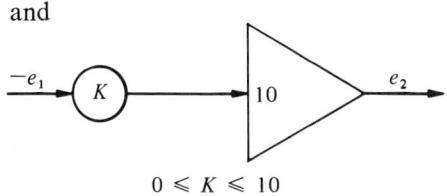

$$0 \leqslant K \leqslant 10$$

3. $E_2 \rightarrow \boxed{\dfrac{1+0.2s}{1+s}} \rightarrow E_3$

The state equations are

$$\frac{d}{dt}x_3 = -[x_3 - e_2(t)]$$

$$e_3(t) = -[-.8x_3 - .2e_2(t)]$$

Therefore, the corresponding analog diagram is

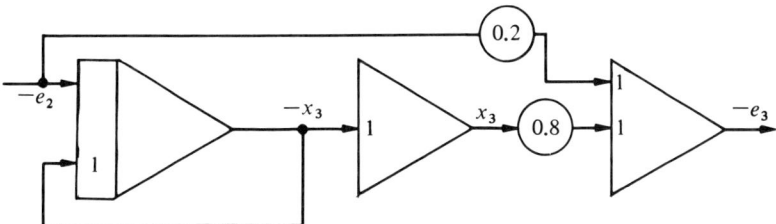

4. $E_3 \rightarrow \boxed{\dfrac{5}{s(1+5s)}} \rightarrow X_o$

The state equations are

$$\frac{d}{dt}\begin{bmatrix} x_1 \\ x_2 \end{bmatrix} = -\left\{ \begin{bmatrix} +.2 & 0 \\ -1 & 0 \end{bmatrix}\begin{bmatrix} x_1 \\ x_2 \end{bmatrix} - \begin{bmatrix} 1 \\ 0 \end{bmatrix}e_3(t) \right\}$$

$$x_o = x_2$$

and the diagram is

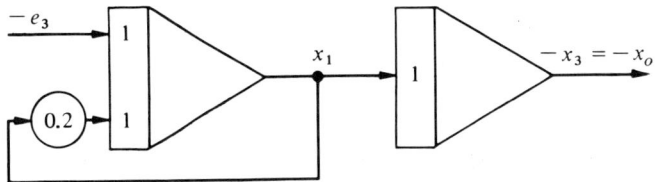

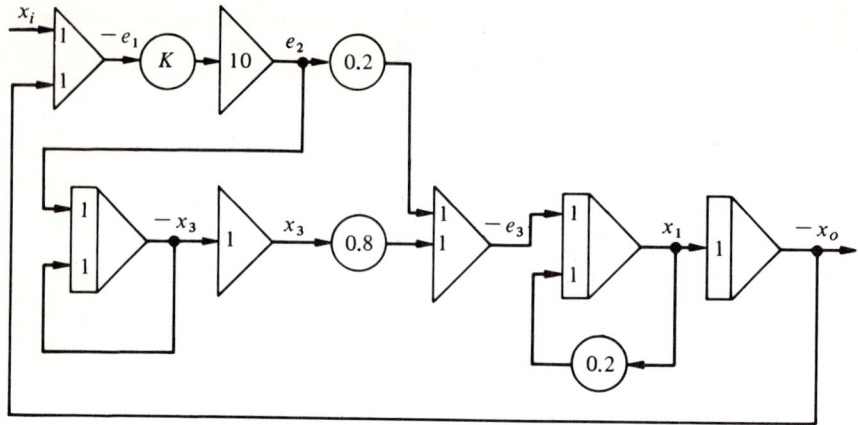

Figure 9.5-2. Analog computer diagram for feedback system.

These four subassemblies may now be integrated into a single diagram representing the entire system. The result is shown in Figure 9.5-2. The analog computer program thus obtained is a simulation of the feedback system. The structural correspondence between the computer diagram and the block diagram of the system is easily recognized. The diagram may be simplified by removing the second and fifth summer-amplifier and combining their functions with those adjacent to them. Figure 9.5-3 shows the simplified

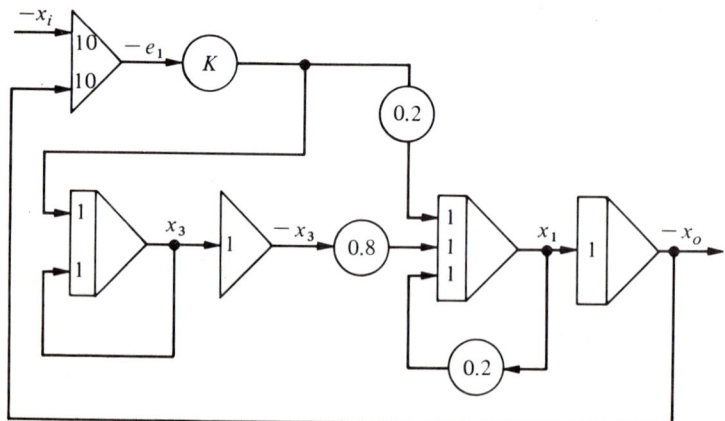

Figure 9.5-3. Simplified computer diagram.

diagram. Although this simplification requires the elimination of the variables e_2 and e_3, it is consistent with good analog computer practice of constructing a diagram of minimum complexity. However, the main structure is still preserved.

In contrast to the simulation approach outlined above, we shall consider a straight state equation solution of the same problem. The combined "closed-loop" transfer function is given by

$$X_o(s) = \frac{K(.2s + 1)}{s^3 + 1.2s^2 + .2(K+1)s + K} X_i(s)$$

The corresponding state equations by direct programming are

$$\frac{d}{dt}\begin{bmatrix} x_1 \\ x_2 \\ x_3 \end{bmatrix} = \begin{bmatrix} -1.2 & -.2(K+1) & -K \\ 1 & 0 & 0 \\ 0 & 1 & 0 \end{bmatrix} \begin{bmatrix} x_1 \\ x_2 \\ x_3 \end{bmatrix} + \begin{bmatrix} 1 \\ 0 \\ 0 \end{bmatrix} x_{in}(t)$$

$$x_o(t) = K[0 \quad .2 \quad 1]\begin{bmatrix} x_1 \\ x_2 \\ x_3 \end{bmatrix}$$

The analog computer diagram corresponding to these equations is shown in Figure 9.5-4.

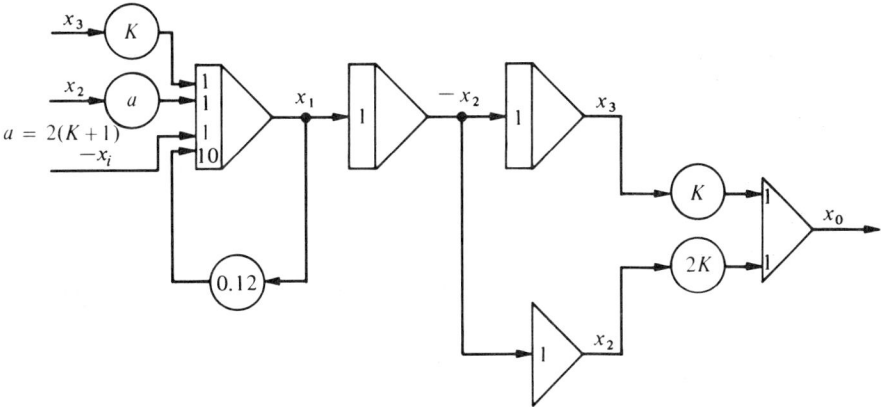

Figure 9.5-4. Direct approach of feedback system solution.

Figures 9.5-3 and 9.5-4 show computer diagrams that represent two approaches to the solution of a dynamic system. Although both will yield identical results in terms of the solution, the simulation approach is usually preferred. Not only is the structure of the system preserved in the computer diagram, but also it is much more readily adaptable for design considerations. For instance, if the parameter K, which represents the gain of the system, is to be selected according to some response specifications, only one potentiometer has to be adjusted on the simulation diagram, against four on the state equation approach.

The example developed above is but a single illustration of analog computer simulation techniques. However, it points out one important characteristic: the breakdown of the overall system into subassemblies, each one of which is modeled individually.

The simulation of systems by analog computer is particularly useful when the system contains isolated nonlinearities. Some subassemblies will be characterized by linear state equations, while others may contain typical nonlinearities that are much more easily dealt with on a subassembly basis than on a complete system approach.

The simulation of discrete systems or sampled-data systems may be most effectively carried out by the use of a hybrid computer if the system operates in continuous time while other parts function in a discrete manner. We shall turn our attention to hybrid computer simulation in the next section.

9.6 Digital Analog-system Simulators

The application of digital computation methods in the solution of technical problems has become widespread. The increased use of digital computation methods is largely due to the availability of larger, faster, and more powerful digital computers and the development of application languages such as FORTRAN and ALGOL. These languages enable the user to express his problem in a computer language closely allied to the language of his own field, thus considerably simplifying the programming process.

During the last ten years an additional class of application languages has been developed which are proving to be very useful. These are the digital analog simulators. Nearly two dozen of these languages have been created, their level of capability roughly corresponding to that of digital computer technology. In an evolutionary manner, two simulation systems that have gained widespread acceptance because of their simplicity and overall effectiveness are MIDAS (Modified Integrator Digital Analog Simulator) and MIMIC. Of these the MIMIC language provides a simple substitute for the hybrid computer, as far as the simulation of discrete and sampled-data systems is concerned. What makes it potentially even more attractive than hybrid computer programming is the ease of programming, the elimination of the need for scaling, and the absence of sign reversals through operational amplifiers.

By education and job experience, the systems designer tends to visualize a system as a complex of subsystems. In this context, a computer control system may well be described by a block diagram consisting of interconnected blocks, each one of which designates a subsystem. These blocks may contain s-transfer functions, z-transfer functions, nonlinear functions, logic functions, time-varying constants, and other characteristics typically found in a large-scale system.

By its very organization, the hybrid computer provides this building block capability. Digital analog simulation languages provide the means of programming a digital computer like a hybrid or analog computer. The problem is programmed on the digital computer in a manner closely approaching an analog computer solution. The user has at his disposal a set of predefined blocks from which he can assemble the system to be simulated. The blocks perform the same functional operations as standard analog computers, such as integration, multiplication, function generation, switching relays, and the like. In addition, hold circuits and z-transfer functions may be simulated. The assembly of these blocks, corresponding to the patching of the analog computer, is performed by a sequence of connection statements.

Before we proceed with the detailed description of MIMIC, we consider a brief history of the development of digital simulator languages.

9.6-1 Brief History and Recent Developments

The first published account of work on digital analog simulation was presented by R. G. Selfridge.* His work was motivated by the need to simulate larger problems than his analog computer could handle and to achieve better accuracy. His program was developed for one of the early computers without the advantage of automatic compilers such as FORTRAN. Adapting the digital computer to block diagram organization was his major contribution and is the basis for all of the subsequent analog simulator programs. Selfridge's program was an interpretive routine, which means that it accepted and executed certain pseudo-instructions without producing a machine language translation. Also, all computation was done in fixed-point arithmetic, and the problem variables had to be scaled to a definite maximum value.

Digital simulator development has paralleled the general development of digital computers. Programming aids such as floating-point arithmetic and automatic compilation are used to great advantage by the newer simulation programs such as MIMIC, MIDAS,† DAS,‡ PACTOLUS,§ and DSL 90.‖ Although these programs differ in format and language, they all retain

---

*R. G. Selfridge, "Coding a General-Purpose Digital Computer to Operate as a Differential Analyzer," *Proceedings 1955 Western Joint Computer Conference* (IRE).

†R. T. Harnett et al., "MIDAS ... An Analog Approach to Digital Computation," *Simulation*, Vol. 3, No. 3, September 1964.

‡R. A. Gaskill et al., "DAS—A Digital Analog Simulator," *AFIPS Conference Proc.*, Vol. 23, p. 83.

§R. D. Brennan and S. Harlan, "PACTOLUS—A Digital Analog Simulator Program for the IBM 1620," *AFIPS Conference Proc.*, Vol. 26, October 1964.

‖W. M. Syn and D. G. Wyman, *DSL-90, User's Guide*, SHARE Library 3358.

the block organization feature. The PACTOLUS program exhibits one of the latest innovations: man-machine interplay. This is a very desirable feature, but it is not practical, since most large computer installations do not allow this procedure.

The DAS (Digital Analog Simulator) is structurally classified as a compiler. In addition to providing a workable and easy-to-use simulation language, it represents the forerunner of the MIDAS program.

9.6-2 MIMIC—A Simulator Language*

The development of simulator languages is an evolutionary process. At the time of this writing it is fair to say that MIMIC is the most versatile and effective simulation language. It represents a direct descendant of MIDAS, with some basic modifications and improvements. In this section a brief description of MIMIC and its use will be given. The description will be only sufficiently complete to permit the reader to obtain a basic appreciation of the operation of the program. For a total understanding only a manual of operation will suffice.

Illustrative Example

The coding of a simulation problem in the MIMIC language is carried out in terms of FORTRAN-like statements. As an introductory example, consider the control system shown in Figure 9.6-1. To code this problem in

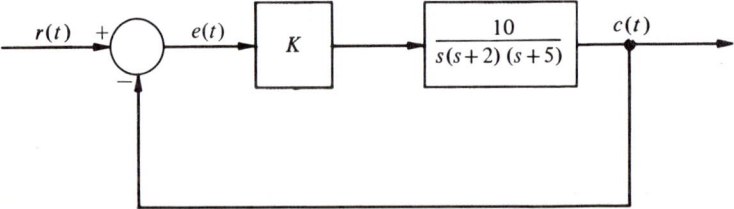

Figure 9.6-1. Simple control system.

MIMIC, we proceed in a manner similar to an analog computer simulation. For each block in the block diagram, a state model is developed. Using direct programming, for the block relating $c(t)$ to $m(t)$, we have

$$\frac{d}{dt}\begin{bmatrix} x_1 \\ x_2 \\ x_3 \end{bmatrix} = \begin{bmatrix} -7 & -10 & 0 \\ 1 & 0 & 0 \\ 0 & 1 & 0 \end{bmatrix} \begin{bmatrix} x_1 \\ x_2 \\ x_3 \end{bmatrix} + \begin{bmatrix} 1 \\ 0 \\ 0 \end{bmatrix} m(t) \qquad (9.6\text{-}1)$$

$$c(t) = 10x_3$$

*H. G. Peterson and F. J. Sanson, "MIMIC," *A Digital Simulator Program*, SESCA Internal Memo 65-12.

The problem would be programmed by keypunching the following statements

Result 10	Expression 19
E	R−C
XM	AK∗E
X1	INT(−7.∗X1−10.∗X2+1.∗XM,0.)
X2	INT(X1,0.)
X3	INT(X2,0.)
C	10.∗X3

The relationship between the coding and the equations is so similar to FORTRAN programming that it should not require explanation.

The six equations may be reduced to three, as follows:

Result 10	Expression 19
X1	INT(−7.∗X1−10.∗X2+AK∗(R−10.∗X3,0.))
X2	INT(X1,0.)
X3	INT(X2,0.)

The block diagram of Figure 9.6-1 may be restructured to contain only first-order transfer functions, such as those shown in Figure 9.6-2. Then it is possible to utilize a MIMIC function that computes the input/output relations for a first-order transfer function (FTR). The operation of this function is illustrated by still another coding version of the same problem. The FTR solves a first-order differential equation.

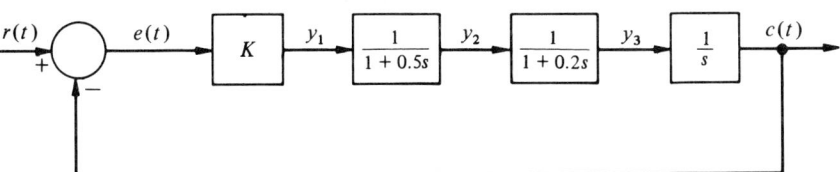

Figure 9.6-2. Alternate block diagram.

Result 10	Expression 19
E	R−C
Y1	K∗E
Y2	FTR(Y1,.5)
Y3	FTR(Y2,.2)
C	INT(Y3,0.)

As is illustrated by the above example, it is possible to code a MIMIC simulation program in various ways: directly from a set of differential equations, or directly from a block diagram, or a combination of both. Obviously, a great degree of flexibility is available to the programmer. Whatever approach is used, the computer will automatically sort the instructions into proper order for sequential solution and then proceed to solve the equations.

9.6-3 Selected Features of MIMIC

Variables

As in FORTRAN, a group of from 1 to 6 alphameric characters constitutes the name of a variable in a MIMIC program. There are six reserved names; they are

T	the independent variable
DT	the amount T changes between printouts
DTMAX	the maximum integration step size allowed
DTMIN	the minimum integration step size allowed
TRUE	a logic constant that always has a "true" value
FALSE	a logic constant that always has a "false" value

Integrator with Mode Control

The mode of each integrator in MIMIC can be individually controlled. These are the RESET, OPERATE, and HOLD modes. A completely specified integrator function is given as

Result	Expression
10	19
R	INT(A,B,C,D)

Here the variable A is integrated with initial condition B. The mode is controlled by the variables C and D, according to Table 9.6-1.

Table 9.6-1 Integrator Mode Control

D \ C	TRUE	FALSE
TRUE	OPERATE	HOLD
FALSE	RESET	OPERATE

Logic Control Variables

MIMIC permits the use of logical variables. Logical variables may assume the value TRUE (1) or FALSE (0). They may be generated by use of the function switch FSW or logical switch LSW. One of the uses of the logical variables is to provide a control over the execution of expressions. If a logical variable is entered in the LCV column (column 2-7), the expression on that line will be evaluated when the control variable is TRUE and bypassed when the control variable is FALSE.

Subprograms

Subprograms in the style of a FORTRAN subroutine may be included in a MIMIC program. A subprogram must first be defined by putting the expressions comprising the subprogram between a BSP (begin subprogram) and an ESP (end subprogram). The subprogram name is entered in the result column of the BSP and ESP cards. The inputs to the subprogram are specified as arguments of BSP, while the outputs of the subprogram are obtained as arguments of ESP. To use a subprogram, two other control statements, e.g., CSP and RSP, are used. The name of the subprogram is entered in the result column of the CSP card; the output of the subprogram will be demonstrated in a subsequent example.

Function Switch

The function switch is designed to generate logical variables. It is used as follows:

Result 10	Expression 19
XX	FSW(A,B,C,D)

The result XX is a logical variable that is equal to B, C, or D, depending on whether $A < 0$, $A = 0$, or $A > 0$, respectively.

Many other functions are defined for use by MIMIC, including, for instance, ZOH (zero-order hold), TDL (time delay), TAS (track and store). In addition to the many operational functions, there are input/output functions to generate outputs in printed or plotted form.

EXAMPLE 9.6-1

Develop a MIMIC program to simulate the computer control system shown in Figure 9.6-3.

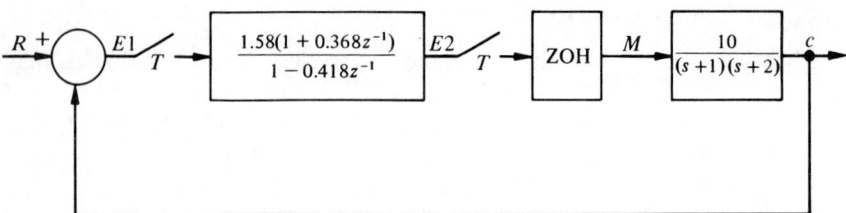

Figure 9.6-3. Computer control system.

It is clear that the operations contained in blocks defining the digital computer program and the zero-order hold are executed only once every sampling period. It would, therefore, be appropriate to introduce a subprogram for these operations and restrict its execution by a logical variable.

A suggested MIMIC program for the entire system which is complete except for initializing and input/output statements is given by

Logical Variable	Result	Expression
2	10	19
	DIG	BSP(R1)
	X	X+TSAM
	C2	C1
	R2	R1
	C1	1.58★R1−.58★R2−.418★C2
	DIG	ESP(C1,X)
	SAMPLE	FSW(T−X,FALSE,TRUE,TRUE)
SAMPLE	DIG	CSP(E1)
		RSP(M1)
	M2	5.★M1
	M3	FTR(M2,1.)
	C	FTR(M3,5.)
	E1	R−C
		FIN(T,5.)
		END

The logical control variable will be TRUE once every TSAM seconds, the length of the sampling period. The program is terminated by the FIN function, when T = 5 seconds.

This introduction to the digital simulation language MIMIC is very brief. It goes without saying that the study of a complete reference manual is required to use MIMIC intelligently. But the brief examples serve to point out the ease with which facility in the use of MIMIC can be reached. MIMIC and other languages like it represent a powerful means of digitally simulating systems.

9.7 Hybrid Computer Techniques and Applications in Simulation

Hybrid computers—computers integrally composed of analog and digital computers—have been developed since the late 1950's. The motivation for this type of computer is founded in the need for high-speed computation and extensive memory and logic capability in large-scale simulations. Utilized to best advantage in this combination are the high-speed simulation capability of the analog computer and the arithmetic, logic capability, and memory of the digital computer.

Cursory references were made to the use of a hybrid computer in previous chapters concerning the analysis of a sampled-data system. This section will give more complete coverage of the characteristics and applications of hybrid computers.

9.7-1 The Organization of a Hybrid Computer

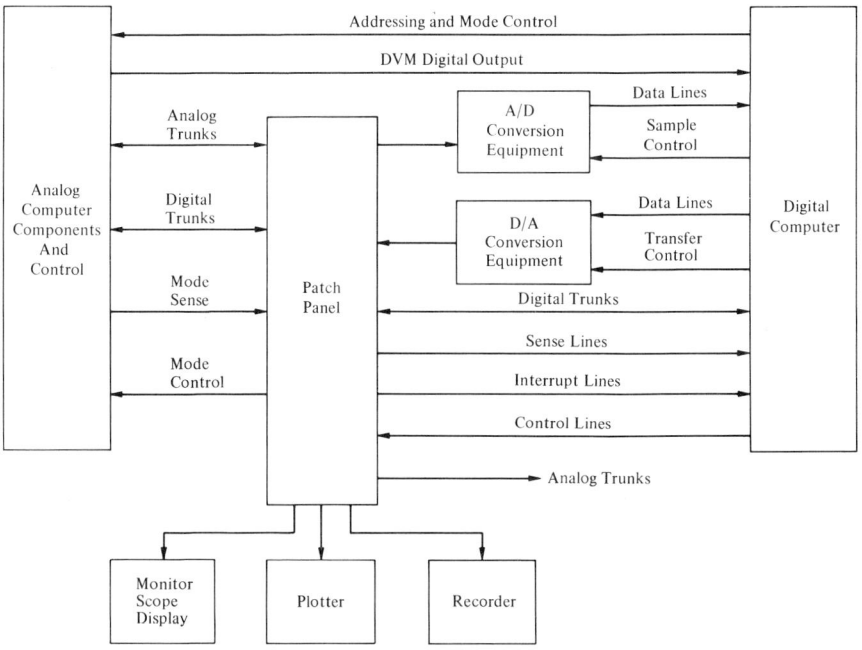

Figure 9.7-1. A typical hybrid computing system.

As is shown in Figure 9.7-1, the hybrid computer is composed of three major parts:

1. A general-purpose digital computer.
2. A general-purpose analog computer.
3. A linkage system to provide for exchange of data and control information between the computers.

We now describe these parts in more detail.

The Analog Computer

The basic computing elements of the analog computer consist of the components shown in Table 9.7-1. The computer is provided with electronic mode control of all time-dependent components, particularly the integrators. All three modes, RESET, HOLD, and COMPUTE, are controllable either by patchable digital logic timing circuits, by commands from the digital computer, or by manual pushbutton control. By means of special forced-

Table 9.7-1 A Selection of Hybrid Components and Their Symbols

No.	Symbol	Description
1		*Integrator:* Separately controlled through its OPERATE (O) and RESET (R) inputs. $(O, R) = (0, 1)$, RESET $e_0 = -e_i$ $(O, R) = (1, 0)$, OPERATE $e_0 = -\dfrac{1}{RC}\int \sum e_k dt$ $(O, R) = (0, 0)$, HOLD $e_0 =$ constant
2		*Comparator:* If $e_1 > e_2$, $E = 1$ ($\bar{E} = 0$) If $e_1 < e_2$, $E = 0$ ($\bar{E} = 1$)
3		*Switch:* If $E = 1$ ($\bar{E} = 0$), $e_0 = e_1$ If $E = 0$ ($\bar{E} = 1$), $e_0 = e_2$
4		*Track-store:* If $E = 1$, $e_0 =$ constant (OPERATE) If $E = 0$, $e_0 = e_1$ (RESET)

No.	Symbol	Description

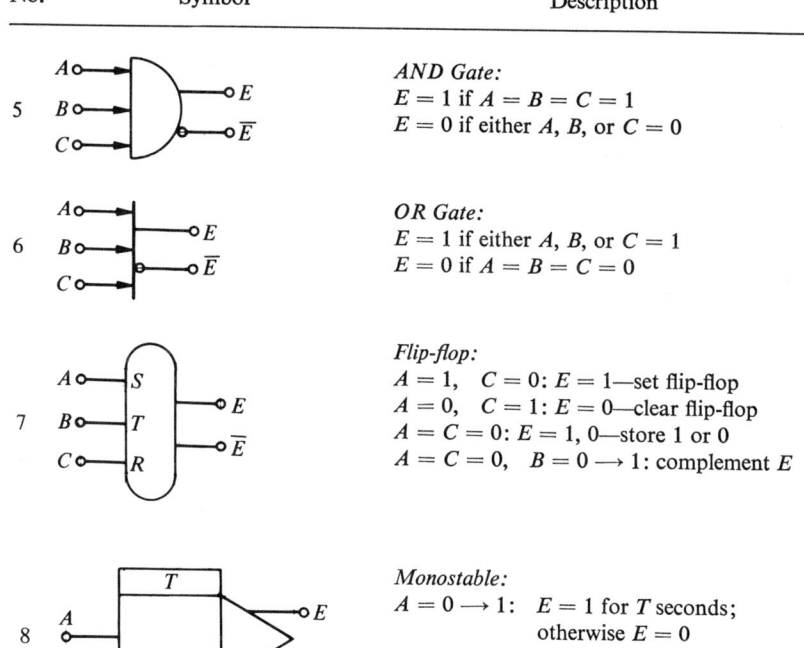

5. AND Gate:
$E = 1$ if $A = B = C = 1$
$E = 0$ if either A, B, or $C = 0$

6. OR Gate:
$E = 1$ if either A, B, or $C = 1$
$E = 0$ if $A = B = C = 0$

7. Flip-flop:
$A = 1$, $C = 0$: $E = 1$—set flip-flop
$A = 0$, $C = 1$: $E = 0$—clear flip-flop
$A = C = 0$: $E = 1, 0$—store 1 or 0
$A = C = 0$, $B = 0 \longrightarrow 1$: complement E

8. Monostable:
$A = 0 \longrightarrow 1$: $E = 1$ for T seconds; otherwise $E = 0$

charging circuits the integrators may be operated at high speed, cycling up to 10,000 times a second between the three modes. The mode selection of the integrators is carried out by logic signals when under the control of digital logic.

In addition to the electronic control of integrators, electronic switches are provided. The operation of such a switch is defined by the description of Table 9.7-1.

A special kind of hybrid computer component is the track-store circuit. Since it is foreign to a conventional analog computer, we shall describe its operation here. The track-store circuit is essentially an integrator under electronic mode control with input solely through the initial condition input terminal. When the integrator is in RESET mode it "tracks" the input, whereas in OPERATE it "stores." To illustrate the operation of a track-store unit, consider the waveforms of Figure 9.7-2. It shows a waveform $r(t)$ connected to the input of a track-store unit that is controlled by a timing signal E. The output $c(t)$ is seen in Figure 9.7-2(c).

Thus when $E = 1$ the unit is in STORE mode and when $E = 0$ the unit is in TRACK mode.

The voltage levels of the logic signals are usually such that logic 0 cor-

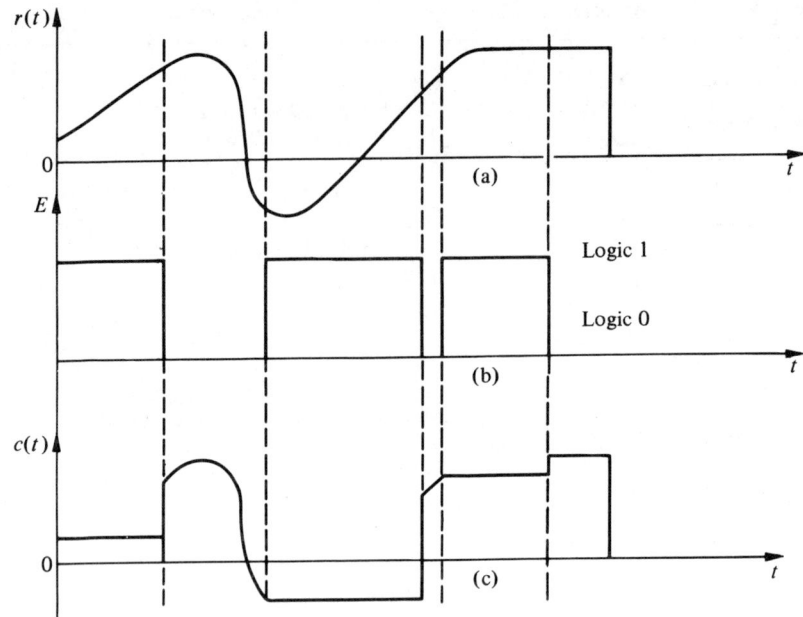

Figure 9.7-2. Track and store operation.

responds to ground level and logic 1 is given by a small positive potential such as 5 volts.

A very useful hybrid computing element is obtained by combining two track-hold units in cascade, as shown in Figure 9.7-3(a). The first unit is driven by logic signal E, while the second is driven by its complement $\bar{E}$. This element functions as an analog memory. Its usefulness can be further enhanced by feeding the output of the second track-hold into the input of the first, as shown by the diagram of Figure 9.7-3(b). In this connection it is an analog accumulator and is usually switched periodically. Thus it functions

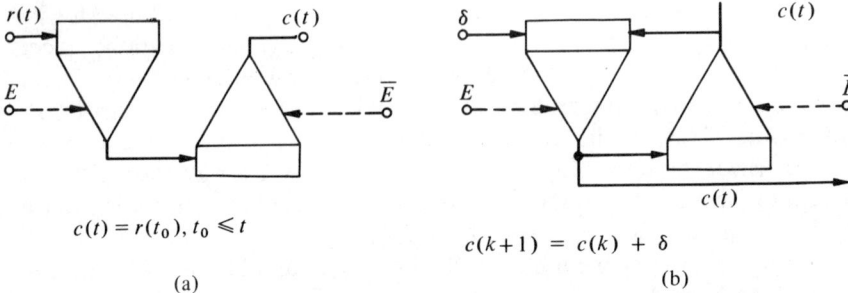

Figure 9.7-3. Cascaded track-store circuits: (a) analog memory; (b) analog accumulator.

according to the difference equation

$$c(k+1) = c(k) + \delta \qquad (9.7\text{-}1)$$

This is recognized as a discrete integrator.

One other feature worth noting is the use of servo-set potentiometers. All potentiometers are set to four-place accuracy by a remotely controlled servomotor. This arrangement makes it possible to use the digital computer to adjust the potentiometers, greatly enhancing the automatic capabilities of the hybrid computer.

The Digital Computer

The digital computer that is used in a hybrid computer installation may range from a small, 4096-word computer to as large a computer as one may require. Typically, however, the computer used is characterized by the following specifications:

1. Memory size of 8192 with 16- to 18-bit words.
2. Ability to interface with a large number of external devices.
3. Multilevel priority interrupt system, internal and external.
4. Extensive hardware capabilities to permit extended arithmetic operations in floating point.
5. Secondary memory such as discs to permit resident storage of library routines and compilers.
6. Medium volume input/output, such as high-speed paper tape reader and punch.

One of the most important factors determining the usefulness of the digital computer in a hybrid installation is the availability of sophisticated software. Of absolute necessity is a FORTRAN compiler. Next in importance are subroutines that consist of (1) arithmetic and mathematical functions, (2) format routines for numeric conversion, and (3) input/output routines to provide communication and control for all peripherals. There should be software packages specifically created for hybrid operation to permit real-time capability, time and delay statements, and complete control of and communication with the interface between analog and digital computer. Further, there are program setup and checkout routines.

Computers that fall into this category are the IBM 1800, CDC 1700, PDP-9, EAI-640, SDS 920, and many others.

The digital computer in a hybrid installation may be used solely in conjunction with the analog computer. On the other hand, recent developments in digital computer technology permit the time-sharing of the digital computer for a variety of uses. Thus, it may be possible to time-share a larger digital computer with a hybrid installation, provided that hybrid computer uses are granted a "foreground" priority, while all other uses during hybrid computation can be handled on a "background" priority. What this means is that the digital computer makes available all the time needed at the right moment to the hybrid installation; all other users must be satisfied with whatever time is left over from hybrid computation.

The main advantage of a time-shared hybrid computer installation would appear to be the availability of a larger digital computer at less cost. But it is questionable whether the priority requirements can be always satisfied.

A hybrid computer may be controlled by digital instructions or by analog patching. Briefly, these two modes of program execution control are described as follows.

Digital Control

The timing of all analog operations, such as HOLD, OPERATE, and RESET is under digital program control. Data transfer from the digital to the analog computer via digital-to-analog converters and from the analog to the digital computer via analog-to-digital converters is triggered through appropriate commands in the digital computer. The entire simulation is initiated and terminated by digital computer instructions. Under digital control the analog computer is completely slaved to the digital computer.

Analog Control

The timing for RESET, OPERATE, and HOLD modes is controlled through patchable logic circuits such as AND and OR gates, flip-flops, counters, delay elements, etc. Data transfer is triggered by counters or differential switches. The program simulating the discrete-time part of the system is executed periodically upon a receipt of a new data set from the analog-to-digital converter. Under analog control the digital computer is slaved to the analog computer.

Program Instructions for Digital Control

For the sake of general flexibility a hybrid computer is usually under digital control. For this purpose we define, then, the following program instructions. They represent only a partial list of all instructions needed to support sophisticated hybrid computations.

CALL   HOLD
    This call puts all integrators into HOLD mode; at the same time, all A/D converter channels sample the output voltage of the amplifiers to which they have been patched.

CALL   RESET
    This call puts all integrators into RESET mode.

CALL   OPERATE
    This call puts all integrators into OPERATE mode for a preselected

interval; at the end of this interval the integrators automatically assume a HOLD mode.

CALL POTSET (KPOT, VALUE)

This instruction sets the potentiometer identified as KPOT to the value VALUE.

HWRITE(I)X

This instruction transfers the contents of location X to the output of D/A converter channel I and holds it there.

HREAD(I)X

This instruction transfers the output of A/D converter channel I into location X.

These instructions are sufficient to develop a simple but complete program. The preparation of this program consists of two parts, the analog computer diagram and the digital computer instructions. An example to be introduced shortly will demonstrate this.

The Interface

The interface provides the link between the two computers. As was mentioned earlier, this linkage must provide for communication and control. As the computers work with physically different signals, digital and analog signals, the primary function of the interface is to provide a number of high-speed analog-to-digital and digital-to-analog channels through which data flow between the computers takes place. The intercomputer information flow is accomplished through A/D converters and D/A converters that operate with an accuracy variable from 6 to 14 bits including the sign bit. Table 9.7-2 shows the relative conversion accuracy that is possible in terms of percentile figures.

Table 9.7-2 Conversion Accuracy

	Bits	Powers of 2	Percentage Accuracy
	1	2	50%
	2	4	25
	3	8	12.5
	4	16	6.2
	5	32	3.1
	6	64	1.6
	7	128	.8
	8	256	.4
	9	512	.2
Normal	10	1024	.1
range	11	2048	.05
	12	4096	.02
	13	8192	.01

The analog computer is capable of an accuracy of approximately .01 percent. Thus to carry a conversion to 13 bits achieves maximum reasonable accuracy.

In addition to conversion accuracy, time required to perform the conversion is often of great importance. Typically, an A/D converter can handle 20,000 to 50,000 conversions per second of words ranging in length from 8 to 13 bits. Digital-to-analog converters operate at a speed of a few microseconds. Normally, A/D conversion of several analog signals is carried out by time-sharing a single A/D converter through the use of a multiplexer. This is shown schematically in Figure 9.7-4. The multiplexer selects the channel

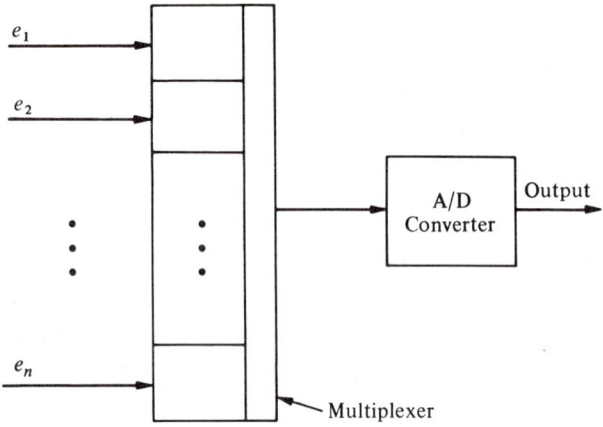

Figure 9.7-4. Time-sharing in A/D conversion.

whose input is to be converted. One problem that is encountered by the use of a multiplexer is that in the case of the conversion of a large number of analog signals, the consecutively generated digital numbers correspond to different moments of time because of the sequential operation. This time difference, called *time-skewing*, could be of significance in a given problem. One effective method of dealing with this problem is to provide sample and hold circuits for a selected number of analog signals and synchronize them. They may be still converted sequentially but may correspond to the same moment in time. The sample and hold amplifier precedes the multiplexer.

The computer linkage provides for intercomputer control for the purposes of timing, interrupt, input and output, and logic decisions during the course of a computer program.

The operation of the linkage can be under control of the analog computer or digital computer. Digital control is usually preferred because of greater flexibility and ease of programming.

Examples of Hybrid Computer Applications

Four examples will be presented here to illustrate the use of a hybrid computer in systems studies.

EXAMPLE 9.7-1

Design for Critical Damping. In the simple control system shown in Figure 9.7-5 the gain is to be automatically adjusted so that the system responds in a critically damped fashion to a step input. To accomplish this,

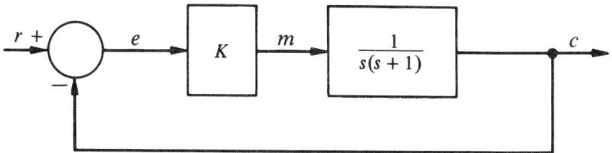

Figure 9.7-5. Gain-adjustable control system.

we shall use a hybrid computer, utilizing only the analog computer as controlled by patchable digital logic. The problem solution is actually fairly simple. The automatic gain adjustment is carried out by use of an electronic multiplexer, whose one input is e and whose other input is K, which is generated from analog accumulator as shown in Figure 9.7-6. It is automatically adjusted according to the iterative equation

$$K_{\text{new}} = K_{\text{old}} + \Delta K \tag{9.7-2}$$

at a frequency synchronized with the OPERATE-RESET cycle of the system simulated.

The parameter a is taken as a variable. A system that adjusts K each time a is changed could operate on the principle that if there is no overshoot during a particular COMPUTE cycle, one should increase the gain, and if there is overshoot, one should decrease the gain. Thus, the gain would tend to oscillate about a critically damped value. When a changes, the system will automatically adjust the gain to reach the new value, giving critical damping. A control circuit operating on this principle is shown in Figure 9.7-7.

At the beginning of each COMPUTE cycle, the storage device is cleared to indicate that no overshoot has been detected in this cycle. If the latching comparator detects overshoot at any time during the COMPUTE cycle, the storage device is set. The condition of the storage device at the end of the compute cycle determines whether the ΔK to be added during the RESET

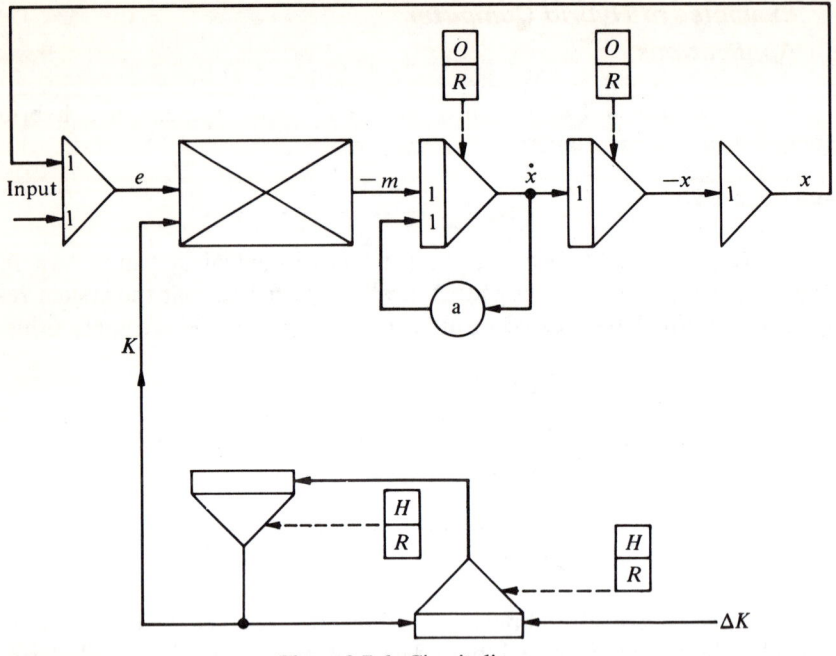

Figure 9.7-6. Circuit diagram.

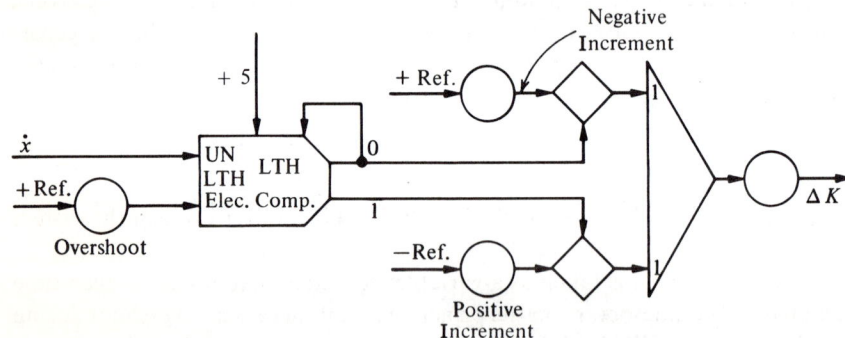

Figure 9.7-7. Control circuit using latching comparator and digital switches.

phase is positive or negative. Thus, the presence or absence of overshoot determines whether K is decreased or increased, respectively.

The OPERATE-RESET signals are timed by decade thumb wheel counters that may be dialed to give the timing for the integrators.

EXAMPLE 9.7-2

Simulation of a Sampled-data Control System. We consider the hybrid simulation of the computer control system shown in Figure 9.7-8. The system

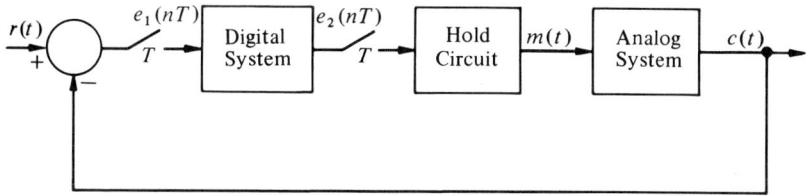

Figure 9.7-8. Digital control system.

is characterized as

Digital system:

$$e_2(nT) = 1.582e_1(nT) - .582e_1[(n-1)T] - .418e_2[(n-1)T] \quad (9.7\text{-}3)$$

Hold circuit:

$$m(nT + \tau) = e_2(nT) \quad \text{for } 0 \leq \tau < T \quad (9.7\text{-}4)$$

Analog system:

$$C(s) = \frac{1}{s(s+1)} M(s) \quad (9.7\text{-}5)$$

or

$$\frac{d}{dt}\begin{bmatrix} x_1 \\ x_2 \end{bmatrix} = \begin{bmatrix} -1 & 0 \\ 1 & 0 \end{bmatrix}\begin{bmatrix} x_1 \\ x_2 \end{bmatrix} + \begin{bmatrix} 1 \\ 0 \end{bmatrix} m(t) \quad (9.7\text{-}6)$$

$$c(t) = x_2$$

The analog parts of the system are simulated on the analog computer, requiring two integrators, two potentiometers, one D/A converter, and one A/D converter. Since the D/A converter incorporates a zero-order hold output, no special provision is made for this system element. The diagram for the analog part is shown in Figure 9.7-9.

The digital computer program is given next. The simulation is set up so as to allow digital control.

```
            DIMENSION
            READ (5,1) A,B,NDATA,TIMEFN
C           PARAMETERS FOR THIS RUN
        21  READ (5,2) GAIN,XIN
C           SET COEFFICIENTS
            CALL POTSET(1,GAIN)
            CALL POTSET(2,XIN)
C           INITIALIZE NEW RUN
            NRUN = NRUN+1
```

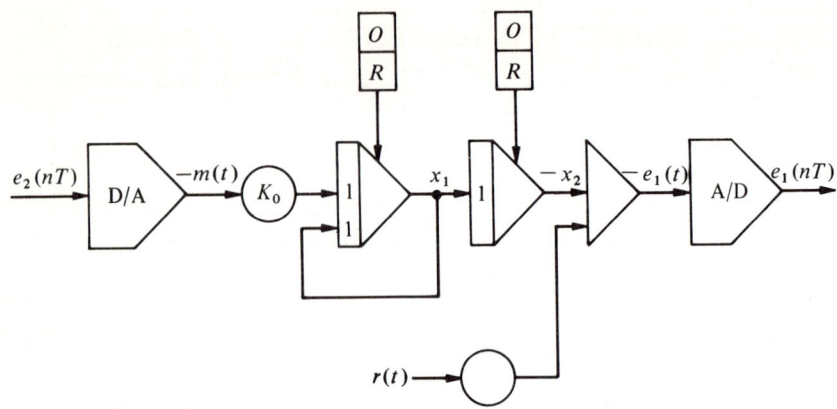

Figure 9.7-9. Analog computer diagram of hybrid simulation.

```
          TIME = 0.0
          DO 11 I = 1,3
          E1(I) = 0.0
          E2(I) = 0.0
    11    X(I) = 0.0
          CALL RESET
C         COMPUTE COMPUTER INPUT
    20    E1(2) = E1(1)
          HREAD(1,E1(1))
C         COMPUTE COMPUTER OUTPUT
          E2(2) = E2(1)
          E2(1) = 1.582*E1(1)-.582*E1(2)-.418E2(2)
          HWRITE(1,E2(1))
C         COMPUTE PLANT OUTPUT FOR NEW INTERVAL
C         NOTE: INTERVAL HAS BEEN SET TO VALUE OF SAMPLING
          PERIOD
          CALL OPERATE
C         PRINT OUTPUT
          TIME = TIME+T
          WRITE(6,3)  TIME,E1(1),E2(1)
C         ALL PERTINENT ANALOG VARIABLES ARE RECORDED BY PEN
          RECORDERS
C         TEST IF THIS RUN IS COMPLETED
          IF(TIME.LT.TIMEFN) GO TO 20
C         TEST IF A NEW RUN IS TO BE INITIATED
          IF(NRUN. LT. NDATA) GO TO 21
          STOP
```

EXAMPLE 9.7-3

Design of Digital Compensator. As in our last illustration of the use of hybrid computers in systems studies, we consider the design of a digital

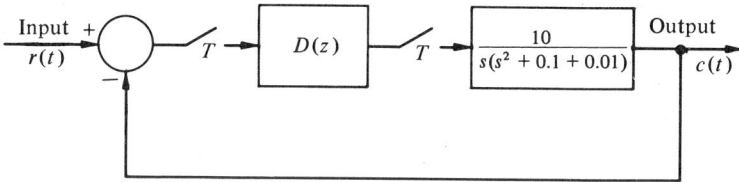

Figure 9.7-10. Digital control system.

compensator. It is required that the digital transfer $D(z)$ shown in Figure 9.7-10 be designed according to a weighted positive combination of the following:

1. Minimum overshoot.
2. Minimum rise time.
3. Zero steady-state error.
4. Minimum settling time.

All conditions are to be satisfied in response to a step input.

There exists no analytical design procedure that may be followed to realize these objectives. Nor is it at all established that the proposed system structure corresponds to the best configuration. Furthermore, the form of $D(z)$ is unknown. A design procedure that performs well under these adverse conditions is based upon a computer experimental approach as outlined below.

The form of $D(z)$ is arbitrarily taken as a second-order digital transfer function

$$D(z) = \frac{b_0 + b_1 z^{-1} + b_2 z^{-2}}{1 + a_1 z^{-1} + a_2 z^{-2}} \tag{9.7-7}$$

The system is simulated with a set of numerical values assumed for the coefficients of $D(z)$. A performance function is determined which measures the quality of the response relative to the design objectives. A good choice for this performance function might be

$$J = k_1(C_{\max} - R) + k_2(T_0) + k_3(C_\infty - R)^2 + k_4(T_{2\%}) \tag{9.7-8}$$

where $C_{\max}$ = output at time of maximum overshoot
R = magnitude of step input
T_0 = time at which output equals input for the first time
C_∞ = steady state output
$T_{2\%}$ = time at which output has settled to within 2% of input
k_i = weighting constants.

The four successive terms of the performance function (9.7-8) correspond to the four design objectives 1 through 4, respectively.

The general objective of the design procedure consists of adjusting the coefficients of the digital transfer function until the performance function reaches a minimum. The class of techniques that may be used for this purpose is generally referred to as *parametric optimization techniques*. During the past decade considerable research effort has been expended in the development of a great number of optimization techniques.* The use of these techniques has become feasible only recently because of the availability of high-speed digital computers.

The minimization technique used in this example is called *pattern search*.† This method begins by changing the parameters of $D(z)$ one at a time, starting at some arbitrary initial choice. The magnitude of the change is arbitrary, but is usually kept small. Associated with each parameter change is a complete simulation of the system and an evaluation of the performance function. Changing a parameter may result in an increase or decrease of J or, possibly, no change in J may occur. Should an increase be the result, the perturbation is repeated with a parameter change of opposite sign. When all parameters have been perturbed once or twice so that the performance function is reduced in each case, then all parameters are changed at the same time in the manner indicated by the individual changes. The process is then repeated. Thus it can be seen that the pattern search method alternates between perturbing the parameters individually to determine a "direction" or pattern and moving in that direction with all parameters. Depending on how well a new direction compares with a previous one, the successive pattern moves may increase or decrease the magnitude of the parameter adjustments. This procedure, therefore, tends to learn as it goes along and is generally quite effective in adapting to problem peculiarities.

In employing an optimization technique such as the one described above, the overall problem solution is under the control of the optimization program, which, of course, is programmed on the digital computer. The organization of the entire program is shown by the flow chart in Figure 9.7-11. It shows the simulation and the evaluation of the performance function as being separate subroutines of the overall program.

The simulation of the system is carried out by employing the digital computer for the evaluation of the recursion equation and the analog computer for the simulation of $G(s)$. The system simulation as indicated by the flow chart of Figure 9.7-12 requires the preparation of a **FORTRAN** program for the execution of $D(z)$ and control of simulation. The detailed instructions of the program are very similar to Example 9.7-2 and therefore are not shown here.

*P. E. Fleischer, "Optimization Techniques," pp. 175–216, *System Analysis by Digital Computer*, edited by F. F. Kuo and J. F. Kaiser, Wiley, 1966, New York.
†R. Hooke and T. A. Jeeves, "Direct Search Solutions of Numerical and Statistical Problems," *Journal Assoc. Comp, Mach.*, Vol. 8, April 1962, pp. 212–229.

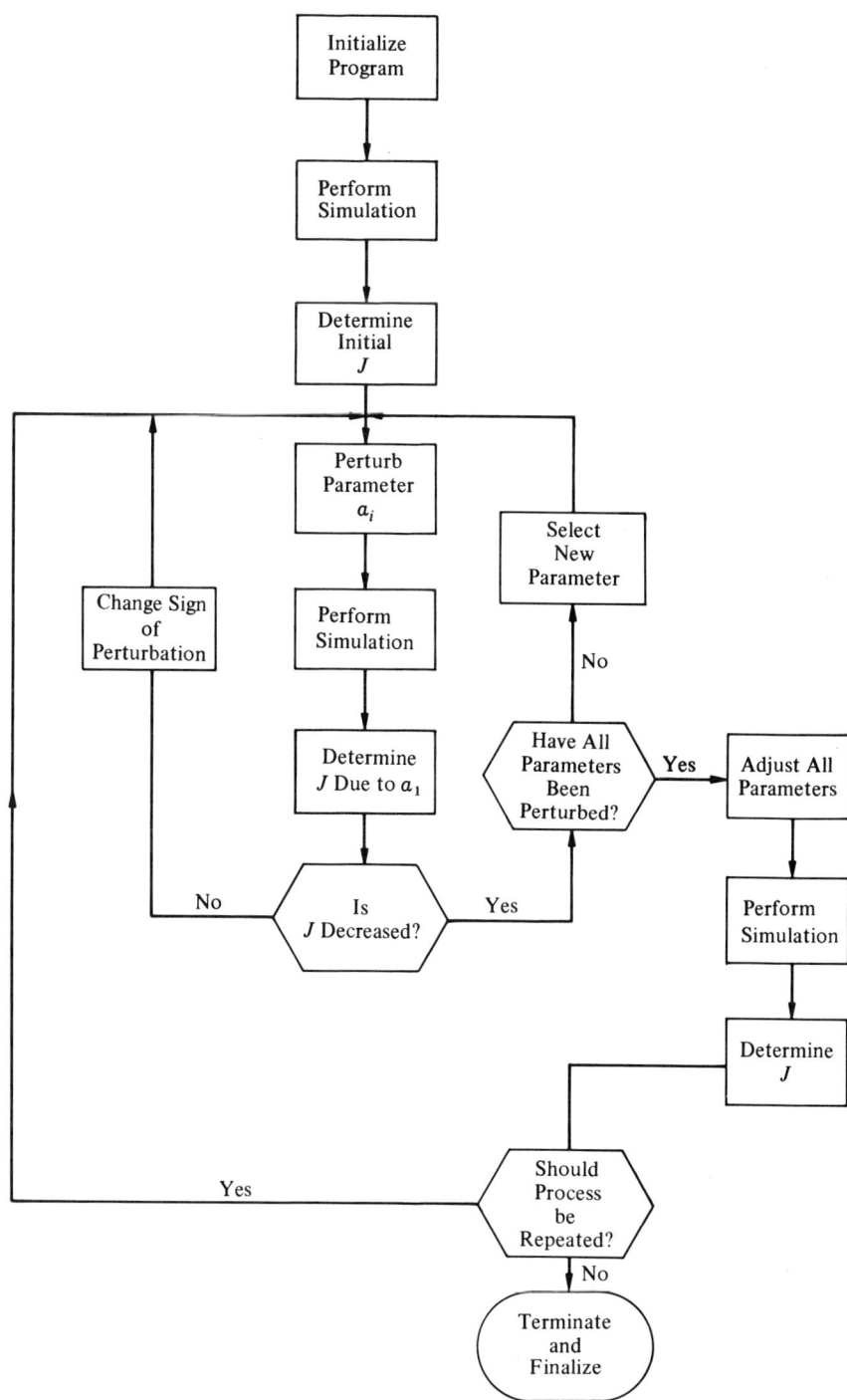

Figure 9.7-11. Flow chart of optimization program.

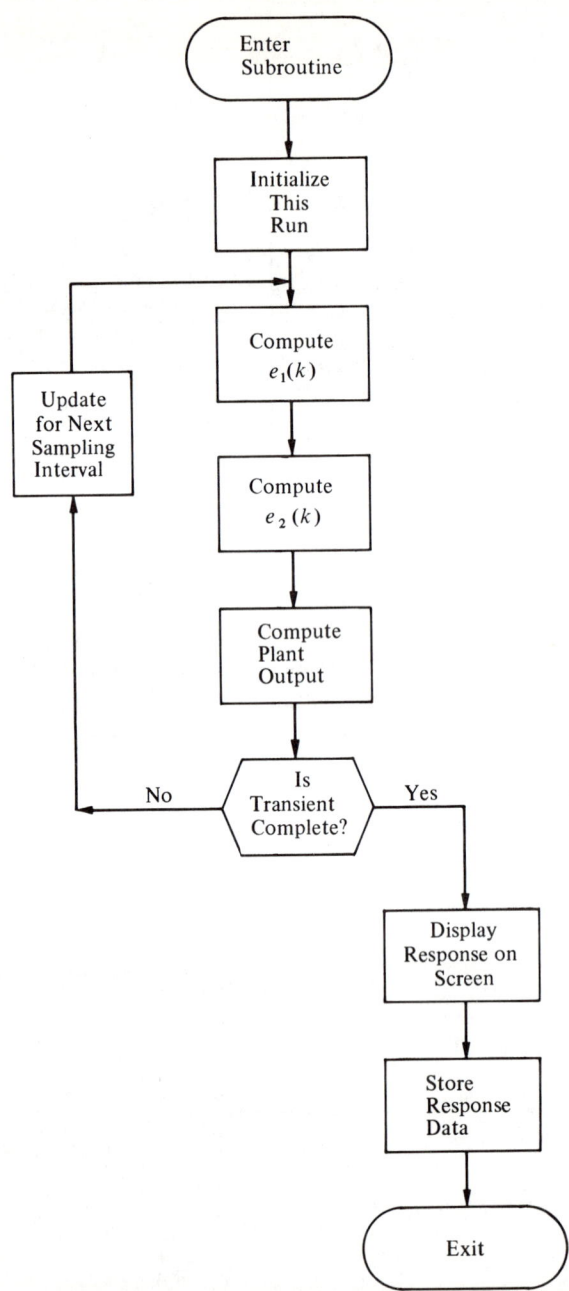

Figure 9.7-12. Flow chart of simulation.

It is reasonable to assume that all digital calculations during the simulation will not require more than 200 microseconds per sampling time. If the sampling time is .1 second and the analog computer operates at a speed of 100 times real time, then it requires 1 millisecond. Thus for each sampling interval 1.2 milliseconds are required. Suppose also that the transient requires 10 seconds or 100 sampling intervals; then the entire simulation requires 120 milliseconds. Thus each parameter perturbation uses up 120 milliseconds. If now eight simulations are required for each major cycle—seven parameter perturbations, one for each parameter plus two extra ones for wrong directions, and one all-parameter adjustment—one full second is used up. During a typical design of the digital compensator possibly from 50 to 100 complete iterations are required. We allow, furthermore, a negligible time of 1 millisecond for other digital computations for the performance functions and general bookkeeping. All factors considered, it is reasonable to expect a complete design every one to two minutes. This amazing speed, of course, is one of the attributes of hybrid computation.

EXAMPLE 9.7-4

A Sampled-data System with Minimum Settling Time. Design and test a digital controller that will permit the system shown in Figure 9.7-13 to respond to step inputs with zero steady-state error in a minimum number of sampling periods. The test will consist of a simulation on a hybrid computing system.

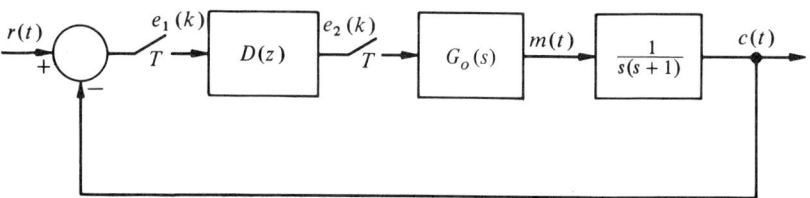

Figure 9.7-13. Computer control system.

The method of design for $D(z)$ is presented in Chapter 7. It is shown that the form of $D(z)$ is a ratio of polynomials of equal degree N. The degree N is equal to the order n of the transfer function of the plant if the plant does not contain a free integrator; $N = n - 1$ if the plant contains a free integrator.

The minimum settling system is simulated on a PDP-8/TR-20 hybrid computing system. The timing for the sampling intervals is under the control of an external timing clock. To carry out the multiplications required in the program in floating point, a number of system subroutines are utilized.

Solution. For the case at hand, the digital transfer function is derived to be

$$D(z) = \frac{F_1 + F_2 z^{-1}}{1 + F_3 z^{-1}} \qquad (9.7\text{-}9)$$

where

$$F_1 = \frac{1}{T(1 - e^{-T})}, \qquad F_2 = -\frac{e^{-T}}{T(1 - e^{-T})}$$

$$F_3 = \frac{(1 - e^{-T}) - Te^{-T}}{T(1 - e^{-T})} = F_1 + (1 + T)F_2$$

For $T = .1$ sec   For $T = 2$ sec   For $T = 1$ sec

$F_1 = 105.2$ $F_1 = .519$ $F_1 = 1.58$

$F_2 = -95.3$ $F_2 = -.0699$ $F_2 = -.577$

$F_3 = .37$ $F_3 = .3093$ $F_3 = .418$

For $T = 1$, the linear recursion equation to be programmed is

$$e_2(k) = 1.58 e_1(k) - .58 e_1(k-1) - .418 e_2(k-1) \qquad (9.7\text{-}10)$$

In order to program the above recursion equation, it is advisable to scale the variables so that overflows and underflows do not occur.

To scale, first find the smallest power of 2 larger than or equal to the maximum value of the variables.

$$e_2(k) = F_1 e_1(k) + F_2 e_1(k-1) - F_3 e_2(k-1)$$

$$e_2'(k) 2^{Q_4} = F_1' 2^{Q_1} e_1(k) + F_2' 2^{Q_2} e_1(k-1) - F_3' 2^{Q_3} e_2(k-1) 2^{Q_4}$$

$$e_2'(k) = [F_1' e'(k)] 2^{Q_1 - Q_4} - [F_3' e_2'(k-1)] 2^{Q_3} + [F_2' e_1(k-1)] 2^{Q_2 - Q_4}$$

Each of the terms in square brackets is now scaled. Multiplication by a power of 2 in a binary computer is handled by a shift to the left by as many times as the power (negative power). We select the scaled factors such that the scaled quantities fall within the range (.5, 1.0). Thus

$$|e_1(k)'| \leq 1$$
$$|e_2(k)| \leq 2^{Q_4}$$
$$|F_1| \leq 2^{Q_1}$$
$$|F_2| \leq 2^{Q_2}$$
$$|F_3| \leq 2^{Q_3}$$

so that

$$F_1 = F'_1 2^{Q_1}$$
$$F_2 = F'_2 2^{Q_2}$$
$$F_3 = F'_3 2^{Q_3}$$
$$e_2(k) = e'_2(k) 2^{Q_4}$$

Using the above definition, we replace the unscaled variables in the equation.

For $T = 1$,

$$|F_1| = 1.58 \leq 2^1$$
$$|F_2| = |-.577| \leq 2^0 = 1$$
$$|F_3| = |.418| \leq 2^{-1}$$
$$|e_2(k)| \leq 2^1$$

$Q_1 = 1$ $\qquad F'_1 = .79$
$Q_2 = 0$ $\qquad F'_2 = -.577$
$Q_3 = -1$ $\qquad F'_3 = .836$
$Q_4 = 1$ $\qquad e'_2(k) = \dfrac{e_2(k)}{2}$

The program is now ready to be implemented on a digital computer. A flow chart of the program is shown in Figure 9.7-14. Note that for $k = 1$,

$$F1*E1(K) = F3*E2(K-1)$$

Therefore, this subtraction occurs first.

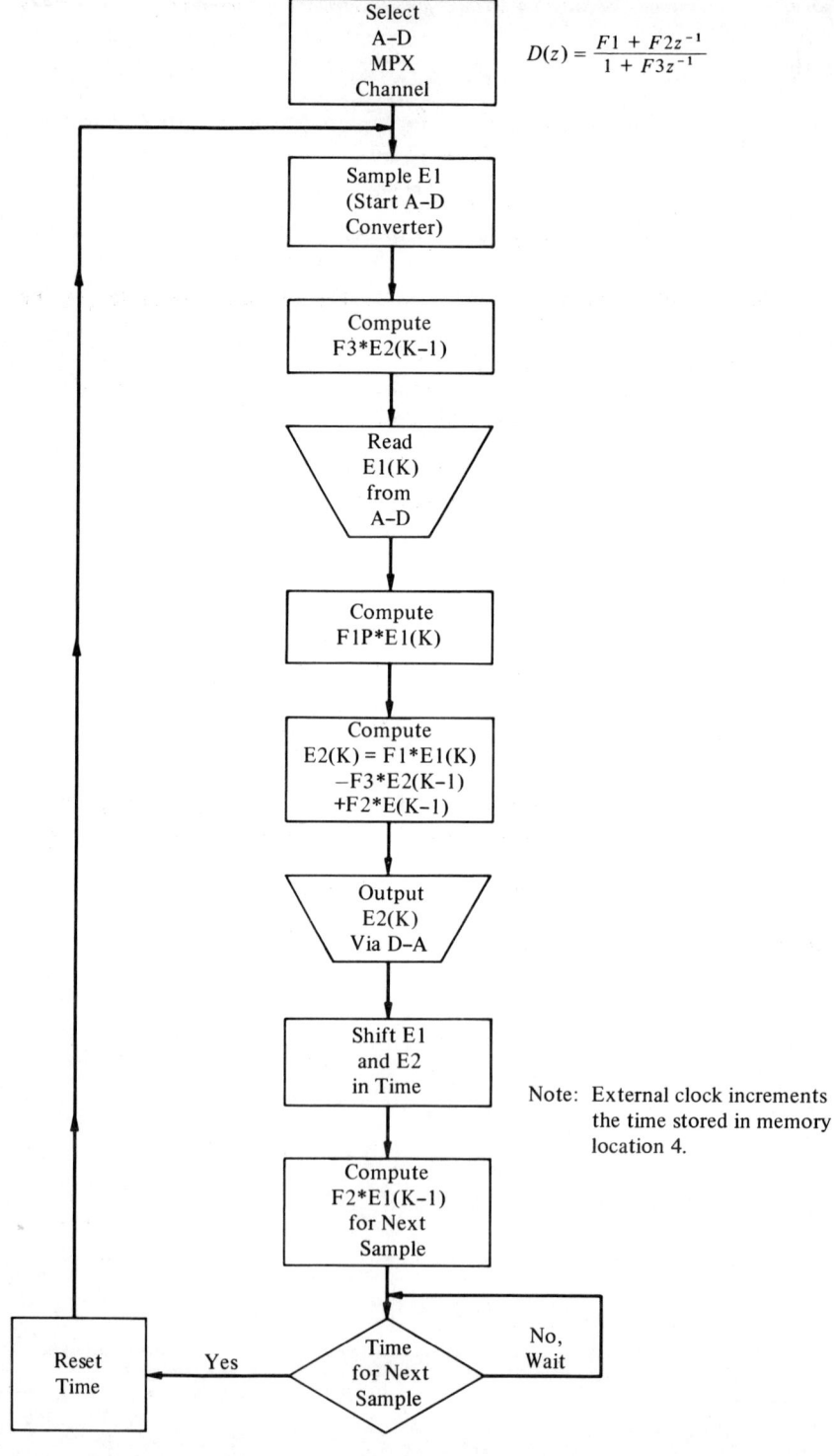

Figure 9.7-14. Flow chart of program.

The program is coded for execution on the PDP-8.* The language shown is the computer's symbolic assembler. The listing of this program is shown in Table 9.7-3. A block diagram of the complete simulation is shown in Figure 9.7-15, where the plant is simulated by utilizing analog computer elements.

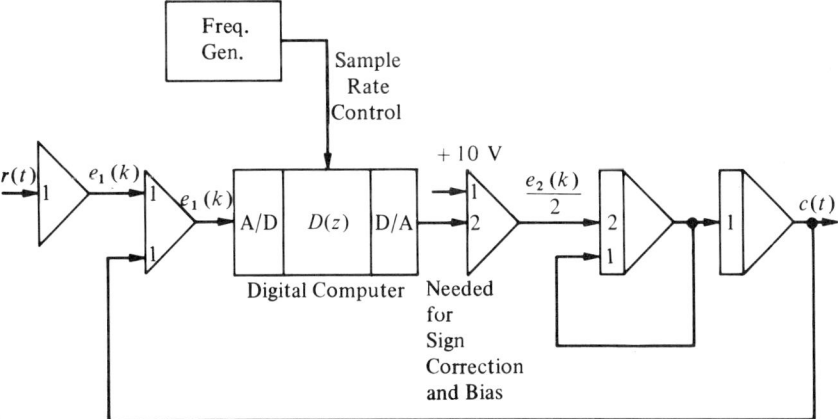

Figure 9.7-15. Block diagram of simulation.

Table 9.7-3 PDP-8 Listing of Program

```
*4
TIME,     -1              /STORE TIME
*1000
STRT,     CLA             /AC←Ø
          ADCC            /CHANNEL Ø
GO,       ADCV            /CONVERT E1(K)
          TAD   E2K1      /E2(K-1)
          JMS   MULT      /F3P*E2(K-1)
F3P,      3260            /.418*2 = F3*2↑(-Q3)
          DCA   SAV3      /MOST SIG. PART
          TAD   MP1
          DCA   SAV3+1    /LEAST SIG. PART
          TAD   M3        /-M3 = NO. OF LEFT SHIFTS
          SMA             /IS IT A LEFT SHIFT?
          JMP   .+4       /NO, M3>OR = Ø, SHIFT IN WRONG DIRECTION,
                           SKIP
          JMS   DPSL      /SHIFT PROD LEFT FOR SCALING
          SAV3
          DCA   SAV3
          CLA
          ADRB            /READ E1(K) FROM A-D CONVERTER
          DCA   E1K       /SAVE IT
```

*See Section 9.8.

Table 9.7-3 (cont.)

```
            TAD   E1K
            JMS   MULT      /F1P*E1(K)
F1P,        3121            /1.58*1/2 = F1*2↑(-Q1)
            DCA   SAV1
            TAD   MP1       /GET LEAST SIGNIFICANT PART
            DCA   SAV1+1
            TAD   M1
            SMA             /IS IT A LEFT SHIFT?
            JMP   .+4       /NO, SKIP
            JMS   DPSL      /SHIFT LEFT M1 PLACES
            SAV1
            DCA   SAV1      /SAVE RESULT
            CLA             /CLEAR AC IN CASE ABOVE COMMAND SKIPPED
            TAD   SAV3      /SAV3
            CIA             /-SAV3 = -F3P*E2(K-1)
            TAD   SAV1      /SAV1-SAV3 = F1P*E1(K)-F3P*E2(K-1)
            TAD   SAV2      /SAV1-SAV3+SAV2 = F1P*E1(K)-F3P*E2(K-1)
                            +F2P*E1(K-1)
            6551            /OUTPUT E2P(K) VIA D-A CONVERTER
                            /OUTPUT SHOULD BE MULTIPLIED BY 2↑Q4 TO
                            /CORRECT FOR SCALING IN DIGITAL COMPUTER
            TAD   E2PK      /SHIFT E2
            DCA   E2K1      /IN TIME
            TAD   E1K       /SHIFT E1
            DCA   E1K1      /IN TIME
            TAD   E1K1      /E1(K-1)
            JMS   MULT      /F2P*E1(K-1)
F2P,        -2235           /(-.577)*1 = F2*(2↑-Q2)
            DCA   SAV2      /MOST SIG. PART
            TAD   MP1
            DCA   SAV2+1    /LEAST SIG. PART
            TAD   M2        /-M2 = NUMBER OF RIGHT SHIFTS
            SMA             /IS IT A RIGHT SHIFT?
            JMP   .+4       /NO, SKIP
            JMS   DPSR      /SHIFT PROD. RIGHT FOR SCALING
            SAV2
            DCA   SAV2      /STORE RESULTS
            CLA             /AC←Ø IN CASE ABOVE COMMAND SKIPPED
            TAD   TIME      /AC←Ø+TIME
            SPA             /IS IT TIME FOR NEW SAMPLE?
            JMP   .-3       /NO. WAIT
            CLA   CMA       /YES, AC←-1
            DCA   TIME      /RESET TIME
            JMP   GO
E1K,        Ø               /E1(K)
E1K1,       Ø               /E1(K+1)
E2PK,       Ø               /E2'(K)
E2K1,       Ø               /E2'(K+1)
SAV1,       Ø
            Ø
```

Table 9.7-3 (cont.)

```
SAV2,         0
              0
SAV3,         0
              0
M3,           0          /-(Q3+1),  Q3 IS LEFT SHIFTS
M2,           0          /-(Q4-Q2-1), Q2-Q4 IS LEFT SHIFTS
M1,          -1          /-(Q1-Q4+1), Q1-Q4 IS LEFT SHIFTS
                         /CONSTANT 1 ABOVE IS DUE TO MULTIPLY
                          SUBROUTINE
/SUBROUTINES MULT AND DPSL AND DPSR MUST BE ADDED
/
PAUSE
```

Hybrid computers, their development and application are just beginning to assume a significant role in the field of systems analysis and design. Examples 9.7-1 through 9.7-4 cited above show but a limited sample of the potential of hybrid computers.

9.8 Computer Control

Throughout the chapters of this text we have presented techniques for the analysis and design of systems containing a digital computer or digital processor as an integral element. It is reasonable to expect that the reader at this point has a fair appreciation of the underlying principles and engineering considerations of utilizing a digital computer in a system. But if some of this textbook knowledge were to be put to a real test, one which involves hardware and software design, chances are only slight that we would be fully equipped to master such a task. This section, then, is designed to help us improve our understanding of the design of a computer control system. We aim to accomplish this by examining three examples of computer control applications where reference is made to a specific computer and its features. This computer is a DEC PDP-8 computer. A brief description of the specific facility used in the examples is given here to present its salient characteristics.

The computer has 4096 12-bit words. Its memory operates on a cycle time of 1.5 microseconds. It has an external program interrupt and a bussed input-output system. Attached to it are an 11-bit plus sign A/D converter and a multiplexer. The A/D converter is capable of performing a 12-bit conversion in 35 microseconds. The analog voltage must be within the voltage range 0 to -10 volts; it must be scaled and shifted to accept a wider range. Also provided is a D/A converter of comparable accuracy with a conversion speed of 3.75 microseconds. All converters are equipped with individual 12-bit registers for use as buffers. This enables the A/D converter to sample and hold an analog signal during conversion, while the D/A converter uses its buffers as zero-order-hold outputs.

The computer has been modified with the addition of a special input-output channel whereby, on a programmable basis, a binary word may be transferred between the accumulator and an external binary register in both directions. This I/O channel is useful in exchanging binary information directly with external devices for such purposes as reading a shaft encoder, sending a binary word to a logic circuit, supplying binary coded control instructions, etc.

The PDP-8 is programmable using an assembler language. Examples of programs will be illustrated by detailed flow charts. It can also be programmed in FORTRAN, although this is very limited because of the small memory.

The PDP-8 is a typical representative of a rapidly growing class of small-scale, almost desk-top size, digital computers, which may be used in a variety of on-line systems and data-processing applications.

The examples to be presented involve the PDP-8 as a control computer. The first case treats the computer as a passive discrete systems element which is to evaluate a linear recursion equation periodically. In this case it is used to provide digital compensation through integration. The second problem places a higher demand on the digital computer by using its computational and logic capabilities in the control of an attitude stabilization system. The third example consists of a digital control system using a digital PID (Position-Integral-Derivative) controller. The PID controller is designed by means of a simulation.

EXAMPLE 9.8-1

Digital Compensation through Integration. The system to be considered is shown in Figure 9.8-1. It consists of a single-loop feedback system. The

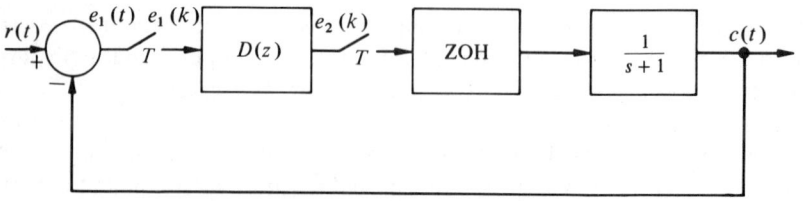

Figure 9.8-1. Digital control system.

digital computer is to control the plant through a zero-order hold such that the error is driven to zero for step input commands. This desired operating characteristic imposes the requirement that the system should be of type 1; i.e., one free integrator should be present in the loop. It is planned to let the digital computer perform this operation. It could, therefore, be programmed according to the recursive relation

$$e_2(k) = e_2(k-1) + e_1(k) \qquad (9.8\text{-}1)$$

We next check whether this recursive relationship will drive the steady-state error to zero. To this effect we apply the final value theorem of the z-transform.

The plant and the zero-order hold are characterized by the transfer functions

$$G_0 G(s) = \frac{1 - e^{-Ts}}{s} \frac{1}{s+1} \qquad (9.8\text{-}2)$$

while its z-transform is given by

$$\mathcal{Z}\{G_0 G(s)\} = (1 - z^{-1}) \mathcal{Z}\left\{\frac{1}{s(s+1)}\right\}$$

$$= (1 - z^{-1}) \mathcal{Z}\left\{\frac{1}{s} - \frac{1}{s+1}\right\}$$

$$= (1 - z^{-1})\left[\frac{1}{1 - z^{-1}} - \frac{1}{1 - e^{-T}z^{-1}}\right]$$

$$= \frac{z^{-1}(1 - e^{-T})}{(1 - z^{-1})(1 - e^{-T}z^{-1})} \qquad (9.8\text{-}3)$$

The error is igven by

$$E_1(z) = \frac{R(z)}{1 + D(z)G_0 G(z)}$$

Since $D(z)$ is given by

$$D(z) = \frac{1}{1 - z^{-1}}$$

$$e_1(\infty) = \lim_{z \to 1} (z - 1) E_1(z)$$

$$= \lim_{z \to 1} (z - 1) \frac{1}{1 + \dfrac{z^{-1}(1 - e^{-T})}{(1 - z^{-1})(1 - z^{-1}e^{-T})}} \frac{1}{1 - z^{-1}}$$

Evaluating the limit yields

$$e_1(\infty) = 0$$

Indeed, we can conclude that the system will behave like a type 1 system with the digital integration. The characteristic equation is given by

$$1 + D(z)G_0G(z) = 0 \qquad (9.8\text{-}4)$$

or, upon substitution and simplification,

$$z^2 - 2e^{-T}z + e^{-T} = 0 \qquad (9.8\text{-}5)$$

It has the characteristic roots

$$z_{1,2} = e^{-T} \pm j\sqrt{e^{-T} - e^{-2T}} \qquad (9.8\text{-}6)$$

It is interesting to show the derivation of the state transition equation for the closed-loop system for the benefit of comparison.

The transition equation for $G(s)$ is

$$x(k+1) = e^{-T}x(k) + (1 + e^{-T})e_2(k) \qquad (9.8\text{-}7)$$

The digital computer equation is given by (9.8-1). To develop the transition equations we must convert the digital recursion equation from its present form into a state equation form. By the methods of Section 2.6 we have

$$\begin{aligned} d(k+1) &= d(k) + e_1(k) \\ e_2(k) &= e_1(k) + d(k) \end{aligned} \qquad (9.8\text{-}8)$$

Combining (9.8-8) and (9.8-7) and using the relation

$$e_1(k) = r(k) - x(k) \qquad (9.8\text{-}9)$$

we obtain the closed-loop transition equations

$$\begin{bmatrix} x(k+1) \\ d(k+1) \end{bmatrix} = \begin{bmatrix} 2e^{-T} - 1 & 1 - e^{-T} \\ -1 & 1 \end{bmatrix} \begin{bmatrix} x(k) \\ d(k) \end{bmatrix} + \begin{bmatrix} 1 - e^{-T} \\ 1 \end{bmatrix} r(k) \qquad (9.8\text{-}10)$$

The characteristic roots of these equations are given by

$$\begin{vmatrix} \lambda - 2e^{-T} + 1 & e^{-T} - 1 \\ 1 & \lambda - 1 \end{vmatrix} = 0$$

or

$$\lambda^2 - 2e^{-T}\lambda + e^{-T} = 0$$

which is identical to (9.8-5).

The investigation of the stability of the system as a function of the sampling time T reveals an interesting fact. Figure 9.8-2 shows a root-locus plot

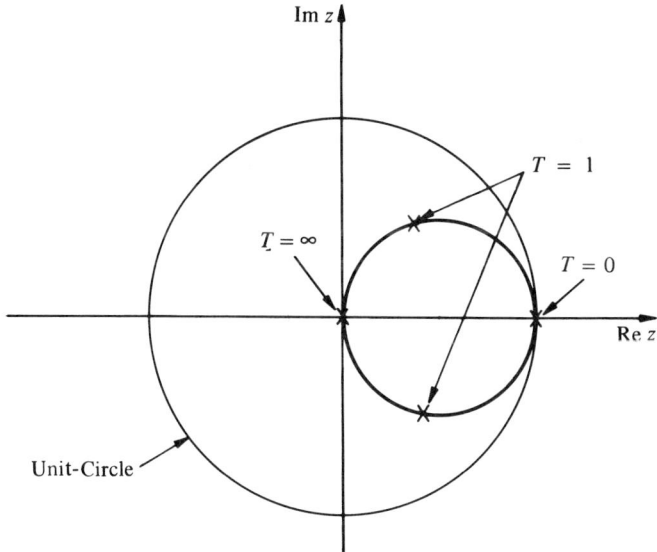

Figure 9.8-2. Root-locus with T as parameter.

of the system with T as a parameter. This is determined by calculating the roots of (9.8-5). This plot shows the location of the two roots to be starting at the origin for $T = \infty$ and migrating to the point $(1, 0)$ for $T = 0$ on two semicircles. This is an interesting result, for it indicates that the response of the system becomes more and more oscillatory as T decreases. This result may be surprising. However, a closer look at the digital integrator quickly reveals that the gain of the system increases in direct proportion to a decrease in the sampling rate. To maintain the same gain, the digital transfer function should be modified to

$$D(z) = \frac{Tz}{z - 1} \qquad (9.8\text{-}11)$$

The analytical aspects of this problem have now been investigated. We next carry out the design of the digital computer program that will implement the recursion equation (9.8-1). A flow chart for this program is presented in Figure 9.8-3. It is sufficiently detailed to show the necessary machine language instructions.

The program begins with an instruction to clear the accumulator and load the multiplex channel number into the accumulator. This may be any one of channels 1 through 24. The next instruction starts the conversion of the analog signal connected to the selected channel to a digital number. The analog number is sampled and held. The conversion requires 35 microseconds when carried to 12-bit accuracy, during which time the computer may carry out

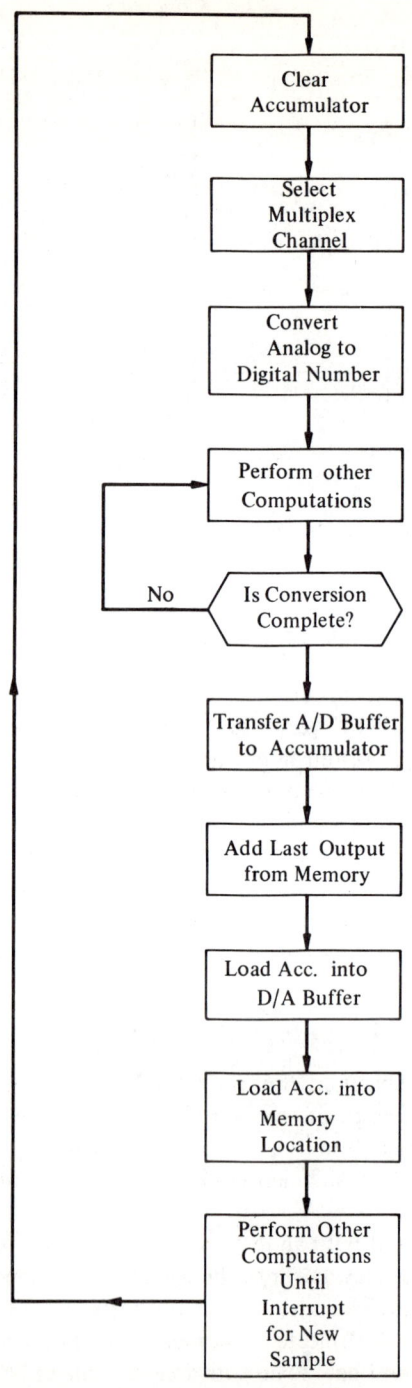

Figure 9.8-3. Flow chart for $D(z)$ program.

other computations independent of the conversion. When the conversion is complete, the contents of the A/D buffer are transferred to the accumulator, representing $e_1(k)$, the new input. To this is added the old output $e_2(k-1)$, resulting in the new output $e_2(k)$, which is stored in memory and also loaded into the D/A converter. Loading the D/A buffer requires 3.75 microseconds, while the D/A output network settles within 3 microseconds. Thus the digital-to-analog conversion is complete as soon as the load instruction is executed. The computer may now perform other functions until an interrupt occurs from an external clock to indicate the need for a new sample and a repetition of the entire cycle. The external clock gives this interrupt command at time intervals that must be larger than the total execution time of the cycle, which is 67 microseconds.

A diagram for the entire system simulation is shown in Figure 9.8-4. An analog computer was used to carry out the simulation of the continuous parts

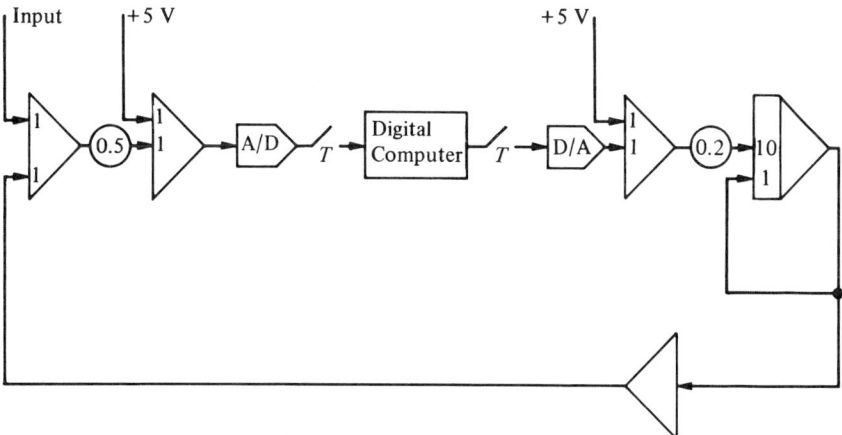

Figure 9.8-4. Complete system simulation.

of the system. Two operational amplifiers are used to perform scaling and shifting of the analog signals before and after the converters. This is needed to change the converter input and output levels (0 to -10 volts) to the analog computer voltage (± 10 volts).

EXAMPLE 9.8-2

Computer Control of Spacecraft Attitude. The second illustration of a computer control system deals with the utilization of the PDP-8 computer (or a computer like it) in the control of the attitude of a spacecraft. For the sake of simplicity we restrict the control requirements to one dimension only; thus the proposed problem can be described schematically as shown in Figure 9.8-5. The principal function of the computer is the determination of the prop-

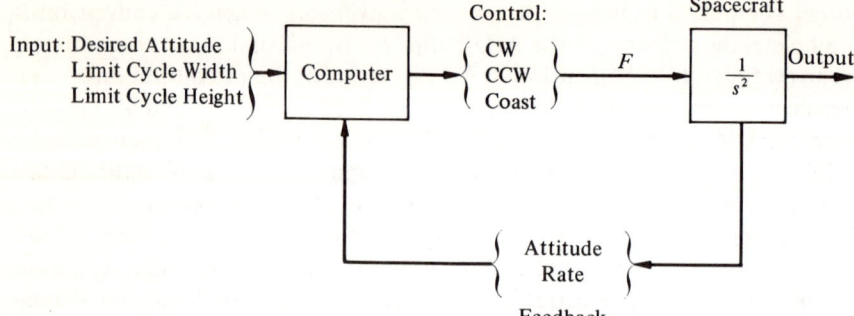

Figure 9.8-5. Attitude control by computer.

er cycling and timing of the control commands, which consist of (1) clockwise ($F = +1$), for which the thrusters are turned on to produce a clockwise moment; (2) counterclockwise ($F = -1$), for which the thrusters are turned on to produce a counterclockwise moment, (3) coast ($F = 0$), for which no thrusters are engaged and the vehicle is rotating at constant angular velocity.

The CW thrusters and the CCW thrusters are controlled by two separate signals which are called FP1 and FM1, respectively, in the computer program. Thus, the control commands are issued by the computer according to the following schedule:

F	FP1	FM1
+1	1	0
−1	0	1
0	0	0

where FP1 and FM1 are names of program variables.

For the determination of its control outputs the computer is provided with input information and feedback information. Since the space vehicle's attitude is controlled by maintaining a limit cycle, the input information consists of the desired attitude angle and rate, the width of the limit cycle, and the height of the limit cycle. These quantities are further defined by the schematic of Figure 9.8-6. A_1 and A_2 define desired attitude angle and attitude rate, respectively. The dimensions of the limit cycle are given by B_1 and B_2. The required feedback information is provided by two sensors measuring angular position and angular rate.

9.8-1 Design of Computer Program

The design of a computer control system requires the development of a computer program capable of satisfactorily managing all demands placed

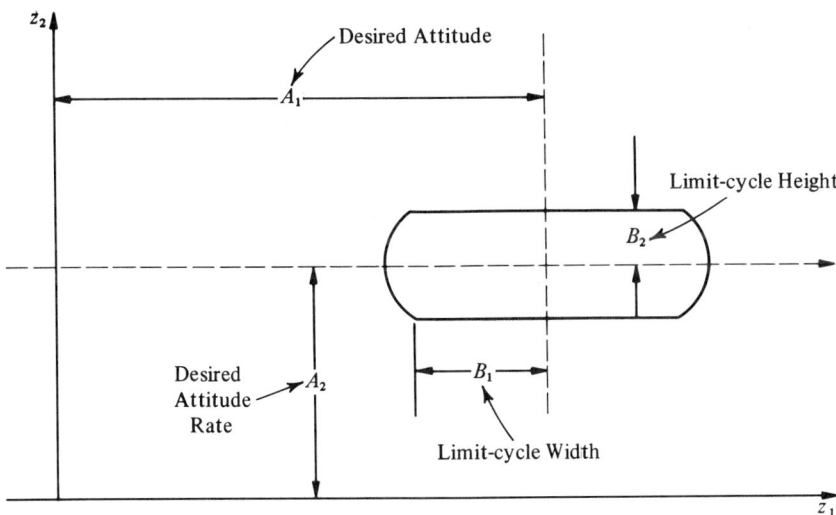

Figure 9.8-6. Limit-cycle operation specifications.

before it. The development of a "bug-free" program is absolutely critical. It is helpful, therefore, if this task can be approached in a methodical manner, although no single programming problem has a unique solution.

In the design of the computer program for the attitude control system two approaches will be demonstrated. The first approach is based upon the technique of flow-charting while the second approach utilizes logic circuit theory. Both programs, of course, yield the same result.

We first consider some relations that are common to both approaches. In considering the requirements of the control problem, it can be established that the choice of the control commands is entirely dependent on: (1) the state of the plant relative to the desired state, and (2) the dimensions of the limit cycle. The two-dimensional state space can be divided into nine regions, as shown in Figure 9.8-7. The dividing lines have the equations

$$x_1 \stackrel{\Delta}{=} z_2 - A_2 - B_2 = 0 \qquad (9.8\text{-}12)$$

$$x_2 \stackrel{\Delta}{=} z_2 - A_2 + B_2 = 0 \qquad (9.8\text{-}13)$$

$$x_3 \stackrel{\Delta}{=} z_1 - A_1 + B_1 = 0 \qquad (9.8\text{-}14)$$

$$x_4 \stackrel{\Delta}{=} z_1 - A_1 - B_1 = 0 \qquad (9.8\text{-}15)$$

For each of the nine regions a unique choice of control exists, as indicated in Figure 9.8-7. Thus the computer programs must be capable of determining the state of the space vehicle relative to these regions, and the choice of the control command is automatically decided. The structure of the overall program is indicated by the flow chart shown in Figure 9.8-8 It consists of the

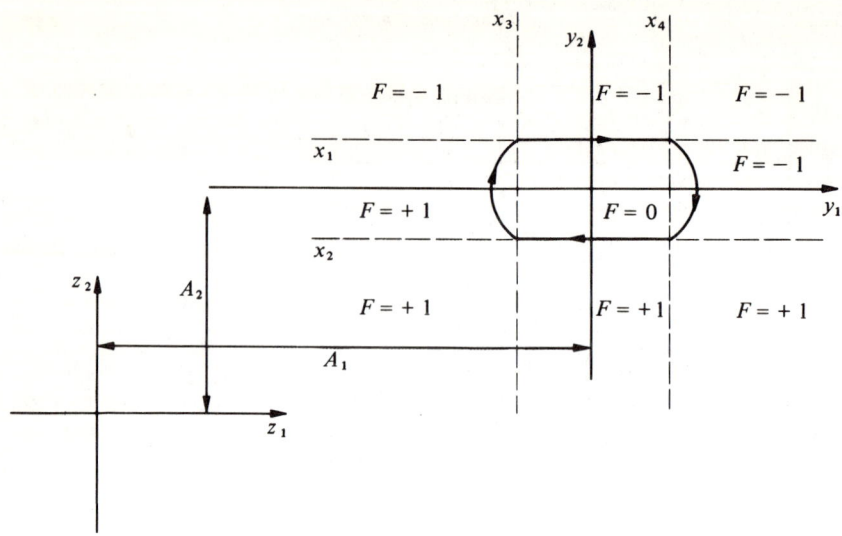

Figure 9.8-7. Regions of control.

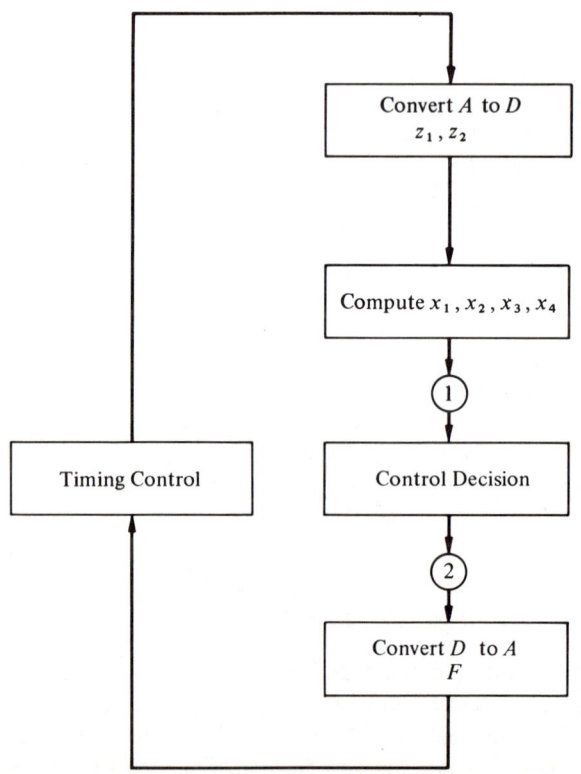

Figure 9.8-8. Gross detail of overall program.

analog-to-digital conversion of angular position and rate, the computation of equations (9.8-12) through (9.8-15), the decision of the proper control commands, and their subsequent digital-to-analog conversion. This sequence of operations is repeated cyclically subject to a timing control. The two versions of determining the control decision will now be presented.

9.8-2 Version I of Control Decision

This version is based upon testing equations (9.8-12) through (9.8-15) as algebraic inequalities. This is carried out by the decision indicated by the flow chart of Figure 9.8-9. A simple program based upon this flow chart is shown toward the end of this example.

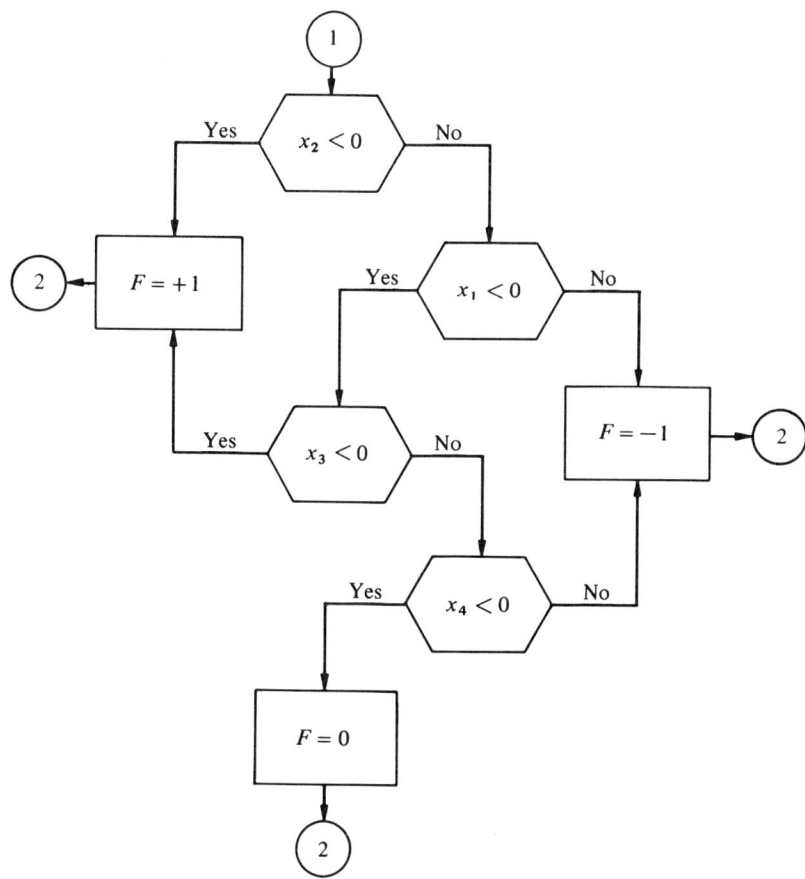

Figure 9.8-9. Flow chart for control command decision.

9.8-3 Version II of Control Decision

An alternate way of implementing the control decision process is by way of combinational logic. If we assign the logic values to the nine regions according to the scheme indicated by Figure 9.8-10, we can set up the table of combinations as given by Table 9.8-1. From this table of combinations,

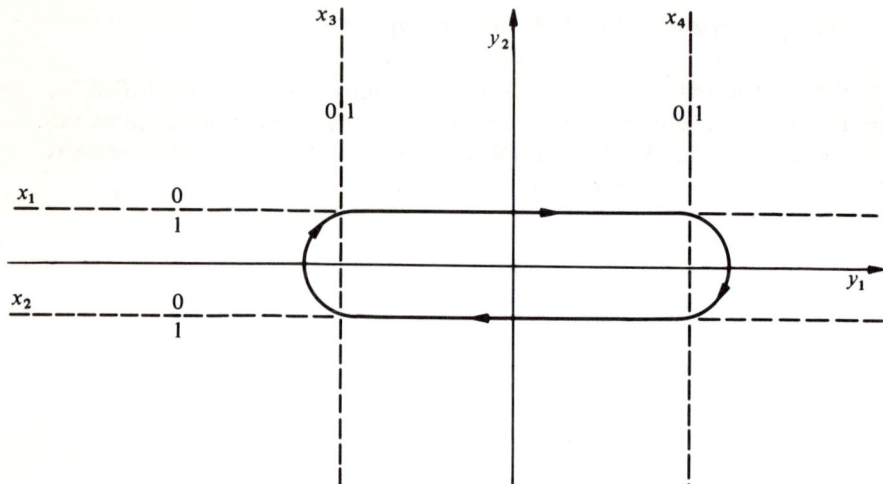

Figure 9.8-10. Logic regions for control decision.

Table 9.8-1 Logic Combinations for Regions of Figure 9.8-10

x_1	x_2	x_3	x_4	$F = +1$	$F = -1$
0*	0	0	0	0	1
0	0	0	1	x†	x
0	0	1	0	0	1
0	0	1	1	0	1
1	0	0	0	1	0
1	0	0	1	x	x
1	0	1	0	0	0
1	0	1	1	0	1
0	1	0	0	x	x
0	1	0	1	x	x
0	1	1	0	x	x
0	1	1	1	x	x
1	1	0	0	1	0
1	1	0	1	x	x
1	1	1	0	1	0
1	1	1	1	1	0

*For assignment of truth values see Figure 9.8-10.
†The x-entry denotes "Don't Cares."

or truth table, it is possible to extract logic equations by the use of a Karnaugh map given by Table 9.8-2. Each entry in this map consists of two parts, corresponding to $F = +1$ and $F = -1$. This table yields the two logic equations

$$FP1 = x_2 + x_1 \cdot x_3' \qquad (9.8\text{-}16)$$

$$FM1 = x_1' + x_2' \cdot x_4 \qquad (9.8\text{-}17)$$

Table 9.8-2 Karnaugh Table

		$x_3 x_4$			
		00	01	11	10
$x_1 x_2$	00	01	xx	01	01
	01	xx	xx	xx	xx
	11	10	xx	10	10
	10	10	xx	01	00

where the prime denotes complement, the plus sign denotes logic OR, and the dot sign denotes logic AND.

These logic equations may be programmed by utilizing the logic instructions of the computer. A flow chart showing this program is shown in Figure 9.8-11. It shows the same entry points as Version I and may be used interchangeably with Version I.

The entire program may now be assembled in detail. It is shown in terms of symbolic machine language statements in Table 9.8-3.

The digital computer communicates with the outside via data inputs and converters in the following way:

Data Input: A1, A2, B1, B2 may be entered into the computer either through keyboard switches, A/D converters, or direct memory access transfers.

A/D Converter: Z1 and Z2, corresponding to attitude and attitude rate, respectively, are entered through channels 0 and 1.

D/A Converter: FP1 and FM1, corresponding to $F = +1$ and $F = -1$ commands, respectively, are exited through channels 1 and 2. Logic signals FP1 and FM1 activate CW thrusters and CCW thrusters, respectively.

Figure 9.8-12 shows a series of phase-space plots for a variety of conditions. Figure 9.8-12(a) shows the limit cycle as approached from different initial conditions. Part (b) shows a response to a change in attitude reference, from $A_1 = 0$ to $A_1 = 30°$. Part (c) shows various limit cycles with B_2 (coasting velocity) as parameter, while part (d) shows the same with B_1 (dead band) as parameter.

450   *Computer Methods in Systems Studies*   *Sec. 9.8*

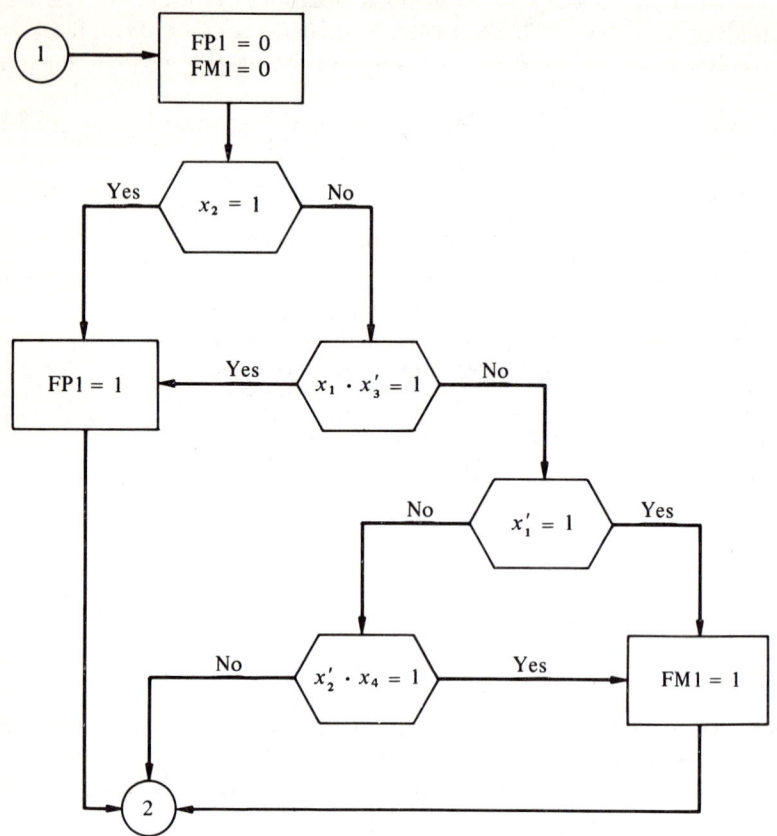

Figure 9.8-11. Flow chart for Version II program.

Table 9.8-3

Symbolic Machine Instruction		Commands
A1,	0	
A2,	0	
B1,	0	
B2,	0	
Z1,	0	
Z2,	0	Assign storage locations to variable and set initially to zero.
X1,	0	
X2,	0	
X3,	0	
X4,	0	
BEGN,	6541	Select Multiplexer Channel 0, assign transfer location BEGN

Table 9.8-3 (cont.)

Symbolic Machine Instruction	Commands
6532	Convert attitude (Z1)
TAD A2	Add A2 to accumulator
TAD B2	Add B2 to acc.
CIA	Change sign of acc.
TAD Z2	Add Z2
DCA X1	Store in X1 and clear acc.
TAD A2	Add A2 to acc.
CIA	Change sign of acc.
TAD B2	Add B2
TAD Z2	Add Z2
DCA X2	Store in X2 and clear acc.
6531	Is conversion done?
JMP .−1	Yes: skip to next instruction
	No: return to previous instruction
6534	Read A/D buffer into acc. (Z1)
DCA Z1	Store in Z1 and clear accumulator
6544	Select MX Channel 1
6532	Convert attitude rate (Z2)
TAD A1	Add A1 to acc.
CIA	Change sign of acc.
TAD B1	Add B1 to acc.
TAD Z1	Add Z1 to acc.
DCA X3	Store in X3 and clear acc.
TAD B1	Add B1 to acc.
TAD A2	Add A2 to acc.
CIA	Change sign of acc.
TAD Z1	Add Z1 to acc.
DCA X4	Store in X4 and clear acc.
DCA FP1	Set FP1 = 0
DCA FM1	Set FM1 = 0
	VERSION I
TAD X2	Add X2 to acc.
SPA	If acc. is positive, skip next instruction
JMP FP2	If acc. is negative, transfer to FP2
CLA	Clear acc.
TAD X1	Add X1 to acc.
SMA	If acc. is negative, skip next instruction
JMP FM2	If acc. is positive, transfer to FM2
CLA	Clear acc.
TAD X3	Add X3 to acc.
SPA	If acc. is positive, skip next instruction
JMP FP2	If acc. is negative, transfer to FP2
CLA	Clear acc.
TAD X4	Add X4 to acc.
SMA	If acc. is negative, skip next instruction
JMP DTOA	If acc. is positive, transfer to DTOA
JMP FM2	Transfer to FM2

Table 9.8-3 (cont.)

Symbolic Machine Instruction	Commands
	VERSION II
TAD X2	Add X2 to acc.
SPA	Check truth value of X2: $X2 = 1$ if negative; $X2 = 0$ if positive; skip if $X2 = 0$
JMP FP2	If $X2 = 1$, transfer to FP2
CLA	Clear acc.
TAD X3	Add X3 acc.
CMA	Complement acc. to generate X3'
AND X1	Logic AND with X1 to generate X1X3'
SMA	Check truth value of X1X3': $X1X3' = 0$ if negative; skip if $X1X3' = 0$
JMP FP2	if $X1X3' = 1$ transfer to FP2
CLA	Clear acc.
TAD X1	Add X1 to acc.
CMA	Complement acc. to generate X1'
SPA	Check truth value of X1': $X1' = 1$ if positive; skip if $X1' = 1$
JMP FM2	If $X1' = 0$ transfer to FM2
CLA	Clear acc.
TAD X2	Add X2 to acc.
CMA	Complement acc. to generate X2'
AND X4	Logic AND with X4 to generate X2'X4
SPA	Check truth value of X2'X4: $X2'X4 = 1$ if positive; skip if $X2'X4 = 1$
JMP DTOA	If $X2'X4 = 0$ transfer to DTOA
FM2, CLA	This is instruction FM2; clear acc.
TAD REF	Add RFE to acc.
DCA FM1	Store REF in FM1: set $F = -1$
JMP DTOA	Transfer to DTOA
FP2, CLA	This is instruction FP2, clear acc.
TAD REF	Add REF to acc.
DCA FP1	Store REF in FP1: set $F = +1$
DTOA, CLA	This is instruction DTOA; clear acc.
TAD FP1	Load FP1 into acc.
6551	Convert DTOA-channel 1
CLA	Clear acc.
TAD FM1	Load FM1 into acc.
6552	Convert DTOA-channel 2
CLA	Clear acc.
6534	Read A/D buffer into acc. (Z2)
DCA Z2	Store in Z2 and clear acc.
TAD TIME	Load time into acc.
SPA	
JMP BEGN	If TIME < 0 begin new cycle
JMP .−3	If TIME > 0 wait until TIME < 0
FP1, 0	⎫ Reserve storage for $F = +1$
FM1, 0	⎬ Reserve storage for $F = -1$
REF, 4000	Reference set at decimal $+0$

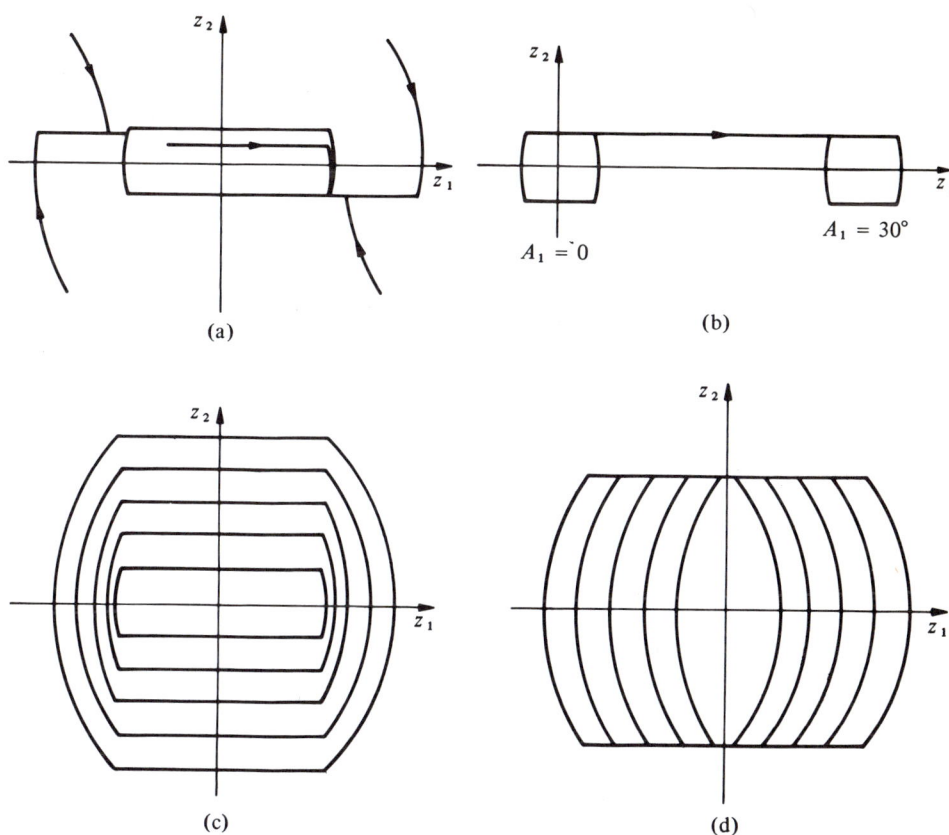

Figure 9.8-12. Phase-space plots of attitude control system.

EXAMPLE 9.8-3

A Digital PID Controller.* Perform a simulation and determine a suitable design for a PID controller that is being used in the system shown in Figure 9.8-13. The PID controller is to be replaced by a digital version, as shown in Figure 9.8-14. The simulation should be sufficiently flexible to permit the adjustment of the following parameters: T, the sampling interval; K_p, the proportional gain; K_i, the integral gain; and K_d, the derivative gain. The system to be controlled is defined by its transfer function in Figure 9.8-14.

The digital computer is programmed so that the output sequence is given by

$$e_2(k) = K_p e_{21}(k) + K_i e_{22}(k) + K_d e_{23}(k) \qquad (9.8\text{-}18)$$

*See also Section 3.7.

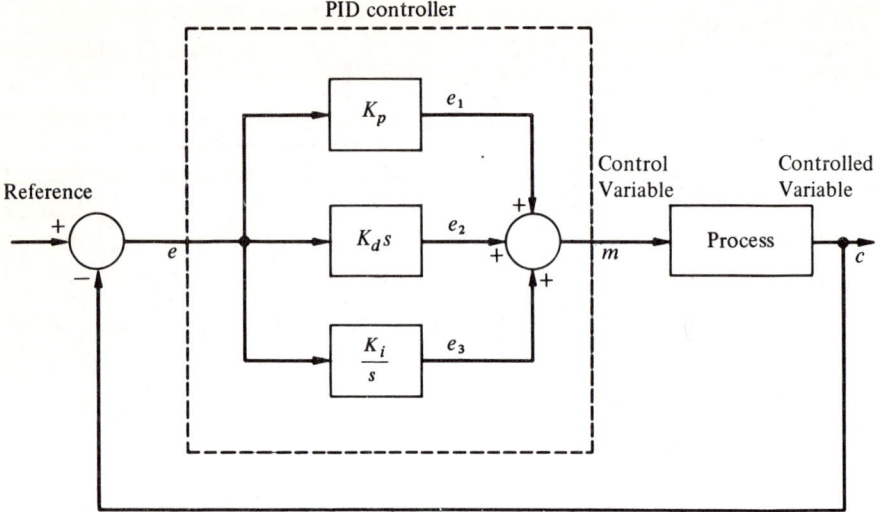

Figure 9.8-13. PID controlled system.

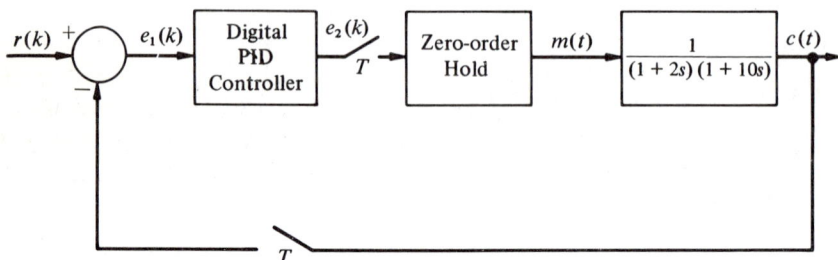

Figure 9.8-14. System with digital PID controller.

The three components of the sum are generated as follows:

Proportional control:

$$e_{21}(k) = e_1(k) \tag{9.8-19}$$

Integral control:

$$e_{22}(k) = e_{22}(k-1) + Te_2(k) \tag{9.8-20}$$

Derivative control:

$$e_{23}(k) = \frac{1}{T}[e_1(k) - e_1(k-1)] \tag{9.8-21}$$

In the application of a digital computer as a PID controller, the quantities K_p, K_i, K_d, and T must be determined. It has been the practice in the process control industry to adjust the gain constants on the job until a desirable response is obtained. Alternatively, a simulation of the entire system may be performed which permits parameter adjustment. This approach is followed in this study.

A PDP-8/TR-20 simulation is chosen as the method of simulation. The process to be controlled is programmed on an analog computer according to the diagram of Figure 9.8-15. A flow chart of the program is shown in Figure

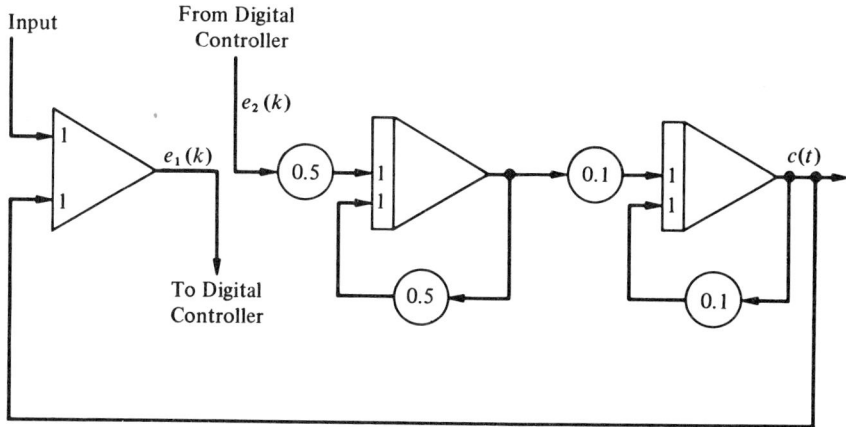

Figure 9.8-15. Simulation of process.

9.8-16. The program is listed in Table 9.8-4. When the program is started, the teletype will ask for the typewriter input of T, KP, KI, and KD in floating point format. In addition to setting these constants, one must set the digital clock to agree with T.

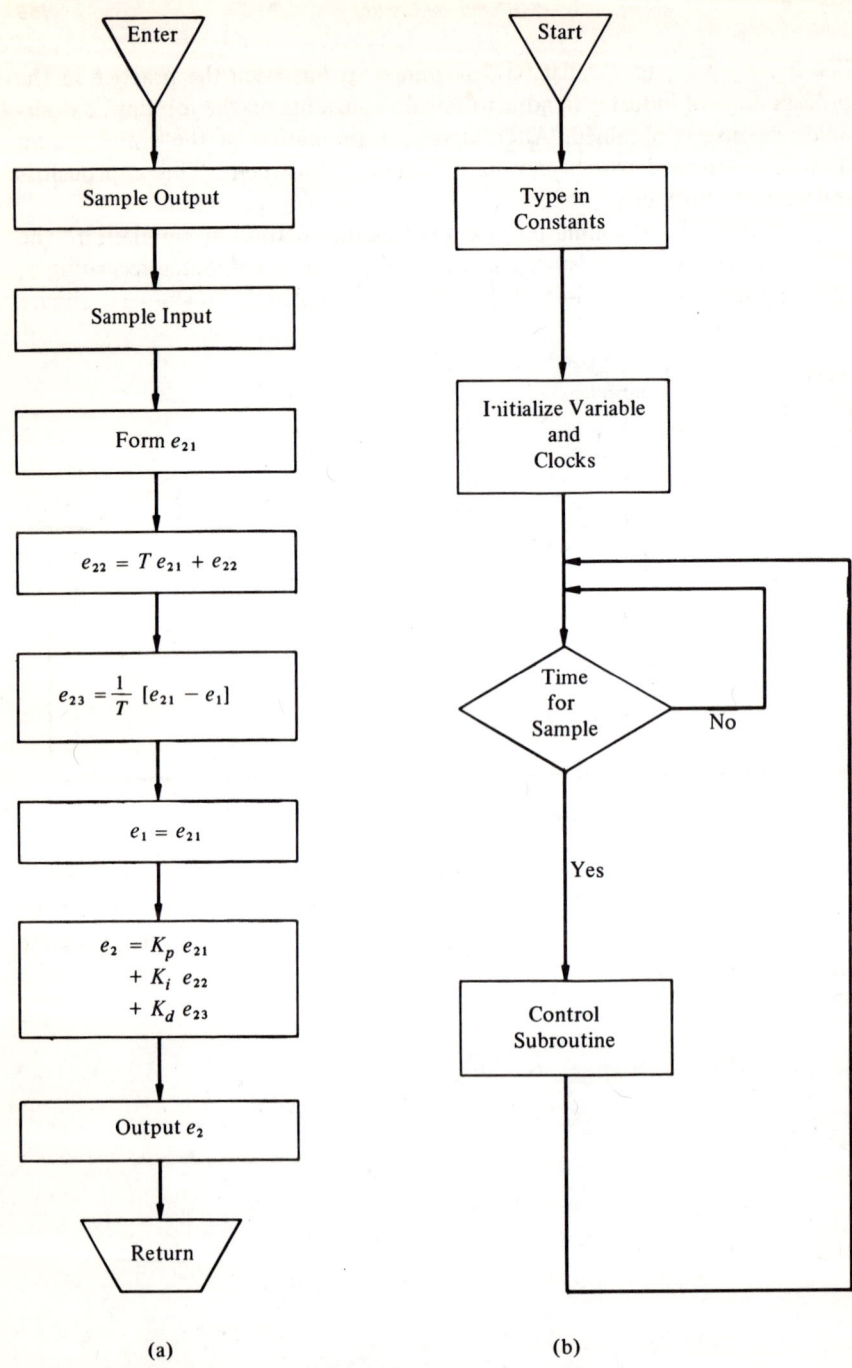

Figure 9.8-16. Flow chart for simulation program: (a) control routine; (b) timing routine.

Sec. 9.8 Computer Control 457

Table 9.8-4 Listing of Program†

```
I CONTROLLER  PROGRAM
/STARTING  ADDRESS  200

INPUT   = 0003
OUTPUT  = 0004
PRNT    = 0005

*5
IN,     7400            /ADDRESS OF INPUT ROUTINE
OUT,    7200            /ADDRESS OF OUTPUT ROUTINE
FLOAT,  5600            /ADDRESS OF FLOATING-POINT SYSTEM

*200
BEGIN,  KCC             /CLEAR READER FLAG AND AC
        TLS             /TRANSMIT 0 TO PRINTER
        JMS I FLOAT     /ENTER INTERPRETER
        FGET ZERO       /FAC = 0
        FPUT E21        /INITIALIZE VARIABLES
        FPUT E22
        FPUT E23
        FPUT EK1
        FPUT E2
        FPUT TEMP
        FGET TOUT
        PRNT            /PRINT LF T =
        INPUT           /READ VALUE OF T
        FPUT T          /STORE IT
        FGET KPOUT
        PRNT            /PRINT KP =
        INPUT           /READ VALUE OF KP
        FPUT KP         /STORE IT
        FGET KIOUT      /PRINT KI =
        PRNT
        INPUT           /READ VALUE OF KI
        FPUT KI         /STORE IT
        FGET KDOUT
        PRNT            /PRINT KD =
        INPUT           /READ VALUE OF KD
        FPUT KD         /STORE IT
        FEXT
        CLA CMA         /AC = -1
        DCA TIME        /INITIALIZE CLOCK
WAIT,   TAD TIME        /GET TIME
        SPA CLA         /TIME FOR NEXT SAMPLE?
        JMP WAIT        /NO, GO WAIT
        CLA CMA         /YES, AC = -1
        DCA TIME        /INITIALIZE TIME
        JMS I PID       /GO TO PID CONTROLLER
        CLA
```

†The floating-point package (DIGITAL 8-5A-S) must be loaded before this program is loaded.

Table 9.8-4 (cont.)

```
              6311            /SKIP IF SW 2 OFF
              JMP  WAIT
              JMP  BEGIN      /SW 2 OFF, GET NEW DATA
PID,          PIDPRO
*400
PIDPRO,       0
              ADCC            /CLEAR MPX TO 0
              ADCV            /START CONVERSION OF C
              CLA
              ADSF            /IS CONVERSION DONE?
              JMP  .-1        /NO, WAIT
              ADRB            /YES, READ OUTPUT C
              CIA             /-C
              DCA  MC+1       /STORE -C
              CLA  IAC        /AC = 1
              ADSC            /GO TO CHANNEL 1
              ADCV            /START CONVERSION OF R
              CLA
              ADSF            /IS CONVERSION OF R DONE?
              JMP  .-1        /NO, WAIT
              ADRB            /YES, READ R
              DCA  45         /STORE IN FAC
              DCA  46         /ZERO LOW ORDER PART
              TAD  C13        /11 IN DECIMAL
              DCA  44         /STORE AS EXP.
              JMS  I FLOAT    /ENTER INTERPRETER
              FNOR            /NORMALIZE
              FPUT TEMP       /SAVE IT
              FGET MC         /GET -C
              FNOR            /NORMALIZE
              FADD TEMP       /R -C
              FPUT E21        /E21 = E1(K)
              FMPY T          /T*E21
              FADD E22        /T*E21+E22
              FPUT E22        /UPDATE E22
              FMPY KI         /KI*E22
              FPUT TEMP       /SAVE IT
              FGET E21        /E21
              FSUB EK1        /E21-EK1
              FDIV T          /<E1(K)-E1(K-1)>/T
              FPUT E23        /UPDATE E23
              FMPY KD         /KD*E23
              FADD TEMP       /KI*E22+KD*E23
              FPUT TEMP       /SAVE IT
              FGET E21        /E1(K)
              FPUT EK1        /E1(K-1) = E1(K)
              FMPY KP         /KP*E21
              FADD TEMP       /KP*E21+KI*E22+KD*E23
              FPUT E2         /E2
```

Table 9.8-4 (cont.)

```
        FEXT              /LEAVE INTERPRETER
        CLA
        TAD  44           /FETCH EXPONENT
        SZA  SMA          /IS THE NUMBER <1?
        JMP  .+3          /NO
        CLA               /YES, FIX IT TO 0
        JMP  DONE+1
        TAD  M13          /NO, SET BINARY POINT AT
        SNA               /11 PLACES TO RIGHT OF CURRENT POINT
        JMP  DONE         /IT IS ALREADY THERE; ALL DONE
        SMA               /TEST TO SEE IF IT IS TOO LARGE
        JMP  ERROR        /YES NUMBER >2**11
        DCA  44           /NO, SET SCALE COUNT
GO,     CLL               /0 TO C(L)
        TAD  45           /FETCH MANTISSA
        SPA               /IS IT <0?
        CML               /YES, PUT A 1 IN LEFT BIT
        RAR               /SCALE RIGHT
        DCA  45           /RESTORE IT
        ISZ  44           /TEST IF SHIFTED ENOUGH
        JMP  GO           /NO, CONTINUE
DONE,   TAD  45
        6551              /OUTPUT M
LEAVE,  CLA
        JMP  I  PIDPRO    /LEAVE
ERROR,  CLA  CLL
        TAD  45           /GET VALUE
        SPA  CLA          /IS IT POSITIVE?
        CML               /NO, L←1
        TAD  MAX          /AC = −2048
        SNL               /WAS ILLEGAL NUMBER NEGATIVE?
        CMA               /NO, CHANGE AC TO 2047
        6551              /OUTPUT MAX VALUE
        CLA
        6301              /SKP PRINTOUT IF SW 1 OFF
        JMS  I  OUT       /PRINT THE NUMBER IN ERROR
        CLA
        JMP  I  PIDPRO
MAX,    4000              /−2048

*100
TIME,   −1                /INCREMENTED BY EXTERNAL CLOCK
C13,    13                /DECIMAL 11
M13,    −13
MC,     13                /STORE −C
        0
        0
E21,    0                 /PROPORTIONAL VALUE
        0
        0
```

Table 9.8-4 (cont.)

```
T,       0       /SAMPLE TIME
         0
         0
E22,     0       /INTEGRAL OF INPUT
         0
         0
E23,     0       /DERIVATIVE OF INPUT
         0
         0
EK1,     0       /INPUT DELAYED ONE SAMPLE PERIOD
         0
         0
KP,      0       /PROPORTIONAL CONSTANT
         0
         0
KI,      0       /INTEGRAL CONSTANT
         0
         0
KD,      0       /DERIVATIVE CONSTANT
         0
         0
E2,      0       /OUTPUT
         0
         0
TEMP,    0
         0
         0
ZERO,    0
         0
         0
TOUT,    212     /LINE FEED
         324     /T
         275     /=
KPOUT,   313     /K
         320     /P
         275     /=
KIOUT,   313     /K
         311     /I
         275     /=
KDOUT,   313     /K
         304     /D
         275     /=

*6547
         7400
         7200
         3000

*3000
PRNTO,   0       /ALPHANUMERIC PRINTOUT
```

Table 9.8-4 (cont.)

```
        DCA  SAVE        /SAVE ACCUMULATOR
        TAD  44          /GET FIRST CHARACTER FROM FAC
        TSF              /IS PRINTER READY?
        JMP  .-1         /NO, WAIT
        TLS              /YES, PRINT
        CLA
        TAD  45          /GET SECOND CHARACTER
        TSF
        JMP  .-1
        TLS
        CLA
        TAD  46          /GET THIRD CHARACTER
        TSF
        JMP  .-1
        TLS
        CLA
        TAD  SAVE        /RESTORE AC
        JMP  I  PRNTO    /LEAVE
SAVE,   0
```

The above examples of computer control systems are perhaps not overly significant from an engineering point of view, but they do serve to illustrate the role that a computer assumes as an active component in a system. In all cases the computer is programmed to periodically execute statements that implement control relationships based upon analytical derivations. The examples are sufficiently simple to permit alternate implementation approaches that do not require a digital computer. However, the use of a computer is completely justified if one realizes that there may be many such control tasks carried by the computer on a time-shared basis. Then there exists almost unlimited flexibility for modification of the computer programs.

9.9 Summary

This chapter discusses a variety of ways in which digital and analog computers, either singly or in combination, may be utilized for work in system analysis and design. Given here are some of the major applications.

Analysis:
1. Computer routines   –Digital
2. Computer simulation–Analog, digital, hybrid
3. Simulation languages–Digital

Design:
1. Generation of control laws   –Digital

2. Implementation of control laws –Analog, digital
3. Time-shared computer control systems–Digital

In each of these categories the capabilities of a computer are employed in distinct ways.

PROBLEMS

9.1 For the following transfer functions, generate digital simulation models using the Euler method, Tustin method, and State Transition method. Put all results into a linear recursion form, and compare the structure of the model and the coefficients. Investigate the stability and steady-state performance. Finally, perform a digital computation with a step input and an integration interval $T = .1$ second.

(a) $G(s) = \dfrac{1}{s(s+1)}$

(b) $G(s) = \dfrac{s+1}{s(s+4)}$

9.2 Investigate the stability of the discrete model as generated by the Runge-Kutta method for the transfer function of Problem 9.1(a) and the following three choices of integration interval $T = .1$, $T = 1$, and $T = 4$. For this purpose it is necessary to obtain a linear recursion equation.

9.3 Devise a scheme by which the Runge-Kutta-Merson method is used to control the integration interval for a numerical integration. Give consideration to the following: The step size should be decreased when the local truncation error estimate exceeds a preselected maximum level, the step size should be left unchanged when the error falls within a range of acceptable error, and it should be increased when the error is smaller than a preselected minimum level. The range of acceptable error should be in some relation to the step size change, keeping in mind that the local truncation error is proportional to the integration interval raised to the fifth power.

9.4 Perform a digital simulation using the State Transition method of the system defined by Figure 3.5-2. Use the computer program to compute $A(T)$ and $B(T)$.

9.5 Construct an analog computer simulation model for the transfer function

$$G(s) = \frac{s+a}{s+b}$$

by giving consideration to the fact that $a > b$ or $a < b$.

9.6 Prepare a MIMIC program for the digital simulation of the systems shown in Figure P9.6(a) and (b).

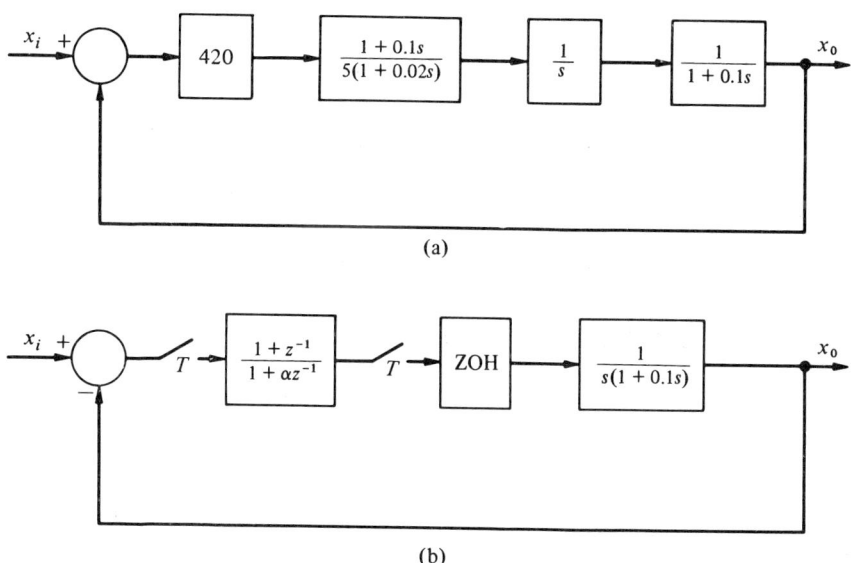

Figure P9.6

9.7 Perform an analog-hybrid computer simulation of the system shown in Figure P9.6(b). Consider α a parameter to be adjusted.

9.8 Repeat Problem 9.7, but use a hybrid computer for the simulation.

Appendix 9A

PROGRAM FOR THE NUMERICAL INTEGRATION OF STATE VARIABLE EQUATIONS

Consider the state variable equations

$$\dot{x}_i = f_i(x_1, x_2, \ldots, x_n, t) \qquad i = 1, 2, \ldots, n$$

The program described in this appendix permits the integration of the *n* first-order differential equations over a preselected time interval. The calling statement is

CALL INTFUN (X, TIME, T, N)

where

X = array of initial conditions for state variables; the end point will be returned through the same array
TIME = running time; must coincide with initial conditions on start of routine; will correspond to time at end point at return
T = time interval
N = number of variables

Program INTFUN will compute a solution of the state equations from X(TIME) to X(TIME + T).

A subroutine supplying the derivative information must be prepared. It has the calling statement

CALL DERIV (X, TIME, D)

where X and TIME are defined as above and D is the array of derivatives.

The integration routine used in this program is the variable stepsize Runge-Kutta-Merson routine as given by equations (9.2-20) through (9.2-22). The stepsize H is automatically adjusted to maintain the local truncation error between EMAX = 1.E-05 and EMIN = 1.E-07. Initially, H is selected as T. Upon subsequent calls to INTFUN H is chosen as the value used in the last call.

The stepsize will be reduced by a factor of 2 if the largest truncation error, ERROR, exceeds EMAX. On the other hand, it will be increased if ERROR is less than EMIN three successive times. If H is reduced to less than HMIN = 1.E 10-4, it is assumed that the system is dynamically unstable and a CALL EXIT is executed. EMAX, EMIN, and HMIN may be altered to suit other conditions.

The use of the INTFUN program is illustrated by the integration of the differential equation

$$\ddot{x} + \dot{x} + x = 1$$

The program uses the printer plotter described in Appendix 3A as output.

The solution is to be calculated for 100 seconds, with printout and plot-out at 1-second intervals.

The calling program is written as

```
      PROGRAM MAIN (INPUT,OUTPUT,TAPE5 = INPUT,TAPE6 = OUTPUT,TAPE1)
      DIMENSION X(2)
      REWIND 1
      N = 2
      X(1) = 0.
      X(2) = 0.
      TIME = 0.
      WRITE (6,9)
    9 FORMAT (1H1,9X,4HTIME,11X,4HX(1),11X,4HX(2),/)
      DO 10 K =1,100
      WRITE (6,11) TIME,X(1),X(2)
   11 FORMAT (3X,3E15.4)
      WRITE (1,12) TIME,X(1),X(2)
   12 FORMAT (7O20)
      CALL INTFUN(X,TIME,1.,N)
   10 CONTINUE
      CALL PLOTX (100,2,1)
      END
```

The derivative subroutine is written as

```
         SUBROUTINE DERIV (X,TIME,D)
         DIMENSION X(20), D(20)
         D(1) = -X(1)-X(2)+1.
         D(2) = X(1)
         RETURN
         END

         SUBROUTINE INTFUN(X,TIME,T,N)
         DATA KUSE/1/
C        X-STATE VECTOR
C        TIME-RUNNING TIME
C        T-TIME INTERVAL
C        N-DIM OF STATE VECTOR
         DIMENSION X(20), D(20), A(20,5),XB(20)
C        INITIALIZE
         KINC = 0
         HMIN = 1.E-04
         EMAX = 1.E-05
```

```
        EMIN = 1.E-07
        TIN = TIME
        TOUT = TIN+T
C       TIN-BEGINNING OF INTERVAL
C       TOUT-END OF INTERVAL
        IF (KUSE.NE.0) H = T
        IF (KUSE.EQ.0) H = HSAVE
        KH = 0
        KUSE = 0
C       INTEGRATION ALGORITHM BETWEEN 22 AND 19
        DO 101 I = 1,N
    101 XB(I) = X(I)
     22 K = 1
     15 CALL DERIV (X,TIME,D)
        GO TO (100,200,300,400,500),K
    100 DO 11 I = 1,N
        A(I,K) = D(I)*H/3.
     11 X(I) = XB(I)+A(I,1)
        TIME = TIN+H/3.
        K = K+1
        GO TO 15
    200 DO 12 I = 1,N
        A(I,K) = D(I)*H/3.
     12 X(I) = XB(I)+.5*(A(I,1)+A(I,2))
        K = K+1
        GO TO 15
    300 DO 17 I = 1,N
        A(I,K) = D(I)*H/3.
     17 X(I) = XB(I)*(3.*A(I,1)+9.*A(I,3))/8.
        TIME = TIN+.5*H
        K = K+1
        GO TO 15
    400 DO 18 I = 1,N
        A(I,K) = D(I)*H/3.
     18 X(I) = XB(I)+(3.*A(I,1)-9.*A(I,3)+12.*A(I,4))/2.
        TIME = TIN+H
        K = K+1
        GO TO 15
    500 DO 19 I = 1,N
        A(I,K) = D(I)*H/3.
     19 X(I) = XB(I)+.5*(A(I,1)+4.*A(I,4)+A(I,5))
C       COMPUTE TRUNCATION ERROR
C       INTEGRATION ALGORITHM BETWEEN 22 AND 19
        ERROR = 0.
        DO 21 I = 1,N
        TE = A(I,1)-(9.*A(I,3)-8.*A(I,4)+A(I,5))/2.
     21 ERROR = AMAX1(ERROR,ABS(TE))
        IF (ERROR.GE.EMAX) GO TO 33
        DO 32 I = 1,N
     32 XB(I) = X(I)
        TIN = TIME
```

```
      IF (TIME.EQ.TOUT) GO TO 39
      TREM = TOUT-TIME
      IF (TREM.GT.H) GO TO 31
      HSAVE = H
      KH = 1
      H = TREM
      GO TO 22
   31 IF(TREM.LT.(2.*H)) GO TO 22
      IF (ERROR.GT.EMIN) GO TO 22
      KINC = KINC+1
      IF (KINC.LT.3) GO TO 22
      H = 2.*H
      KINC = 0
      GO TO 22
   33 K = H/2.
      IF(H.LT.HMIN) GO TO 35
      TIME = TIN
      DO 34 I = 1,N
   34 X(I) = XB(I)
      GO TO 22
   35 WRITE (6,36)
   36 FORMAT (1H1,22HH IS LT HMIN,TERMINATE)
      GO TO 40
   39 IF(KH.EQ.1) RETURN
      HSAVE = H
      RETURN
   40 CALL EXIT
      END
```

Index

A

Adams-Moulton method, 385
Analog accumulator, 418
Analog computer simulation, 404
Analog-hybrid computer, 106
Analog signal, 12
Analog-to-digital converter, 203, 356, 422, 437
AND gate, 417
AND operation, 340
Anticipative system, 44
Attitude control, 443
Augmented matrix, 395
Automatic gain adjustment, 423

B

BCD counter, 346
Beckett, J. T., 337, 344
Bidirectional counter, 337
Bifilar windings, 360
Binary complement, 341
Binary number, 326
Binary operators, 340
Binary up counter, 343
Bit capacity, 14
Block diagram, 2
Brennan, R. D., 409
Brush encoder, 334
Buffer register, 346

C

Cascaded systems, 45, 141
Cayley-Hamilton theorem, 70

Characteristic equation, 70, 229
Clear, 342
Closed-loop discrete state equations, 88
Closed-loop sampled data systems, 86
Codes, 332
Comparator, 416
Complement, 342
Complex poles, 228
Complex variable, 225
Complex variable theory, 212
Compound interest, 30
Computer control, 437
Computer-controlled systems, 51
Computer control system, 91
Computer generated plot, 95
Computer program:
   exponential matrix, 399
   integration, 465
Computer program design, 444
Constituent matrix, 72
Continuous signal, 2
Controllability, 274
Convergence, 162, 165, 167
Conversion, analog-to-digital, 13
Conversion accuracy, 357
Conversion time, 354, 356
Convolution integral, 35
Convolution summation, 40, 45, 210
   general, 44
Critical damping, design for, 423

D

DAS, 410
Data hold device, 77

DCD gate, 341
DC gain, 308
Dead band, 449
Decimal number, 324
Derivative control, 100, 454
Difference equation, 33
   linear order, 34
Difference equations, 5
Differentiation, 96
Differentiator, 375
Digital analog simulator, 408
Digital comparator, 337
Digital compensation, 438
Digital compensator, 426
Digital computer:
   input-output relationship, 34
   transfer function, 215
Digital encoding, 14, 331
Digital integration, 438
Digital modules, 339
Digital PID controller, 453
Digital process controller, 100
Digital processor, 12
Digital recursion equation, 55
Digital signal, 12
Digital simulation, 390, 398, 402
Digital-to-analog converter, 19, 77, 204, 350, 422, 437
Digital transfer function, 216
Dirac delta function, 23, 36, 206
Direct programming, 56
Direct reading encoder, 336
Direct transmission matrix, 49
Discrete signal, 3
Discrete state equation, 22, 33
Discrete state model, 55
Discrete step input, 42
Driving matrix, 176

E

Eigenvalues, 98
Encoding ambiguity, 336
Encoding error, 337
Error-sampled feedback system, 219
Estimation, 27
Euler method, 16, 378
Exponential matrix, 20, 387
Exponential matrix program, 399
Exponential series, 388
Extrapolator, 77

F

Feedback, 91
Feedback control law, 306
Ferritic core, 336
Final-value theorem, 136, 222
First difference method, 16
First-order function, 411
First-order hold, 78
   transfer function, 207
   z-transform, 217
Flip-flop, 342, 350, 351, 417
Floating point subroutine, 455
Fredriksen, T. R., 367
Function of a matrix, 70
Fundamental matrix, 197

G

Gaskill, R. A., 409
Geometric series, finite, 42
Gray code, 337, 332

H

Harnett, R. T., 409
Hexadecimal number, 325
Hold circuit, 15, 204
   equivalent system, 209
   z-transform, 217
Hooke, R., 428
Hoppe, S. G., 389
Hybrid computer:
   digital, 419
   analog, 416
Hybrid computer organization, 415
Hybrid computer simulation, 415
Hybrid interface, 421

I

Idempotent matrix, 73
Impulse response, 35, 36, 53
Impulse response method, 17
Impulse sampled function, 209
Impulse sampler, 23, 204
Impulse sampling, 207
Incremental encoder, 336, 349
Infinite sum, 163
Initial condition, 6
Initial value theorem, 136

Input, 2
Input matrix, 49
Input variables, 173
Input vector, 174
Integral control, 100, 454
Inverse z-transform, 222

J

Jeeves, T. A., 428

K

Kaiser, J. F., 428
Karnaugh table, 449
Kronecker delta input, 147
Kronecker delta response, 38
Kronecker delta sequence, 36
Kuo, B. J., 224
Kuo, F. F., 428

L

Lagrange multipliers, 303
Laplace z-transform pairs, 211
Left inverse matrix, 289
Leverrier's algorithm, 68
Limit cycle, 448
Linear-difference equations, 146
Linearity property, 39
Linear system, 6
Liou, M. L., 388
Logic control circuit, 370
Logic input, level pulse, 341

M

Magnetic encoder, 336
Maneuverability, 27
Mapping, 225
MIDAS, 408
MIMIC, 408, 410
 integration, 412
 logic variables, 413
 programming, 411
 variables, 412
Minimal prototype design:
 time domain, 249
 z-domain, 268
Minimum energy controller, 286
Minimum settling time, 431

Mode control, 417
 hybrid, 420
Modern control theory, 19
Modified two's complement binary, 354
Modified z-transform, 234
Modified z-transform pairs, 238
Monostable multivibrator, 348, 417
Multiplexor, 104, 450
Multiplying DAC, 354

N

Nested programming, 56
Noise suppression, 27
Nonanticipative system, 12
Nonlinear system, 8
Number system conversion, 328
Number systems, 324
Numerical differentiation, 96
Numerical integration, 5, 9, 16, 17, 402
 accuracy, 391
 comparison, 390
 speed, 391
Numerical methods, 377

O

Observability, 280
Octal number, 327
Open-loop sampled data system, 80
Optical encoder, 335
Optimization techniques, 428
OR gate, 417
OR operation, 340
Orthogonal matrix, 73
Output, 2
Output matrix, 49, 177
Output variables, 173
Output vector, 174

P

PACTOLUS, 410
Parseval's theorem, 158
Partial-fraction expansion, 215, 223
Pattern search, 428
PDP-8 computer, 431
Periodic strip, 225
Permanent magnet motor, 359
Peterson, H. G., 410
PID controller, 100, 453

Piecewise constant inputs, 51
Poles, 226
Pole-zero configuration, 226
Polynomial extrapolation, 77
Power series, 165
Program, recursion equation, 35
Proportional control, 100, 454
Pulse rate comparator, 349
Pulse rate measurement, 338

Q

Quadratic performance index, 302
Quantization, 13

R

Radar tracking, 27
Radius of convergence, 165
Radix, 325
Rational function, 229
Ratio test, 164
Realizable system, 44
Real pole, 227
Recursion equation, 35
Regulator, 283
Reset, 342
Residue, 235
Response between sampling instants, 106
Response between sampling periods, 397
Riccati transformation, 305
Right inverse matrix, 289
Ring counter code, 333
Root locus, 225
Root locus program, 234
Root locus techniques, 229
Round-off error, 380
Root test, 164
Runge-Kutta-Merson method, 385
Runge-Kutta method, 383

S

Sample and hold circuit, 85
Sampled-data system, 22, 51, 203, 216
    simulation, 424
Sampled function, 23
Sampler, 23, 203
Sampling, 3, 12
Sampling effects, 82
Sampling interval, 3

Sampling switch, 207
Schmidt trigger, 349
s-domain and z-domain, 225
Selfridge, R. G., 409
Sequence:
    input, 34
    output, 34
Shaft encoder, 76, 334, 369
Shifting theorem, 217
Shift register, 338
Signal, 2
Signal conversion, 12
Simulation, 15
    analog computer, 404
    digital computer, 390
    hybrid computer, 415
Simulation of PID control system, 106
Stability, 97
Stability analysts, 223
Stability criterion, 224
Stability regions, 99
State space representation:
    direct programming, 187
    iterative programming, 185
    nested programming, 190
    partial fraction expansion, 178
State transition equations, 12
State transition matrix, 49
    program, 64
    properties, 52
State transition method, 387, 395
State variable representation, 45, 378
State variables, 173
State vector, 174
Steady-state response, 151
Stepper motor, 76, 357
    applications, 365
    digital feedback, 367
    driving circuit, 364
    dynamics, 362
    slewing, 364
    speed, 359
    translator, 369
Step size, 384
Subroutine, 377
Subroutine MATEXP, 65
Switch, 416
Switch-tail ring counter, 333
Sylvester's expansion theorem, 72, 98
System, 173
    continuous, 4, 173

System (cont.):
  discrete, 4, 173
  hybrid, 4, 173
  matrix, 176
System response, 150

T

Table of combinations, 448
Time invariant system, 10
Time-sharing, 104
Time-skewing, 422
Trace of a matrix, 69
Tracking, 27
Tracking test inputs, 298
  acceleration, 302
  ramp, 301
  step, 299
Track-store, 416
Track-store unit, 85, 106
Transfer function, 126, 139, 147
Transient response, 151, 226
Transition matrix, 67, 177, 199, 200
  properties, 74
TR-20 computer, 431
Truncation error, 379
Tustin method, 381
Two's complement binary, 330, 353

U

Unstable system, 43, 89

V

Variable, discrete-time, 3
Variable reluctance motor, 358

Variables:
  input, 2
  output, 2
Vector, 9

W

Weighting sequence, 33, 36, 42, 58

Z

Zero-order hold, 14, 24, 78, 355
  transfer function, 205
  z-transform, 217
z-transform, 26
  complex differentiation, 128
  complex integration, 129
  convergence region, 120, 165
  divergence region, 120
  final value theorem, 136
  initial value theorem, 136
  inverse of, 142
  inversion integral, 146
  linearity, 122
  one-sided, 119, 162
  Parseval's theorem, 158
  poles, 121
  product of two functions, 157
  properties of one-sided, 132
  transfer function, 126
  two-sided, 119
  weighting, 148
  weighting sequence, 147
  zeros, 121
z-transform analysis, 204
z-transform of Laplace transform, 210